Homesick for a World Unknown

Also by Miriam Horn

Rancher, Farmer, Fisherman

Earth: The Sequel

Rebels in White Gloves

Homesick for a World Unknown

THE LIFE OF
GEORGE B. SCHALLER

Miriam Horn

PENGUIN PRESS NEW YORK 2026

PENGUIN PRESS
An imprint of Penguin Random House LLC
1745 Broadway, New York, NY 10019
penguinrandomhouse.com

Copyright © 2026 by Miriam Horn

Penguin Random House values and supports copyright. Copyright fuels creativity, encourages diverse voices, promotes free speech, and creates a vibrant culture. Thank you for buying an authorized edition of this book and for complying with copyright laws by not reproducing, scanning, or distributing any part of it in any form without permission. You are supporting writers and allowing Penguin Random House to continue to publish books for every reader. Please note that no part of this book may be used or reproduced in any manner for the purpose of training artificial intelligence technologies or systems.

PP colophon is a registered trademark of Penguin Random House LLC.

Illustration credits appear on pages 607–608.

Map illustration by Daniel Lagin

Designed by Alexis Sulaimani

Library of Congress Control Number: 2025018683
ISBN 9781984881342 (hardcover)
ISBN 9781984881359 (ebook)

Printed in the United States of America
1st Printing

The authorized representative in the EU for product safety and compliance is Penguin Random House Ireland, Morrison Chambers, 32 Nassau Street, Dublin D02 YH68, Ireland, https://eu-contact.penguin.ie.

For Don MacNeill, Almira Kutzer, Rose Weiss Horn,
and the many others who opened my eyes to the wonders

"Whatever there is in him of the scientist, the poet, the primitive seer, the watcher of fire and shooting stars, whatever obsessions eat at the night side of his mind, whatever sweet and dreamy yearning he has ever felt for nameless places faraway, whatever earth sense he possesses, the neural pulse of some wilder awareness, a sympathy for beasts, whatever belief in an immanent vital force, the Lord of Creation, whatever secret harboring of the idea of human oneness . . . whatever burning urge to escape . . . the circumscribed and inward-spiraling self . . . all these are satisfied."

DON DELILLO

"To write honestly and with conviction anything about the migration of birds, one should oneself have migrated. Somehow or other we should . . . feel the feel of feathers on our body and wind in our wings, and finally know what it is to leave abundance and safety and daylight . . ."

WILLIAM BEEBE

Contents

Homesick for a World Unknown

PROLOGUE

He would be torn limb from limb.

All the Great Men of Science said so, if this rash young man went to live with the real King Kong, Africa's monstrous mountain gorilla. Known the world over for its crazed chest-beating and hooting, slavering scream, this ape was savagery incarnate, or worse: a creature—as the first Westerner to ever see one had warned—of that "hideous order, half man, half beast." Fearing that not even bullets would stop the demon, Paul du Chaillu had ordered his men to load their flintlocks with ragged hunks of cast iron. None ever forgot the goliath's dying groans, "full of brutishness" yet "terribly human."

George Beals Schaller had read Du Chaillu's account when he set off for the Belgian Congo in 1959. He knew that Carl Akeley, who had killed five gorillas to stuff for the American Museum of Natural History—pausing along the way to strangle an attacking leopard with his bare hands—had called any white man who got close to one without shooting "a plain darn fool." He had listened to famed paleoanthropologist Louis Leakey argue with other "old Africa hands" over whether Schaller should wear armor or a disguise, what weapons he should carry, whether he would survive at all.

And then, without retinue or pith helmet, George went to the gorillas. Alone. When charged by a roaring silverback, he stood his ground. He couldn't have shot the brute if he'd wanted to. Certain that it would change his bearing in ways the apes would sense, he had refused—then and forever after—to carry a gun.

What emboldened a twenty-six-year-old graduate student to defy the scientific magi, in these jungles and for the rest of his life? How did he step free of the fear that has driven so much of human history, and into relationships no scientist had yet ventured, including with Earth's most reviled creatures?

History offered scant precedent for approaching wild animals in Schaller's utterly unguarded way. Most zoologists had until then studied only specimens on cold tables, or broken creatures in shackles or cages. C. R. Carpenter had observed free-living howler monkeys in Panama and gibbons in Thailand, but even he soon retreated to the controlled environment of his rhesus macaque compound in Puerto Rico. Konrad Lorenz's studies of animal behavior would win him the Nobel Prize, but he had met his featherweight geese and jackdaws as near pets within the menagerie he made of his own home.

The animals Schaller would spend his life with were far bigger than any of those, and sometimes truly dangerous. But they could be known, he believed, only by entering their world. So he transformed scientific practice: committing months or years to winning the trust and acceptance of lions and tigers and bears, assimilating to their rhythms and rules, endeavoring to see the universe through their eyes.

Though risky for a young scientist, that meant breaking all prevailing norms. For much of the twentieth century, the natural sciences, extending a lineage that ran from Descartes to B. F. Skinner, had insisted on a "disinterested" stance toward creatures seen to be of a distinctly lower order: insensate automatons, animated only by reflex. For the serious scientist, they were data; by "twisting the lion's tail" (in Bacon's apocryphal phrase) the "objective" investigator could answer a question, test a hypothesis, advance his own grand theories.

Schaller stood nearly alone in claiming a radically different conception of nature and science, this one descended from the nineteenth-century Romantic understanding that the "expression of the emotions in man and animals," as Darwin titled his final book, were variations on an evolutionary theme. If animals' interior lives were closely akin to our own, empathy and intuition became critical instruments for understanding them, and lyricism for expressing their complexity. Nature held out a path not to cold mastery but to communion, a stepping free of the self into something far larger and more powerful.

Schaller's quest to know these others would take him into six continents and a life of high drama and romance. Unmoored from his earliest days, he never quit moving, crashing into all the darkness and light the century had to offer, piling up enough brushes with death, transcendent encounters, and grand adventures to fill a dozen lives. Born in Berlin in 1933 to an ambitious young American socialite and a pliant diplomat working for Hitler, he survived aerial bombings, American soldiers' rough searches, and a narrow escape from the Russian zone—fending for himself much of the time, evacuated again and again as the front closed in. Completing high school in Missouri as an enemy alien, then college in the territory of Alaska, he joined the expedition that gave birth to the Arctic National Wildlife Refuge, writing the trip report that ultimately helped sway Eisenhower. He stayed through the violence that erupted when Belgium abruptly left Congo, witnessing the tragedy that unfolds when desperate animals and desperate people collide, and taking mad risks to protect the gorillas. Again and again, for the sake of the giant panda or Marco Polo sheep, he navigated historic upheavals—from the aftermath of China's Cultural Revolution to the flood of Kalashnikovs, Apache helicopters, and opium dealers unleashed in Afghanistan by its series of imperial wars.

Despite all the wrecked countries and tense borders, the informants, corrupt officials, and opportunistic expats, Schaller completed an astonishing run of first-ever studies—of Indian tigers, Serengeti lions, Himalayan snow leopards, Brazilian jaguars, Chinese pandas, Gobi Desert wild camels, and the grand mountain goats, sheep, antelope, cats, and bears of the Tibetan Plateau, to which over forty years he was granted a near-unfettered access no other Westerner has ever had. As David Attenborough, one of his myriad acolytes, once said, "I thought he must sit back and think: What other impossible creature can I imagine that nobody else could get near?"

Coming of age just as air travel became commonplace, Schaller reached far more of the planet than Alexander von Humboldt, Charles Darwin, or any of the earlier generations of adventurer-naturalists. And, in this age of specialization, likely more than any since. Focused by his thirties on the unprotected and unstudied expanses where most life on Earth remained, he walked thousands of miles alone or with local companions, often traversing landscapes outsiders had never seen.

In many of those places—Pakistan's scorched deserts, Sichuan's iced bamboo forests, Brazil's sultry wetlands—he stayed, sometimes for years. Settling into leopard or panda time, he endured alongside them the shrapnel of sandstorms, freezing blizzards and blistering heat, hunger, thirst, leeches, and larvae that burrow under the skin—and also their wild freedom, griefs, and loves. On occasion, he traveled even across time. Driven to know what life was like when we ourselves were still wildlife, he once spent a week living as a hominid near the site of our African origins, reliant on tools of his own making and scavenging what he couldn't kill.

Like devotees across centuries, Schaller embraced the mortifications of the flesh and extreme tests inflicted by these pilgrimages—from abandonment in a whiteout at seventeen thousand feet to venomous snakes in malarial jungles. He came to know many kinds of emptiness and aloneness.

All was made bearable by his love, from age nineteen to the end of his life, for the slight, acerbic, ever-undaunted Kay Morgan. It was she who kept their sons from toddling into the tiger ravine, she who ran after the pet baby warthog and bottle-fed their rescued lion cub, searched out vegetables to stave off the boys' rickets, shepherded them through middle school in Lahore while their father vanished for months into Asia's highest mountains. Though she loved wildness as much as he, she paid her own high price for his commitment, superseding all others, to the largest idea of family.

Like the Horatii swearing their oath to Rome as wives and children wept, Schaller's code of honor seemed of another age. Those asked to describe him often reached for archaic phrases. He was a "pillar" in the world, a man of "sterling character" and deep humility, a companion one would entrust with one's life. He was not without powerful internal drives—the explorer's to be first, the scientist's to know and name, the artist's to faithfully convey, the missionary's to save. But his own psychology never much interested him, far less, certainly, than the mystery of these other beings. Much of his power lay in all he didn't need: not wealth, nor fame, nor even adulation—at least from humans. Lacking (as his younger son put it) the "selfie emotions," he steadfastly refused whatever might steal his time or dispirit him, including solipsism.

Instead, it was to the world that he brought his attention, as intense and devout as prayer. His patience and openness were the very opposite of the shallow, inflamed voyeurism we call our attention economy, so bent on searching out all the ways we are not "them." What kept Schaller going through decades of acute danger and misery, political chaos and heartbreak, was an urge few beyond children or poets remember: to climb inside the other's skin, to live unalienated in the manner of a bird or caribou. To be recognized as someone others needn't fear.

If ultimately on a spiritual quest, Schaller had an immeasurable impact on the world. It's hard to find a wildlife biologist on any continent untouched by "the most important animal researcher in the twentieth century," as admirers from Attenborough to Michael Crichton have called him. It was Schaller who guided Jane Goodall in the first months of her chimp study. It was his work that Dian Fossey took up a decade later: She learned from him how to use noseprints to identify individuals and even worked with his tracker Sanwekwe, though she never forgot Mary Leakey's withering dismissal: "'So you're the girl who's going to out-Schaller Schaller.' It was an intimidating thought to carry with me." The first chapter of her memoir, *Gorillas in the Mist*, is titled "In the Mountain Meadow of Carl Akeley and George Schaller"; in the film version, after Fossey (Sigourney Weaver) has fled headlong down the mountain pursued by a roaring male, the character based on Sanwekwe (John Omirah Miluwi) asks her, "What does Schaller's book say when a gorilla charges?" It says, she answers, "Never run."

Schaller's nearly two dozen books, the finest of which zoologist Desmond Morris, author of *The Naked Ape*, ranked with the works of Darwin, Lorenz, and W. H. Auden, inspired many: Robert Sapolsky became a primatologist after reading *The Year of the Gorilla*; E. O. Wilson drew heavily on Schaller's gorilla and lion studies in *Sociobiology*. His exacting methods—which included a "micro-documentation [of the] frequency, weight, section count and position excreted of the creature's faeces" that one admirer called "astonishing"—set the standard for all who followed, and cemented the enduring value of his studies. Half a century later, his monographs on gorillas, tigers, and lions are still relied upon by scientists and conservationists as their bible.

Most fundamental was his remaking, or recovery, of our understanding of animal consciousness. In the way that tunnels now give wild animals safe passage under highways, Schaller carried older, even ancient, values safely through the arid wastelands of modern science. His vivid rendering of these creatures—their singular personalities, memories, aspirations, friendships, and rivalries—laid the ground for generations of scientists who followed. Emulating his methods, they have reached across to elephants, octopuses, owls, hawks, crows, wolves, and myriad others, taking as given his insights on those beings' complex societies, communication, and minds.

His influence reverberated far beyond science, marking some of the most important cultural artifacts of the past half century. Schaller is the "GS" who led Peter Matthiessen in search of the snow leopard. Director Stanley Kubrick based the Dawn of Man sequence that opens *2001: A Space Odyssey* on his gorilla work, and Crichton wrote him into the sequel to *Jurassic Park*, offered to a diffident young scientist as an exemplar, a renegade who proved expert after expert to be "wrong . . . just wrong."

His impact on the global political landscape was just as profound. From his walks across continents, he grasped before others, and more comprehensively, the radical changes unfolding in even the remotest places. Over seven decades he saw the before and after, crossing terrain largely unaltered since the Pleistocene, but also lands once vibrating with millions of animals now emptied of life. Witnessing firsthand our modern Great Dying, he led his peers in turning from pure science to learning what animals need to survive, setting in motion what in the 1970s would become the new field of conservation biology. And despite a reserve that borders on the hermetic, he proved masterful at diplomacy—persuading governments to protect parks in China, Afghanistan, and a half dozen other countries that together span more than two hundred thousand square miles, an area bigger than France. Were it not for Schaller, we would likely have no mountain gorillas, no wild pandas. As *Mannahatta* author Eric W. Sanderson put it: "George saved more nature than anyone, ever."

That accomplishment depended on yet another, if far slower, transformation. When Schaller began his work in the Belgian Congo, conservation was still infected with ideas of the white man's burden. The men in London and Brussels who ruled the fate of most of the world's wild animals asserted the same authority over the people who lived alongside those animals. They were "*Naturvölker*" to be managed, or intrusions on Eden to be cleared away.

Perhaps opened up by his rebellion against the scientific establishment to other kinds of knowledge, Schaller was able, stepwise, to see more. From the Gwich'in in Alaska and Batwa in Congo he learned how deeply people with millennia of tenure know the animals and land. In India he saw that tigers would be safe only if their human neighbors were too; in Nepal, that villagers and wild sheep depended on the same forests and grasslands, and were equally doomed if the trees and grasses were stripped by loggers or overgrazing. Though it would take decades, he would come to see in China and Afghanistan that it had to be the local communities who shaped and enforced protections, because it was right, and because it was the only way they'd endure.

As George grew into these insights, he helped pull the world along with him, leading a process still underway of decolonizing conservation. As head for many decades of global field science at the Wildlife Conservation Society (WCS), the nation's longest-running wildlife conservation organization and the first to hire field scientists, he advanced community-led conservation and nourished what he counts as his most important legacy: three generations of Indian, Chinese, Brazilian, Rwandan, and Afghani scientists, many of them women, creating new possibilities in their home countries for animals and humans together.

Though he is a magnificent writer, winning many honors including a National Book Award, Schaller always counted himself a "biographer of animals." Far from the "awful lot of 'I'" that characterizes much nature writing, he offered only sparing glimpses of himself. Subsumed in these other beings' lives, the existential crises he wrestled with were theirs. He had practical reasons, too, for his self-effacement: In places like China, he was often most effective when least visible.

Having turned down many biographers, he submitted to my scrutiny only late in life, hoping (it seems) to leave one more useful bit of himself behind. Because I had spent half my life working in conservation, for the US Forest Service and Environmental Defense Fund, and written in previous books about the challenges and synergies in reconciling human and nonhuman lives, he thought I could be trusted, he later told me, to be true to the complexities.

Over the course of nearly six years, George and Kay allowed me to spend many days in their New Hampshire home, interviewing them individually and together and wading through family albums, letters, diaries, and crates of slides, fossils, horns, and other treasures. George often loomed about as I worked, answering any scientific questions precisely, even ones he'd not considered for half a century. Twice, we traveled together. In early 2020 we went to India where, more than fifty years after his landmark study of tigers, acolytes traveled many days and nights to meet him, and locals still sought his guidance on how to resolve life-or-death conflicts between humans and wildlife—pressed into closer proximity there than anywhere else on Earth. Though I saw how ill at ease he was in the throngs jamming India's streets and airports, he clearly loved animal teemings, content to watch for hours as sambar, wild pigs, chital, and finally a tiger wandered through the maidan. A second trip to the US–Mexico borderlands in summer 2024 found him still intent at ninety-one to understand how the border wall had doomed Arizona's historic population of jaguars, and how Mexican and American scientists were working together to recover them.

Like all who spend time with him, I crashed into his world-class taciturnity. Most people, when invited to talk about themselves, go on and on. George answered as tersely as possible, then stopped; there was no warming to the attention nor pleasure at the sound of his own voice. "Not easy to know," Peter Matthiessen once said of him.

Silence is an obligatory habit for those who watch animals. But Schaller's went further. Dropping again and again into new countries—thirty-two in all—each infinitely complex in its history, social structures, and politics, he learned to listen, speaking on what he knew (ecology), quiet on all he recognized that even years in a country would not suffice to understand. Running deeper still was an awkwardness born of his perennially exiled and perilous childhood. Cat biologist Alan Rabinowitz, who overcame the isolation imposed by a childhood stutter by visiting the jaguars in the Bronx Zoo, saw in his mentor a familiar hesitancy. George himself saw that "animals tend to be shy because people have made them afraid" with our noise and menace, because he had been made shy that way too.

My task, in short, often mimicked Schaller's: to peer into an opaque creature.

I stumbled into a second, stranger parallel when George and his siblings put seventy-year-old Bundesarchiv documents into my hands. Like my father, who had escaped from Germany in 1939 but returned in 1941 as an American spy, Schaller's father spent the last years of the war in Bavaria, interrogating high-ranking prisoners of war. Our fathers had the same job, that is, on opposite sides.

It was the Schallers' openness, matched by their gift across generations for preserving artifacts, that ultimately made this biography possible. I was permitted to read a century of letters saved through war, exile, and many transcontinental moves; journals kept by Kay and the boys; "lost" materials including Schaller's proposal to Pakistan's Zulfikar Ali Bhutto for a national park; talismans saved by others, like his 1973 packing list for Peter Matthiessen. Colleagues and friends across sixteen countries were just as generous with their vivid memories, dating as far back as the early 1960s.

Richest of all were the two daily field journals Schaller kept for seventy years—one for data, one narrative—recording in real time everything seen, heard, smelled, tasted, and felt by one of history's most acute sensoria. It was through this writing, said colleague David Western (called Jonah, as he will be here), that "George thinks and reveals himself." As the first to read the more than twenty thousand pages, now archived at the Yale Peabody Museum, I found "that real poetry," as Nabokov called it, "with which the live experience of these receptive, knowledgeable and chaste naturalists endowed their research."

Yet even in these stunningly precise accounts, the intensity cannot be felt vicariously. It is only through presence, our own animal body, that one can know the peace of being allowed near a wild creature or of having it meet your gaze; the primal anxiety of standing alone in a boundless landscape, ill-equipped with a human's weak senses to know who else is alive nearby. That I share with Schaller a call to these other beings posed a paradoxical obstacle. It is hard to explain a love one can't imagine being without, to describe this deepest antidote to loneliness offered by the more-than-human world.

George Schaller's Global Footprint

Taken together, the parks Schaller helped create encompass an area greater than that of France. Pictured here, some of the parks and other places he worked; at right, one of his many unprecedented treks.

Arctic National Wildlife Refuge
ALASKA
Pico da Neblina
BRAZIL
Pantanal Conservation Area
KEY
Parks Schaller Helped Create
Places Schaller Worked
Chang Tang Traverse Route
N
Kilometers 0
2000
Miles 0
2000

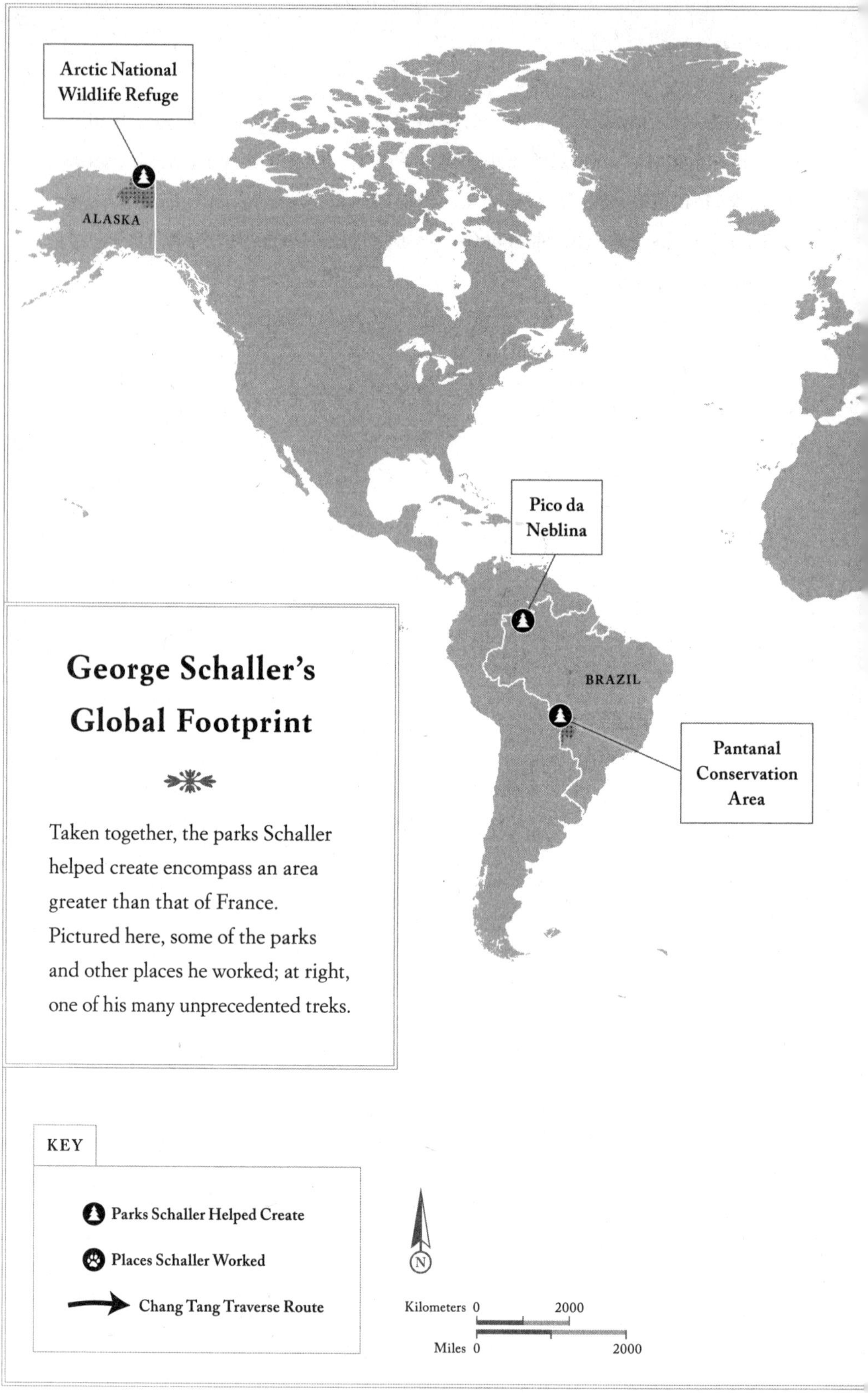

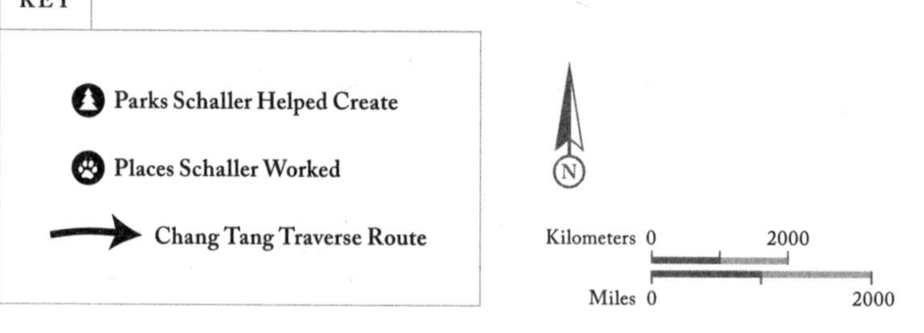

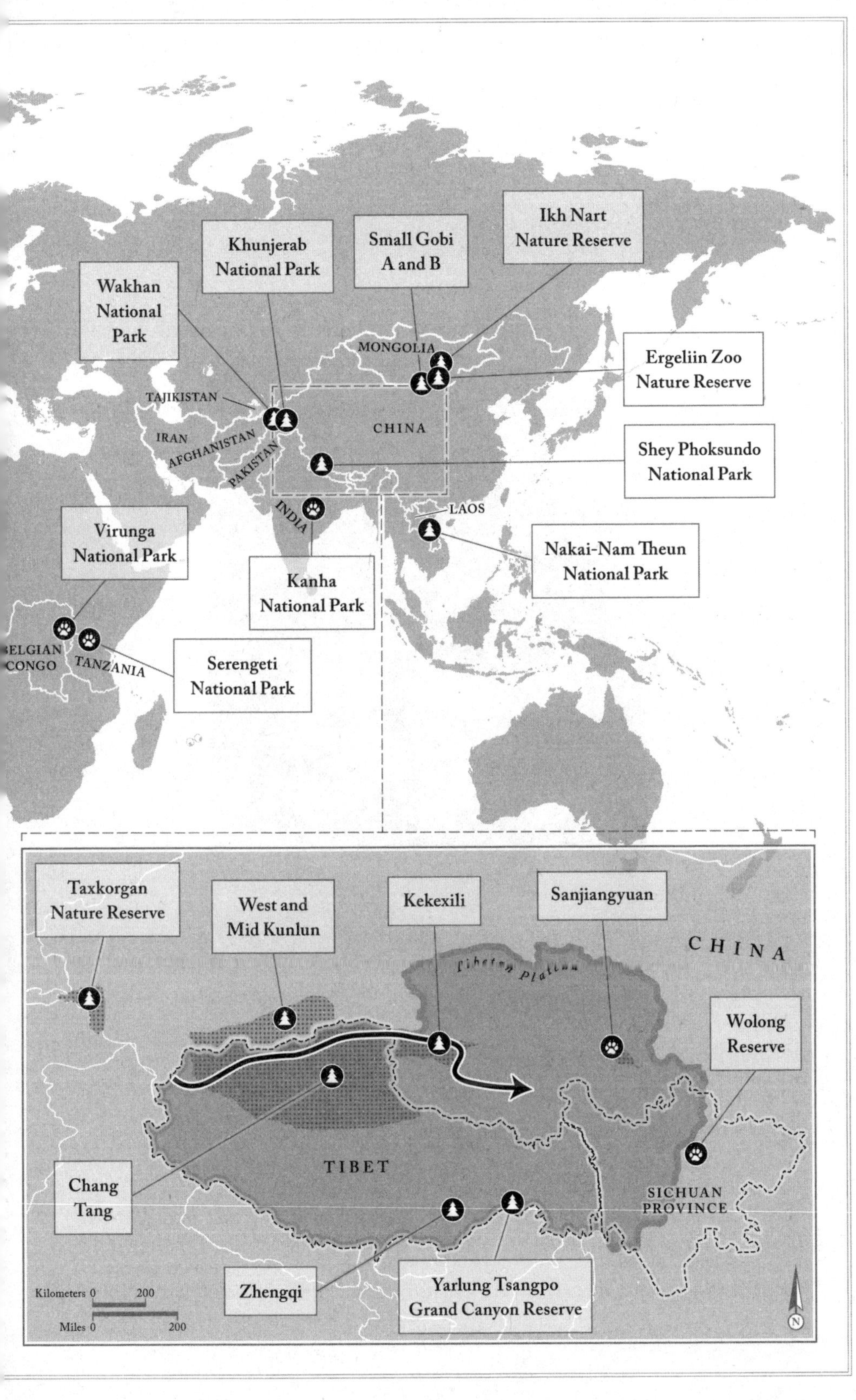
Ikh Nart
Nature Reserve
Small Gobi
A and B
Khunjerab
National Park
Wakhan
National
Park
MONGOLIA
Ergeliin Zoo
Nature Reserve
TAJIKISTAN
CHINA
IRAN
AFGHANISTAN
PAKISTAN
Shey Phoksundo
National Park
INDIA
LAOS
Virunga
National Park
Nakai-Nam Theun
National Park
Kanha
National Park
BELGIAN
CONGO
TANZANIA
Serengeti
National Park
Taxkorgan
Nature Reserve
West and
Mid Kunlun
Kekexili
Sanjiangyuan
CHINA
Tibetan Plateau
Wolong
Reserve
TIBET
Chang
Tang
SICHUAN
PROVINCE
Zhengqi
Yarlung Tsangpo
Grand Canyon Reserve
Kilometers 0 200
Miles 0 200
N

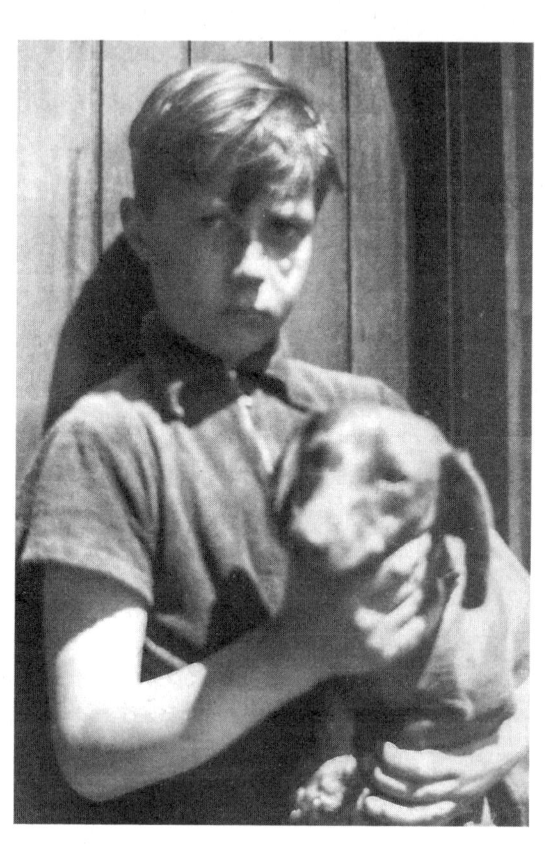

1.

A Box of Eggs

Europe, 1933–47

Oscar began to demonstrate in such a painful way around 4am. 1000s of years later, Georgie said something funny. I laughed. Rap: the water broke. Patients two floors down were inquiring who was being killed. I pulled out my hair, endangered a doctor's eye, split my front tooth. . . . Georgie sound asleep on the couch throughout. At 1:30 a.m., there was the furious face of George Beals Schaller. 8 1/2 pounds. Cute with a largish nose; pointed face and flat ears like my father; my forehead and high eyebrows; and eyes of his own, large grey-blue with curly lashes. He is strong, with a beastly difficult disposition.

When Bettina Byrd Beals Schaller wrote to her dearest friend Marie Kuehn to announce her firstborn's arrival, she was just twenty years old and still dazzled by her new life as the wife of a promising German diplomat. Dancing all night in palaces lit by a thousand candles, switching among five languages to bewitch aging princes and dashing ambassadors: This was the life this Midwestern socialite had been groomed for. But it was also May 1933, and Bettina and her husband had recently settled in Berlin. "Georg has joined the N.S. [National Socialist] forces," the letter continued, "as we are well informed as to goings on. It looks as if Bealsie will only grow up to be cannon fodder; I guess many mothers are thinking the same."

Bettina was eighteen and had only just returned from six years of boarding school in Europe when she met George's father, then the German vice-consul in Chicago and ten years her senior. Georgie, as she would call him, fell immediately for the dark-eyed beauty with fine arched brows. Their engagement, in June 1932, was covered in three Chicago papers, illustrated with photos of Bettina in her fox stole (head and tail intact) or the jowly Consul Schaller in double-breasted jacket and fedora, helping her into the rumble seat of his car. To prepare for the wedding, the bride had gone to her aunt Paula's mansion in La Salle, Illinois, where Bettina's maternal grandmother, Helen Conkley Barnes, had settled following the death of her husband, US Congressman Lyman Eddy Barnes. "I would like to kiss you as many times as there are stars above," Georg wrote her there. "Sometimes words are a poor medium. . . . Read the waves of love between the lines. Late at night, I send millions of such waves, knowing that you are a fine, sensitively vibrating antenna."

Married August 30, 1932, at St. Chrysostom's Episcopal Church, the Schallers left the same day on the much-storied 20th Century Limited. (Among its many glories, the elegant train had invented the red carpet for passengers that included J. P. Morgan and Enrico Caruso.) The German consul general had invited them to dinner at the Waldorf Astoria, then took them to the dock to board the SS *New York*. Six days later, they landed in Bremerhaven and boarded the train for Berlin.

After a "heavenly" crossing in the bridal suite, Bettina wrote Marie, "It's awfully nice to wait for the key in the door . . . there always the same good-natured little man, busy and reliable." They were pregnant almost immediately. "We visited the doctor and after thumping me and prying around he gave the verdict that the Blessed Event will put in an appearance end of May." Miserable with morning sickness, she arrived for a weekend at the country mansion of the former president of the bank of Germany with "a face that would have dispersed a herd of gorillas." By Christmas, they were calling her baby bump Oscar. "Cute little Georgie seems thrilled to death about the baby and I guess I will love the little thing too. Oscar is very small but determined and lively. If a girl she will be Helene Alexandra-Marie. If a boy, he will be George Beals."

Bettina did not share the news with her family, not even in a long Christmas letter to Grandmother Barnes. Instead, she recounted their visit to Radebeul, near Dresden, to meet her parents-in-law, Edmund and Helene. "They are just darling. Papa Schaller met us at the station with a bouquet of flowers held straight in front of him." She made fun of herself, her favorite way to beguile: A feast of hare goulash had "added ten pounds to my buxom physique" and her sniffles kept poor Georgie from sleeping through "my constant blowing and honking." She finished with the material obsessions that by temperament and soon necessity would preoccupy her. The diamonds big as Easter eggs in Dresden's Grünes Gewölbe were "enough to make a lady commit murder." And might she have the family silver? "We need it badly. . . . But oh well, do as you like Dear. If worse comes to worse, we can always eat with our fingers."

George Beals was barely a month old when the family was transferred to Prague. As was customary for her class, Bettina left his care mostly to servants, often shipping him off to *Opa* and *Oma* with his baby nurse so she and Georgie could keep their calendar of parties and grand tours. Such was the life she had known as a girl. Her much-decorated father, Major Frank Beals, was rarely home, busy with such duties as preparing officers to lead the American Expeditionary Forces against the German Empire. Her "angry hornet" mother, Alice, abandoning him for Europe, brought along twelve-year-old Bettina but handed her straight off to boarding schools, settling her finally at Pensionnat Roseneck in Lausanne. "You would like the sweet directress Mlle Ecuyer," Bettina wrote to Aunt Paula, "tiny with very pink and white skin. She has a way of reading people. She said that when she shook hands with mother, she had a peculiar sensation of no contact at all."

It was at Roseneck that Bettina had met Marie, a steel heiress four years her senior to whom she would write hundreds of intimate pages from Europe between 1928 and 1947, preserving a remarkable chronicle of George's childhood world. Bettina had fallen in love at first sight of Marie's closetful of shoes, "wondering how anything human could navigate on shoes that diminutive." She sustained her swoon over the gifts Marie regularly sent from home. "Cinderella has nothing on us now," she wrote of mules with silver stitching on satin heels. "If we *marcher sur la tête* it is to show off our *piedis* who are *d'une elegance incroyable.*" She shared bits of the sparkling life into which the school was initiating them: dining at the Savoy, shopping at the famous watchmaker A L'Emeraude, "blubbering so" at *Madame Butterfly* at the Opéra-Comique "we almost drowned those next to us."

The many gifts Bettina would bequeath to George are manifest in these letters, her acute eye and skilled hand prefiguring the drawings he would later use to illustrate his correspondence and journals. With great economy she captured the long-faced scowl of a stately lady or the bulges of middle-aged bathers. Her strongest suit was satire, especially when telling of the hearts she broke. The most tragic was Dr. Maurice, the dentist who while fixing the seventeen-year-old's braces suddenly froze, gaping. As he stood trembling, huge tears dashing down his cheeks, Bettina thought it a violent attack of apoplexy. Instead, "like poor Thomas Diafoirus in [Molière's] *Maladie imaginaire,*" he blurted out a proposal, "said he was crazy about me and could give me everything I want (which means he's rich!)." She told him she didn't love him but "felt cruel saying so and started weeping. He bundled me in his arms and I must have fainted for I woke up feeling suffocated." He took her monogrammed handkerchief to carry to his *service militaire*, which fortunately began the next day.

•••

Returned to Europe as a bride, Bettina resumed chronicling her glamorous life. Finding "herds of bachelors (counts, barons, and whatnots)," she waltzed through grand ballrooms strewn with lilies of the valley, spooned up caviar with the Japanese legation, flirted with Polish industrialists and Italian dignitaries. She and Georg holidayed in the Alps, visiting castles and gathering mushrooms and wild raspberries.

"This is the first time in my 21 years," Bettina wrote Marie at Christmas, including in the envelope a highly staged picture of herself gazing adoringly at her baby boy, "that I am truly happy. Georgie is ideal for me and the baby; he would make anyone happy. Beals is bright and lively, moves all the time like an eel and goes off in wild explosions of laughter over nothing. Eats like a horse." By eighteen months, grown "perfectly enormous," he careened about "like a mad hatter"; they had to tie the Christmas tree to the radiator with her corset strings to keep him from pulling it over. Called *Pilzie*, little mushroom, he is "so much of a person. His hands are sweet, his fingers long and tapering and so dexterous. He loves anything tiny; affords us no end of pleasure." The other diplomats' wives coveted the pink Snuggle Bunny ("a scientifically-designed crib cover") he slept under. Could Marie perhaps send more? Just once, Bettina betrayed awareness of what her absences cost George. Returning from a trip she found him, not yet two, "in a frenzy, hugging my knees and trilling 'Mama.'"

As to the history unfolding around them, Bettina often seemed naive, no doubt willfully so. In 1932 she assured Marie "there is no depression here at all"; two years later, urging her to visit, she guaranteed "at least ten peaceful years."

Only rarely did she let slip glimpses of the growing darkness. On March 14, 1933, seven months pregnant, she began a letter apologizing for her déclassé stationery and for having "nothing of importance" to tell. Then, without mention of the Reichstag fire two weeks earlier, she said how glad they were to have left central Berlin.

> *Things are badly upset in the city, lots of demonstrations and shootings. The Jews are having a hot time of it and many are leaving as fast as their flat feet can carry them. The new government is shedding most high officials . . . so Georgie and I are glad that he isn't much worth getting rid of at present. . . . The doorbell rings all day with beggars, decent looking people, clean but with pinched desperate faces that would melt an anvil. They all but kiss your feet if you offer them dry bread. PS: if you could send me a second-hand book on the care of children I shall need it badly, I'm afraid.*

In June, a month after Bealsie's birth, she sent pictures of them in disarray. "Someday when Georgie is a big ambassador, dictator or such, you can haul these out and blackmail us."

It was a perverse joke for the time, but even as the storm gathered, Bettina's ambitions for her husband seemed to exceed his own. When in early 1934 he was appointed secretary of the Prague legation, she was thrilled to be the next Lady after the ambassadress. Her own charms, she knew, were crucial to his advancement. Since "our next position depends on what the department head thinks of me," she asked Marie to send tools of her trade: an eyelash curler and Princess Pat brilliantine, with which she created a sleek wave in her lustrous hair.

Her continuing parodies suggested little unease with the ascendant ideology. Sitting at a show behind Irish girls, "I'd have given my kingdom for a bottle of *eau de corps*." On seeing *The Jazz Singer*, she wrote that "it's about a Hebrew so of course all the Jews were there and just to see the prize-taking profiles made me split with laughter." In April 1935, she responded to Marie's account of her reunion with Roseneck girls: "Rozie is . . . about the only Jew I have ever liked. I'm glad to see her husband is not the kinky-haired type. She has none of the characteristics of her race, but Hortie's just a damned little kike if there ever was one."

In 1937, soon after George turned four, his father was transferred again, becoming vice-consul to Katowice, Poland. As the world careened (Bettina wrote of gathering at the consulate to hear *der Führer* speak, and of her relief that Marie—who had finally visited—had not gone home on the Hindenburg) the Schaller revelries grew boozier, edged with an end-of-the-world decadence. Bettina sketched their sotted friend Schueller "in an advanced stage of syphilis" greeting them in his gaping bathrobe, champagne glass in wobbly hand being tipped of its contents. She gave news of the diplomat who had flirted with Marie: Worn out by his "mouse-like" Polish *maîtresse*, he'd passed her on to a friend. Their gravely ill pregnant friend was now at the mercy of Polish doctors, since "no German doctors are allowed to practice here anymore."

George (now become "the Buzzard" or "the Buzz") continued to delight, relishing any adventure or present. Some Sundays, they took him to the Beskid Mountains to gather armfuls of primroses. "You are so full of life and into everything," Bettina captioned one photo, "that these sedate people did not know what to make of you." He talked endlessly, though only in German, which his parents believed a safer choice than English. One August afternoon, the nurse carried him home white in the face; he had fallen off his scooter. Bettina "got the whole consulate in an uproar searching for Zozzle" (another nickname for her husband), who returned from the doctor with a long face. George had broken his thigh straight through, requiring a week in the hospital in terrible pain, and a cast from hip to ankle. "He has a fever and is hoarse from crying. I felt wretchedly sorry for him."

By Christmas, fully restored, the Buzz was chortling over Marie's present of popcorn; "he simply couldn't get over the way it exploded." Bettina marveled at how he "talks so fast and says such logical things. I took him to see a play about pirates and he has been whooping around ever since. *'Froh dass meine Eltern kein Piraten sind.'*" (I'm glad my parents are not pirates.) He was particularly grateful for the presents from Bettina's father, *Opa aus Amerika*, who on a visit brought him two telephones linked by a wire and a real canary. "He is overjoyed to have found a playmate, is a whole lot less pestiferous and enjoys all his toys, finding a use for everything." Photos from the time tell a more complicated story. Perhaps rebelling against the formal jackets and short pants he was made to wear, George's handsome little face is almost always in a scowl.

It was becoming increasingly difficult to shield him from the chaos outside their door. He wanted to know why there was "patsy" (shit) smeared on monuments honoring Germans. When in March 1938 Hitler marched into Austria, he noticed the cars "held nearby to rush us across the frontier" should Poland erupt in reaction. The Schallers were dining with Polish and Czech nobility ("pro-Hapsburg, as they all are") the night Hitler reached Vienna. "We all steered around the burning subject most carefully. The only casualty of the evening was my fur-lined boot, chewed by the host's pugs." The couple did enjoy some victor's spoils, traveling soon after the Anschluss for a private viewing of the Albertina museum's famous Dürers. "I was just reveling," Bettina wrote of her purchase from a ruined French banker of a 1777 painting in its original frame—of Marie Antoinette.

In August 1938, with George ready for kindergarten, Bettina asked for Marie's help in securing her Ahnenpass (ancestor passport), proving that "back to our great grandparents' parents we were Aryan . . . which Beals needs for entering school."

Her letters increasingly mixed bravado with black humor. She'd paid a visit to an Italian friend, "to strengthen the Berlin-Rome axis, don't you know." Of the Munich Agreement, which gave Hitler a chunk of Czechoslovakia, she complained that

> *everything was alright until the Poles and Czechs got huffy with each other. It's a right coy life, it is. In Wisła district, we drove past 28,000 soldiers and by gum it looks business-like. On Friday night we expected an air raid from the Reds but they did not put in an appearance, so at dawn we put the Buzz and nurse on the train to Dresden.*

Having for the first time to do her own domestic chores, she kept up the self-mockery. Trying to restuff a feather bed, "I dragged my prey into the bath and cut it open. Stepped on the animal and it sent great gusts of feathers flying. A custard pie would have made the scene complete."

Another Christmas, this one marked by air raid rehearsals: lights-out at 6:00 p.m., sirens, planes dropping fake bombs. "Buzzwuzz had the time of his life." He was equally excited by his first circus, deciding to "become a lion-tamer when he grows up!" But with anti-German sentiment mounting—"conditions here are such that I could not allow him out in the Park alone with the nurse"—they decided that he should be sent to his grandparents to start kindergarten in Dresden. "The Jews are leaving in droves. The Germans here are starving because the Poles won't give them work. The French and English have left. We have orders to sit tight."

By April 1939, Bettina had also left Poland, joining George in Radebeul. A picture finds him dressed for school in a dark jacket and shorts, white knee socks, and a leather pack. Though his arms are full of the traditional first-day paper cone of candy, nearly as tall as he is, his face is somber. Georg Sr. had stayed behind, and on September 1, 1939, "watched the arrival of our troops." Within days, the German Army had burned the Great Synagogue of Katowice to the ground, and Britain and France had declared war.

When later that year, Georg Sr. was again reassigned, this time to Copenhagen, Bettina rejoined him, enrolling six-year-old George in St. Petri school. They had only enough coal to heat two rooms, no hot water, and difficulty finding coffee, rice, and vegetables. But Bettina—unearthing a stoicism unpredicted by the spoiled little girl she'd been—knew it was "paradise. . . . In this beautiful city I shudder thinking back to the soot, mud, and hopelessness of Katty." Though she and Georgie were never in bed before 2:00 a.m. ("I am getting the greenish bar-room tinge") there was no "trip-ing" now. But "it is good to be healthy again and to have a real live husband."

The repeated moves into communities that feared and loathed his family isolated George. Though he was soon fluent in Danish, neighbors forbade their children to play with him; he was similarly shunned in Germany as an American. As Kay would later reflect, zoologists understand play as essential to the young of any species; without it, George "fell behind" in his socialization. Although the repeated uprootings may also have honed his interpretive acuity. Trying to safely navigate a world that fears you, in which you understand neither the culture nor what anyone is saying, is almost like dropping into another species.

The alienation certainly required George to cultivate a capacity for creative solitude that would serve him all his life. In December 1940, confined to his bed by pneumonia, "he is horribly busy reading and writing and drawing, the cutest little critter." One of those early drawings survives: a cartoon somewhat in the style of *The Katzenjammer Kids*, accompanied by a poem in Danish. Both depict a candle burning through a rope that restrains a dog with sharp teeth, who will soon be free to attack a bulbous-nosed onlooker. Feral dogs will be one of the few beasts that ever frighten him.

Bettina's letters now rarely made it to America. "I got your letter saying Christmas was the last you heard," she wrote Marie in October 1941. "I've written at least five since so some dirty work is going on." George, now eight, was in his third year of school. "You would like him, a lovable sort of boy. He keeps my mind taut answering a never-ending stream of questions. He would be a congenial travel companion except for his talking so much. The display windows downtown have him all agog." Again, a picture of him holding his dachshund, Erna, and glowering at the camera, tells a less lighthearted story. "Somebody's cross about something!" reads the caption in Bettina's hand.

On December 4, 1941, four days before the US entered the war, Bettina wrote her last letter to Marie, or the last to arrive, until 1946. She was finally pregnant again. "Baby is progressing nicely and with his squirming and kicking promises to be another edition like Pilzie." Chris arrived on April 9, 1942. But with Georg soon ordered back to Berlin, Bettina was left to manage wartime life with two small boys on her own. Two of her portraits of nine-year-old George survive: In one, in pencil, he looks directly at her, his face a mix of defiance and defeat; in the second, a watercolor profile, he appears full of sorrow.

They never did join Georg in Berlin. In a family history Bettina wrote for her sons forty years later, she explained that when they refused to accept an apartment that had been expropriated from Jews (a story her sons suspect is revisionism), they were told they would be offered no other housing. Stranded, in January 1943 Bettina took George and Chris to live with their grandparents again. Georg Sr. moved into a room at the Hotel Adlon and visited once a month.

Settled in Radebeul, George began painting and drawing in earnest. A charming watercolor of "the animals a fox hunts" includes a wild-eyed hare, an elegant stag, and a broken-necked

Schaller's first predator-prey study.

goose dangling from the predator's mouth. With colored pencils he drew Indians in headdresses and teepees, perhaps under the influence of Radebeul's most famous son, Wild West–adventure writer Karl May, part of whose home was by then a museum. For Christmas he made his *Vati und Mutti* a *Tierbuch* (animal book) that included a chimpanzee, lynx, Grant's gazelle, and baboon. He began looking carefully at plants, painting three kinds of cherry and four types of plum. Having learned to forage fungi, he frequently signed his nickname, often just drawing a little red-crowned mushroom with white spots followed by a lowercase *i.* One Easter, he turned it into a little house, with a mama bunny waiting at its door.

The ten-year-old frequently knocked heads with his grandfather, who was not happy to have a rambunctious boy tromping through his small house and cherished garden. ("Opa would come yapping out of the house," Bettina later wrote, "if Pilz even stepped off the path.") "The rigid Germanic type," as George remembers him, gruff Edmund believed in strict training and harsh punishment for children—all of which inspired in George only insubordination. He deliberately bungled his English lessons. "If someone's pushing you and you make a little joke and he blows up, then you make another joke. I'd purposely write down the wrong word. He'd hit me wherever he could reach, with his hand or a stick. There was a wall he forbade me to climb so naturally I did. He chased me around the yard but I could run faster." George didn't yet appreciate the strain and danger his grandparents had taken on by hosting these Americans.

George had scarcely seen his father for eighteen months when the family was splintered further still. On October 22, 1944, Consul Schaller received a hand-signed letter from Foreign Minister von Ribbentrop.

> *Hitler has decreed that positions in the government, party or Wehrmacht can no longer be filled by men who through marriage or other reasons have a special connection to enemy countries. I must therefore share with you today my particular regret* [besonderen Bedauern] *that*

> *your further service is no longer possible. By the Führer's official document from 28 September, 1944, enclosed here, you are thus transferred into retirement. . . . On this occasion, I would like to convey again that the loss which the departure of your person means for the foreign service is very regrettable to me, and to express my heartfelt thanks for your longstanding faithful service.*

The letter, both sacking and lauding him, further muddies the question of how committed a Nazi Georg Sr. may have been. Documents preserved in the Bundesarchiv (which include papers he submitted to the foreign service after the war, when he sought unsuccessfully to be reinstated and pensioned), suggest a middling career, with no promotions after 1935. But they also record several honors: the Sudetenland medal for his service in the occupation of Czechoslovakia; the War Merit Cross and an honorary post as a *Sturmbannführer* (battalion commander) in the SS in Poland; training in the SD, the state security service under Himmler, in Denmark. In his application for reinstatement, Consul Schaller said that he had joined the Party and accepted the merely "honorary" appointments to protect his career and family—as he was married to an American, he had to show his support for the regime—but that he never did any SS or SD work nor wore the uniforms. One document indicates that while in Katowice he helped smuggle out two journalists from Upper Silesia.

Demoted to a common soldier, by Christmas 1944 it appeared that Georg would be sent to the Eastern Front—a near-certain death sentence—had an influential friend not managed to instead secure him a post near Munich as translator at Stalag VII-A, Germany's largest POW camp.*

The Russians were now just fifty miles east of Radebeul. "We could hear cannon fire and allied planes flying over to bomb Berlin," Bettina wrote in

* This is according to official documents. Bettina wrote that he was at Stalag IX-B, which was infamous for searching out the Jews among the American forces and sending them to a forced labor camp.

her retrospective chronicle. “As thousands of refugees poured in, fleeing the Red Army, the 10- and 11-year-old boys were being sent down at night to the Dresden *Hauptbahnhof* [Central Station] to help. Father and I decided to get George away.” They arranged for him to be accepted at one of the Hermann Lietz boarding schools, where landed gentry sent their sons to be trained in estate management.

George was sent first to the school’s Schloss Ettersburg campus, in a castle in Weimar, though not for long. SS officers soon arrived, with the rare treat of chocolate bars and orders to evacuate. Only later would George understand the significance of the school’s proximity to Buchenwald, which by then housed tens of thousands of slave laborers as well as a major munitions plant.

Twice more the students were moved, as the front kept drawing near. In the only bit of autobiography George ever wrote, he described going by military train to Schloss Bieberstein, another of the Hermann Lietz schools. They were on a siding in Halle when an Allied air raid “demolished part of the city as we crowded into a railway underpass. A bomb hit one entrance and a hurricane of debris blasted over the moaning crowd. The train survived, battered and tattered but functional, and we continued.” Looking back years after writing this, he downplayed the drama. “You get bombed; well, so it goes. People all over the world have been bombed by the Americans.” And getting caught in violence, he learned, focuses the mind. “When bombs are hitting the bunker you’re in, do you lie down and hope you aren’t trampled, or push through a door?”

Moved again, to the Grovesmühle school on a farm near the Harz Mountains, the boys spent long hours digging potatoes and (soon) trenches.

> *One night the ground shook and distant explosions thundered. From the trenches we observed flares arcing across the sky and a flaming horizon: the retreating German army had blown up an ammunition dump. Within days two American half-track vehicles filled with combat troops pulled into the school courtyard.*

Ordered by their principal to lie in their beds, the boys froze as the soldiers went cot to cot searching for hidden snipers, shoving their weapons into the scared young faces. A few days later, the school was overrun again, this time by Serbs freed from a German labor camp, who looted all the food and clothing they could carry, taking even the school cats. Again, George recalled seventy-five years later, each boy sat alone. "We were there too briefly to make friends. And as a boy you don't think about the depths. You think only 'How am I going to get food?' You just survive day by day."

By now Soviet tanks had reached Radebeul, as Grandfather Schaller described in his own account, driving in front of them more refugees carrying suitcases "like a cat her kittens." He and Helene had decided to stay in their house come what may, but soon "felt it unwise for a young woman to remain where Russians troops were sure to pass." How Bettina might go, however, was not clear. It was impossible to get a travel permit; rail travel was reserved for soldiers. She was preparing to go by bicycle with Chris in a baby seat when she heard that a friend's sister-in-law had a railroad pass to return to the estate her father managed, where they had access to food. The woman agreed to take Chris until his family could retrieve him. Bettina would pay with money, coffee, and soap.

Late on the brilliantly clear afternoon of February 13, 1945, Bettina, Opa, and Chris took the winding electric train across the Elbe River. Bettina had written often to Marie of the beauty of Dresden, especially at night, when its grand old buildings and illuminated fountains looked like a magical stage set. Now, as the sun set and the evening star rose, "we caught one last glimpse of the copper roofs and greenish domes." They went to the apartment of an elderly Swiss friend, Frau Schoch, who lived two blocks from the station. The plan was to put two-year-old Chris to bed until 1:00 a.m., when Bettina would deliver him to the woman who would take him to safety.

Opa said goodbye and headed home, leaving Frau Schoch and Bettina chatting over tea. Suddenly, air raid sirens howled; no warning alert, just straight to a full alarm. Rousing Chris and putting on his shoes, snowsuit, and cap, Bettina led him one toddler step at a time down four steep flights to the shallow basement already overflowing with terrified neighbors. “No sooner had we pulled the door shut than we were shaken by a tremendous thud, the walls swayed and soot blew out of the chimney base blackening our faces. Despite our great fear we had to laugh. Then detonations, frighteningly loud and close.” (The stomping footsteps of giants, Kurt Vonnegut called them.) “The building heaved back and forth and we heard loud cracking. Mothers bent over to shield their children. After several explosions the deafening noise abated.” When Bettina climbed out with Chris in her arms, the five-story building was gone, destroyed by a “blockbuster.” All they could hear were live wires crackling and terrible cries for help from beneath the rubble. “All we could see were the stars.”

From their home, the elder Schallers had watched the sky over Dresden turn bloodred. At dawn, Grandfather went by trolley in search of his family. Wading through husks of burned buildings littered with “naked, blackened bodies,” he found Frau Schoch’s home, at Sedanstraße 15—an incinerated shell. A strong wind blew dust in his eyes and, feeling as though he were sinking into the earth, he began to cry. When US Air Force bombers roared over “to destroy what the British hadn’t” (leaving by the end of three raids an estimated thirty-five thousand dead) he hid under a bridge, then headed home in despair.

Bettina, meanwhile, had determined to walk out of the city with Chris on her back. Wading through several inches of glass splinters, past streetcars tipped on end, she smelled gas and felt the air to be “strangely hot.” Looking back, she saw gigantic pillars of flame. At the city’s edge, she rested at a friend’s until a female warden knocked to warn of another wave of bombers. Bettina stuck her fingers into Chris’s mouth to keep his eardrums from bursting and told Frau Schoch to remove her dentures lest she swallow them. As phosphorus bombs set houses on both sides of them afire, “we were so frightened that we ceased to be frightened.” Chatting, pretending for Chris, they were soaking sheets to crawl out in when panicked animals from a nearby farm raced past, making a fearsome noise, "their horns, wings, tails and fur aflame."

Finding a small cart in the cellar, Bettina piled in their baggage and set Chris on top. Then, like Mother Courage, she pushed and pulled the cart nearly four miles to the village of Gorbitz. Their eyebrows were singed and their nostrils covered with scabs, but "in the distance a lark was singing." Reaching the home of another friend, Bettina managed to get a message to Radebeul.

When Grandfather arrived home from his desperate search, his wife met him at the door. She had just received the message via the neighbor's phone. "Escaped death—they live." At six the next morning, the doorbell rang. There stood Georg. "'Dear God what's happening' he said, as we fell around his neck." Fearing what the Russians might do to his family, Georg had won a leave of absence and a spot in an open car of a military train crossing Bavaria and Bohemia. Ordered off the train when it began taking fire, he had walked six hours through the night (stopping to check on their stored belongings, in Bettina's version) before continuing sixteen miles to his parents. After a few hours' rest, he went by bicycle to fetch Bettina and Chris. They walked many hours home, pushing Chris on the bicycle.

Two days later, Georg, Bettina, and Chris said goodbye to the elder Schallers, not knowing if they would ever see them again, and headed west to what would become, per the agreement just concluded at Yalta, the American zone. The bombers, fires, and cries of the dying followed the family throughout their three-hundred-mile journey. In the Spessart Mountains, Bettina wrote, "enemy" (by which she meant Allied) reconnaissance planes droned overhead.

Twelve-year-old George had gone months without word of his family when one day he spotted a "stocky man with a rucksack" walking alone up the lane: his father come to find him. Wanting George nearer the family, now settled in Bad Orb near Frankfurt, his father was here to move him yet again, to a fourth school in a seventeenth-century castle. George's lessons at Schloss Buchenau were desultory affairs, taught by ruined men: "a shell-shocked air force officer; a math prof ancient and abstruse . . . an athletic director with part of his skull missing who played soccer even though an errant kick could have penetrated his throbbing brain." When his history teacher, a Catholic priest, decreed that he would grade based on church attendance, George made a point of not going. He still worked in the fields, now paid mostly in rutabagas. Served for breakfast, lunch, and dinner, they became one of the rare foods George would forever after refuse to eat.

But it was also here, at Buchenau, that George "discovered nature and the joy and contentment it brings."* With the son of a forester who became his first real friend, he searched the surrounding woods for the nests of buzzards, mistle thrush, and collared doves. Collecting an egg from each, he blew them clean, then tucked them into a wooden box padded with cotton wool. He waded into streams to grab trout by the gills, impaling them on a stick to roast over a fire. "He is so unlike the other boys," Bettina wrote to Marie, "you would die at him. He is a combination artist and fist-fighter. His teachers have never run into such a specimen. I guess he is rather unusual. He has a perfectly beautiful face. I wonder what on earth he will do?"

For his family, meanwhile, the end of the war had brought new travails. Georg was taken into custody for several weeks, in handcuffs. For his freedom, his father wrote, "he can thank his wife who had stayed an American citizen."

Food was ever harder to come by. "I guess my starvation training came

---

* Nature as an antidote to war is a recurrent theme in literature. George's great WCS predecessor William Beebe called his 1918 book on British Guiana, which George would read in Alaska, *Jungle Peace.* "After creeping through slime-filled holes beneath the shrieking of swift metal . . . I turn with all desire to . . . the great green wonderland . . . to slip quietly and receptively into . . . this age-old fraternity. . . . The peace of the jungle is beyond all telling." In *H Is for Hawk,* Helen MacDonald describes an English craze for long country walks born of the trauma of the Great War. The character *Dr. Dolittle* first appeared in Hugh Lofting's letters home from the trenches of that same conflict.

in handy," Bettina wrote Marie, "because I have kept up my weight and most people look like overstuffed chairs or coat racks, according to their connections. Though when we were nearest to starving, a grippe almost sent me flying. I just didn't have reserve strength." The family received aid from the United Nations Relief and Rehabilitation Administration (UNRRA) and the Red Cross, and Bettina's father, for a time, sent care packages of candy bars and peanut butter, vitamin pills and toothbrushes. Whenever George was home from school, he and his father rifled through those packages for the most coveted items, to trade at farmhouses for bread. George also took a big tin can to the garbage bins behind the American troops' mess, intercepting bits of pancake and omelet to take home for the family meal. With Chris, he gathered cigarette butts for Bettina, who scraped together any remnant tobacco into smokes for herself. On one of their reconnaissance trips, Chris fell into a creek. George fished the four-year-old out and then got hell for letting him fall in, which Chris thought exceedingly unfair.

Since it was clear that Bettina's petition for repatriation would not be granted soon (if ever), and with food growing scarcer, in November 1945, Georg decided that he and George would return to Radebeul, to retrieve the family silver and other heirlooms to sell. His parents had fared passably well so far. They had seen the Germans' last desperate moves: first blowing up the Elbe bridges and shooting any villager who dared call it a lost cause, then rushing to replace swastikas with white flags. But Edmund—who wrote in his memoir that he always wore his velvet skullcap so the invaders would know he was a real grandfather, "even if a capitalist"—had buried their valuables, alcohol, even jams in the garden. When wagonfuls of raggedy Soviet soldiers came demanding schnapps, he gave them his cheap stuff; they then insisted he join their toasts, shaking his hand and offering cigarettes. Four times armed men rifled the house, stopping only when they came upon a painting in the library of Mary and Jesus.

Getting east to his grandparents was easy, but George would later, in a sketch written for high school, memorialize their harrowing return to Bad Orb. At 5:00 a.m., his grandfather helped them load their heavy suitcases onto the streetcar to the train; as it wound its way through bombed-out Dresden, George saw people sheltering in crates and cardboard boxes. A local train with missing windows and a shattered roof took them to the mobbed Erfurt station. There, they pressed into a train so packed that George had to fold himself into a windowless cubby meant for dogs, then roll through the pitch-black hours not knowing whether his father had made it on. (He had, but only onto the roof, where he held on through the freezing night.) Having made it to the western edge of the Russian zone, they sought passports out but were told no one would be allowed to leave for at least fourteen days.

Georg decided they would go anyway, in the dark of night, and found a potato farmer who for cigars and whiskey would take them across the three-mile neutral zone. They feared most the German ex-soldiers the Soviets had hired to patrol, who were so poor they would, wrote teenage George, "take all your things away if they catch you." Crouching in bushes in the pouring rain whenever the border patrol passed, they made their way through the deep mud of unplowed fields. "I fell down very often and looked like a pig climbing out of its mudhole."

Then their potato farmer ditched them. It had taken four hours to go a mile and a half; they could see the lights of the American zone but George was so tired he could barely stand. After a half hour's rest they set out again, pushing and pulling their suitcases through mud that "stuck like glue," finally risking the road. Twice they jumped into the ditch to hide from Soviet patrols, but at last they reached the final obstacle: a river to cross, its bridge guarded by Russians. They would have to wade through chest-deep water, twice, carrying the suitcases one at a time above their heads. Fearful of being arrested as American spies, they went in silence and without lights. The first crossing left George's feet numb. On the second, he stepped into a hole, felt the water rise to his nose, and with his feet still sinking and the suitcase slipping from his grasp, cried out softly for help. Seeing his distress, his father rushed to toss the case he was carrying onto the far bank, then splashed back to take George's load so the boy could swim to shore. Dragging themselves the last hundred yards to the village, they found a deserted hut with a corner of straw. George had the best sleep of his life.

It would be two more years of waiting in Bad Orb, sometimes in a single room, sometimes crowded into a displaced persons (DP) camp with Jewish refugees, George coming and going from boarding school, before the US government would make its determination on the family's future. They managed in those years only halting contact with the Schaller grandparents, who at first tried to send more of their things, though the Russians capped parcels at a pound "so mother-in-law is dismantling our clothing to send piece by piece. Every so often she includes a relic like that can of pore-grains, that I'm applying every evening." By Christmas 1946, the grandparents were the ones in dire need. "They have no fuel or light this winter, almost no food, though they don't complain. No one in the Eastern zone dares to." Georg bitterly joked that it took just five days to get a letter from the US but a full month from his parents because "the Russians took away the second rail and can't decide which direction to send the train. The dear Bolsheviks have a fine talent for organization."

Though it seems unfathomable given all that was coming to light, until their moment of separation the couple seemed to adore each other. "Good old Zozzle is lots of fun, and unperturbable as always," wrote Bettina. "With no telephone, no radio, not many 'musts,' we've had lots of time for one another, an advantage of these times." Both were grateful for the packages Marie still sent: the children's first lollipops, Nescafé, Pall Malls, boxed pudding.

We had Pilz come over for Zozzle's birthday, and had the wieners and crushed pineapple. The nylons and pink soap are divine, though no

> *coal means no baths, just piecework which I loathe. To have new shoes is so rare, people stare at my feet. Though it was sadistic of you to mention ice cream and hot butterscotch sauce.*

Georgie added his thanks for the suspenders Marie had included. "My old ones were a dilapidated accordion. The new ones lift me in body and spirit. Along with the candy, I gradually take shape from inside out."

In a sweet note to his German grandparents, ornamented with a flower and the *o*'s colored in pink and green, George described playing ball on his visits home with nice American soldiers, who frequently gave him chewing gum. At school, he focused on staying warm, often waking to thick frost on the inside of the castle's stone walls. Now "tremendous, all arms and legs, I can no longer brush his hair without straining," as his mother wrote, he was resigned to wearing her corduroy pants. ("At first he rebelled but after doctoring them he decided there was no alternative.") And "when he saw the sheepskin-lined boots you'd sent me, he lost his heart, and mama that I am I figured he needs them worse than I do. So they went with him to Buchenau." Her selflessness did have its limits. "We consumed the last tin of fruit for his 14th birthday. He was on a school hiking trip but the peaches tasted mighty good even without him."

As the months dragged on, Bettina's hopes sagged. "Discouragement seeps into my vitals. Maybe I sound down in the dumpsy—maybe I am. I try to keep it in the box and sit on the lid but sometimes it manages to stick a leg out." She fortified herself briefly by recalling, if flippantly, a moment soon after Chris's birth, when she mustered an utterly unexpected act of generosity. "Speaking of cows, did I tell you about the conglomeration I fed in the hospital? I took on twins from the next room, at home sent a quart to the baby clinic, and gave the dog a daily bowl full."

As the wheels finally began to turn for their transit to America, Bettina also acknowledged the gratitude she owed her aunt Paula, who until now she had only lampooned—for her reliance on the ten-cent store, or habit of getting "sidetracked by some book" and leaving domestic tasks undone. A mining heiress married to Alice's brother Talcott, Paula had tapped the help of Wisconsin Senator Alexander Wiley, who had finally sprung Bettina and the boys by pressing the State Department with "vigor and decision." "What a mess it would have been," wrote Bettina, "without Paula."

Though prepared by the DP camp for the equally packed troop transport they would sail on, Bettina worried over the embarrassment their ragged state might cause the family. "Pilzie has nothing left. I have applied to the UNRRA as an interpreter; maybe they can help me secure clothing." She herself would have to accept looking "foreign," with everything too short or too long. "Style here is whatever you have left. Women wear coats made of US Army blankets—a giveaway since no GI gave his blanket for nothing! I saw an ad in the Paris *Herald Tribune* for a coat made of feral cat." When warned by Marie that in America she'd not have the grand life of a diplomat's wife, she joked that if they wished to make her feel at home, they could provide "sawdust bread, boiled cabbage, straw mattresses (don't forget the bedbugs), and a score of people to rush in without knocking."

Finally, in September 1947, Bettina, George, and Chris made their way to Bremen and the SS *Ernie Pyle*, their father promising to join later. The three tried to avoid bottom bunks, George later told his own sons, because that's where vomit from the top bunks landed, a useful precaution given the storms they hit, violent enough that the crew once readied the lifeboats. Allowed to bring just one treasure, George chose his box of eggs. He knew that the tiny, speckled sky-blues were song thrush, the marbled ones crow, and the turquoise blackbird, but they weren't yet specimens. "I didn't yet know what questions to ask." Rather, each egg was "a memory of a day, my excuse to wander."

Arriving in New York Harbor after eleven days at sea, the three were interned on Ellis Island for several days until Grandfather Beals secured their release. George went ashore an enemy alien. With the twenty-five cents his grandfather gave him, he bought a can of pineapple.

George's childhood would prepare him well for the rigors of his long life in the harshest places on Earth, where "one settles at times for mere survival": the days without food, months cut off from those he loved, opaque political dangers. Its surfeit of violence and human darkness would shape his own renunciation of weapons and his complex view of the relative nobility of species. Having no fixed vantage of his own, he would find the rarest kind of empathy, the ability to inhabit the perspective of the ultimate other, making a kind of contact with furred, clawed aliens that he would achieve only rarely with human beings.

2.

Coming Into the Country

US, 1947–59, Raptors, Ravens, Grizzly Bears, Caribou

Though neither of Bettina's long-divorced parents were eager to take her family in, she and her boys found unhesitating shelter with Uncle Talcott and Aunt Paula in the St. Louis suburb of Webster Groves. It was not an easy gesture. Though once fabulously rich, the Barneses were now struggling and had little room to spare. Bettina's first cousins Ed (twenty-three) and Bill (twenty-one)—who she found "stupendous in size, looks and intellect"—were often home, and their friend Charles had moved in. To add to that crowd a sullen teenager and rambunctious five-year-old was an act of generosity for which George remained forever grateful, even more so when he came to understand that Paula was already sick with the cancer that would end her life. The leaden legs, persistent fevers, a breast wound that would not heal—she kept hidden from all but a few.

Bettina stayed just three months, still writing Marie Kuehn. On a trip downtown with George "our eyes popped out, as if we dropped in from another century. Even a candy store is to be marveled at." She worried about "what's to become of poor old Zozzle," to whom she had sent a box of Christmas treats. "It's going to be dull without his mental hustle-bustle."

But such glimpses of her old sparkle were few, her tone more curdled than ever. Her expressions of gratitude to "wonderful" Paula were cut with shudders at her unkept house and "high-water pink pyjamas and flying grey hair. . . . She is not particularly female. She has a man's brain." Bettina loathed Charles, who spent all his time "scuttling about . . . goading the family into saying nasty things about me and then taking great joy in telling me. . . . He thinks it fascinating but I do not. He and Paula seem to be continually going over my this and my that. It is high time I clear out." She even lashed out at Marie. "Don't think I haven't been thinking of my children. What in heaven's name do you think I came here for? For the pleasant little trip we had?" Gone were the zest and wit that had sustained her through the war.

> *My inside has gone quite deadpan. I cannot seem to feel or show joy or sorrow. I don't know what has happened and the terrible thing is that my two boys seem to be the same way. Prisoners of war often get into this condition, I have noticed. But don't worry, I will snap out of it, or into it, however you wish to put it.*

Her father and his third wife had offered to take Chris for a time, but "one of those ten-minute-on-edge-of-chair visits" changed their minds.*

> *So, I am bogged down here but will soon head East and take myself a neat little job as a second maid. Can't you just see me in cap and apron, scrubbing toilets and washing dirty whiskey glasses? By gum, many is the time I've done just that for no pay. What fun it will be to have roles reversed for a change. And don't worry about my hopes being too high. I got over that years ago.*

---

* In a letter written to his grandfather from Pakistan twenty-five years later, George remembered this as their last-ever meeting.

In late December 1947, leaving the boys in Paula's care, Bettina moved into her mother's "very chic" Greenwich Village apartment. Their ten years of estrangement, begun when Alice insisted the Schallers return her gift of Louis XIV furniture, seemed forgotten. "Mother knows everyone and is terrifically amusing; we spend our time chatting and giggling and shopping in the Italian food stores on Bleecker St." Bettina occasionally wondered how "my two hyenas are enjoying life without mother." But she soon landed her dream job, designing textiles at the venerable J. H. Thorp & Co. Eighteen months later, she brought seven-year-old Chris to New York, settling him, to his lasting sadness, in the Wartburg Orphans' Farm School in Bronxville.

For all of high school, Paula became for George a "kind of second mother," worrying over his loneliness, and combativeness, and uncertain future. She knew what it meant to live through a precipitous change of fortune. Her grandfather Edward Hegeler had founded a zinc company and grown rich producing arms for the Civil War. But the Depression had brought an end to all that. Talcott borrowed money he could not repay. "He came home for lunch one day," Bill recalled, "and said, 'It's over.'" Moving the family to the country, he supplemented their fresh eggs, milk, and home-cured bacon with deer he shot, without a license, by flashlight late at night.

Talcott was a severe presence, a "believer in the belt," in his younger son's words. At age ninety-two, Bill still remembered watching his enraged father throw teenage Ed against a riverbank and knock a guy out in a Chicago bar. When the FBI came to him ahead of the Schaller family's arrival and said, "'What if we discover that your niece was a loyal Nazi?' He said, 'Then take her out and shoot her,' and he meant it. If she was a Nazi, to hell with her."* Talcott's toughness, in Bill's view, influenced Ed's "limited

* Bill recounted this in an interview with the author. Chris believes Talcott said "If she's a Nazi, shoot her," not about Bettina but about his sister Alice, Bettina's mother, who had settled in Vichy when it was the center of French collaboration.

availability" to other people; "he preferred being out alone in the wild to the city and human relations." As the man George would emulate and, together with Bill, be "the closest thing he's had to lifelong friends," Ed would introduce his young cousin to the solace of unpeopled places.

For his first year and a half in Missouri, George wrote to his mother in German, still a sweet boy signing "your Pilzie." He began every letter with earnest thanks for whatever note or gift she might have sent, and ended with a plea for her to visit, or at least to write again soon. "We made cherry and vanilla ice cream. It was so good; a shame you were not here."

His letters soon filled with animal tales. If Uncle Talcott was rigid on most things, he was remarkably tolerant when it came to the menagerie he allowed George to keep in the basement. First came an aquarium. "I bought myself Black Mollies for $1.50," he wrote his mother, "and even nicer water snails. My fish are feeling good. And it is awfully fun." Again finding a friend to scour the woods with him, he added a terrarium for his skinks and ring-necked snakes. Then came a raccoon (who liked to sit on his lap but not walk on a leash), an opossum, and a turkey vulture; George still bears the scar on his lip left by its beak. As he often would, with these charges, he faced sad turns. "I got young fish today, tiny as a pinhead. Unfortunately, I could only rescue three because the mother ate the rest. And the turtle disappeared without a trace. I think Christopher took it out and forgot to put it back in and Daxi [the family dog] ate it up."

As in Germany, the woods offered respite from an alienation teenage George sometimes took out on the world. "When he is on his good behavior you couldn't ask for a better boy," Paula wrote in a long confiding letter to Charles. "He and Chris are both good with Talcott in the house, and seem happier too, behaving. Strange! If only they were always like this, how much heartache would have been avoided." She didn't like to leave them together. "Chris adores him but George cannot say a kind word unless he wants something out of him. Just now he wants to sell Chris cherries for 15 cents, so is being rather pleasant to the little fellow. But his continual sarcasm will only do them both harm." Most distressing to Paula was George's treatment of the family housekeeper. "He enjoys teasing and tormenting Mrs. Haley, as well as sometimes talking very cross. I can't leave the boys alone with her. They don't obey, just fight with her and each other."

Of Mrs. Haley, Bill had a warmer memory, of shared affection with George. "Both on the periphery of the family, they found each other." But he also recalled serious transgressions. Knowing that their German neighbor Ella Dietz was terrified of snakes, George presented her one day with a venomous copperhead. "She screamed and ran home. Mr. Dietz soon appeared at our house with a club and let George know if he ever did that again he would get a good beating."

School brought its own kind of troubles.* Thrown into ninth grade with little English and a German accent that he would never completely shake, George was challenged to fights by boys determined to avenge on him their family's wartime sufferings. "One of the he-men would say, 'Come out in back.' That's typical of the whole world; you beat up the other that's different. Most of them, when they saw I wasn't just going to lie down . . . well, you'd mutually decide this isn't going anywhere." A mediocre student, getting mostly C's, George was nudged toward vocational school. An aptitude test showed him suited for interior decorating.

For all the separations he'd endured, this uncrossable distance from his parents seemed hardest to bear. His first Christmas without his mother, Paula wrote, was "really sad." She pleaded with Bettina to be in closer touch. "We've not heard from you in a long time. George asks every day when he gets home from school if there has not been a note from you. . . .

* Webster Groves High has had an uncommon measure of publicity: A 1966 Charles Kuralt documentary painted the town as the epitome of sheltered privilege: "To be sixteen in Webster Groves is to be insulated against all the cold winds of life—war, death, and poverty." (*16 in Webster Groves*, produced by Arthur Barron, reported by Charles Kuralt, aired February 25, 1966, on CBS News). Alum Jonathan Franzen wrote a riposte to this portrayal in his memoir *The Discomfort Zone*. George's future boss, Bronx Zoo director William Conway, was also an alum.

And I do think he gets lonesome for his father at times. Ed and Bill are not quite the same, and Talcott is too stern and severe."

Bettina responded with her own troubles. With neither "old Schaller," as she now called her husband, nor her own father helping, she was struggling to find the ten dollars a week her mother required for rent. "Many boys George's age [fifteen] must help support their parents. Sometimes I think it would have been a whole lot simpler for everyone if those bombs had been direct hits. I don't see any point to a life being like this."

At some point, George found that letter. Underlining those last lines, he wrote, "Such unbounding confidence." He continued his annotations on Bettina's next letter, sent to Paula a month later. "My original plan was to leave George on the other side to finish school. That would have divided the burden with my husband." ["*News to me!*"] "Father agrees that the Navy would be a good place for the boy. I think he would like it." ["*Shows how little you know me.*"] "I am eager to resolve the Big Georg situation, having been through just about everything a human can go through." ["*You had it easier than 90% of the people in Europe. You greatly exaggerate your hardship and you know it.*"]

Shortly before George's sixteenth birthday, Bettina wrote Marie about his changing voice (like a foghorn!) and the "good job" Paula had done with him. "I can't think of anyone else going to such trouble." Certainly not her own father, "who doesn't want to be bothered with any of us." Her stepmother, she added, had just miscarried. "Trust Father to cook up little brothers or sisters for me."

Paula's worries deepened when Bettina began to talk of divorce. "George Jr will be very resentful," she wrote Charles. "He is jealous of anyone whom he thinks wants to take his father's place in his mother's affections. Perhaps this is why he pestered you, and still makes drawings of you in embarrassing situations." Georg Sr. may not have been doing his share; he sent thirty dollars a month, which the spreadsheets Bettina liked to create showed fell short of her own contribution. "But I don't think Bettina has given him much chance."

As late as September 1949, Georg was still signing letters to his wife "love, Georgie" and lamenting her long silences. "Is he upset that you did not bring him here when you could have?" Paula asked Bettina. A few months later, George wrote that he'd "bought two pounds of chocolate for Dad's birthday. I hope the package will arrive on time." He was excited to learn that his father had sent him money—"I didn't know he could!"—as well as a book on monkeys. In May 1950, a German court granted Georg a divorce on grounds of desertion.

"Have you told George Jr?" Paula wrote to Bettina. "He should hear it from you or his father, so he can keep his faith in you both. It's important he be able to talk about it so it does not get buried and make trouble for him. Or perhaps you've told him and he won't mention it? Also not good." When Georg Sr. remarried, Paula again asked if anyone had told George. He was now a junior, so she added a question she would ask many times. "Do you think Georg Sr will help young George with his college education? Should I write him to see if he could find some jewelry George could sell? He is doing better at school. He really should have at least two years of some kind of training."

George *had* by then begun earning his own keep. Writing now only in English, he boasted to his mother of his jobs—at a fur farm (fox and mink) near his uncle's summer house in Fort Qu'Appelle, Saskatchewan, and as a bounty hunter with the .22-caliber rifle Talcott had given him. "I get 25 cents for every crow I shoot and 10 cents for every rat." (Seventy-five years later, he still had his junior marksmanship diploma, awarded by the NRA.) He had also begun catching baby hawks, possums, raccoons, and mice to raise, tame, and sell. He needed nothing, he wrote, wanted no toys or candy for Christmas. His only desire was for a book, *Animals of the World*, though if four dollars was more than Bettina could manage, "never mind then, because I could not find any other book I want."

His disinterest in material things stood in stark contrast to his mother; that was perhaps its point. When his father wrote asking if he might hold on to a few relics of their lost life—"I would like to keep the blue coffee pot (I have none) and one Meissen figurine. Please tell me what you want"—Bettina replied with a thirty-one-page list (a marmalade pot, a Venetian shaving mirror) noting which had been a gift from her grandmother and which "I bought with my allowance." The latter included the painting of Marie Antoinette. Alice, too, had staked her claims, growing angry when some things got broken in shipping. Georg tried to calm her. "What's missing is not so much. I lost much more."

That sadness sometimes slipped into the holiday letters Georg and Bettina exchanged over the years. Three years after their parting, he thanked her for sending a picture, though "I did not recognize Christopher at first. Poor me!" He had sent money via Alice, he added, which "I hope you will get before Christmas so you can give the children something from me."

•••

From George's vantage, at least in hindsight, the important event of those years was not his family's final splintering but a trip in the summer of 1949 to his cousin Ed's little island on Rainy Lake, Ontario. George helped fell, peel, and notch logs for a cabin, turned his box camera on a porcupine, slept under the stars, and heard wolves howl, falling for the first time under the spell of the wild.

The next summer brought more adventure. Ed took George to the Yukon Territory, to paddle the Big Salmon with an ancient prospector who claimed to know where to find a lode of gold. Youth is typically callous to age. But George saw the man's struggle to recognize anything through cataract-clouded eyes, as well as his shame at sitting helplessly in the boat as the young cousins spent hours in the icy water pushing it upstream. Pulled up on gravel bars to camp, drying their clothes next to driftwood fires, George felt free as he never had.

Paula saw the salutary effect these wild sojourns had on George. "He is working hard to get better grades and learning to live with other people in a more peaceful manner. Many of his difficulties seem to be smoothing out. He is gradually becoming a helpful and pleasant boy." At school, George dove excitedly into Mr. Barnett's biology class, joined the wrestling and track teams, worked on his senior yearbook, and became a member of the Scribblers (the writers club). On weekends, as he wrote his grandmother—first complaining that he never knew where to send letters, as she went from Positano to Capri, Casablanca to Madrid—he went swimming with friends, diving to catch baby snapping turtles. He closed one letter with the kind of detail he came to love: a sighting of "a black raccoon eating a dead dog by the side of the road." Paula saw that "he is simply no good if he has to do work he is not interested in. You cannot make him into something he is not."

One worry still hovered. Bettina had been working to secure George citizenship, a move his father feared. "I've thought a lot about Pilzie," he wrote her in February 1951, three months before his eldest's eighteenth birthday. "Pilzie can and should be loyal to the USA, but why a citizen right away?" As an American, he could be drafted, "to fight and maybe die in China or some other place. I think we all learned our lesson in the last war . . . please don't throw him into the arms of the war machine."

Seeing that the prospect of the draft "sent chills up and down George's spine," Paula urged Bettina to heed Georg's guidance, though she wondered, if he stayed German, would he have to go back? Ed had a better idea: George should apply to the University of Alaska and get a deferment.

Graduation at last brought Bettina to Missouri for a day, which pleased George greatly, though he "sweated straight through" his heavy borrowed suit, got seven blisters from his also-borrowed shoes, and found "the robes and caps with tassels the biggest nonsense, of course he would," as Bettina reported to Alice. The next day, he left for South Dakota to work on a US Geological Survey team, but that job lasted just seventeen days. Word came down from Washington that the "enemy alien" had to be fired. The hundred dollars he'd earned was already gone, so he spent the summer pumping gas, sweeping out a movie theater, and shocking wheat.

The years of fending for himself left a lasting mark. "George never dwelled on his childhood," said cousin Bill. "He never dwelled on himself. From those earliest years, he had to survive emotionally on his own. I think he became resigned to loneliness, made up his mind to it. That's the way it was, so he didn't live in pain. He just never focused on himself but on the world out there."

•●•

Though the university never replied to his application, in September 1951 George packed a duffel and flew to Fairbanks anyway. In his pocket was the thousand dollars Uncle Talcott had pressed into his hand as he'd gone out the door. The registrar took his sixty-dollar tuition and let him stay.

Alaska was then still a territory and the campus little more than a dusty frontier camp, which pleased George. Though his first Christmas away from family was "naturally not very nice," as he wrote Talcott, the school gave stranded students "all the turkey we could eat." And for the first time, he found lots of kindred spirits: kids up for playing Ping-Pong all night, greeting the new year on ice skates even at fifty below, or hunting moose, though with only three hours of daylight "it got dark before we got close enough to take a shot." He wrote to Bettina about a dance with piles of orchids and the girls in sarongs ("It was wonderful! Though I don't dance, just walk around in a certain fashion") and to Chris of various kinds of fun: a snack bar job that gave him endless ice cream; dog races, especially entertaining when "a bunch get tangled up and fight"; and how he'd busted his skis "going like a shot out of hell. I'm not very good but can dodge the trees and that's what counts." He began illustrating his letters with cartoons. For his brother's tenth birthday he sent a satirical self-portrait, bald and knobby kneed, with stuck-out ears and a corncob pipe, delivering a giant cake.

George found another kind of home in the zoology department, volunteering to clean muskrat skulls (320 in one particularly productive week) and skin wolves killed in the name of "predator control"—their tongues still greenish-yellow from cyanide. Any that had been shot, he carved up for the grill. "Old wolf meat is stringy," he wrote Grandmother Alice, "but I had steaks off a yearling female which were delicious. Fried in beaver fat. My whole window is packed with wolf steaks, a soothing sight." His most onerous task was checking the animals' decaying internal organs for parasites, unleashing a smell so overwhelming that he often worked until dawn to be done with it. In one night alone, he boasted to Bettina, he earned eighteen dollars. Brina Kessel, a remarkable young biology professor who had studied with ecologist Aldo Leopold, described him fondly in the school paper as "the most active manufacturer of incense in the lab, greeting you with a merrily boiling kettle of skulls."

The once-middling student was now devoted to his studies, eavesdropping on graduate students as they debated the need for what he called the "wolf pogrom" and poring over scientific papers like Swiss ethologist Rudolph Schenkel's 1947 study of canine packs that originated the (now largely discredited) idea of "alpha males." George did sometimes skip class to go work in the woods. Once during May exams, he went caribou counting and got trapped by a snowstorm. He made it back to campus three days later at 7:00 a.m., in time for his chemistry final at eight and zoology final that afternoon. He made four A's and three B's, the second-highest average in the freshman class.

The new horizons had not erased his longing for past ties. When, in the spring of his freshman year, his father made his first trip to North America representing a German trade group, George traveled two thousand miles to meet him in Canada. His father stayed just two days. In a nine-page letter to his mother ("You sure don't write often," it began) George barely spoke of the visit, except to imagine that it must have been hard for Chris, when his father stopped in New York, "to act like a son to a dad he can barely remember." Bettina's reply was absorbed with her own vindication. "Father had written that he was very sad that Chris had to be in an orphan's school." But after a visit to the campus, "he knows that Chris is in good hands, and that I always did the right thing for him to the limit of my ability." As for his father's praise for George's hard work, she assured him, *she* had praised him still more. "I said that I am a very proud mother to have such a nice and serious-minded son and that your sense of responsibility has always been a great help and consolation to me. He said that you did not seem too interested in money, as you flatly refused uncle Talcott's offer of further help. Father was pleased with your attitude and rest assured that I am just as pleased." She told him the news she'd shared with Georg and Chris, that she was remarrying. "Your father left three pairs of German socks for you from grandmother Schaller."

It was Dr. Brina Kessel who launched George's life in the field by sending him that first summer to the Arctic Slope to survey birds. Though Kessel had won the grant, the Navy barred her from entering the petroleum reserve without a husband to accompany her. So in her place went freshman George, who was dropped off by a floatplane on June 4, 1952, with graduate student Tom Cade and their collapsible canoes. Over the next two and a half months, they would descend first one tributary and then another north into the Colville River and on to the Arctic Ocean.

Commencing the relentless note-taking that would eventually fill more than three hundred hardcover notebooks with his tight, neat script, George kept this first-ever journal in pencil and a careless scrawl. He recorded their daily sightings: a red-spotted bluethroat, visiting from Asia; a just-hatched peregrine falcon, still on the half shell; white-crowned sparrows and Lapland longspur singing in the midnight sun. He also recorded the birds they shot, his marksmanship paying off: On one occasion, he got three red phalaropes—a two-ounce shorebird—with a single shot. All told, they "collected" 178 specimens of 45 species. Taking that many lives was then standard scientific practice, as George explained to his mother—while he did not like shooting a goose off of her seven eggs, "geese are confusing to scientists so we have to collect many"—but the body count is startling, nonetheless, given where both men ended up.

Each of the fifteen dozen birds they killed had to be skinned and stuffed with cotton. It was not easy sitting for hours over small birds that tore at the slightest mistake; if the carcass was too badly shot up, they made it into a skeleton. They ate the meat, cooked over willow branches or the coal they found in seams on the bluffs. Glaucous gull was too fishy, but George relished fried goose breast and a two-ptarmigan stew (willow and rock). They ate eggs, too, when collecting the shells, sometimes shooting both parents off the nest.

Tom was a falconer and schooled George in the art of capturing wild birds, rappelling down a bluff to take a young gyrfalcon and two peregrines, which he tied to his canoe. Wanting a pet, George tried the same approach with a raven's nest but slipped and wound up hanging from the rope, growing dizzy from cutoff circulation. He had scared one raven out of the nest and into the river below, so once Tom lowered him to a ledge, he jumped in to fish the baby out. They then caught five young Canada geese to use as training lures.

Several of their captives met untimely ends. One peregrine "hanged himself dead." A second raven jumped from the canoe "and drowned, damn it." They fed it to Tom's gyrfalcon, a species famed for its ferocity, though this one proved to be afraid of prey: It took the smallest gosling only after they tied raw meat to its back. Still, George was fascinated. He copied Tom's templates for a hood, then knocked his raven out with paraldehyde to fit it with jesses (thin leather tethers) and a leash. Nineteen days after its capture, he let it fly free for the first time; it made a big circle and returned. Even through nine July days of continuous rain and snow, the birds stayed with the men, all sitting dejectedly in the canoes. Lost in a fog so thick they could see nothing ahead, George and Tom finally pulled up to sleep on a sandbar. They woke to a wall of sea ice and immense relief; any farther, and they'd have been adrift in the Arctic Ocean.

Returning with his companion to the university, the raven lived up to his species' magical reputation. One day Kay Morgan, a willowy anthropology student from Anchorage,

> heard terrible shouting outside my dorm window and looked down to see this tall young man shaking his fist at the sky. I thought, "This is the Arctic Madness I've always heard about." They said people went mad up here and, heavens, here was someone who absolutely had! The bellowing continued for several minutes. Then suddenly this big black bird came down and landed on his arm. He bent and spoke to it. The next day in class I said, "Are you the one who has the raven?" and he said yes, and we just went on from there.

The raven continued to work its spell, sometimes landing on a light pole to peer in Kay's window, which felt "like George was near." It soon vanished, but left George in thrall to the "vibrant girl with golden hair." He began inviting Kay to the cafeteria, to play horseshoes, even to help him clean skulls. Though by then he had a colony of assistants: beetles to eat off the meat.

Kay could not take seriously the "downy-cheeked" nineteen-year-old, three years her junior. "Good heavens, I was much more interested in dating older graduate students." (And they in her; one remembers her as the center of much male attention.) "But he was teased for trailing after me all the time." George was the cartoonist for the student paper, *The Polar Star*, but soon also landed in its gossip column. "Ouch. That tall skinny blond just sat down to pester me while I ramble on—yes Kay Morgan. Seems she just finished a lab with George called Outdoor Techniques. Wonder if it's called Indoor Techniques when the weather is cold?"

Alaska suited George as no place ever had. Between semesters, he camped out on campus in a leaky tent. He replaced the raven with a groundhog, which he managed to housebreak before it went to sleep for the winter, and a few reptiles, though they froze to death when George accidentally left his window open on a minus-thirty afternoon. He joined the hockey team, regularly updating his mother on his latest stitches. ("A stick laid the top of my head open. We tied the game.") His college scrapbook includes a news clipping on Schaller's two goals in a game they won, and four-minute penalty for slashing in a game they lost. When prankster mining students dynamited the ice rink, he joined the skaters trying to jump the holes, then watched the girls storm the men's dorm to carry the perpetrators out in their sleeping bags and little else. "Everybody howled as they tried to hop, crawl and roll back into the dorm." George's own defiant streak had not entirely faded. He flunked PE. And when a modern art show came to campus, featuring work George thought nonsense, he dashed off his own painting, signed a French name, titled it *Infinity*, and sneaked in to hang it alongside the others. When the exhibit closed, it was carried off with the rest.

Food remained an adventure. Joining the rifle team, George hunted almost every weekend for grouse. "We throw a big dinner and eat four or five each." At his department's annual wildlife dinner, he tasted polar bear, fox, raven, muktuk (raw frozen whale blubber), and pickled walrus skin. He sent his mother recipes: To cook a ptarmigan over a fire, put it in a pot with a lid and in half an hour it will be roasted golden. For lichen flour, use genus *Cetraria* or *Cladonia*, boil it in baking soda, dry and grind it. For pie: one cup lard, three cups flour (half lichen, half wheat); fill with every kind of wild berry. Bettina, in turn, sent him seeds of something called Vitacress to grow on damp cotton. The first batch froze, he told her, and his pet chickadee ate the second. But he now had them screened near the radiator and when they sprouted would eat them on wolf steak. He became known for going off to the mountains with no food. Out for three days with just a sleeping bag, camera, and gun, he wrote Bettina, "My delicious menu included ptarmigan stuffed with wild herbs and mushrooms, rabbit meat rolled in lichen flour and fried in porcupine fat, berries and leaves and choice morsels of ground squirrel." In summer, he sometimes left even the sleeping bag behind. "He'd just take naps since it stayed light all the time," a graduate student named Dave Klein recalled. "We'd ask him what he had eaten. He'd say, 'I ate a grouse.' 'Well, did you build a fire?' 'No, I just ate it raw.' He never had trouble going without food. He'd make up for it when he got back to the cafeteria."

A marvelous illustrated letter to his little brother dated December 1952 captures George's newfound joy. He began by asking Chris what he wanted for Christmas; in the accompanying cartoon, a grinning George with one bucktooth sneaks a red truck beneath the tree. "It sure is cold," the letter continued, his cartoon self now in scarf and hat with a big red nose, then padded up for rough hockey. "–40° last night. When it gets this cold, you can take a glass of very hot water outside and throw it into the air, and it will not come down. All the water evaporates in a white cloud before it can come back down." He illustrated a successful hunting trip (cartoon George shooting a grouse). And another, less successful, on snowshoes: "I saw a rabbit but I couldn't hit him. [They're] completely white . . . So you just can't see them against the snow. All you see are 2 black eyes jumping along, leaving big tracks behind." He signed off with his avatar in tails at an all-night dance, arms around a girl in pink and a sly smile on his face, and then flat on his back, cutting *z-z-z*'s. Some letters offered advice. "Learn to swim real well, you never know when you might need it." And when shooting birds, "aim for the head so you don't spoil the meat." Others included small gifts such as a chip of a mammoth tusk, thousands of years old. "I heard you got $2.50 for being a good student. Well, I can do twice that good; here's $5."

George's relationship with Bettina remained complicated, veering between resistance and tenderness. Though almost twenty, he'd come late to

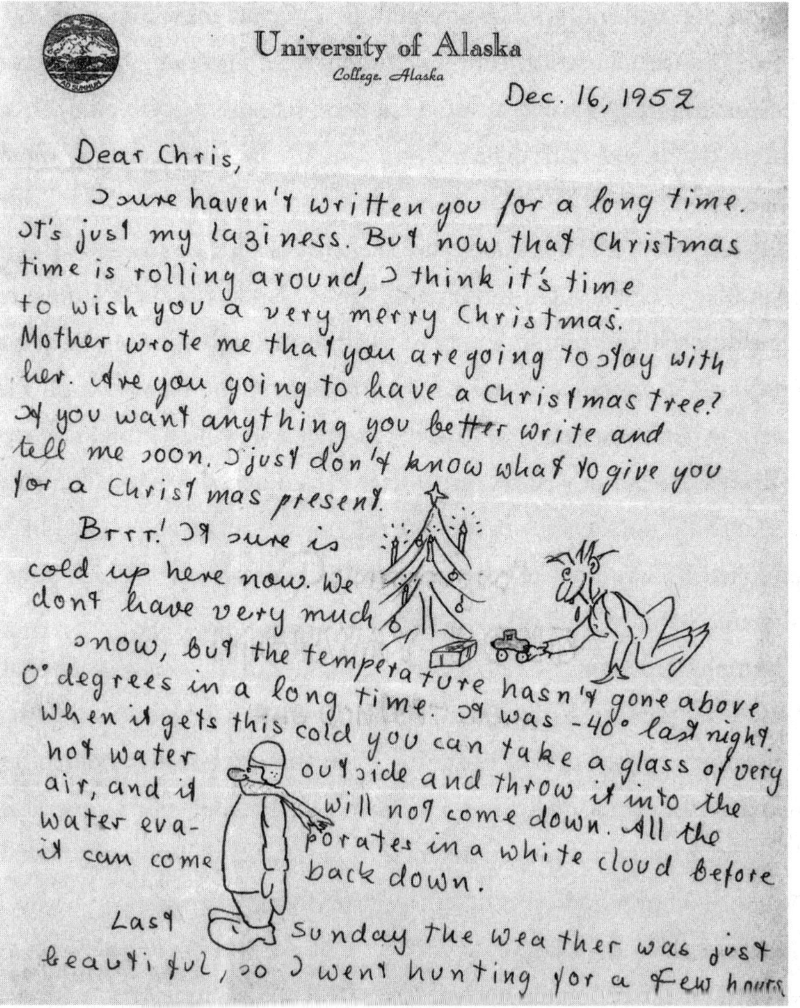

University of Alaska
College, Alaska

Dec. 16, 1952

Dear Chris,

I sure haven't written you for a long time. It's just my laziness. But now that Christmas time is rolling around, I think it's time to wish you a very merry Christmas. Mother wrote me that you are going to stay with her. Are you going to have a Christmas tree? If you want anything you better write and tell me soon. I just don't know what to give you for a Christmas present.

Brrr! It sure is cold up here now. We dont have very much snow, but the tempetature hasn't gone above 0° degrees in a long time. It was -40° last night. When it gets this cold you can take a glass of very hot water outside and throw it into the air, and it will not come down. All the water evaporates in a white cloud before it can come back down.

Last Sunday the weather was just beautiful, so I went hunting for a few hours

By nineteen, George had become a sweet big brother.

teen snark. On a gift she'd sent: "After several days of no response to my genuine Audubon bird call, I sat on it and broke the tail. No point replacing it. I can do better with a blade of grass." Another note began, "I was just writing a theme on how to shrink human heads when someone brought me your letter, so I'll answer it right away."

Some cut deeper. George's criticism of Chris's school—"he sure is getting religion pumped into him"—brought an angry defense from Bettina. He replied, "rather amused by your long tirade over the goodness and virtue of the Warthog," his derisive nickname for the school. "I gave you my honest opinion but if you don't like my opinions, I can always change them to please you."

His decision to visit his aunt and uncle in Saskatchewan rather than see her in New York brought more pain. "Your visit to the Barnes must have seemed more important to you than to grant my wish," she wrote. It was a question of distance, he answered, and of pride. "Find Fairbanks on a map sometime. And you threw $400 at me in such a way that I could not accept it." To his complaint that he rarely heard from her, she insisted she wrote religiously. "I don't want you to write religiously, but when you feel like it," he replied. "I write out of motherly love," she answered. "Which your ill-mannered letter robbed of all pleasure."

Bettina's growing agitation showed up in her drafts—page after page of crossed-out paragraphs and scribbled amendments, which along with her finished letters she kept for posterity. "You think you're a wonderful great big man," she wrote, "bred in the wilds of Alaska, and your mother a fool. My own wishful thinking led me to believe you *had* grown into a man. But you're just a boy who thinks he knows everything."

George, learning arts of de-escalation that would crucially serve his work, yielded. "You made me feel like a heel and a skunk. You take everything I say as disrespectful or presumptuous. [But] I hope that you will forgive and forget all my rudeness, which was never intentionally directed at you but just my own worries." Of Bettina's own peacemaking gesture, a holiday gift of sausages, dark pumpernickel, and sweets: "Your packages are my whole Christmas," he wrote. "Let us start the new year as it always was and always will be, I hope."

Whether out of innocence or guile, George applied the summer after sophomore year to be field assistant to Dave Klein, his chief rival for Kay's affections. George's "tremendous vitality" had already impressed Klein.

"Most of us grad students had served in the war and were older. George was the young person you couldn't help but like, even though you might feel he was going to outshine you. He was the kid who hung around, persistent. Mind wide open, not ashamed to ask questions, though always in the most appreciative way. Eager to learn about the whole world."

Klein found those qualities only magnified in the field. After long hours of work, a wolfed-down dinner, and more than his share of chores, George would ask if he could go out again in the midnight sun. After another fifteen miles and an hour or two's sleep, he would recount what he'd seen. "He knew he could learn just by being there, trying to think like the animals. When that bear is searching, what is he searching for?"

Outlines of the pathbreaker were coming into view. "Nobody has ever done this before for alpine tundra," George wrote Bettina of his effort to determine how many caribou the Nelchina range could sustain. Beyond gathering the first data, that is, he had to figure out *how* to do this novel work.

His writing, too, became a concerted practice. He occasionally went overboard: "Only the alarming splash of a beaver's tail [and] melodious trill of a hermit thrush interrupted the stillness of the drowsy meadow smelling of fern, resin and grass." But already, he was beginning to pare away the excess that afflicts much nature writing, developing—in his second language—the precision that would prove essential to his impact on the world. A bear with a stomach round and hard from a night of gorging on moose looked at him from fifteen feet, smacking his lips. Though his mother served as audience—he even sent her copied-out pages from his field notes—he was not writing for her, exactly. "I've been rambling on."

Most important, he began to learn from those who had been on this land a long time. Some of the Native people he interviewed in the field, asking about both furbearers and the Copper River salmon, were wary. "How many muskrats did you get?" he'd ask. "Plenty," they'd answer. But most liked to share stories, of the wolves they'd killed or the huts they built atop graves for their ancestors' hunting and cooking things. Some invited him to dinner, sharing cloudberries and dried raw salmon caught by kids in a way George admired enough to describe for Chris. Standing in a deep pool, they'd slip a wire noose over the neck of a passing fish, then "the battle was on." A thirty-pounder put up a fight but when "he" tired (that pronoun a carelessness he would soon outgrow) was dragged to shore and cut open. "You should see all the eggs. Two quart-jars full." Finding eight-thousand-year-old hunting tools, he was moved to realize just how long such traditions had been carried on.

While his singular fortitude and curiosity were already evident, George had not yet shed the habits of the old-school Alaskan biologists. He remained attached to his gun, still often seeing animals as menace or meat, and imagining they saw him the same way. A long letter to Bettina recounted "a bad experience I still dream of." He was out hunting marmots, "a real sport, you have to hit them in the head or they run into the rocks." Having bagged two giants—together weighing forty pounds—he'd walked eleven hours with the rodents in his wood-frame pack when he stumbled upon a grizzly just a hundred feet away. It sat on its bottom like a teddy bear, hind legs stretched in front, forepaws digging out willow roots to devour with snorts and drools. "All I had for protection," George wrote, "was a .410-gauge shotgun, though luckily, I'd brought rifled slugs [ammunition that enables the use of a shotgun for large game], said to be as effective as a rifle at 50 feet. I loaded the gun and backed away. Without warning he charged."

George had aimed the gun right between the grizzly's eyes when it dawned on him that "I was not frightened. When he saw I didn't run, the bear's own courage faltered. I noticed that and didn't shoot. He trotted to within 20 feet of me, slid to a stop and stood on his hind legs. Seven feet tall." The bear began moving away, but when George saw that he was only circling to stalk him from behind, fear finally flooded in. Having not yet learned that fleeing is the worst response—triggering an animal's chase instinct—George ran for a knoll. As soon as he did, the grizzly came after him.

Again, George's wiser instincts kicked in. "Like a rabbit" he ran zigzag through the tall brush. Running on four legs, the bear couldn't see where his prey had gone unless he stopped to stand, losing ground. The grizzly finally abandoned the chase, at which point George realized that his own body was soaked with sweat and shaking.

A moment later, he reconsidered the whole encounter. "I think he was mostly curious," he wrote Bettina. That altered perspective—granting the beast a glimmer of an inner life—carried him through his next meeting, with an animal far more likely than bears to harm people. "I nearly collided with a moose. She and I were both walking down a small path thinking our own thoughts and not looking where we were going. We were just ten feet apart when we finally looked up into each other's faces. I apologized and stepped off the path."

That toggle—between the old ways and the new relationship to wild animals George was inventing—would persist for a few more years. A professor studying why rabbits ate telephone poles asked George to collect as many as possible. "I took Irene along because she likes to shoot, and we got a car full. I shot 85. Most turned out to be pregnant. Probably needing some mineral." Going by dogsled up the Chatanika River with a friend and five huskies, "everything seems pretty dead and silent until you really look and listen," he wrote Chris. "Then you see fox tracks following the river and tiny fish swimming deep beneath the ice, hear the chatter of red squirrels, the thud of snow falling. At 30 below you can stand a quarter mile apart, talk in ordinary voices and hear every word." But that patience and acuity again gave way to a more brutal encounter. Sleeping curled up with the dogs on spruce-bough beds, the young men heard a scream. A trap George had built had caught a lynx, who looked quietly at them with yellow eyes until "my friend got too close and wham! The cat hit him in the leg leaving big gashes. So he took a stick and broke its neck. I have the hide and it's beautiful." Frustrated on a moose hunt to find only an off-limits female—a big bull could have meant seven hundred pounds of meat—he was also glad of it, writing for the first time: "There is no thrill in killing."

At the end of the Nelchina study, George doubled down on the audacity that had taken him into the field with Klein. Waving goodbye to his mentor and rival, he traveled more than a hundred miles to arrive unannounced at Kay's front door. It was not an entirely easy meeting. Both her brother, Wayne, who had been shot twice in the war and helped liberate Dachau, and her father, who had served in the first World War and was now critically ill, were leery of her taking up with a German. George didn't help his case by arriving in his usual wild state, nor by blasting the stereo; Mussorgsky's fevered "Night on Bald Mountain" was a favorite. But if Kay was still undecided, George was not. Bettina did not miss his glancing references to the nights he'd stayed up until 4 a.m. and then slept until noon, or the crab dinner and swim he had with "that girl." She'd graduated that spring, he explained, and "so far I haven't found another one that I can stand for a steady diet."

Junior year brought big family changes. Bettina married Emilio Iwersen, a Mexican-born US citizen who in 1937 had moved to Germany to manage a factory for Henckels ("probably making bayonets," said Chris) and in 1939 had applied for German naturalization. This nonetheless seemed a warmer stretch for mother and son. To George's avowal that he'd been at her wedding "in spirit," Bettina replied, "I assure you I felt your nearness and affection." When she sent pictures, he asked, "May I please keep them? You sure look young." After finishing off an eighteen-pound Thanksgiving turkey with three friends, "I came home to find a box from the girl I used to go with. A big cake. So I started eating again. The rest of the evening I thought of the people I love, wishing you could come to Alaska, and that I see you and Chris soon." He'd shot a bull caribou and wanted to share with her the delicious tenderloins.

Although after five years without seeing Chris "I don't even know what you enjoy anymore," he held out similar enticements to the twelve-year-old: a rattlesnake thick as his bicep that George had skinned and eaten, the rainbow trout he caught with just a handheld line and a horse fly ("here we call them moose-flies"), the eagle who watched from his aerie until George shared.

He was most overwhelmed by the fine camera Bettina sent for his birthday. Having urged her to spend her "hard-earned money" on Chris or herself, he thanked her in letter after letter, even imagining traveling to New York to thank her in person.

When Aunt Paula died, he grieved openly. "I loved her very much." Of his father, who worried what his eldest could do with a zoology degree, and that he might marry too young, George wrote Bettina, "I can't blame him for not being able to make sense of what I'm doing here. He doesn't know me anymore."

Developing skills he would draw on all his life, in the summer of 1954 George attempted a first ascent of Mount Drum—a steep, glaciated volcano in the Wrangell Mountains—with three companions, two of whom he respected. In his journal, he mocked the third, an instructor who had never been on an expedition but insisted on leading. "A true cub-scout, wanting to take compass readings all the time," Dr. Swanson led them into a blizzard and across deadly hidden crevasses and avalanche chutes to a false summit where, despite all, George enjoyed being in the "beautiful void" between cloud banks. The next day's blowing sleet kept them in their sleeping bags, soaked and frozen. Swanson "griped as usual." George settled in to read *The Life of the Spider* by Jean-Henri Fabre, whom Victor Hugo had called "the Homer of insects"; like Hugo, Fabre saw that even the lowliest creature was on a hero's journey, a lineage George would continue. "And oh, I have a birthday today, I guess." It was his twenty-first. "A restful day anyhow." A picture from their hike out captures him in all his wild handsomeness. Standing in a scrubby meadow, one hand resting on an ice axe and the other on the rifle slung from his neck, his dark hair tousled and high cheekbones burnt by the sun, he looks straight at the camera, the free adventurer feeling his full young power.

George got back to campus just in time to be asked to back up a rescue team on Mount McKinley: a nine-hundred-foot fall had killed one climbing friend and severely injured another. A week later, he attempted Drum again, this time with Austrian alpinist Heinrich Harrer, recently returned from his famous seven years in Tibet. They drove together to Chitina, a copper town gone bust where "the buildings are collapsing but so are the whites that live in them," then flew to base camp. As they climbed the crumbling rock, hacking platforms in glaciers to sleep on, George allowed himself a note of triumph over his "unpleasantly Germanic" companion, who sixteen years earlier had made a first ascent of the notoriously deadly Eiger north face. "Until Heinrich became used to the heavy packs (Sherpa and Indians usually carried his loads), our going was slow." They summited on day four, enough of a feat to be covered in the *Anchorage Daily News*; on the trip down, Harrer told stories of Tibet and Peru while George ate all their candy. Impatient for their plane's return, George went exploring and found a hawk owl nest. As he climbed to it, the parents attacked, crashing full speed, claws first, against his head; he was left with twelve puncture wounds and great respect for their fearlessness. Back in Chitina, they stopped for a two-dollar all-you-can-eat breakfast. George had sixteen hot cakes, six eggs, uncounted slices of bacon, and two bowls each of cereal and fruit.

George spent the rest of that summer in Katmai National Park's Valley of Ten Thousand Smokes, exploring the forty square miles of barren ash laid down in 1912 by the century's largest volcanic eruption. With National Park Service biologist Victor Cahalane, he was there to see what vegetation had returned four decades after all was incinerated by one-thousand-degree-Celsius lava flows. Feeling as if he were present at the planet's formation, the molten Earth just cooling, George played with the elemental forces. Finding active fumaroles spewing sulfurous gases, he tied a can of stew to a rope and lowered it into the steam until it boiled. "No cold lunch for us today!" At a cooler vent, he climbed in for a sauna.

Scrambling in and especially out of the deep gullies proved a Sisyphean task, the flowing sand carrying him down again. He spent hours searching out his first glimpse of life, a creeping willow "hanging tenaciously to the shifting sands." It is hard not to hear echoes of his own history in his ardent response. "This small plant, so tender yet so strong, fighting the forces that tried to crush it in this alien country, received all my admiration." He was still more amazed to hear the *sik-sik* of a ground squirrel, which led him to an isolated patch of brilliant green moss. How had the two-pound creature crossed fifteen miles of sterile land without starving?

Living where other people dream of going even once in their lifetimes, George met in Katmai another of the many notables who would cross his path over the years. Adlai Stevenson had brought his son to fish the famous sockeye run up the Brooks River, and asked George to take them bird-watching. As George gorged on fresh salmon, he watched bears do the same. One big "brownie" would hold down two fish at once, sticking his muzzle in the water to bite first one and then the other thrashing spine, or cradling each like a fat baby to crunch its head. For animals, too, he saw, killing could be sport. Caught up in the "playful destruction"—snuffing fifty in an hour, taking just a few bites from each—the bear "quite forgot" himself, getting so close that when he shook to dry his thick fur, he splattered water over George's camera and lens.

As he developed as a scientist, George's writing deepened in tandem. To both, he brought an unwavering attention few can sustain, finding insight and art not in embellishment but in precisely rendered detail. A baby red squirrel uncurled cottonwood leaves to eat the worms inside. Meadow mice played until a sudden shriek—a short-eared owl's talons landing like the scythe—stopped everything alive for a moment of listening. Then all resumed. Even in assembling data, he seemed incapable of artlessness. In an unpublished article recording observations of eighty-eight species of bird, he noted that the "naked and weak" hatchlings of double-crested cormorant were ten days later gorged and covered with black down and lice, their nest turned to a mess of rotting fish.

Back on campus for his final semester, George tried his hand at fiction—and at rendering people—publishing a story in the *Farthest-North Collegian* about two German soldiers. The maimed men returned from the Eastern Front on a windowless train like the one George rode with his father. When sirens blared, one retreated as Bettina had into a cellar, emerging after the concussive bombs quieted to a sky bloodred like the one his grandfather had seen through tears. Young people in this world were "made old by terror" and hunger. Shocks ceased to touch their consciousness, "the inner self walled in."

Graduating with degrees in zoology and anthropology, George again endured the ceremony. "We look like so many undertakers," he wrote Grandmother Alice. "Too bad mother can't be here; it has been two years since I've seen her." Having been accepted into graduate school at the University of Wisconsin, he gave the eggs he'd carried from Germany to the University of Alaska natural history collection, then rejoined his Missouri family. Traveling to Yellowstone with Uncle Talcott, he found the bears pathetically deracinated, running toward cars to beg. He returned to the Kenai Peninsula with Ed and his wife, Arden. Kay, who had also come to graduation, soon joined; she led George off to explore what remained of the barabaras, the Aleuts' traditional dwellings. Soon after, or maybe the following year, Kay had a dream about marrying someone else, and realized she loved George. She had been stirred to tenderness by the air of loneliness he carried. "I thought, 'If I could just make him realize someone cared about him.'"

In late July, George made his long-promised visit to his mother. They saw a Broadway show (*Plain and Fancy*) and dined at the Waldorf, enjoying Xavier Cugat's orchestra. But just a few weeks later, Bettina sent George a ten-page letter that she knew might burn their relationship to the ground. What had set her off was a note from George saying that, because his father was saving to bring him on a visit to Germany, they'd agreed he would quit sending George twenty-five dollars a month. Ah, another of his ruses to escape responsibility, Bettina shot back. She had always had to press him for child support, even as he criticized her for putting Chris in Wartburg and claimed he'd have done better for the boys had they stayed with him. And, George should know, his father had saved nothing for them. "In other words," she concluded, scorching the earth, "he did not think of you; I can prove it." She hadn't told George all this before, not wanting to burden his soul. "Your father is a man who says untruths, and that is why I could not live with him any longer."

George's response, at twenty-two, was remarkably calm. He told Bettina that he would deal with his father directly regarding money, but "I do not judge anyone based on one letter." As to her worry that she might have cost herself his affection: "I don't know what you imagine me to be, but my feelings do not change that easily."

A day later, he wrote her again, working to layer softness over the rough feelings like a snowfall smooths a ragged landscape. He assured her that he would save his money for when he needed it badly, "like getting married." He told her that Kay loved the scarf he'd brought back from New York. "How glad I was to have had you, with your good taste, to select it!" By now in Madison, he riffed on his first forays into teaching, telling her how he'd shrunk his classes by giving hard tests and then waiting in his office "for the patter of anxious little feet." He complained about the handful of girls and a football player who wouldn't dissect frogs, then made fun of himself chasing insects all over campus with a net and trying to find the "6 to 8 erect bristles on the second trochanter" that would identify his catch. "You are lucky if you can find one trochanter. If you can then find bristles, you've achieved lasting glory. By afternoon's end you've become a raving idiot, if you aren't one already." He asked if she had read the book he'd recommended, by physical anthropologist Harry Shapiro, an early critic of scientific racism. He told her he'd sent stories to *Holiday* magazine and *Sports Afield*, because the "more rejection slips I collect, the more determined I get."

•••

George began graduate school in ornithology, under the guidance of John "Doc" Emlen. Raiding wild bird nests, he hatched the eggs in his room, raising ducklings and owlets to study whether fear is inborn. He also got his first pieces published, in *The Alaska Sportsman* and *Ford Times.*

It was the summer of '56, however, that set the course of his life. Having "heard by the grapevine" that the president of the Wilderness Society was planning the first biological survey of the Brooks Range, he'd written (from "George Schaller, College, Alaska" to "Olaus Murie, Moose, Wyoming") asking to come along. He was six foot and strong, he said, and would do all the manual labor for the same wages Olaus might pay the Native Gwich'in, or for no wages at all.

It was in those two glorious months on the tundra with Olaus and his wife, Margaret, or Mardy, plus Brina Kessel and fellow graduate student Robert "Bob" Krear, that George discovered that life could hold both wild freedom and love. In sixty-seven-year-old Olaus—who was never happier than when digging in warm scat to see what a grizzly had eaten—he found a model of undimmable curiosity; in Mardy, a devotion and depth of knowledge that Kay said she tried to live up to all her life. The Muries had built their marriage on shared effort, exuberant love for the living world, and unending adventure. After a candlelit wedding at 3:00 a.m. (Olaus was late getting back from a waterfowl survey), they had spent their eight-month honeymoon on dogsleds; their dances "in the moonlight . . . where the wild things are known" were romantic enough to move John Denver to write them a song. In their company George found the first sense of safety and untroubled family warmth he had ever known. Sad and serious in nearly all his childhood photos, he blossomed here: In one picture he sprawls loose limbed in a low-slung camp chair, exchanging a frank, charmed look with a ground squirrel.

From the first moments, George's unleashed spirits set the tone; "the tall warm-blooded one," Mardy called him. Having gone with Krear a day before the Muries to set up camp—working through the night on a log bridge so Olaus and Mardy could cross from the solid-ice landing strip to solid ground without having to wade the lake's slushy edge—George greeted them with a jubilant shout. "Two ptarmigan nests with seven eggs already!" And then "Grizzlies!" which he saw across the river, chasing and rolling about in the pale yellow grass. In Mardy's telling, George could sound like a bobby-soxer: "Oh boy, those flowers—they send me!" She loved the chatter of the "three young ones, the family, holding a little infirmary." In a hard rain, all five crowded into the cook tent, sitting on Chevron Blazo fuel boxes to share what they were reading—about the Karakoram mountains or explorer Fridtjof Nansen. Sitting with the group quietly on a ridge, George filled up with a sense of completeness he struggled to describe but knew they all shared.

Eager to learn all he could about the logistics of expeditions, George apprenticed himself to Mardy, copying her lists of food and gear. These included tents and stoves from Recreational Equipment Cooperative (now REI), pounds of Sailor Boy pilot biscuits and French's powdered potatoes, and twenty-dozen dried eggs, building a "suggested diet" of five thousand calories a day. He copied her recipe for Logan bread, a mix of flour, cornmeal, wheat germ, brown sugar, honey, molasses, margarine, powdered milk, and raisins. Though Mardy oversaw camp, the little team scrambled gender norms. When the women returned wet from a hike, the boys waited with hot stew. The day after a close encounter with a grizzly, Brina went back out alone. Another day, George heard a shot and found her thigh-deep in mud, collecting sandpipers. At the end of each day, she gave George her notebook to copy, "a source of information far surpassing my own."

George's energy, already legend, grew to match this near-boundless landscape. On three solo trips, exploring unmapped reaches of the vastness the Muries hoped to protect, he covered nearly one thousand miles, meeting wolves, lynx, and Dall sheep. The very first day, he went thirty miles on a pulled muscle, the pain worsening at each crossing of a turbulent stream. By nightfall his muscles were too taut for sleep. But having promised to be back the next day he rose at 3:00 a.m., ate his last four raisins, and set off across the tussocks, the Arctic's notoriously exhausting bunch grasses that force a walker either over their unstable tops or through the icy furrows in between. Slipping and falling, mosquitoes swarming till he wanted to scream, he finally limped into camp at 8:00 p.m. All of which just whetted his appetite for more. Ten days later, when their seaplane landed them at a new camp under the midnight sun, George had to summit a nearby limestone peak, climbing 2,500 vertical feet in just over an hour, before he could settle down at 2:00 a.m. He returned from his longest solo journey, in mid-July, bone thin and ragged. The talus had peeled the soles off his boots, so he'd reinforced them with rubber cannibalized from his sneakers, tied on with string.

His eating habits remained those of a predator's—starving for days and then gorging. "I am ravenous," he wrote in his journal. "I can't get enough to eat, maybe an after-effect from not eating last school year." Going off for twelve days alone, he took just nine meat bars (an army concoction resembling cold hamburger), 390 grams of canned Multi-Purpose Food (another atomic-age invention made by General Mills for fallout shelters), and two pounds of buttered bread. Mardy always saved a feast for his return: wieners, macaroni with cream of mushroom soup, *knäckebröd*, Jell-O, and once even a watermelon flown in with the mail and chilled in the permafrost. No matter how gaunt he'd grown after days afield, she marveled, he always skinned or pressed and labeled whatever animal or flower he'd returned with before he would eat.

The wild energy, uncomplaining toughness, insatiable yearning to see and understand—in all those ways, George was, as Roger Kaye would write in a book about the group's efforts to protect this *Last Great Wilderness*, "Olaus reborn." Though lately recovered from tuberculosis, O. J. (as George sometimes called him) was out every day and even in the tussocks was "magnificent," Brina recalled, "because he was wobbly but managed to wobble between them." Tipping out the tin cans he spent his days filling with treasures, he showed George how to tell ground squirrel hair from caribou in wolf scat and what plants the grizzlies subsisted on, alongside an occasional bit of bone, eggshell, or even bee. George's own eye became ever more acute. Peering beneath the skin of ice on the lake, he spotted a muskrat swimming full speed from below to batter it open with his head. He noticed the hollow sound caribou hooves make on sand.

Olaus modeled other qualities that would become essential to George, particularly his relationship with what he called (cognizant that other species had been here millions of years) the first *human* settlers. Nicknamed "little-bird white man" in honor of his obsessions, Olaus knew several Native languages and many locals. When Gwich'in tribe member Abraham John came along on a mail run from Arctic Village, bringing sugar, flour, and tea for his three kinsmen camped across the river, George spent an hour crossing the cold, swollen, muddy waters to let the campers know. He took note of their eight skinny dogs tied to trees, a wolf hide pegged out to dry (the Gwich'in had come to depend on the territory of Alaska's fifty-dollars-per-head bounty), a tin can on the fire brewing rhododendron tea, and that they had little to eat beyond a well-gnawed porcupine paw and a scrap of bread fried in its grease. Ambrose Williams, who George soon discovered knew the Sheenjek valley "better than any man," shook out a tobacco pouch to roll and smoke one cigarette after another. David Peter, spotting a moose, went after it with his rifle. The youngest of the three, Peter

Tritt, crossed the river with George to retrieve the new provisions. On a quiet word from George, Mardy added beans, rice, and dried applesauce.

That meeting began a string of reciprocal generosities. When their bush pilot returned, this time with Margaret Sam, a shy Gwich'in girl who became a lifelong friend, George spent a happy hour with her gifts (candy bars and comics), then waded the river with letters she'd brought Ambrose. David's hunt had been successful and he offered moose meat, which George said they'd accept only if the men had any left when they were ready to move on. A few days later, Ambrose and David brought "twenty-five delicious pounds," staying several hours to teach Olaus the Gwich'in names for local birds. When the pilot brought a ten-pound salmon, George carried half to the men; when David got a caribou, he brought the Murie team the liver and hind leg. "What wonderful neighbors," George wrote, experiencing for the first time the world's oldest and most universal value, a value wholly desecrated in the dark world of his childhood: the welcome offered to the stranger.

A six-thousand-year-old hide scraper found on a walk again testified to the antiquity of these civilizations, as did a set of carved posts marking what seemed to him a perfect burial ground. "All that view and nobody next to you and no slab holding you down." In the trip report, which it fell to George to write as his master's thesis, he included a translation of local names for the months: May the "month of fawning," July "of mosquitoes." He wrote of the aftermath of the Hudson Bay Company's 1847 establishment of Fort Yukon: Within twenty-five years, "the Birch Creek Kutchin were annihilated by an epidemic of scarlet fever." He quoted Aldo Leopold's *A Sand County Almanac* on the two great unfolding tragedies: the exhaustion of wilderness and the "worldwide hybridization of cultures." The wildlife preserve the Murie team proposed, he wrote in a long feature in *Outdoor Life*, "would not in any way interfere with the lives of these natives. They will still be able to trap, hunt and fish in the ways of their forefathers."

George's trip report cataloged eighty-five species of birds, tallying for each their number, coloration, habitat (elevation, terrain, vegetation), hunting and feeding behaviors, courtship, nest construction, song, and temperament, as well as whether their eggs were warm to the touch and (for those he dissected) the weight and size of their ovules and testes. He recorded the parasites found in the pectoral girdle of a vole and the "greatly enlarged cloaca and fully developed egg in the oviduct" of a female sandpiper he'd shot. And though he apologized for his superficial insect collection—"I swept no foliage nor did I sift litter"—he brought back 40 species of lichen, 138 of flowers, and 23 kinds of spider, many of which—he determined with the help of the twenty-seven experts he conscientiously consulted—had until then been entirely undescribed.

The report developed ideas then new to conservation. On one solo trip to the Sheenjek headwaters, he'd been awakened at 5:00 a.m. by what sounded like a freight train. For five hours, the Porcupine herd flowed all around him in a "river of life." He sat counting: 1,992 caribou—bulls with heads bent under massive antlers, newborns struggling to keep up—while the others, in camp, counted thousands more. Layering those observations atop reports by Native people, bush pilots, US Fish and Wildlife, and even the Royal Canadian Mounted Police, he compiled the first comprehensive map of this migration, laying out an early case for the vast scale at which a landscape had to be protected if migratory animals were to survive. Drawing on Leopold's and Frank Fraser Darling's findings on the damage done when, in the absence of predators, caribou or deer overpopulate, overgraze their own habitat, and thus eventually starve; and Adolph Murie's on how wolves improve herd fitness by culling the weak, George further pressed for managing not individual species but whole ecosystems, predators included. Though he was also already confident enough to correct his elders, noting that under cover of dense willows, wolves also take healthy animals.

Beyond meticulous data and these pioneering arguments, George captured the intensity of living close to these animals. He seemed particularly moved by avian acts of maternal bravery and loss. A pintail feigned a broken wing to lure him from her nest, a cup of sedge she'd lined with soft, warm down. Despite her valor, when he checked a week later her six eggs were gone. A willow ptarmigan refused to move off her clutch even when George touched her. He finally pushed her aside, but she beat her wings at him and climbed right back on. Another ptarmigan, her nest menaced by a ground squirrel, jumped on the "surprised rodent, who retreated in utmost haste."

Still experimenting with his writer's voice, George tried drollery: a too-close encounter "illustrates that care should be taken not to force one's attention on a grizzly." He tried the pulpit. "Is there a person that for a few more trophies would forever erase the track of the wolf?" He kept his thesaurus close: robins scolded, warblers warbled, loons moaned, juncos trilled. He quoted Longfellow on trees bearded like Druids, and Kipling's exhortations: "Something lost beyond the Ranges. Lost and waiting for you. Go!" But he was learning the force of understatement. Spotting a caribou cow on a talus slope, he noted that her calf's front left metacarpal was broken and dangling useless. The cow ran, then waited. "Ten minutes later I saw her without the baby," he wrote, leaving the rest unsaid.

Beneath all else, George found in this summer with the Muries his core beliefs. Across the decades to come, he would often invoke Olaus on the need to honor "precious intangible values." The Sheenjek's trees and animals were not natural resources, quantifiable in their commodity or even ecosystem value, but something both less and more purposeful. As Olaus had written George in their earliest exchange, "I am just as vague as anybody else about what I am going in the mountains for . . . I simply had the urge to get into that country."

Olaus no longer carried a rifle but left the collecting to Brina and George, who learned even to fish with his gun. Shooting into the water to stun Arctic grayling, he grabbed them when they went belly up, saving their stomachs for a fisheries biologist.

His view of guns shifted again, however, the day he and Brina followed

Bob through shadowy spruce, their footsteps muffled on moss, and startled a grizzly. Bob threw his tripod at the charging bear, who wheeled and ran. But George realized that a gun would have been useless. "No man is fast enough to shoot when not expecting the attack." Beyond that practicality, fear's grip on him had grown ever weaker. Though the wind tore at his clothes as he stood transfixed by a grand Dall ram silhouetted against high peaks, "I had never been so free."

The freedom George found in the Sheenjek was nothing like the individual license it is now often confused with. Quite the opposite: It lay in protecting the freedom of others, human and nonhuman, to be fully themselves. That the caribou moved without noticing him, repeating a journey they had made since long before humans first walked here, gave him a glorious feeling. As their plane lifted off on their final day, Mardy saw a grizzly with two cubs stand to watch them go. "Good-by, grizzly bear, your valley of the Sheenjek is all yours again." Even the free-running river, she saw, was "blessedly itself."

It took four years after their trip's end for Olaus, bearing George's report, to persuade Eisenhower to sign the executive order protecting 8.9 million acres (later doubled) in the Arctic National Wildlife Refuge. He had help from Justice William O. Douglas, who with his wife, Mercedes, had joined the merry band for a rainy week, cheerfully crawling through deep mud to find fairy forget-me-nots and a moss that grows only on moose dung and sprouts delicate yellow parasols that George identified for him as *Splachnum luteum.* To their meals of caribou steak and angel food cake, Bill (as he insisted on being called) added his alder-smoked grayling and jam boiled from the tiny cranberries he spent hours in the bog gathering. He told stories George remembered forever—of the four hundred plants he'd brought back from Persia, his travels in Pakistan's Hunza Valley, the brutality of British colonials in Tasmania—and later wrote his own tribute to the arctic's "loneliness that is joyous and exhilarating." Mercedes, who spent much of her time perched on an overturned bucket knitting argyle socks, pleaded with George to stop gutting his trapped mice. She "couldn't stand the sight of all the tiny skins."

Sheenjek would be the Muries' last expedition. Olaus, to whom George would dedicate his first book, died of cancer in 1963. Mardy carried on their work for another forty years, standing with Lyndon Johnson as he signed the Wilderness Act, which she and Olaus had helped craft and pass, and before Bill Clinton at age ninety-five to receive the Presidential Medal of Freedom.

•••

Returned to Wisconsin, George resumed his bird studies, now with an assistant to help care for his "surfeit" of babies, many of whom needed 2:00 a.m. feedings. Though she had a possible museum job awaiting her in Copenhagen, George had persuaded Kay to stop in Madison on her way. When at year's end she was still there, with a job at a bank, an apartment, and a Christmas goose to cook with George, her father let her know he approved. Kay helped George with his experiments, showing the nestlings such potential terrors as colored cardboard and a rubber hawk. Bronzed grackles, they found, crouched at the sight of a strange object three days after first opening their eyes.

Konrad Lorenz had just published *King Solomon's Ring*, describing ducklings imprinting and following him about, and George was delighted when his feathered charges did the same. None were more possessive than a great blue heron hatched from one of two eggs he had collected high in an elm. After watching the young heron grab and twist his sister's bill until she fell over, then jab at her with his own "flashing" bill, George named him Siegfried for "the hero of German legend who had a magic sword."

Looking with his gangly neck "like a fuzzy gourd," Siegfried spent weeks trying to stand. Thwarted by his heavy bottom, he could stay upright "only by leaning far forward and balancing like a tripod on his bill." Deeply attached to George, the bird danced when he walked in, lay in his lap to preen the hair on his arms, and roared at strangers, though he grudgingly tolerated Kay. A picture from those days shows her in plaid Bermuda shorts, hair carefully curled, crouching with a mirror to show an alarmed Siegfried his own reflection.

Their story ended with a poignant parting. After three months and several excursions to practice fishing and flying, George took Siegfried to a marsh where he immediately began hunting for minnows. "When I walked away, he flew up and circled before landing at my feet. Again, I carried him into the marsh, then hurried away." Though he knew from the bird's leg band that Siegfried flourished and covered many miles, George never heard from him again.

Although George found lab work useful for learning quantitative science, after Sheenjek he knew that his future was not with tamed or captive animals. That resolve deepened when he crossed campus to visit the labs where Harry Harlow was conducting his infamous experiments inspired by John Bowlby's study of war orphans. Harlow isolated infant rhesus monkeys to test their responses to a "warm" surrogate mother—made of terry cloth heated by lightbulbs, with a face and sometimes milk in her breast—versus a cold "rejecting" mother, made of wire and mined with spikes to repel a clinging infant. Nearly all the babies wound up staring and self-clutching, or huddled in a tight ball on the floor. Harlow also did captive breeding, with an apparatus he "affectionately" termed the rape rack. "It was awful," George recalled. "This was a time when people would say 'these animals have no emotions, feel no pain.' They were after an answer to their question, but went about it in a way deeply harmful, especially to the animal's psychology. After visiting a few times I wanted nothing to do with any of those labs." Even were the animals not suffering, to view them as objects generating data to elucidate a theory was the opposite of the kind of unmediated communion George would spend his life seeking.

In January 1957, Emlen interrupted a meeting on their bird experiments with a startling question: "Would you want to study gorillas in the wild?" George thought it a whim, or a joke. But his professor had in fact been mulling the idea for years, not giving up even when the leading ape experts in Africa, Europe, and the US deemed it crazy. He had won tentative support from Harold Coolidge at the National Academy of Sciences and, two years earlier, had gone so far as to ask his former student Brina Kessel if she knew any young scientist intrepid enough to spend years in the jungle with its most feared inhabitants. She did have the ideal candidate, she'd answered, though "he was still an undergraduate and could be highly impulsive, so I should approach him cautiously until I had the arrangements in the bag." She then urged George to apply to Wisconsin, without explaining why she wanted him to go to a school that had thwarted her own ambitions.

Without a moment's hesitation, the twenty-three-year-old answered Emlen. "Yes!" He then rushed home to write Coolidge, who at the very same age had made his own trip to Congo, shooting a gorilla for Harvard's zoology museum. Again George offered to work for nothing, or even pay his own way. Olaus wrote, too, from his hospital bed, to assure Coolidge that George was up to the task. After his hard childhood, "he can travel on next to nothing [and] has a complex about wasting anything."

Olaus did echo Brina's caution. "He is a tiny bit impulsive and may take vigorous action if it seems interesting or valuable. He's still very young and his judgment will be developing."

It took two years and dozens of letters "buttering up the right people" to get the requisite grants and permissions. While waiting, George managed to get himself invited to a primate seminar at Harvard Medical School. Among his pictures is one of the lowland gorilla he dissected in the comparative anatomy lab; its thrown-back head and arched ribs look unbearably human. Since he would be closer in Cambridge to New York, George wrote Bettina suggesting he visit her afterward. While she'd enjoy that, she replied, she "could not possibly" put him up in their small apartment.

Weeks later, he sent a curt postcard: The seminar had gone long, so he had to rush back to Madison. A month later, he let his mother know of his plan to be married in August on Cousin Ed's island, whenever the First National Bank gave Kay her annual week's holiday. (He had proposed via cartoon, asking under a portrait of the two of them bedraggled, potbellied, and swarmed with babies: "How about it?") When he next wrote, the event was just three days away. They had no money for a ring, and no one but Ed attended from either family. Bettina, whom Kay would meet for the first time two years later, sent a telegram offering "affectionate regards."

Returned from their honeymoon—a week of canoeing on Rainy Lake—bride and groom both wrote to Bettina. The ceremony had been "short and painless," George reported. "The preacher was lazy but got it over with in eight minutes." He added wry thanks for her wedding gift: a set of Henckels knives. He still carried the one Emilio had given him the previous year, with which he'd "cut off a bear's head in two slices last summer." And the new set was most welcome, though the grapefruit knife arrived bent, so he'd straightened it.

"Don't believe him!" Kay wrote soon after, in the first of many letters charming her mother-in-law with a flirtatiousness not unlike Bettina's own. She thanked Mrs. Iwersen for raising such a "sweet and helpful" son, who had done an excellent job scrubbing and painting their new apartment. "Not at all haphazard. He is a *very* thorough person. His one fault is that he loves rock and roll on the radio!" When a Henckels manicure set arrived for Christmas, for which they roasted one of George's "former turkeys," she thanked her again. "George is a great believer in cutting fingernails. I must pass inspection every other week. When we first met, I wore them not only long but painted. He still shudders." The set of daggers Bettina and Emilio had sent, part of the Henckels wartime repertoire, had pleased George no end. "He is like a small boy playing soldier. And I am in fear of my life! I suppose he'll be wearing them to school soon." George's own letter sounded like a Christmas carol: Twelve turkeys hatched, three pairs of Japanese singing quail, one grant application "that will probably be turned down."

George had been asking for years for a visit from Chris, and at last the sixteen-year-old came. He loved going out early mornings with his big brother to watch the famed mating dance of the prairie chicken, and even more that George—impressed by his varsity jacket—dubbed him, as Chris recalled many decades later, "a rock, like the Fonz." Kay wrote Bettina that it was wonderful to have Chris, "except that he sometimes sided with George not me! Two weeks wasn't long enough for me to train him." Bettina responded in kind, resurrecting the wit that had lain dormant for a decade. When Emilio had come down with the flu she had "followed him loyally," winding up feeling as if a steamroller had flattened her. "But don't worry. *Unkraut vergeht nicht.*" You can't kill a weed.

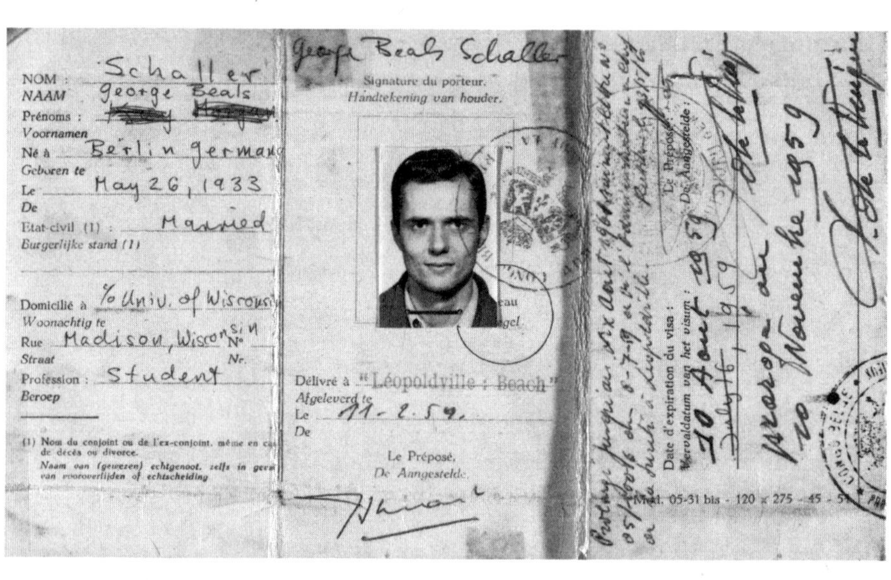

NOM : Schaller
NAAM
Prénoms : George Beals
Voornamen
Né à Berlin germany
Geboren te
Le May 26, 1933
De
Etat-civil (1) : Married
Burgerlijke stand (1)

Domicilié à % Univ. of Wisconsin
Woonachtig te
Rue Madison, Wisconsin N°
Straat Nr.
Profession : Student
Beroep

(1) Nom du conjoint ou de l'ex-conjoint, même en cas de décès ou divorce.
Naam van (gewezen) echtgenoot, zelfs in geval van vooroverlijden of echtscheiding

George Beals Schaller
Signature du porteur.
Handtekening van houder.

Délivré à "Léopoldville : Beach"
Afgeleverd te
Le 11-2.59.
De

Le Préposé,
De Aangestelde.

Date d'expiration du visa :
Vervaldatum van het visum :
10 Août 1959
Le Préposé,
De Aangestelde :

Mod. 05-31 bis - 120 x 275 - 45 - 5

3.

Gorilla Honeymoon

Congo, 1959–60, Mountain Gorillas

Having lost all fear in Alaska's expanses, George found it anew in his first encounter with the crawling intensity of the Central African jungle, an "eerie shadow-world, full of grotesque shapes, monsters that appeared for a moment only to vanish again." An accidental ambush by a red forest duiker, an antelope no bigger than a dog, had been enough to set his heart pounding. So had getting lost, on a solo ascent of 14,787-foot Mount Karisimbi, in the ravines that wheeled around its flank, as snow threatened and dusk set the leopards and buffalo in motion. Still, nothing had raised the hair on his neck like this: the *pok-pok-pok* of massive hands slapping a monstrous chest, a musty sweet odor. And there—just two hundred feet away—a four-hundred-pound silverback risen to his full height, huge, shaggy arms whipping out the rapid tattoo and blackened teeth bared as the alarmed animal let loose an explosive *roar, roar, roar.*

George would not be this afraid again until a year later in Uganda's Budongo Forest when a troop of unseen chimps, hooting like demons as they encircled him, suddenly fell silent; he knew they were closing in only by their crazed pounding on ironwood trees. He would not feel this powerful an urge to run until, in this same forest a bit later still, he was chased by men with machetes, bent on killing him for revenge.

But here, with this silverback, his fear quickly dissolved into something else: a yearning to make contact. "He was the most magnificent animal I had ever seen. . . . I felt a desire . . . to let him know . . . that I wished only to be near." And to glean, if he could—in the curious, soft, melancholy gaze he attests leaves no one unchanged—whether the gorilla, too, "recognized the kinship that bound us."

While George had been "scaring chickens" (as Chris called it) in Madison, Coolidge had been gathering financial, political, and scientific support. He persuaded New York Zoological Society (NYZS) president Fairfield Osborn Jr. to fund the gorilla fieldwork, though Osborn had little sense of its demands;* he proposed they send a zookeeper. As US directors of the Institute of the National Parks of the Belgian Congo, Coolidge and Osborn also prepared the way with colonial authorities. When Coolidge then reached out to his "old Cambridge friend" Louis Leakey, already wildly famous for his finds in Olduvai Gorge, George found himself suddenly at the center of rivalries spanning three continents. Leakey was opposed: He pre-

* Though himself a retired banker, Fair, as everyone called him, had been exposed via his scientist father, Henry Fairfield Osborn, to wildlife conservation, including its darkest strand. As head of the American Museum of Natural History, H. F. Osborn had embraced eugenics, joining the like-minded Theodore Roosevelt and Madison Grant (whose 1916 book, *The Passing of the Great Race*, became Hitler's "bible") in such groups as the Boone and Crockett Club, dedicated to preserving the continent's noblest wild creatures as quarry worthy of their equally fine breed of men. The same group founded NYZS in 1895, opened the Bronx Zoo four years later, then backed the decision by its first director, William Hornaday, to display a Congolese Batwa named Ota Benga. Though Osborn Jr. had his own complicated relationships with the colonial powers, he did provide essential correctives. In his prescient 1948 text *Our Plundered Planet* he directly rebuked his father's scientific racism, and his entire class for their imperialism: "Africa illustrates perhaps more vividly than any other continent the ill effects of transplanting European cultures, and specifically European methods of using the land, to other regions." At NYZS, he hired William Conway, the estimable zoo director who over time supported its shift in focus to global conservation led by local people.

ferred young, unschooled women for studies of the biggest apes, believing them to be unthreatening to male animals. Perhaps, suggested Leakey, George would rather study chimps in Gombe?

That exchange roused another titanic ego. South African Solly Zuckerman, who had been tapped as a military advisor by Winston Churchill following on his study of homicidal baboons, warned that Leakey was "not beloved" by colleagues in Africa. By the time the National Science Foundation (NSF) assembled a dozen experts to resolve the disputes, "we had trouble holding the whole idea under control," Emlen recalled. "The enthusiasm from primatologists was almost alarming."

George, meanwhile, had kept up his campaign, writing Coolidge to say that he would soon be a citizen and had married, which "in no way affects my desire to go. My wife, a trained anthropologist, is quite willing to remain here or to work at some job in Africa for the duration."

At last persuaded that "Schaller is a first-class field man [and a] good bet for the two-year pull," Coolidge approved the selection on condition that Emlen help George get started, including by coauthoring his NSF grant application.

Emlen and Schaller began by laying out the great void in knowledge. Every gorilla study to date had been of dead animals or, in one case—by Yale psychobiologist Robert Yerkes—of a lone individual (Miss Congo) in captivity. The few studies of free-living primates had focused on chimps, monkeys, or gibbons—all, not incidentally, a fraction of the gorillas' size. The men discussed methods, including techniques from the lab: They might stimulate the immense apes with a live snake or a leopard skin, a mirror, even a caged gorilla infant. They proposed that Kay join the team to explore ethnological questions, like the gorilla's role in local folklore, and for her secretarial skills and knowledge of botany. (The ideas of both artificial stimulation and an official role for Kay were soon jettisoned.)

The heart of the document made direct appeal to the growing obsession in postwar science with how primates might illuminate man's nature, especially the violence that had defined the century, the proper structure of the family, and male and female roles. "To understand man stripped of the masks of culture," they wrote, the mountain gorilla promised the truest window, having diverged far less than other great apes from man's own evolutionary course.

To bolster that claim—reflecting the degree to which scientific racism was still, as historian Donna Haraway put it, "normal, authoritative practice"—Emlen and Schaller quoted Harvard anthropologist Earnest Hooton. "More is to be learned about . . . community life in early man by the study of contemporary infra-human primates living under natural conditions than by the studies of retarded human groups." (Even four decades later Hooton, known for measuring the skulls of his three subjects—criminals, the Irish, and Harvard freshmen—had not been disavowed. In 1995, the National Academy of Sciences published a tribute granting him a stature equaled only by Franz Boas and Aleš Hrdlička and noting without irony that this man who believed intelligence to be inscribed in the body was disqualified from military service for nearsightedness.)

Even as they submitted the application, the question of how to protect against gorilla attacks remained unresolved. When George would not yield on his refusal to carry a gun, certain that it would introduce an edge of aggression the animals would sense, Coolidge suggested a tear-gas pistol. "Though if you do shoot a gorilla, it will in all probability mean the termination of permission. And if you don't it might also spell termination." Emlen rejected another Coolidge idea: shooting a drugged dart should the gorilla charge. "It might work on ungulates, but I can't see that it would . . . against a charging male gorilla. And I can't imagine tagging a doped gorilla while the troop watched and then expecting normal behavior after he came to . . . anthropoids don't quickly forget." Perhaps instead they might contrive a disguise? In any case, "George and I are convinced that an appeasement posture would stop an attacking gorilla faster than an aggressive bluff, though neither of us would like to put the hypothesis to a test." In the end, mostly to appease Kay, George agreed to carry a little track-meet starter pistol, though he would find cause to use it only with people.

Equally nettlesome was how to manage the jockeying now in full swing among researchers in Africa and Asia. Not wanting anyone to steal their thunder, Emlen campaigned to keep a Japanese team, experienced with habituating macaques and chimps, from making a simultaneous gorilla study. And worried that "without Leakey's wholehearted cooperation the project will be unthinkable," he wrote the anthropologist a deferential letter. "We feel humble in our lack of direct experience . . . you know the animals and problems of field work better than anyone. . . . I enclose our NSF application which I'm sure will seem naïve to you in places. . . . [Do] gorillas accept pygmies more than white men? Is it possible a man could protect himself with light armor? A wild idea to be sure but that's the type of thing we need."

With Zuckerman continuing his anti-Leakey campaign, Emlen wrote despairingly to University of Chicago anthropologist Sherwood Washburn, who had been with Coolidge on a primate expedition to Southeast Asia and studied free-living baboons in Rhodesia.* Given the feud between Leakey and Zuckerman, "the only course is to make a choice accepting one and excluding the other."

"They all feud," Washburn replied. "Our policy has been to pay no attention to the rows. Leakey is a dramatic, brilliant, original guy . . . but I'd let no part of the scheme depend on him."

With just six weeks till departure, one last hurdle landed. "Just to add to the excitement," Emlen wrote Osborn, "Mrs. Schaller (George is in Columbus studying the gorilla pair there) received a notice from George's draft board—1A!" A flurry of letters to Selective Service followed. The NSF and NYZS argued that George's special abilities and training made him

* Now Zimbabwe

irreplaceable. And it is through this work, the university added, that "the welfare of the United States will best be served."

His deferment granted, George could at last turn his full attention to gorillas. He read everything he could find, back to Hanno's account of his fifth century BC expedition from Carthage, concluding that "no animal has [so] fired the imagination of man"; in these creatures "biology and myth combine." That was particularly so since the 1933 release of *King Kong* (and its rerelease in 1952, when *Time* named it Movie of the Year). Its success had set off a craze for gorilla films, played for lewd comedy or terror; actor Charles Gemora built an entire career in a gorilla suit opposite Bela Lugosi and Abbott and Costello. Pulp magazines churned out tales of the beast ravishing white women; the racial overtones of the helpless blonde in massive black arms could hardly be called veiled. War propaganda continued the theme: In one British poster, a gorilla in a swastika armband carried off the maiden Liberty. For anyone seeking to "animalize" other humans, apes proved an ideal trope. That included Jews, a tradition parodied by Kafka in "A Report for an Academy."

George found hunters' accounts to be the most useful. The best since Du Chaillu's was a 1956 book by Fred Merfield, who had killed 115 gorillas for European museums. Having hyped the apes' monstrousness to burnish their own daring, these men often had awful epiphanies. To Du Chaillu the gorillas "looked fearfully like hairy men . . . running for their lives." Merfield saw that troops never abandoned the wounded but tried to get them to safety. Carl Akeley, who had shot a female and then speared her infant to fill out his diorama's "hierarchically ordered" family tableau, saw a "heartbreaking look of piteous appeal . . . Like a little child sick and lost, he needed help. I knew that he would have come to my arms for comfort."

George learned less from the handful of researchers who preceded him.* Of the scheme hatched by professor Richard Garner—who in 1893

* One notable exception was Raymond Dart protégé Jill Donisthorpe, who made contact with mountain gorillas several times, held her ground through two bluff charges,

set up house for 112 days and nights in an iron cage, planning to learn some gorilla words, lure them in with conversation, and then record their language "of a hostile or amatory nature"—George wrote, "as might be expected, his success was as restricted as he." Garner did bring an infant to the Bronx Zoo, wheeling her about in a pram and lace cap, but she soon died, likely due to his insistence on "correcting" her vegetarianism. Louis Leakey protégé Rosalie Osborn, whom he'd dispatched to Uganda when their affair in Nairobi became untenable, mostly hid when she saw gorillas and found it all too much. To avoid exhaustion, she advised, George must spend two days at low altitudes for every three on the mountain, and buy insurance. "The strength of a male is ten times that of a human and they delight in dislocating bones."

The trip began, as George diligently recorded, with the Emlens and Schallers on a Pan Am jet from Idlewild, "cruising at 37,000 feet, outside temperature -88 . . . arrived London 9:35a . . . cab a very old-fashioned thing with luggage tied to the outside, fare <2." The next morning, after early birdwatching (he listed the eleven species), "we sit on bed eating good breakfast (price incl) of tea, fried eggs, sausage and toast." In London, he and Doc met hunter-collectors from the previous generation, including ornithologist Beryl Patricia Hall, who had spent two decades in sub-Saharan Africa for the British Museum, and a fading Merfield, "whose cold flat had nothing but an elephant foot" to recall his thirty-five years in Cameroon. After stopping in Brussels to get letters of introduction from Victor van Straelen, president of the Institute of the National Parks of the Belgian Congo, George continued on with Kay to Germany on a train "quite improved from the last I'd ridden here in 1946."

Though George had seen his father just once in twelve years, and Kay was meeting him for the first time, they spent only one night. George met

and discovered the value of remaining visible. (J. H. Donisthorpe, "A Pilot Study of the Mountain Gorilla (Gorilla Gorilla Beringei) in Southwest Uganda, February to September, 1957," *South African Journal of Science* 54, no. 8 (1958): 195–217.)

his stepmother, Annalise, and six-year-old half sister, Renate, who loved his silly pranks, including stuffing her shoes full of paper. They walked through still-bombed-out streets to a carnival, then spoke till dawn. Kay had met George's mother on their passage through New York, dining with her and her new husband at the Starlight Roof. "I admit I can see why George has no relationship with her," she wrote afterward to her own mother. Bettina was elegant but with a superior air, charming but insincere. "She thought you'd married beneath you," Kay told George, "though was relieved that I was blond and not an Eskimo." With George's father, Kay was "completely captivated," finding him "a warm dumpling, short and smiling and a little round, one of the most thoroughly nice people I've ever met. He was interested in many things . . . felt so dreadful about his separation from George and Chris, wrote us wonderful letters. A good diplomat, he knew how to make me comfortable."

On February 9, 1959, George and Kay and Doc and Jinny reached Africa at last, though it would take five more flights and a crowded ferry across the Congo River to get from Kano, Nigeria, to the gorillas' home. George's journals from those first days preserve glimpses of a continent on the cusp: He noticed officials' tattooed faces, an emir with a black staff adjudicating disputes, a healer selling charms of crocodile skin and vulture wings—and also that towns were carved into quarters for Africans, whites, Arab traders, and Indian merchants. He saw his first topi "with black thighs and a curious rocking gait" and wild elephants hovering as a newborn crossed a creek, only its little trunk above water.

At first sight of their destination—the eight Virunga volcanoes that traverse the Albertine Rift and rise to almost fifteen thousand feet—George was madly impatient to begin. "Fortunately, Doc . . . had learned the wisdom of having the proper support" before rushing to the work. For a month, they visited officials and scientists in Congo, Uganda, and Rwanda, paying deference and gathering advice and permits for the six-month survey Doc would stay for—to map the "perilously small area" that was the mountain gorillas' entire range and find the best place for George and Kay to settle in.

Their passports were not always warmly received. The CIA had been busy dispatching or propping up regional allies to parry Soviet influence and secure access to valuable minerals. And the rare American tourist had done little more to win local hearts. George met one "bloodthirsty" trio in Uganda's Queen Elizabeth National Park who "had to shoot several of everything." His Belgian colleagues told him of another American who, when he didn't see lions, "complained, 'but I have money.'"

The Belgians added to the tensions, imperious even as the clock wound down on their brutal reign. Marc Micha, the warden of Albert National Park* and a former army officer, demanded military discipline of his 250 African guards. (And offended young George, who pouted that "Kay and I, as Doc's assistants, always get ignored.") Heading at last into gorilla country with Rousseau, the Belgian chief of guards, George watched the porters twist grass straps that, looped across the crown of their heads, helped bear the load, and savored feeling "out of my century." But after wondering why for just a week they needed sixteen men, he saw that they were carrying Rousseau's mattress, linen sheets, porcelain tableware, and baskets of fresh fruits and meats.

On this first reconnaissance trip with Doc, to the cabin in the remote study area called Kabara that would become the Schaller home, both men made errors of inexperience. Though Doc had been to Africa before, he found it hard to stay quiet. The first to spot gorillas, he burst out with a "Hey!" that sent them crashing away. A few hours later, Doc was enjoying a postprandial nap when George had that frightening first encounter of his own, with the roaring, black-toothed, chest-pounding silverback. It hadn't taken long for the animal to settle, calming the others enough to resume the lazy feeding that fills a gorilla's days, watchful but not worried by

* Now Virunga National Park

George. Then Doc woke up. Not knowing how long he'd slept or where George was and whether he was safe, he shouted his name. "Ten minutes later George returned, furious!" Doc had again chased the animals away.

George's own rookie embarrassments began with the hike in—five hours up four thousand feet—which left him collapsed in the grass and the porters laughing. Worse was his reaction when a sudden wheeze sent him scrambling up a tree, realizing too late that he was no higher than the back of the passing elephant, who could easily push the tree over anyway. Doc had to wave his sheepish student down.

Still, they caught glimpses of a sweetness and placidity contrary to everything they had been told to expect. A female gorilla, using a male as a pillow, held a baby so spidery and wet they realized it was brand-new; they later found the afterbirth in a nest. Another male clowned: ambling past a female, surprising her with a yank on the leg, then cantering off. A big group—twenty-two, including four silverbacks—finished eating and lay down like sated picnickers, arms folded on their chests. These sightings often required stumbling up and down steep slopes through a relentless assault of nettles and spiny vines and brush so thick George sometimes had to crawl on his belly or lie face down on the tangle, packing it with his body so he could walk on top. Frequently tripping, he sometimes just rested where he fell, cheek pressed on the cool ground.

Their beginner's luck notwithstanding, George realized that he needed teachers. He found his first at the Traveller's Rest Hotel in Kisoro, Uganda, which became the Schallers' "home away from home." Proprietor Walter Baumgartel had hired Reuben Rwanzagire, who was unequaled as a gorilla tracker and would train several generations of scientists. For their first lesson, he took the two couples to Walter's mountain camp, where Osborn and Donisthorpe had done their surveys. Speaking only in low whistles, "like pensive birds," Reuben showed them how to look for trees peeled back to the tenderest bark and holes dug to reach borage root. Coming suddenly upon gorillas sheltering under an overhang from a downpour, George startled at the male's roar. Reuben urged him to go closer but he demurred, preferring for now to stay at eighty feet. When another silverback surprised them, rearing up to beat his chest just ten feet away, they learned Reuben's most important lesson. Don't Run. "I felt a rush of adrenalin," wrote Doc, "but glancing at Reuben, saw he was standing firm." Later, visiting hospitals, they saw the price of panic: half a dozen patients with severe wounds, always on the back of the calf or thigh. With few exceptions, a gorilla attacked only if someone fled.

Jinny did not feel well and returned to the inn, but Kay settled happily into her first African campout with George. She helped him skin and stuff a mouse and together they pressed plants; sixty-five years later, his journals sometimes fall open to reveal a mid-century flower or the feathers of a cinnamon-chested bee-eater, still a vivid orange and green. Kay brought George treasures, delighting him so with a bulging-eyed chameleon that he devoted an entire page to it, describing it risen up on its prehensile tail to wave him away with flipper feet. His notebooks filled up with pictures, often the quickest way to capture a posture or moment. Taken with warthogs, who shoot their tails straight up when excited and "run with a certain erect dignity, their upturned tusks like a big mustache" as if they were "British military circa 1900," he sketched them in high grass, the only visible bit those stiff tails.

George was now used to the altitude and summited Mount Muhavura three days running, because he'd seen gorilla sign at 13,500 feet but mostly because up there "everything was great and free." On a rare sunny day, he took Kay to climb Mount Gahinga, one of the smaller volcanoes at 11,400 feet. There she had the first encounter she had mostly dreaded: Eighty feet away, a silverback stood to beat his chest. As a child, Kay had been scared by both a live monkey and a cartoon gorilla; even as an adult, her nightmares invariably featured gorillas. Now, when this real one opened his mouth, "I had to shut my eyes. George's professor was there, and the Africans watching the memsahib. I tried to hide just how terrified I was."

For his next apprenticeship, George sought out the aboriginal Batwa, who—unlike the Hutu and Tutsi,* who raised crops and cattle in the lowlands—lived in the montane forest and knew it as no one else. He and Kay drove at night into Uganda's Impenetrable Forest, through hordes of hippos heaving and rearing in the headlights. Getting lost, they arrived to find the Lake Edward Hotel dark; when a man roused by their banging said there were no rooms, they rolled out their sleeping bags on the dining room floor. The next morning George woke to a sight he'd dreamed of since boyhood: the Ruwenzoris, the glacier-topped Mountains of the Moon.

Though the Batwa had lived here for thousands of years, their world had been steadily shrinking, as first Hutu and then colonial planters cleared lower-elevation forests for fields. By the 1920s, their situation was precarious enough that when a remorseful Carl Akeley—seeing the gorillas' survival threatened by the collecting craze he'd help fan—persuaded King Albert to create Africa's first national park, it was conceived to protect these "*Naturvölker*" too. Here, scientists would be able to study two endangered groups: the "most advanced" primate and "most primitive" humans. Julian Huxley, who was director-general of UNESCO when it established the International Union for the Conservation of Nature (IUCN), praised the move in language that suggests how quickly reverence can slide into something else. The Belgians had treated the Batwa properly, he wrote, as "fauna rather than as tribes to be civilized."

The protection afforded by the park was far from complete. Some Batwa had already been compelled to abandon their traditional life; the first George met, at Walter's hotel, was dancing for coins. But taken to visit a village, George saw how at home most remained in the forest. Their huts, built quickly of bent saplings and ferns, were intentionally transitory. When the pigs and monkeys they hunted grew scarce and shy, they moved on, their huts reverting to tree and frond.

---

* The Tutsi were then more commonly known by the plural ("wa") form and a slightly different spelling. George himself uses variants but most often "Watutsi."

George and Doc hired four trackers including a Batwa chief named Bichumu and a bright twelve-year-old boy, Ephraim Busingye, who would use his mission-school English to interpret. From the moment they set off, George felt the clumsiness of his own height and gear; he had to break vines the Batwa easily slipped through. He struggled to keep up with Bichumu, who though white-haired and wrinkly kneed trotted tirelessly from dawn to dusk carrying two spears, a panga (machete), and a goatskin bag to gather edible roots and leaves. In this tangled, buzzing forest, in light so dim as to seem subterranean, George watched in awe as his guides found an invisible pig track in matted leaves or jabbed a spear into smooth ground and unburied a hidden tuber. He recorded all they showed him, including the native names: the *ornukungashebeya* tree used for dugout canoes; a vine, *olushuri*, they wove into bags. But when he asked too persistently about the gorillas—Do they eat this plant? Do they nest on the ground?—he was met with evasion or silence. Finally a distressed Bichumu quieted him. "If you call the animal you are seeking by name, you will never find it."

George was fascinated by the Batwa technologies, drawing them from several angles and using colors to distinguish parts. A few of their snares had a Rube Goldberg complexity: A passing animal's hoof compressed one stick, releasing a second, releasing a bent tree with a snap to tighten a noose around the animal's leg, or levering a log-fall to crush it. He was particularly delighted by their skill with wild bees, sketching a man high up a tree with a burning hunk of wood. Having smoked the bees out, the man reached in shoulder-deep to haul out cream-colored combs dripping with golden honey. Gleefully devouring both the honey and white bee larvae, the Batwa spit out the wax and licked their sticky arms, yelping and laughing whenever a mired bee stung their tongue. George enjoyed their noisy exuberance: the call and response as they worked, the chatter that erupted from nowhere in the middle of the night.

At times, writing about the Batwa, George edged into his own essentializing. "The gurgling of water . . . the swift flight of a flock of grey parrots, had ceased to be received by their senses for they formed the heart of the senses themselves." But he also nodded to the strategic way they used their knowledge: to right the balance of power with him—they sometimes deliberately lost the trail—and to retaliate against the Hutu porters who refused to share their huts or food and ordered the Batwa about. When George asked Bichumu why the Hutu believe gorillas catch and throw back flying spears, he smiled. "We tell such tales . . . and they believe them," a mockery and also a canny way to keep their lowland neighbors out.

Before leaving the Batwa, George wished to see one last bit of their forest, most easily reached by car. The young ones were frightened of his blue VW bus but climbed in. Bichumu refused. Turning his back, he sat down in the road. When he wouldn't budge, they left him there.

•••

As he began to know the animals and their human neighbors, George also saw the drastic changes reshaping everyone's world.

Gorillas were still regularly killed: by white miners for sport, and by collectors for circuses and zoos, who typically slaughtered whole groups to get babies. For a species that numbered in the hundreds and reproduces slowly (beginning at age ten, a female has just one baby every few years, a quarter of whom die), such losses were by themselves unsustainable and, by disrupting stable social structures, often led to more deaths. Additional pressure came from the decimation of other animals vital to the ecosystem beyond the montane forest, much of which Schaller explored. In a misbegotten effort to eliminate tsetse flies, the Belgians had ordered the extermination of waterbuck and bushbuck. And with globalization advancing, subsistence was giving way to market hunting, for meat, ivory, rhino horn, and even giraffe sinews for bowstrings. These lucrative commodities were often cruelly secured. Accompanying a warden tasked with shooting an injured elephant, George saw how the snare's tightly wound steel cable had cut its hind leg to the bone. Though the elephant was still alive, he wrote in his field journal, "vultures had begun on its anus and neck," summoning the affectless voice he often used to protect himself—and devastate his reader.

Most threatening to the gorillas was the ongoing destruction of their forest home. Fair Osborn had condemned Africa's colonizers for forcing "native peoples . . . from the fertile valley lands so that they were compelled to burn and cut forests" to make space to grow food for their families. He had also warned of the damage done in replacing local practices that did not take "more from the land than it could give" with cash crops (coffee, pyrethrum) that mined out soil fertility and set it on course—as Aldous Huxley wrote in a fervent appreciation of Osborn's book—to become another Dust Bowl. George saw this all unfolding on the ground. Outside of the national parks, including in areas the Belgians had taken out of the protected boundaries to restore to local use, he found the land empty of game. Some places had turned to desert, useless for sustaining humans or animals. As Hutu plantain groves and European tea plantations and Holstein milk cows pressed into the forest, the gorillas were encircled in ever smaller islands, isolated from contact and genetic mixing with the others. Climbing one day to nine thousand feet, George found a family of Flemish homesteaders raising three daughters and—because their pigs had all died—turnips.

George had read deeply into the history of Congo, as traumatized as any place on Earth by the most barbarous of imperial despots, whose "sole ownership" of the country had been ratified in 1885 by the other colonial powers. When the treasure of elephant ivory had failed to sate his greed, King Leopold II had turned to forced labor, ordering his private army to sever the hands of any man, woman, or child who fell short on their rubber quota. He hung the heads of tribal chiefs alongside his animal trophies and exhibited 267 Congolese people in his "human zoo."

Ranging across the region's ecosystems, George could read traces of that brutal history on the land. Hiking through a hot and nearly uninhabited region of the Congo basin, he gleaned from the rows of palms marking vanished roads that this had once been a farming community. Enslavers and smallpox had taken most of the people. The few who remained, he learned from a missionary there since 1928, had seen their land sold out from under them to white planters. Arriving in the "most wretched village I had ever seen," he saw the consequences for communities of stolen and ruined land—the children's bloated bellies and mucus-caked eyes. Waking in the morning, he found everything "very quiet, as though exhausted by tears."

Reading back still further, George saw that Indigenous hunting and farming had been part of this ecology for millennia, reshaping it in ways that sometimes benefited the gorillas.* While too little light penetrates undisturbed equatorial forest to grow much gorilla food, centuries of swidden farming—using fire to clear small bits of forest, plant bananas and manioc, then after a few years abandon it for new land—had created a patchwork of forest in different stages of regeneration in which, George realized, the gorillas thrived. In those abandoned fields, the apes found lots of their favorite ferns, herbs, and leaves.

The coexistence was not uncomplicated; it never is. If the gorillas strayed into an active banana grove, he'd learned from Merfield, villagers would beat the silverback to death, then turn on the mothers and babies. "They don't even try to get away, and it is most pitiful to see them putting their arms over their heads to ward off the blows." If these farmers obtained guns, George feared, such rare killings might become commonplace. "Improved" farming methods, similarly, would unmake this habitat. He would eventually recommend the creation of a new reserve in these basin forests that preserved traditional shifting cultivation.

---

* In 1965, William Allen would publish *The African Husbandman* summarizing four decades of studies of the profound ways farmers and herders had reshaped these lands. Schaller's findings, that those changes sometimes benefited other species, anticipated by more than twenty years William Cronon's *Changes in the Land* and the full flowering of ecological history.

With Jinny gone to visit friends in Rhodesia and Doc heading off on his own explorations, George and Kay set out with local guides to a little-visited region of the park. Laboring up the steep rift escarpment toward Mount Tshiaberimu, wet leaves and river rocks slippery under their feet and thick bamboo showering them with icy drops at every touch, they got a foretaste of the miserable two weeks ahead. Day after day, in unceasingly beastly weather, Kay stayed in camp, tediously drying the blotters they used to press plants while George headed off through knifelike sedges that slashed his hands and arms, bitten by ants that marched in columns so big he could hear them rustling the grass. Nearly done in by a trap set for elephants—a vine that, had he not dodged it, would have tripped a six-foot spear-tipped log right at his head—he often heard the tuskers' frightening crash. But he never once saw a gorilla.

Still, both found satisfaction in the hardship, appreciating the heightened savor of pleasures forgotten to those of us forever wadded in comfort. Returning frozen and exhausted to Kay waiting with hot cocoa by the fire was a kind of bliss for George. Having finally mastered the art of sleeping in a jungle hammock—getting undressed in the small suspended space, turning over at night without getting entangled or tumbling out—Kay loved the feel of the sleeping bag growing warm with her body's heat, the hammock holding her tightly, raindrops plucking out a melody on the tent ropes. She described her life of luxury in letters to their University of Alaska friends Thelma and Dick. "I am growing spoiled from this new experience of sitting while others set up the tent and bamboo hut. My housework consists of laying a new floor of ferns each day."

Kay was also enjoying the fellowship with their companions, "wonderful people, and so fun. When they see the camera, they laugh and ask to have their picture taken, then rush to change clothes and comb hair." Though George had taken an immediate dislike to their "major-domo" Christoph, "for the swagger with which he pushed other Africans around and for his cowering and whining whenever he talked to us," both had warmed to shy, sweet Kieko, who taught Kay a song in Swahili, which she wrote down. "*Tumechoka kutafuta ile ngagi*," they sang. "We are tired of searching for those gorillas."

After three months in the park, George now felt confident enough to leave Kay in camp and head out for a night alone. Knowing that a solitary white man presented a strange spectacle, he made a discovery that would serve him in the coming decades: the bridge offered by curious children. Passing through a village, he found himself trailed by a giggling band, who screeched with delight and ran off when he dramatically wheeled about, inching back in hopes he'd do it again. He began carrying Band-Aids, seeing how the little ones liked having him put them on their scrapes and sores.

George's only real moment of fear, on this first solo night, was sparked by a pile of elephant dung. Confirming its freshness by shoving his fingers into the soft, warm mound, he strained in the thick fog to locate its depositor but heard only the hard pounding of his heart. Finally, he resorted to a strategy he would later use with tigers: direct beseeching. "Elephants, hello," he said out loud. "I am only a human being, a weakling without weapons . . . Please leave the trail and let me pass." To his relief, they did. His good fortune continued with discovery of a bamboo lean-to where he could stay the night. He built a fire to dry his clothes and buried a potato in the coals. When it was crisped, he pierced the skin, "which opened with a steaming puff. . . . I could imagine no more perfect evening." The shelter, he knew, had been built by poachers. "The only thing I really miss," he wrote in his journal, "is a good hunk of meat. I can now well understand the African hunger for it."

One afternoon George found Leopold III—dubbed "the traitor king" after surrendering to the Germans and forced to abdicate in 1951—drinking in Goma's Palm Beach Hotel. On another he met a Mr. Fangoudis, here to seek his fortune. After years of prospecting for tungsten, he had turned to killing elephants: The colonial administrators got the tusks; he got the meat to smoke and sell. Fangoudis introduced his Cypriot mail-order bride, now pregnant, showing George the ring he'd bought with the money he had saved for her return passage, in case he hadn't liked her.

Many of these expats shared George's yearning for intimacy with wild animals, though less as self-possessed inhabitants of their own world than as surrogate children or exotic ornaments. He sometimes called on Mrs. de Munck and her deadly vipers, puff adders, and green mambas. Or on Baudart the coffee planter, who had paunchy eyes and a red face but was "extremely nice." Living alone with a yearling buffalo and two bristly warthogs, he'd trained the two-hundred-pound pigs to sit and beg like dogs, which George found hilarious.

Most remarkable were Charles and Emy Cordier, who'd built an elaborate outpost to supply the rarest animals for Western zoos and films.* Kay especially adored their baby gorillas: a toddler in diapers and two-year-old Mugi, an "absolute devil" who would play at tripping or biting her, "then run to me for protection when about to be punished, knowing I forgave him everything!"

With Kay happily settled in this menagerie, George took off on his longest survey yet, an eighty-mile hike with several of Cordier's men. By now he was used to feeling awkward. Having watched his porters, loads balanced on their heads, skip across a broad raging river on a slender, slippery, swaying log, he "humped along on my belly like a worm, grasping . . . with arms and legs." Each night at camp, as the others combed their hair, washed, changed into clean clothes, sometimes even polished their shoes and brought out hymnbooks, he felt ever more aware of his own stink, having brought neither extra clothes nor soap. He was ashamed to decline their offer of a shared meal, not yet able to overlook the graying skin and bloat of a wild pig left in a snare too long.

But he was also finding ease with his companions, the mix of intimacy

* The most famous film featuring the Cordier animals was *Masters of the Congo Jungle*, narrated by Orson Welles, which George and Kay had watched with excitement on their way to Congo, not realizing it was all artifice.

(physical, comic) and distance in which he would always feel most at home. Having forgotten a spoon for his canned corned beef, he whittled one of which he was "right proud" (his occasional Huck Finn colloquialisms a reminder that he learned English in part through books), and used it until the Africans laughed at him and gave him a tin one. He enjoyed their desultory talk by the fire, about immediate things—"the affairs outside the forest had lessened in our minds"—and found it peaceful to lie in his tent listening to a language he didn't understand. On the night of a full moon, he let the others look through his binoculars at the craters on its surface, explaining that they were similar to the volcanic craters in the park. "They understood but did not believe it."

By August, with the survey complete and Doc and Jinny returned to Wisconsin, George and Kay settled at last in the cabin at Kabara. Kay pushed herself on the arduous hike in, again "fearing the Africans would think badly of me if I couldn't manage." But she found the cabin dispiriting—impossible to keep warm with the wind finding every gap, its roof leaking, and the three small plastic windows so clouded they let in almost no light. She and George did their best to brighten things, setting out a turquoise washbasin and yellow water pitcher. But it was the flower-strewn meadow, the *Hagenia* trees shaggy with flaky bark like "kindly unkempt old men," the weird senecios—huge gnarled succulents—and especially the view of Mount Karisimbi, the volcano used by Akeley as the backdrop for his AMNH diorama, that buoyed their spirits. George's descriptions, even in his field journals, sometimes call to mind paintings by John Constable or Caspar David Friedrich: Clouds pouring off the frosty summit are "torn into pieces by an updraft and tossed up like flames."

The first weeks at Kabara tested George's fortitude. His journals record not a single cloudless day but page after page of a bleak refrain—rain, cold, mud, hail, scary buffalo, scary elephants. And nettles, so many nettles. Next to "gorillas," that may be the word that appears most. His knees are forever red, swollen, throbbing, and itching, his face welted where they slap him.

He sometimes has to push through solid fields of them, leaving his whole body burning. When he grabs at something to stop his frequent falls on the steep trails, if it isn't nettles, it's a spiky palm frond or thorny bramble. And for all that misery, at first, came scant reward. On the rare days he did find gorillas, the encounters were glancing—obscured by thick vegetation, or cut short with frightened screams and their quick retreat.

At last, after weeks, his persistence paid off: He made the kind of sustained, intimate contact he had dreamed of. It began with a silverback's stroll past the branch George had climbed onto, displaying heavily muscled arms and thighs and a tremendously broad back to their intended intimidating effect. Next came a peekaboo game with some adventuresome females who had carried their babies into an adjacent tree. Bravest was a young male who once and for all broke the ice. Coming within ten feet, beating his palm on the ground, he looked up at George like a mischievous child awaiting an adult's response, then kept looking, as if "searching for a clue to my being." This youngster would become George's closest gorilla friend, regularly approaching to pass long stretches sitting contentedly close by. He would be the first gorilla George named, breaching with "Junior" a scientific protocol he found nonsensical. "When my notebooks began to refer to gorillas by name, the animals ceased to be anonymous members of a species and became individuals, each with foibles, sensitivities, problems, family ties, traditions and past experiences." The kid would inspire some of George's sweetest notes, like this study in gorilla languor.

> *Junior sits and peers intently at the vegetation . . . stretches far out and with a quick twist decapitates a Helichrysum . . . stuffing the leafy top into his mouth . . . The sun appears briefly and Junior rolls onto his back . . . holding the sole of the right foot with the right hand. Then he switches and rests on his belly. After about ten motionless minutes he . . . slides his hand up the stalk of a Carduus afromontanus, thus collecting the leaves in a bouquet which he pushes with petioles first into his mouth.*

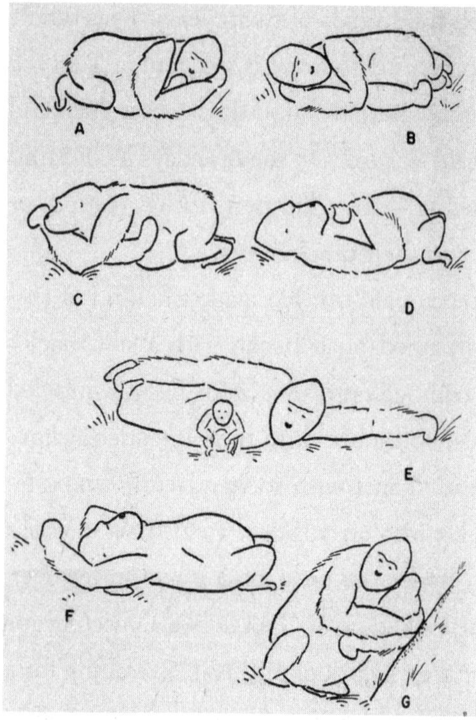

Sketches were often the quickest way to capture a silly antic or tender moment.

On eight occasions, George slept at the foot of whatever trees the gorillas had chosen for their night's rest, nestled into the little caves made by thick roots that reminded Kay of "what I always thought Peter Rabbit's house must have been." Though she worried when he went out overnight, she never once asked him not to go. "I just had to wait until he returned. He'd say, 'I'll be back tomorrow. Maybe in the afternoon.'" Over time, these glossy blue-black animals, utterly unlike the drab, cement-scuffed creatures he'd seen in captivity, granted him witness to their tenderest family moments. Babies played, sliding on rump or belly down a mossy tree, then pouncing on an adult female who'd come with folded arms and a pointed look as if to say, "Time for bed." A "demure little fellow" donned the leafy crown of a lobelia like a hat. Caught in what seemed unquenchable grief, Mrs. Wrinkle carried her dead infant for days. When it rained, George and the gorillas all hunched their shoulders, turning their backs to the cold drops. Their features were as distinct as humans'. "I tended to forget that I was looking into the faces of apes."

With George often gone for days and nights at a stretch, Kay was making her own peace with life in the forest. For company, she wrote long letters to her mother, the pages accumulating like a serialized adventure over the weeks between the changing of their guard, when they could send and receive mail. When the aerograms reached Alaska, the family gathered to read them aloud.

Living in a world of men and badly missing "women's conversation," Kay entertained with a running account of the trials and triumphs of a 1950s American homemaker tossed up in the Congo jungle. The stove became a character in its own right: its too-narrow pipe wrapped with cement and cheesecloth "as if it has plaster casts on," but still dripping black tar on everything; its leaky oven leaving everything stewed on the bottom and dried on top. But oh, to be delivered "fresh meat, 40 eggs, 2 Life Magazines and a Saturday Evening Post. Of such things, heaven is made. I made Swiss steak and added a tiny cauliflower. And peach tarts, with ground almonds and a touch of almond flavoring, absolutely one of the most delicious things I have ever tasted! George never appreciates me more than on baking days."

Kay played at being the most conventional of new brides. In the morning, she saw George off to work as if with briefcase to the train, standing on a rise to watch him cross the meadow, waiting until he turned to wave before heading back to what George called her "housewife's chores." After a day of sweeping, "conveniently down the cracks between floorboards," mending his packsack with a thick needle that punctured her fingers, or using socks as rollers to curl her hair, she was by four waiting again, peering out the door every few minutes, bread in the oven timed (she hoped) to come out just as he appeared. Greeting him with a cup of hot bullion, she helped him peel off his sodden clothes and sat with him by the fire, eager to hear about his day. On the days he insisted on leaving her again, to be alone on the high flanks of Mount Mikeno, she sulked until he reminded her that a day with the gorillas could be as taxing as a day with a human crowd. She teased him for puttering in the storeroom rearranging supplies—"they made him feel like such a good provider"—and for taking on a few silverback habits of his own, unwittingly slapping his chest when he stretched in a yawn.

Though they had hired a twenty-three-year-old from the village named Andre Batinihirwa to live in one room of the cabin and help with chores, Kay was sometimes left completely alone.

> *It seems strange to be by myself with no human beings within miles, only the chickens and wild animals for company. I rather think I enjoy it! Except when I got up this morning and had to build my own fire in that miserable stove. . . . I cheated and used a bit of kerosene and soon it was smoking merrily. The sun came out and I sat in my chair watching our small duiker feed at the edge of Akeley's grave. So far (9am), I've had only coffee and stewed prunes. Our radishes and lettuces are coming up, and the watercress is showing.*

Nights alone were harder to bear. She read Agatha Christie to divert herself from the "vulgar chewing and snorting" of the buffalo outside her door, past whom she had to creep to reach the outhouse, and from the thoughts of all the accidents that might befall George.

Off all day with his gorilla friends, George worried over Kay's loneliness. More than once, out walking together, he caught a baby duiker for her to hold in her arms. Then one day, he brought what he thought would be a more lasting cure. "George came home with a present, a baby hyrax! It is furry, the size of my hand, and just darling. It ran across the trail and he quickly put his foot on it. It is probably semi-independent but may still require milk." Soon the little creature—an unlikely cousin to the elephant, with tiny tusks and an unearthly scream—seemed to settle in. He sat quietly in their arms, "ate 3 raisins and a carrot slice and tried to explore our laps. We put him in a shoebox lined with moss and he kept me awake most of the night with his squeaking and moving."

Alas, despite all-night feedings of warm honey and egg yolk, he survived just two days. In his journal, George told the story with the brevity and pragmatism he often relied on to corral pain—summoning a scientist's distance, or perhaps something older, learned in childhood. The little hyrax wandering over his arms was "tremendously appealing with its long soft brown fur and small velvety ears." But when its neck went limp, he went steely. "I may have to kill it." Two days later, he recorded that "the hyrax died. He got better and ate some droppers full . . . but by morning got steadily worse. Kay baked blackberry tarts and I had a bowlful with sugar. Delicious."

Kay had better luck with a pair of white-necked ravens, who showed up at the cabin the day after they moved in. She fed them until she could whistle them down onto her arm—one at a time, the two together were too heavy—recalling the first moment she'd been fascinated by George in Alaska. The birds became family: crash-landing on the metal roof at dawn, sliding down with a scraping and *click, click* of their claws, banging the loose gutter about as they wiped their bills. When George came out to check mousetraps, they flew in to grab any unwanted specimens he tossed their way. Kay found the male quite funny. "He looks at me very closely then darts up for his piece of cheese."

Her courtship with the gorillas took somewhat longer. One afternoon, when George sent word with the guard that a troop was at the end of the meadow, she went with a book and sat under a tree. "I heard one roar three times and one beat his chest. This is the second time they've come in hearing distance and it always gives me an eerie thrilled feeling."

A few days later, she ventured farther, walking for hours with George to visit his favorites, over muddy trails and sunny slopes of wild celery and jewelweed. They came upon the animals sooner than expected, startling two big males, who headed their way. "George shoved me up a tree," she wrote, "not a good one. I couldn't sit but had to crouch . . . and try to see past a branch that kept hitting me. George got up another tree and looked so comfortable. He had the lunch. He had the pack so could pull out a plastic to cover himself when it began to rain."

With the animals roaring and a sleety wind blowing, Kay began to shiver so hard she feared she'd shake herself out of the tree. Then a rustling took her attention to the next tree over. There, stretched along a branch on his belly, was a young gorilla watching her, his chin propped in his hands like a kid watching TV. She knew it had to be Junior. "I stared back and he ducked his head, broke off a piece of wood and dropped it, took lichen between his teeth and dropped that, beat his chest, then peered at me again—presumably to see if I'd been impressed. I was—I thought he was darling!" From then on, Kay's fear of gorillas was gone. When Mrs. Wrinkle lost her baby, she wrote Thelma. "They seem so human and are such good mothers, I can't help but mourn with her."

For both, these days without time (possessions, news of the world, other people) were paradise. On the rare dry morning when they knew the gorillas would sleep in, they lay in the meadow baking in the hot sun, Kay on a "cushion of moss and mint," George shirtless at her feet. "You have no idea what a wonderful, gorgeous day we are having!" she wrote home. "My face is a lovely brown with reddish highlights—not burnt as I had Sea & Ski on." Food, especially sweets, stirred similar raptures. Hers: "George was home all morning, so we had delicious pancakes with wild mountain blackberry syrup (my jam didn't jam.)" And his: "Kay baked a loaf of molasses oatmeal bread. I ate 3 slices. Delicious with honey." As a photo he took of Kay—in their metal tub, her naked, elegant back turned to the camera—recalls, these were two beautiful young bodies alone in the jungle.

Though it can sometimes feel, in their stories, as if Laura Ingalls Wilder, or perhaps O. Henry, had set up house in central Africa. At Christmas, George headed up Mount Mikeno to cut them a five-foot branch of giant heather, carrying it down at a run through the dark groves and fog, "almost flying." They decorated it with yellow blossoms they'd gathered, silver stars cut from the foil lining of a cookie box, and Kay's favorite ornament, secreted in by George: a wooden chickadee. Having sent Andre and the guard home with gifts—canned oleo and meat—they exchanged their own small gifts. "George was disappointed to find his stocking mostly stuffed with paper," Kay wrote Bettina. "Nothing to the look on his face when he opened an unmarked package from my mother and discovered a box of home permanent curlers!" They nibbled at nuts and oranges until George, remembering a fable that on Christmas humans can talk to animals, had a sudden urge to see the gorillas. Off to commune with them, he lost track of time and arrived home after dark to find Kay waiting in the fog with tears streaming down her face. "And when I held her close, she told me that it was very late, it was Christmas, and she thought that I had been injured because I had not come home." (George would later commemorate this night in a cartoon holiday card for the Emlens. He and Kay—drawn dumpy and disheveled, with stray whiskers on their noses—are on their front porch serenaded by four gorillas, come to sing them "O Tannenbaum.")

As his hours and days with them mounted, George began to live by gorilla principles. They were happiest if he came alone—as vulnerable without weapons or retinue as the scrawniest among them—and sat in plain sight. "Much like us," said Kay, "one is more afraid of the unseen." If he inadvertently surprised one or unsettled them by staring, he learned to shake his head, a courtesy they sometimes paid him in return. Assimilating their languid rhythms, he became both calmer and more alert. Although at first they heard and saw things he didn't, he found with each passing week his own senses recovering an acuity lost in civilization's din; soon, anything they startled at, he did too. That included their roar, which he never got used to; it always made him want to run. But he took satisfaction in seeing that it jolted the "other gorillas" too. As he found his way back into his own animal self, he began to both apprehend things closed to most people and be accepted in turn as just another forest creature. In those moments he escaped what John Berger called the "loneliness of our species," gaining fleeting readmittance into the more-than-human world.

•••

Except for the three times they left the cabin to resupply and meet visitors (first Fair Osborn,* then Kay's mother), George and Kay saw in nine

* Osborn's visit in February 1960 was not a success: George loathed, and was awful at, the clubby chitchat his urbane visitor was used to. Nor could he fathom why Osborn declined every outing, including to see gorillas. Instead, his patron fussed at George for

months only Africans, and only men. Kay's most constant companion was Andre. She joked that her days were "evenly divided between keeping the stove and Andre going—both preferring to smoke leisurely." Half exasperated that he soaped and scrubbed each item of laundry far longer than seemed possible, she nonetheless enjoyed working alongside him, pinning the clothes on the line as each tried to learn about the other's life. She was charmed that he liked the typewriter and often stood quietly behind her while she typed.

> *"Almit, Madame," he says, with an air of wanting something. "Almit to you too," I answer as my hand creeps to the Swahili dictionary. I can't find Almit. Maybe it's French, almée, which means Oriental dancing girl. Why should he come to me for that? Finally he points to matches; ah, allumette. Why he stays with us I don't know.*

George was tougher to please. He appreciated that Andre was "honest and congenial" despite missing his family terribly; when George found a dead duiker, he carried it to Andre to take home. But he also gave vent to frustration—at least in his journals—at the youth's lack of initiative and "blank" face. He railed even more at his guards' recalcitrance. He had little respect for cocky Donati, who did not get along with Andre, griped constantly—"When are we going home? The gorillas are far. I'm sick."—and raced screaming down the trail when a surprised buffalo came crashing their way. Nor for Desmos, who always hid behind George, blew his nose loudly when he saw gorillas to scare them off, and fled at the first roar, vanishing for hours. "The name 'guard,'" he would later write, "seemed, on the whole, to be honorary."

For these derelictions, George occasionally exacted petty revenge. When

---

"pitch[ing] too freely into the lettuce and raw fruits. You are far too important to fool with the g.d. microcosms that hang around" (Osborn to Schaller, March 25, 1960, WCS). As she often would, Kay stepped in with the necessary social graces, charming George's future boss and handling most of their subsequent correspondence.

Gerard announced that he was out of food after ten days (the guards were supposed to bring enough for a three-week posting), George sent him to the village alone. "They hate to go alone in the forest," wrote Kay, "so that in itself is punishment." When he'd had enough of Joseph's requests for aspirin, which the dandyish young man ate "like candy," George played a mean trick: substituting Epsom salts, a laxative.

But he also understood why the guards hated this life none had volunteered for. All came from warm lowland farming communities; they had no experience of the forest or its frightening animals and dreaded the bone-biting cold. Away from family and friends, they had no one to talk to. And still, some rose valiantly to the occasion. George admired Philip, who never complained or asked for *dawa* (medicine), and seemed to enjoy their wanderings. He and Kay both grew attached to N'sekanabo—"a huge fellow with tremendous feet [George wanted to make casts of his footprints], excellent in the field, a quiet personality but such a good man." The two men regularly spelled each other beating a trail through head-high nettles. And N'sekanabo never abandoned his post, even on a day when, as he guarded the base of a tree George sat in, Junior began showing off just twenty feet away. When George looked down and saw that the guard's lips had gone gray with fear, he motioned him up the tree. And "we had wonderful looks together."

George soon became aware of another reason for his Hutu guards' fear. On December 11, he found several hundred cattle in the park and three corrals. Knowing that Tutsi herders must be near, his guard panicked. George held off lighting a fire until he was certain their smoke would not be seen.

With no radio at the cabin, George did not understand until his *Time* magazine came weeks later that these were Rwandan refugees, the first of thousands who would come across the border to hide with their prized cattle in the forests of Congo's Kivu province. For decades, the German and then Belgian colonial authorities had found it useful to sustain a feudal arrangement in Rwanda, with the minority Tutsi—possessed of greater wealth and status by virtue of their livestock—ruling over the Hutu majority. Now, as colonial structures teetered, the Belgians decided it served their interests to foment an uprising against that arrangement. The brutal methods deployed by the Hutu would launch decades of retaliatory cycles. They included, as George wrote, "pruning the tall Watutsi down to size, so to speak, by cutting off their feet."

If George understood the refugees' desperation, he also knew that cattle and gorillas could not both survive in these mountains. The cows would soon trample and graze the remnant forests into ruins capable of sustaining no one. And this was, for the gorillas, their last redoubt; they truly had nowhere else on Earth to go.

He therefore decided—establishing a habit that would often land him in danger—to go toward the trouble, heading with a single guard to the Congo-Rwanda border. Camped at the foot of Mount Visoke, the two men heard cowbells and then caught sight of cattle accompanied by two Tutsi men in long overcoats bearing spears. They again sat silently until enough fog rolled in that they dared build a fire. The next morning, seeing the smoke of two fires below, George ventured a bluff crazier than any silverback's. Shooting off his cap pistol, he shouted in Swahili and French as if directing troops: "Surround the Watutsi! Quick, quick! Catch them." His guard kept saying *kwenda bas* ("let's go"), but the bluff worked: The fires went out and the herders fled.

Leaving Kabara a week later for the first time in five months, George got word to Van Straelen, who passed it on to *The New York Times*. "As these cow worshipers who measure their social status by the number of cattle they possess" enter the park, John Hillaby wrote, "about 350 mountain gorillas, the biggest anthropoids on earth and the last survivors of their species, are threatened by extinction . . ."

Driving the crowded streets of Goma and Rumangabo in January 1960, George and Kay saw that the yearning for independence roiling Rwanda was also beginning to sweep Congo. Everywhere they went, young men and children flashed the *V* for Victory and shouted "Uhuru"—"freedom."

Less than a year in, Schaller was creating a sensation, not only as a source of frontline news but with his field reports reverberating across the global scientific community. Typical was the response of Raymond Dart, then at the top of the anthropological charts thanks to his "killer ape" theory. He wrote to George Merck, the pharmaceutical heir who was for many years NYZS secretary: "How indebted science is to this remarkable work."

Though Schaller would be a primatologist only briefly—dropping in for these few years, then moving on to the great predators—he managed to disrupt the discipline more than most who worked on primates their entire lives.

To begin, he demonstrated to all who had deemed the study impossible that a scientist could habituate even the most "savage" animals to their presence. As Robert Ardrey would write in *The Social Contract*, George's 466 hours of close observation "broke some kind of ethological record."

Also radical was his revelation, through that sustained intimacy, of these animals' rich interior and social lives. The decades of marvelous work since on the complexities of sentience and culture in species from squid to starlings are built on the foundation George laid here, at a time when most still viewed animals as beings without reason, feeling, or purposes of their own.

For primates in particular, George's work upended nearly every dogma. Sherwood Washburn had been using studies of baboons* to prove that male dominance maintained through aggression was "natural." Solly Zuckerman had asked why, in a world driven by a competition of each against each, apes lived in groups at all: His answer was that it gave males access to

* With Harvard evolutionary biologist Irven DeVore

action whenever they wanted it. (He'd shot and dissected twelve female baboons to show that each was in a different stage of her estrus cycle, so that at least one would be receptive at any given time.) Carleton Coon, at the University of Pennsylvania, had argued that young males were driven out of the troop by the silverback, fearing rivalries "in the well-known Oedipus fashion."

George found the last so egregious—blackbacks were not, he saw, exiled—that he called Coon out by name as an example of how many ideas about gorillas "might be quite wrong." He then went on to topple the rest. Far from the violence, territorialism, and sexual jealousy virtually every expert described as innate to primate character, he documented an abiding gentleness and lack of possessiveness toward both territory and females. Where he did see dominance behaviors, they were on the order of summer camp. Females and young males might wrangle noisily over who got the best place to sit under the tree or (like Robin Hood with Little John) had to yield a narrow trail—but rarely came to blows. The reigning silverback decided when everyone got up, napped, and went to bed, and where they went for the day—and interposed himself between the troop and any external danger—but otherwise was exceedingly tolerant. George described watching "as many as 4 babies . . . play all over him [and] hitch rides on his rump."

Sex, too, was nothing like the driving preoccupation or droit du seigneur it had been made out to be. Contrary to Zuckerman, George found that big males often went a year or more without it if all the females were pregnant or lactating. And twice, he watched a silverback lounge indifferently as one of his juniors noisily enjoyed a female nearby. George described these events with brio. "A new sound, *ö—ö, ö, ö*; she on her knees, belly and elbows, DJ mounted behind, holding onto her hips." The couple had chosen a steep slope for their assignation, so George watched them slowly slip downhill, she using her hands to part the vegetation. "He thrusts, she screams, until a hoarse trembling sound, almost a roar, escaped from DJ's parted lips." He also witnessed a female make the first move: repeatedly mounting the male of her choice, who paid no attention until—finally yielding—he swiveled, pulled her into his lap, closed his eyes, and after a time let out a loud *aaahhh.* George captured their various postures with winning drawings, a kind of *Kama Sutra* for gorillas.

There had been a strong measure of self-justification in the social dynamics described by his predecessors: a world run by powerful males with compliant harems was a "natural" hierarchy they could embrace. That George could see a different reality, beyond what they'd told him was there, may have come from having two powerful female mentors in Brina Kessel and Mardy Murie. He in turn would champion female scientists: from primatologists including Jane Goodall and Shirley Strum to three generations of women in China. He was, in a sense, one of them, practicing field study in a way some would characterize as "female": interested in individual variation more than generalization and in the connective tissue of relationships more than taxonomic distinctions, and recognizing a scientist's emotional engagement not as failure but as necessary to deep understanding. The aim was not to tie up the story, prove a theory, but to open it wider: "If Washburn saw big male baboons running the show and Strum saw cooperation, both were useful."

Even among the gorillas, despite an occasional reference to a hierarchy like "the human family, with the father as the master," George saw himself not in the silverback's "brute force" but in the females, who might also slap their chests but in whose faces he found sympathy and kindness. In a brief essay that never made it out of his journal, he compared the warm nurture offered gorilla babies to the more barren terrain some humans had to navigate, awkwardly using the pronoun "it" for both, though he'd long since broken another scientific taboo by referring to the gorillas as "he" and "she." While the gorilla baby was always near its mother, riding on her back or chest and sleeping with her at night, "human babies lack this close contact with their mother. It is put in nurseries during the day. It is put in its crib for long unattended hours. When weaned it is put into large and unknown social units, like schools. It is little wonder that the average child feels insecure not only of its mother's love but of its general standing in the social unit." That human mothers in east Africa carried their babies for two years, he added, undoubtedly worked to good psychological effect.

David Attenborough has described another of the transformations George led. "In my youth, you were a scientist if you worked in a laboratory—with dead . . . or captured things. To go into the bush and watch animals was not regarded as science, but a sort of natural history a gentleman might do." The wild world was simply too full of complexity and contingencies; it could not be pared down to the few controlled variables seen as essential to authoritative science, nor did super-rare animals allow for replicable experiments. "The notion that you could get accurate, rigorous scientific data simply by watching in the field scarcely existed, until Schaller. He produced documents of such intellectual rigor that they had to be taken seriously. And he did so in conditions of such physical rigor that few could match him. He had a stoicism and patience that take your breath away."

While George had turned his back on lab-based primatology, the labs did pay close attention to him. In Frederick Wiseman's 1974 *Primate*, filmed at the Yerkes National Primate Research Center, scientists chat, as they insert electrodes into a chimp's rectum to stimulate ejaculation, about his observations of copulation in the wild.

Schaller's challenges to accepted wisdom pressed at fundamental boundaries and hierarchies. He wrote of "approaching [the gorillas] in utter humility with the knowledge of being in many ways inferior"; of granting them a say in the project: "Assess[ing] their willingness to be observed" via the language they spoke to him "with their expressive eyes." Assimilating into their rhythms, he felt a diminishing need for the language long considered the essential divide between our two species. Even with each other, said Kay, "we learned to speak silently."

George never tried to insert himself into the troop; he never fed or touched the gorillas. But in what would always be his peak moments, he sometimes ceased to be the investigator and became instead the object of their curious study. Sitting as usual one day on a tree branch to allow the gorillas to see him, he was suddenly joined by a female with an infant on her back, who swung up alongside; like strangers on a park bench, they sat exchanging furtive glances. Another female later delighted him by treating him as just another gorilla, tapping him on the leg to let her pass. Like the people of Beringia, who esteem whales as their most important moral judges, George cared what the gorillas thought of him.

While George's reports to NYZS were rattling global scientists, the romance of these beautiful young adventurers captured the public imagination. *Sports Illustrated* got first crack, publishing a nine-page feature in June 1960 that struck a newsreel tone to convey the import of the Schaller mission. "For the first time in history a man and woman, alone and unarmed, have gone among wild gorillas to live with them on a basis of friendly coexistence . . . result[ing] in scenes which have never before taken place between the world's two largest primates." That set off a bit of a frenzy. Michael Korda at Simon & Schuster asked them to write a book "in view of the recent success of *Born Free* . . . I know the pygmies and wildlife are threatened; we envision a series to preserve what we can, if only in print." *The New York Times* asked for dispatches from Congo. Tell them "maybe later. I'm too busy now," George instructed Kay, who had by now assumed her lifelong role as go-between. "When my thesis is published, we'll get renewed offers so don't need to take them all now."

That certainty proved well-founded. The 1963 publication of his dissertation—*The Mountain Gorilla: Ecology and Behavior*—immediately opened doors, beginning with a central role in the "Primate project" at Stanford's Center for Advanced Study in the Behavioral Sciences, and an invitation to make his next study, of tigers and their prey. It also greatly extended the reach of his revolution in field biology, with his nuanced, evocative language—rare in scientific monographs—igniting many imaginations.

The full fourteen pages he devoted to the nine moves that make up the chest-beating sequence—a leaf pressed between the lips, one-legged kick, sideways run and slap of the ground, all accompanied like a big Broadway number with slow hoots building to a crashing finale—became legend. In *Sociobiology*, E. O. Wilson reproduced the sequence in its entirety.

Top journals across several fields paid homage, beginning with eight pages of reviews in *Current Anthropology*. The rare swipe among them came from D. F. Owen, who found the book overly "Germanic" and insisted that "contrary to popular opinion, it is not difficult to collect information on wild primates." Though his own dissertation had been on owls and he was currently studying snails, Owen knew so, he explained, because he lived in Uganda, "where primates and primatologists are common." A similar complaint was made by Zuckerman in *Oryx* (until 1950 the *Journal of the Society for the Preservation of the Wild Fauna of the Empire*), who tried to deflect the gutting George had delivered his work with his characteristic condescension: George's prolonged attention to such matters as excretory habits suggested a "lack of intellectual priorities."

All the rest were simply awed. C. R. Carpenter, whose studies of howler monkeys in Panama had set the standard, now handed the crown to George for completing "the most difficult of all primate studies." Several rightly admired him for his restraint: refusing to exaggerate or "profiteer" from his subject's spectacular nature. Anthropologist Walter Goldschmidt noted how disappointed all the scientists must be who'd premised their theories (that human culture, for instance, emerged to repress our animal lust) on violent competition for females.

Attenborough had been captivated as a boy by the adventures of George's predecessor at NYZS, William Beebe, as he told an interviewer filming him at home. "Then," he continued, holding up a well-thumbed copy of *The Mountain Gorilla*, "another great explorer-zoologist star appeared on the horizon . . . with this epic . . . one of the great books that first drew the world's attention to this marvelous species. . . . Both those figures have loomed very large in my life as models of excellence and pioneers who I would do my best to emulate." Having described his own meetings with gorillas as the peak experience of his life, Attenborough particularly admired Schaller's debunking of their reputation "as the most ferocious, aggressive creatures in the universe, representing everything that was terrifying." It was the crashing of that myth against George's portrait of tenderness that lent the book its "huge piquancy and buzz. This wonderful creature that had been thought to be a demon, turned out to be not an angel exactly but something almost as large."

For Dian Fossey, who began her work at Kabara in 1966 before moving, after being kidnapped by Congolese soldiers, across the border to Rwanda, *The Mountain Gorilla* was the wellspring. For anyone working with her, it was required reading. "Everyone had a copy," recalled Ian Redmond, who began at Fossey's Karisoke Research Center in 1976. "It was what we referred to most, to know if what we were seeing was new." Amy Vedder—who arrived in 1977 and with her husband, Bill Weber, helped create the ecotourism program that gave Rwanda's gorillas a path to survival—thought of George at every bluff charge. "A silverback will come roaring down and stop a yard away, glare into your eyes, and stomp off. Others just shot them, thinking they'd be torn to pieces. George was willing to wait to see what would happen."

Of greatest value to his successors was George's ecological study, one of the earliest efforts by anyone to answer the question fundamental to conservation: What does this animal need? By the time Vedder arrived, half of the park on the Rwanda side had been carved off for farms, and mountain gorilla populations had crashed by 40 percent. World Wildlife Fund (WWF) told her she was wasting her time; the gorilla was doomed. But no one since George, she knew, had asked the key questions: What are the gorillas eating? Where do they travel? Is there enough habitat left? His findings from twenty years earlier remained the only data anyone had but were fortunately, as Vedder found when she set about updating them, still "phenomenally accurate. He picks up fantastic detail, is so meticulous, captures it in real time, and then conveys it with incredible fidelity. He built us an incredibly strong foundation of knowledge. Every gorilla scientist feels that we follow in his footsteps."* His insistence on recording everything, even phenomena he could find no meaning in, bore fruit in the way he'd hoped: allowing this next generation to glean insights he himself had missed. Though his work is sometimes mind-bogglingly quantitative ("the animal moved 3.5 meters"), says Vedder, "he isn't there to test a hypothesis but just goes out with this big, wide-open mind."

In 1964 George published a second book on gorillas, for a general audience—a pattern he would repeat for many places and animals. For all his renegade ideas, his dissertation had largely hewed to scientific norms. Only rarely, amidst the seventy-three tables and sixty-nine figures, had he let slip the animals' names; his descriptions of their temperaments and interactions retained a measure of impersonality. (Though he did make an exception for baby Max, who, after wriggling free when an auntie pretended to pin him down, stood with arms overhead to fall straight backward like a proper goof.) *The Year of the Gorilla* allowed Schaller greater freedom and reach, and effectively invented a new genre: the popular field narrative, unfolding the intimate dramas of a tightly woven cast of nonhuman characters and their human friends. This genre, as Ian McEwan would write, has "all the major themes of the English nineteenth-century novel: alliances made and broken, individuals rising while others fall, plots hatched, revenge, gratitude, injured pride, successful and unsuccessful courtship, bereavement and mourning." The courtship here included George's own of the gorillas, full of coy overtures, rejections and rages, doubt and fragile harmonies.

---

* Vedder never forgot their first meeting, when in 1977 she came to the Bronx Zoo to seek Schaller's support for her study. He was by then NYZS's head of global conservation. She had reached out her hand, awestruck. "And the first thing out of his mouth was: 'So, you're interested in the digestive physiology of gorillas. Do you think the pig is a good analogue for looking at passage rate time?'"

The same gifts that set George apart as a scientist—his ability to sit still and listen to whatever his subject had to tell him, his intensity of observation and commitment to capture all that he saw and heard in the world—now fully revealed themselves in his writing. His gorillas take on Dickensian identifiers: The Outsider has a "seaman's gait and glowering temper"; Mrs. Blacktop stares morosely at her paunch. Where there is pathos, he lets the quiet accretion of detail do the work. For two days, Mrs. Wrinkle gazes down at her feeble baby, its gums closing around nothing. For two more days, after it stops breathing, she drags it about by the hand. When she finally leaves it, face down, limbs askew, George retrieves it for an autopsy. "As I stripped the soft meat from the fragile bones to preserve the skeleton, it was like dissecting a human baby."

Having already transgressed the twentieth-century norm that science should be an exercise of pure, cold reason, he now went further, edging into the metaphysical. He described Kay's ravens as "black messengers from Thor," plummeting "out of the wind-tattered clouds, rushing downward with folded wings, until abruptly they veered skyward and disappeared, only a disembodied *krrrua-krrrua* echoing." Crawling to the rim of a volcano, he found "a sight so beautiful that my heart wanted to cry out": the lake of lava a portal into the center of the ever-changing Earth, a full moon's soft silver light lifting him into the eternity of the cosmos.

George broke with norms in another crucial way: His reliance on local knowledge, if not greater than his peers', was more openly credited. He included in his dissertation Reuben Rwanzagire's findings on the gorilla group that he had himself habituated and come to know as individuals. In areas George couldn't survey directly, he interviewed residents and owned up to doing so. Some passersby were so shy they dove into bushes at his car's approach, but most were willing to show him where to find gorillas and nests and "always proved to be correct. . . . We came to have a high regard for our informants' intelligence and integrity." Having carefully recorded each person's name in his journal (a habit he would maintain all his life), he included in his text or acknowledgments many of his African teachers and guides.

The book met with rapturous reviews. Naomi Bliven at *The New Yorker* had read George's scientific monograph, loving its "uncompressed field observations . . . so full of delights one hardly dared expect more." To that she now added pages of admiration for his refusal of the "vanity" that leads human beings "to regard the universe as a collection of metaphors designed to describe themselves," offering instead a sensitive portrait of gorillas in their own right. The *Times Literary Supplement* warned that "for those who have been brought up on King Kong and gigantic rapist apes . . . the real gorilla may well come as something of an anti-climax"; for those who would like to "blaze away . . . it is hard to see how anyone could ever have considered it a valiant opponent." To study gorillas required a truer valor. "One must be prepared to sit with them, hour after hour, in the forest wetness, watching and writing down every move they make. If after months of this you are still keen enough to leave your wife . . . on Christmas Day and crouch, soaked to the skin in the torrential downpour, staring silently at the huddled forms of the great apes as they stare back at you . . . then perhaps you may one day be able to produce a book as compulsive as Dr. Schaller's. [And] you will be a rare specimen."

The impact of the work went further still. The very premise for the bone-throwing that opens Stanley Kubrick's *2001: A Space Odyssey* came from Schaller's speculation that the gorillas' habit of throwing things might have given rise to weapon use. In his *Moonwatcher's Memoir*, choreographer and lead man-ape Dan Richter detailed other templates provided by George: that chest-beating "can be done on a thigh or a log [and] be a nervous, tentative thing"; "how the gorillas also observe him . . . tentatively approach . . . then scurry back to the safety of the band"; and how the gorilla's "gentle, delicate grace . . . works in counterpoint to [their] power." He described the thrill when Kubrick "dug up some footage" shot at Kabara in 1963 by Joan and Alan Root. "This is fabulous, Stanley. We're actually seeing Schaller's gorillas." Having read Schaller carefully, "I get an eerie feeling of closeness."

George was even the featured mystery guest on a December 1965 episode of the game show *To Tell the Truth*. After the off-camera introduction—"I spent two years studying the mountain gorilla [in] the dark continent. Armed with nothing more lethal than a notebook, I tabulated and recorded with sketches and photographs every aspect of gorilla life from birth to death twenty-four hours a day"—came questions reflecting the jumbled ideas still in play about apes, naked or otherwise. Panelist Kitty Carlisle stumped one impostor asking them to name a book; she had in mind the bestselling *African Genesis*, the first of many in which Ardrey drew heavily on Schaller's work. Peggy Cass asked if gorillas are monogamous. Orson Bean asked a question of particular resonance given that journalists invariably described Kay as George's "pretty blond wife":

> **BEAN:** "I saw *King Kong* the other night and what he did to Fay Wray, I don't want to go into the details, this is a family show. Now, for years I grew up seeing gorillas carrying blondes off into the jungle. It ain't true, right?"
>
> **SCHALLER:** "That's right. It's false. They don't carry off blondes."

•••

As they returned to Kabara with replenished supplies in February 1960, George and Kay's two weeks in Goma and Rumangabo had left them alert to both the ongoing influx of Tutsi herders fleeing Rwanda and the accelerating momentum toward Congo independence. News trickled in with the mail—of violent power struggles, miners' strikes, Belgians (and their money) fleeing. One April afternoon, Kay stepped out of their cabin door and saw at the end of the meadow three tall figures in long dark overcoats holding spears. The men stopped in surprise at the sight of her. One, hat in hand, made an exaggerated bow. Only when George stepped out of the cabin behind her did they go.

That same month, Raymond Dart sounded an alarm in the pages of the *South African Journal of Science.* Reproducing a report from George mapping the sizeable area already made useless for the gorillas, he urged an exercise of "imperial responsibility" on behalf of the "strange beings of gigantic bodily proportions . . . whose disappearance would have a criminal quality." Protecting the forests, he added, was essential to sustain water supplies and climactic equilibrium for the entire region.

The mounting tensions spilled into the Schallers' journals and letters. George, arguing with porters over what they were owed, vented something darker than anger: "I hate their sullen faces." When a guard sat down without a word when they hit nettles, forcing George—who continued to break trail for a time before realizing it—to retrace his burning steps, he scribbled: "Lazy, stupid fool." The usually burbling Kay sent a glum note to Dick and Thelma. She'd had it with the Belgians, who are "rarely pleasant or friendly; aren't interested in a couple of Americans," and when asked about native customs, "seem to know nothing and are even less interested, except to make derogatory remarks." Uncharacteristically, she followed suit. "I've come to the conclusion that the people here have little in the way of art, music, folklore—they are unusually dull and have been a terrible disappointment. I have surprised our camp-boy dancing, and occasionally he and the guard will sing some dirge-like phrase over and over [or] drum on our oil can. Otherwise, they seem more interested in looking at ads in American magazines and asking how many francs the watches and radios cost. They never ask George, only me. I think they're scared of the big Bwana."

With independence officially set for June 30, George and Kay decamped to Uganda, planning to wait for a few weeks to see how events unfolded. From the newspapers they learned that the road from Goma (their supply center) to Kisenyi* (the Rwandan resort town five miles to its east) was blocked with barbed wire and patrolled by choppers and troops with machine guns. Nonetheless, when on July 3 George read in the *Uganda Argus* that "tribesmen were celebrating freedom by hunting game in Albert Park," he resolved to go back. Six days later, he and Kay drove to Queen Elizabeth National Park, stopping to see the warden, Captain Frank Poppleton. As they watched a long convoy of cars pour across the border from Congo, Poppleton explained that soldiers were in mutiny there and these were the last of the whites, fleeing.

All the more reason, George decided, for him to head back in: He had to retrieve some field notes and was "curious . . . the flight of the Belgians had all the earmarks of a panic reaction." On July 10, leaving Kay with the Poppletons but taking along Richard Clark (another graduate student he had just met), he crossed the border and drove the deserted road to Rutshuru at the edge of the park.

Only two of the Belgian park staff had not fled. "Baert was mighty glad to see us for he had been isolated for days with no word from anywhere." And biologist Jacques Verschuren recounted "with waving arms and many cries of '*terrifique*' and '*mon Dieu*'" how the Force Publique (FP) had searched house to house for guns.

George's experience surviving in a war zone now came in handy. When a jeep forced them to the curb and armed rebel soldiers leapt out, Richard asked to take their picture; lining up and pulling George into the frame, the soldiers puffed out their chests, mugging and grinning, and let the men pass. Spotting a cordon ahead, they just bore down, compelling the police to "jump nimbly." At a more ramshackle barrier, George wrote Emlen, "I slipped a fellow a pack of cigarettes and he moved it before the drunken crowd came too near." Finally heading back, with George's notes, to

* Now Gisenyi

Uganda, they had a flat and then a second, limping the last twenty miles into Elizabeth Park on the rim. "On the whole, however," he continued, "the population is friendly and I believe it will be possible to work there again. So far as anyone knows, no whites have been injured. My next step is to find quarters for Kay in Kampala. I don't want to take her back to the Congo—probably ever."

As George would later learn, they had gotten out just in time. Half an hour after they left Baert, soldiers joined by a few park guards had arrested and beaten him, standing him against a wall to face a firing squad before releasing him. That same night, he wrote Kay, "13 drunken soldiers waved revolvers under Verschuren's nose and wanted to rape the nurse who lives with him. He had to talk for two hours before they left."

Returning Kay to Kampala, George found her safe haven in the girls' dorm at Makerere College; neither guessed that in the next two months she would see George just once. Determined to get back to the gorillas, he went first to Walter Baumgartel (by now "chaotic and bellowing"), pausing there to pick off number six of the nine volcanoes. "Climbed Sabinio with Reuben (2 3/4 hours)," he wrote Kay. "Situation bad. All officials are being rounded up and held hostage. Verschuren left word for me not to go into the Congo because I'm considered a Parc official. I miss you but am glad you are where you are. love, love George."

A few days later, with the all-clear from Verschuren ("Everything normal," read his telegram. "Come Here"), George headed to Kisenyi. He spent the next several weeks slipping into Congo by day, out again for the night, driving back roads to bypass razor wire and land mines. Visiting government offices—to secure road passes and permission to return to the park—was like stepping into a double exposure. "New flags flying," George wrote in his journal, "but in the administrator's office they still speak French, portraits of Leopold still on the wall." The Congolese authorities had doubled down on Belgian bureaucratic formalities, inspiring the kind of droll riff George sometimes spun to avoid going mad. (This one featured a dawdling clerk with many visitors, each of whom had to be elaborately introduced to George on arrival, then go through an equally theatrical farewell.) The arrival of the UN added to the surreal comedy: Bagpipe-playing Irish forces confronted the FP, then climbed into their trucks to "ride around in the friendliest way."

The state of the park was more alarming. One of the officers, Le Comte Claude Cornet d'Elzius, had left headquarters in chaos, "personal bills and even his last will and testament jumbled in with Parc business." Guards were warning that if they weren't soon paid, they'd begin shooting animals to sell. Verschuren assured them that they were not abandoned but now worked for the Parc of the Republic of Congo; he even managed to scrape up a few months' pay. The guards said they would not accept anyone who had fled—not park chief Micha, nor Cornet, nor chief of guards Rousseau—but that as far as they were concerned, George was welcome to go to his gorillas again.

"Keep this under your hat," he wrote Kay. "It may be possible to go to Kabara for a week or two. At least everyone here black and white thinks so." He added that he'd dented the car, "where a marabou flew smack into me. Killed him," and asked that she save his stamps, which were rapidly becoming collectors' items. That was not so easy. Kay's dorm mates sometimes took them before handing her the letter, thrilled to see depictions of Congolese replacing Belgian royals. "I finally at dinner had to tap my glass and say, 'I will share them with you but please let me see them first. Because sometimes my husband doesn't date the letters and that's the only way I know when he sent it.'"

Nights in Kisenyi were surreal. George found the "drawn, haggard" Micha and Cornet packing, "giving up, panicked. They took all their stuff out of the Park even though Verschuren told them their presence, even a token visit, was critical to the Park's survival." Rousseau, meanwhile, played tennis. With little to do besides "gathering rumors," George worked—writing up a report on mice, labeling plants—then went to the lake to float on his back watching fruit bats feed on figs. "Dearest Katie," he wrote, "right now I'm sitting on the beach and I'll sleep here tonight. Lots of women and children are sunbathing and swimming. I'll swim too, after dark since I don't have pants. I see no sense in rushing back to Kampala because I'll have no place to live and it will take time to get visas; do try to hurry them through, just in case. I closed the accounts, have a bag of 29,000 Congo francs I can't exchange. Hope you're having fun. Love George, and love again."

Returning to Walter's to await permission to go back to the gorillas, George wrote Kay nearly every day. He told of his visits with Junichiro Itani, a primatologist he had sometimes treated as a rival but now loaned a cot to use in Burundi; they spent hours talking about Japanese efforts to find the (Abominable) "Snow Man." They'd called on a rich old bachelor Swiss planter. "He said he wanted us to see his artifacts but the real reason, he is lonely. He talked incessantly." Lunch was in the toolshed, so crammed with crates of heirloom furniture they could barely sit down; the planter was living there while his house was being finished, an arrangement now in its twentieth year.

George then prepared her for the weeks ahead.

> *I've set up shop, keeping forty days of food and selling Walter the rest. I will remain in Congo for two or three months more. This will be a long separation and I will hate it but I must finish the work and you are safer there. Please write my uncle and other family. You will hear less from me now and not at all when I go to Kabara so don't worry. I will take care. And I will not forget our anniversary. I hope you can stand it in Kampala. I know it's been rough on you and I love you for taking it so nicely.*

As weeks without word of the gorillas wore on, Verschuren got another story into *The New York Times*. This one quoted Van Straelen, lamenting the disruption of the "sensational" study by the Schallers, who "were practically adopted by the gorillas" and claiming that "unfortunately, the native guides have turned aggressive. . . . At this point it seems everything there connected with the white race must be got rid of." Alarmed by the story, Emlen immediately wired George: "Osborn, Merck, Emlen advise abandon gorillas." The dispatch reached him but, as with several sent over the next weeks, he ignored it.

In fact, George's hopes had been lifted by his first meeting with the twenty-three-year-old "thin, efficient" agronomy student Anicet Mburanumwe, who had just been appointed *conservateur* in charge of the park. Gathering his staff, Mburanumwe told them "that only Verschuren and this American remain, and all help should be given them." The next day, when the three men went together to check on the Rwindi plains section of the park, Mburanumwe confirmed that Schaller could resume his work.

George did not immediately go to Kabara. Instead—his instinct, when the world is in chaos, being to climb mountains alone—he took time to conquer two last volcanoes, the still-active Nyamuragira and Nyiragongo. With Andre, two porters, and a guard, he spent four days scrambling over sharp, twisted lava rock, the demonic scent of sulfur in the air. Coming upon two rondavels (traditional drum-shaped huts), the five men had to chase off elephants drinking the rain barrels dry: Only by tipping them and squeezing the leaves at bottom did they collect enough water. George urged his companions to use the furnished tourist hut, but "tradition was too strong and independence too recent," and they refused. He camped alone on the summit anyway.

As in Alaska, the proximity to Earth's molten interior transported George to the dawn of time. Catching sight on this "just-formed planet" of a tiny, tender, bright-yellow flower, he admired its insistence on beauty "amidst desolation all around." But he felt unsettled, too, looking down 1,200 feet into roiling black lava, fissured red as if by wounds. Just as "a human who lays his emotions bare makes others uneasy," so was Earth's heart too nakedly revealed. Beginning his own laying bare—"I'm so lonesome without Kay it hurts"—he cut it short with stilted syntax. "I can feel the chest contract."

With two months elapsed since George and Kay had left their cabin, George finally brought Fair Osborn up to date. "As of today," he wrote on August 9, 1960, "I am the only white man living in Albert Park. Micha, Cornet, Baert—all are in Belgium or Rwanda, and have lost face completely." Mr. Mburanumwe "is honorable and intelligent [and] the Park is completely safe at present with no more poaching than before independence, [but] even the most brilliant person could not straighten out the mess left by the Belgians." He laid out his two-month plan, beginning with his imminent return to Kabara. "Kay, I am sorry to say, has to stay at Makerere. It should be interesting—only two whites and the rest are Indians, Arabs, etc. I am sad not to have her with me, for she has been a wonderful companion and I admire her greatly for sticking it out at Kabara for 9 months. Any success that may come from this project is due to a large extent to her presence."

Though George had endured his share of painful journeys, his return to his gorillas was "the saddest walk I had ever taken." Accompanied by Andre and a conscientious guard who called himself Bonne Année, he found deep-rutted cattle trails everywhere. The sight of the fragile soils and plants lacerated by their hooves and teeth nearly brought him to tears.

When he then found the offending herds, Olaus's prediction that George's youthful judgment might fail him was fulfilled. As he watched the "dull creatures" wreak destruction, "I was mad with fury and with a heavy stick hit a large calf on the neck and it collapsed." (In a letter to Kay, he added that "she dropped and kicked like Ingemar Johansson," referring to that year's world heavyweight boxing champion.) Farther on, coming upon a hut with a boy out front, "I bellowed and charged and the boy and two Watusi men ran away. I called after them 'this is the independent republic of Congo now.'" With his Hutu companions, "we smashed their spears and burned everything."

At Kabara, they found the cabin wrecked: furniture and windows smashed, beds slashed, stove upended, dead duikers rotting in water barrels. George bought two spears from the porters before sending them to Mburanumwe with a brutal note: "Help must come fast. Only by shooting a person or two and killing cattle will they be driven out." He set up house, banging together a table and dipping green water from the lake; adding bullion, he discovered, precipitated the particles out in a frothy foam. Then eleven Watusi bearing spears stepped into the meadow. Andre and Bonne Année hid in the cabin. George, left alone, hollered, "You better get the hell out or I'll kill your cattle." Mimicking him, they laughed, then vanished. "There was nothing to do. I hid money, food, and film under the cabin floor, loosened a board in back as an escape hatch. We slept with all sorts of booby traps, a hammer and my cap pistol next to me." When any of them went to the outhouse, he carried a spear.

George did not dare leave either of his companions in the cabin alone, so the next morning he went out by himself. He was glad to find that "all the raving I did had some effect." The fleeing Rwandans had moved their cows out. Walking to Rukumi meadow, he found that cabin also destroyed but grazed the brambles, heavy with ripe blackberries. Returning to the hut he had burned, he retrieved a wooden pot, "a nice one that smelled of milk, that the Watusi left behind. Or rather, yesterday I threw it into the brush and retrieved it today. I burned their mats and clothing. I must be popular." His heart sank at the discovery he'd feared—a dead gorilla. Maggots had cleaned the skeleton, so he carried it home. Finding the radius and ulna bones in the left forearm severely deformed, he was "glad to know that I had never met him." Happy, too, he wrote Kay, to find "your ravens" still here. "Old friends, jumpier but still daring."

The next day, George had the reunion he'd hoped for, with a troop that included Junior, who came and sat close to him for an hour. Though he couldn't focus, the animals offered a brief respite. "It was peaceful to sit with the gorillas again."

Not so the night. "We are in a bad position," he wrote in his journal. "From the forest we are watched unseen and never know if an attack will come. We are only 3; if the Watusi decide to take revenge, they would come 20 or more. And we are so isolated. I only hope help comes tomorrow."

Help did arrive, in the form of five armed guards and two policemen sent by Mburanumwe and bearing a letter from Verschuren on *Institute des parcs nationaux du Congo belge* stationery, the word *belge* crossed out. "The guards are allowed to do the maximum," he advised George. "They are allowed to kill the cattle, for example." (Also in the mail, a feat of direct marketing, was his new Wilderness Society membership card.) George was pleased that Anicet had the "courage and foresight" to take measures the Belgians would not. He watched as the guards shot a dozen cows, some immediately dropping, others "lowing as the bullets ripped through them." One guard tossed aside his rifle and "plunged his long knife to the hilt again and again into the stolid cows, until his arm was red with blood and his hate against the Watusi vented"—neither the first nor the last time that protecting animals would become the pretext for violence that had other motives or ends.

When the guards headed back to town, herding sixty cattle ahead of them, they carried a letter from George to Osborn as amped-up as any he would ever write. "Things have been popping at Kabara," he began, "and I better give you up-to-date news." He described finding hundreds of cattle, "mutilated" ground cover around Akeley's grave, and the forest destroyed. "First thing, I killed one of their cows, burned two of their houses with belongings, and told them if they didn't leave, I would kill them too. The next day they left which made me feel quite good since I am armed with a spear only. As it was, eleven Watusis armed with spears did not dare anything." He described the "reinforcements with rifles," sent by Mburanumwe. "This is the first real strike against the Watusi. Harroy in Ruanda has not dared anything to help his side of the Park in 2 years." He laid out his plan. "Since the lost cattle represent an investment of 7–10,000 dollars, we expect Watusi reprisals. I never thought I would watch gorillas armed with a rifle, machete and 2 spears. I have asked for one or two more guards so I don't have to go alone or leave the cabin untended. I hope for the best and will try to finish the filming. My gorillas are as tame as ever but the weather is terrible."

By the next morning, George had accepted the impossibility of staying longer: Looking for gorillas, he realized he was being "hunted" through the forest. The Watusi wouldn't come in the cabin—they didn't know if he had a gun. But walking alone, "I could hear them running after me, shouting. They would probably have killed me because, up there, who's to find you?" Although the next day was sunny, he didn't dare go out. "This waiting gets me down. I read the account of Scott's last Polar expedition. How those men stood being away 2 years or more from their family is beyond me." He wrote to Kay, deflated, distracted. "I am happy you did not come. You can remember Kabara as it was. Though the cabin sure is empty, cold and dreary. And I wish you were here in a way besides for company. I have to do all the chores myself. You are a good girl to fix my green coat so nicely." On day three, abandoning caution, he went out again. Hearing voices, a long whistle, and a gunshot, he hid in a cave. Only hours later, when he snuck back to the cabin, did he learn it was Anicet's men come to escort them to park headquarters. He finished his letter to Kay with a foretaste of the remorse he would come to feel for actions that decades later he would call "stupid and rash":

"The chief who lost the most cattle is rumored to have committed suicide."

•••

That so many of these letters arrived is remarkable; that they were often delayed left Emlen and Osborn lagging in their grasp of events. While all hell was breaking loose at Kabara, Emlen was writing the Schallers with a mix of petulance and good cheer. "We finally feel that we are caught up with what you people have been doing. We're very pleased George has been able to fit in and contribute importantly to the rescue of the Park." Osborn wrote of his "joy to learn you are well, and such vivid detail of the doings in the Congo, an almost incredible story." With Coolidge, they continued to try to sway events from afar. Emlen wrote in French to Jean Miruho, leader of the province in which the park sat. "Mr. President, my colleague George Schaller has apprised me of the admirable manner in which you took a position on the celebrated Park Albert and its irreplaceable troops of wild animals. Your reaction will be applauded for many years to come." Coolidge urged UNESCO to create a mission from a noncolonial country to assist the new government, and to include George, though in a secondary role, given the "Communist propaganda about U.S. economic imperialism." Osborn opposed that idea, and any further attempt by George "to get the Park situation straightened out." He wanted him out of Congo. "And this will be better for Kay who, after all, must be considered." Emlen pleaded with Osborn to let George stay.

> *He is in a unique position to rescue the Park from destruction. The fact that he and Verschuren are acceptable to the guards and respected by them creates an opportunity that simply must not be bypassed. George is, as you know, an impatient soul who becomes exasperated when things go wrong or his work is interrupted. But I'd be surprised if he wouldn't be eager to make personal sacrifices for this cause.*

Osborn consulted with Fraser Darling and decided. "Enclosed find a copy of the letter that has gone forward to Schaller today," he wrote Emlen on August 24. "This was not an easy letter to write. All I can say is that it expresses our best judgment here." It acknowledged conflicting views but advised that NYZS would end its sponsorship and responsibility for George should he choose to stay on.

It didn't matter, since he had already left the park and written Osborn to say so, their letters crossing again. And he had by now learned to interpret correspondence as it suited him. Ignoring the larger directive—to leave Congo—he assured Osborn his time would be dedicated only to gorillas.

George did leave Congo, briefly, driving to Kampala to surprise Kay. He stayed just two nights, leaving after a farewell dinner at her dorm to return to Congo under cover of darkness. He first tracked down the Cordiers, who, when the Force Publique began arresting whites, had hidden in the forest. Charles, pecked by a heron, was now one-eyed. But he "talked of nothing but finding his man-apes." (The search for the Yeti, here called *Kakundakari*, will be a recurring motif in George's life.) Cordier was determined to stay until he could ship out his animals—though soon even the $5,000 he was paid for each gorilla ceased to seem worth it.

Reunited with Andre, George resumed work as if the world weren't upended. Exploring Congo's lowlands, he saw that "poor, shy Andre has it rough here," as he wrote Kay. "He is not of the tribe so they won't let him sleep with them." George offered the comfort of a red sweatshirt, "which he wears all the time," and the meat of a mona monkey, bought from a local. For himself, he poked fun at the terrifying gendarmerie. "Bought a nice blue monkey today. Bloody up to my elbows cleaning it when the FP banged wanting to see my papers. Ground his cigarette out on my floor. Pointed to his epaulettes, '*Je suis un capitaine*.' Obviously just been promoted."

By September 1, 1960, Osborn was frantic. Again he wrote George. "It has become abundantly clear that you should get out of the Congo—and stay out! While your purpose and courage in helping to clear the Watusi and their cattle are admired, you are becoming involved in risks not related to your assignment. I must put it on the line that your primary responsibility, indeed your only one, is to come out with a maximum of information." He wrote Van Straelen in Brussels. "We are concerned for the safety of Schaller. He is utterly fearless and could get himself into critical difficulties.

I trust you will concur that we have done the right thing in requesting that he withdraw immediately." He wrote Coolidge. "An endeavor was made to get you on the telephone. He is getting himself into a great deal of trouble from which he may not be able to extricate himself." He sent a telegram to Kay. "Please send word to George leave Congo without delay."

CONGO BELGE — *BELGISCH-CONGO*
SERVICE DES TÉLÉCOMMUNICATIONS
DIENST DER TELEVERBINDINGEN

NUMERO *Nummer*	ORIGINE *Oorsprong*	MOTS *Woorden*	DATE *Datum*	HEURE *Uur*	VIA *Via*
12/167	Kampala	19	2	1205	mbi

Arrivé à : *Aangekomen te :* Rumangabo 3/9/60
Heure : *Uur :* 1450

Indications de service taxées
Betaalde dienstaanwijzingen.

TÉLÉGRAMME
Telegram

george schaller
c/o parc national
albert rumangabo
via goma

Explications des abréviations admises pour les indications de service taxées :
Verklaring van de afkortingen toegelaten voor betaalde dienstaanwijzingen :
RP = Réponse payée. *Antwoord beta...*
LT = Télégramme lettre. *Briefelegram.*
CR = Accusé de récep. *Kennisgeving van ontvangst.*
TC = Collationnement. *Te collationneren.*

La Colonie n'est soumise à aucune responsabilité en raison de la correspondance privée par voie télégraphique.
De Kolonie is niet verantwoordelijk wat betreft de private correspondentie langs telegrafische weg.
(Ordonnance législative n° 254/Téléc. du 23 août 1940.)
(Wetgevende ordonnantie nr. 254/Telev. van 23 augustus 1940.)

osborn orders immediate departure

Kay

Although uncertain where George was as Congo erupted, Kay kept her cool head.

Kay's response was a marvel of self-mastery. "Dear Dr. Osborn, I received your cable this morning and immediately sent cables to George telling him to leave the Congo immediately. As he might be in any of several places, I sent a cable to each. How long it might be before it reaches him, I have no idea." George received the one Kay sent to park headquarters, and answered her. "Tell Fair that telegrams sometimes take several weeks to reach me. In other words, I'll judge when to leave."

George did write to Osborn, imploring him to finance essential areas until the new government could take over—"if not now, it may well be too late"—and slamming the Belgians for failing to train Congolese as wardens and otherwise prepare for independence. Emlen also wrote Osborn, "distressed that George took it on himself to correct a bad situation in such a short-sighted way. I agree that the Park cannot be kept alive by force. At longer range, I'm sure George would have seen the prudence of holding off in favor of the frustratingly slow procedures of diplomacy."

Amidst all these matters of consequence, the trivial also demanded attention. Dozens of letters and telegrams went between New York and central Africa about their VW combi, impossible to sell in a market glutted by fleeing Europeans. (It eventually netted fifty-five pounds.) The NYZS comptroller wrote everyone at length about a missing freight refund of $21.22. George's letters to Kay devolved into randy teasing. "I got your latest letter with your bust measurement. Have you grown? You have a damned nice figure. Wish you were here. For once I could give you 10,000 francs and tell you to buy yourself clothes." He gossiped about Verschuren's nurse ("they probably sleep together") and Julian Huxley, whom he'd taken Anicet Mburanumwe to meet. Or rather Lady Huxley, who made the more lasting impression. "She's a wildflower specialist and asked me what the mating positions of gorillas are."

Having sent off a final report to Van Straelen—"We found no cattle in the forest. Some poaching but no extensive damage in 3 months of independence"—George was at last ready to go. "Project finished leaving Congo in two days stop," he cabled Osborn. Still carrying 6,000 francs, useless beyond Congo's borders, he followed a European acquaintance to a shop in someone's home and bought dresses for Kay. On October 3, reunited in Kampala, they shipped out their crates and headed to Nairobi.

Though *The New York Times* would credit the gorillas' survival to the "tact, common sense and bravery" of Verschuren and Schaller, the biologists laid most of the honor to the Congolese guards. Even when vastly outnumbered, they had put their lives on the line to turn back poachers from Uganda. Some met gruesome ends; the hacked-up body of Valère Kurubandika was buried with military honors.

Historian Samy Mankoto Ma Mbaelele deemed Schaller "instrumental in saving an entire sector" but also joined George in praising the park administrator, whose manifesto anticipated a debate that would last into the current century. The parks are "national heritages belonging to all," wrote Mburanumwe, "not to a single clan or tribe pretending to have rights of first occupancy." Forty years later, when Mburanumwe was still protecting gorillas as head of L'Institut Congolais pour le Conservation de la Nature, journalist Alex Shoumatoff met him on assignment for the UN Foundation. He carried with him, Shoumatoff reported, "an affectionate letter Schaller had sent him in 1988."

Despite the promising beginning, life soon turned as grim for the gorillas as George had feared. Returning in 1963 for *Life* with photographer Terence Spencer (soon to be famous for his pictures of the Beatles), he found no cattle but ample evidence that something drastic had happened. Though he recognized one old friend—a puckish female transformed by stress into a stooped old woman—just ten of her troop had survived and at the sight of George all fled, a sure sign they'd been hunted. As he often would, he absorbed the pain to offer plain witness, the more devastating for its quiet. He described gorillas without hands, lost as they tore themselves in panic from snares, or beheaded for sale as tourist trinkets or meat. "It is difficult," he wrote in a postscript to *Year of the Gorilla*, "to reconcile the congenial nature of gorilla society with the brutality of my own species."

As they prepared to leave the continent in September 1960, George received a telegram from Louis Leakey, asking if they might stop to see him in Nairobi. As Kay would later recall, they found the white-haired anthropologist at his desk at the Coryndon Museum* "looking very big-browed

* Now the Nairobi National Museum

and grumpy. [He] looked up at us as if we were the most unwelcome people in the world, [but] as soon as George said who he was, he was just extremely kind." George remembered a "bulky man dressed in a sort of jumpsuit delightedly showing me fossils he'd found." Leakey explained why he'd asked them to come. After the "spectacular breakthrough" of his intimacy with the gorillas, George was uniquely qualified to advise the young woman Leakey had just sent to study chimps, who they therefore must go see. George agreed, and on October 8, he and Kay drove to Kigoma.* Checking into the Lake View Hotel, they were joined for tea by Mrs. Vanne Morris-Goodall and "Miss Jane. The mother is freckly, short, fairly thin with upturned nose. Jane too is freckly, 2 prominent upper incisors, golden hair and rather cute. And she has lots of get up and go."

The next morning, all took a boat to the Goodall camp: a tent at the mouth of a canyon, with a fresh stream behind and a sandy beach "curving moon-shaped to a lone palm. Fishermen dugouts and grass huts. Quite friendly people. Mrs. Goodall treats their sores." While Kay visited with Vanne, George and Jane climbed into the dry hills to look for chimps. They heard them once but did not see them. They tried again at dawn, finding bathing baboons but still no chimps. The women envied George's own baths in the cool, clear lake; they could not swim lest "the Mohammedan natives object." In the evening all watched a three-foot monitor lizard scavenge in their fire pit and the fishermen row out by lamplight, their silhouettes black against a red sky.

Though Goodall had been in Gombe for three months, as her biographer wrote, "nothing Jane had done or seen so far contradicted or surpassed [Yerkes lab director] Nissen. . . . In most ways, she was that summer and autumn still busily confirming the conclusions of her scientific predecessor. Then George Schaller showed up." George found her unprepared: "She wasn't sure what she was looking for. I helped her understand what to

* In present-day Tanzania

record." He also offered what she later recognized as life-changing guidance. "George said he thought that if I could see chimps eating meat, or using a tool, a whole year's work would be justified."

On departing Gombe, leaving Jane a polyethylene sheet for shelter, George sent a report to Osborn commending her work but noting the challenges. "She lives alone with her mother, isolated by 16 miles of lake from the nearest white settlement. The heat is terrific and there are many diseases." He noted the pressures on the reserve—home to one of just two known groups of chimps in Tanganyika—in consequence of its troubled history: Residents forcibly evicted at its creation in the 1940s had been readmitted in 1955 to dry their dagaa (a sardine-sized staple) on the beach, but were now "agitating" to farm and cut wood inside its boundaries, which would threaten the ape's existence. He asked if NYZS might buy Jane a hut. "Miss Goodall's tent is not ideal during the rain. Even more, she needs a spotting scope . . . she has only a poor pair of binoculars."

Jane wrote home in a state of great excitement.

> *I dare say Mum has told you about the visit of George & Kay Schaller? The American who has just done a superb study of the gorilla. It really was nice talking to someone who understood what I was doing, and why, & who didn't think I was completely crazy. He did not envy me my mountains or my heat. He, poor man, had the most freezy wet climate. When in their little cabin they wore layers of sweaters and shivered, & he ate porridge for breakfast. It rained regularly every day—usually 4 times—often more. He said he sometimes used to sit on a branch and watch the gorillas below. The branches were all covered in long moss—it looked lovely & soft, but when you sat on it it went "squelch" & was cold, soggy and nasty. He used to get very embarrassed at presenting the boy, time after time, with pants that were a queer greeny brown colour!!*
>
> *It was a pity all my chimps departed on his arrival, not returning till the day after he went! But he had only come to look at their habitat;*

the possibility of the Reserve and, mainly, to spy on me! He admitted as much before he went! Which proves, I think, that he was not too horrified by my work.

Jane remained deeply intrigued, writing home in May 1961 that "they have a baby in a month & George is most anxious Kay keeps it with her from the moment it's born—he feels the early mother/child relationship is particularly important—after studying his gorillas!!" When his books came out, she read them carefully, even correcting Dian Fossey when she wrongly insisted that George had overlooked the gorillas' taste for wild blackberries.

A bit of rivalry did creep in. When chimp David Greybeard climbed a palm to be near her, Jane crowed: "Better than George's 'Junior.'" And over time, she muted George's role in her story. George didn't mind: He was nothing but grateful for Jane's impact as the celebrity face of primate conservation, a role he would have both failed at and loathed. And while he shared the view that she went too far in habituating her chimps—not only altering their behavior ("she didn't get normal group life anymore; they just sat around camp waiting to be fed"), but also putting them at risk of human disease—he said: "You make mistakes, and you learn."

George and Kay made one last stop in Africa, heading to Olduvai Gorge to meet Leakey. He never showed, but they wandered happily around the dig and spent a beautiful cool night alone. Kay liked to believe it was where their firstborn, Eric, was conceived. Sailing across the Indian Ocean, they paused for a holiday in Ceylon* before continuing on to Malaysian Borneo, where George would spend his final months as a primatologist. After a few weeks, discovering she was pregnant, Kay went on to Alaska to be with her mom.

George was lucky in Borneo to again find a deeply knowledgeable local guide. A year earlier, Gaun anak Sureng had completed the first survey of

* Now Sri Lanka

Sarawak's orangutans, the least known of the great apes. Wading behind Sureng through chest-deep peat bogs, George had his first intimate encounters with leeches. But in two months they met the rare "man of the forest" just four times. The only way to estimate their numbers was by counting the nests they build in trees, though Sureng shook his head when George began measuring one, showing him the claw marks left by the nest's actual owner: a bear. George did make the most of his few encounters, capturing in an article in *Zoologica* how the orangs used their intuitive understanding of physics to bend branches with their body weight so they could walk onto an adjoining tree, and made vocalizations that sounded like kissing, chugging, and burping. He particularly admired how this "least-gregarious" ape dropped branches on visitors, or swung and released them at just the right moment to hit their target; he himself had to dodge some thirty branches one female flung his way. He did think it odd that while both Alfred Russel Wallace and Attenborough had seen this behavior (in 1869 and 1957, respectively), his own NYZS predecessor William Hornaday had not, "probably because he shot the animal before it had a chance." (In one visit, Hornaday shot forty-three.)

The two years that followed, in Wisconsin and at Stanford, gave George time to write his books and Kay time to edit them while also getting on with what she called their very planned children, who had to be born and sufficiently grown in time to be vaccinated for their next adventure. (It also gave Kay the chance to take driving lessons, financed by a cheque for a million francs George wrote on the Banque Belge d'Afrique.) George missed his eldest's birth. Though he had repeatedly postponed a trip to Europe to present a paper, "Kay kept delaying," he said, with the feigned pique he often used to tease her. (In their late eighties, the two still flirted constantly.) She fully supported his going.

> In those days, you married your husband and he was in charge, and his profession was in charge of him and of you, and you went along

> with whatever was required for your husband. You didn't have that need to be independent and equal. You belonged to your husband essentially, and to his life. If I had been equally trained, I still would have put his career ahead of mine. I liked having George there in charge and taking care of me.

That she had already accepted his vagabond ways was plain in her July 24, 1961, letter announcing George Eric Schaller's arrival. She reported the frontline details she knew her husband would relish, without a hint of resentment. "I labored from Thursday morning to Saturday night, 42 hours. With the doctor's help we staged some pictures, from the time of his head appearing to the cord being cut. His eyes are swollen and squinty, and he has your vulture lip. He looks like Mr. Magoo."

4.

Burning Bright

India, 1963–65, Tigers and Their Prey

A royal son of Gujarat, M. K. Ranjitsinh was just twenty-five years old but would soon be named "collector" (administrator) of the Mandla district of Madhya Pradesh. An avid seeker of big cat trophies, he was thrilled that his new dominion included one of the last refuges for India's vanishing tigers, which in December 1963 he went to see for the first time. It was early in the morning and he was driving fast, alerted by a swamp deer's alarm call that a predator was nearby. Roaring up out of the ravine, "I was greeted with this tableau: In the tall winter grass sat a jeep, on the seat of which stood a lanky white man." The man's binoculars were trained on the group of barasingha, a rare species of deer named for its unique twelve-tined antlers. The stag and his six hinds were staring at a patch of grass, braying like donkeys, when Ranjitsinh's arrival "disturbed the dramatis personae." Out of the grass stepped the source of their terror: a tigress in the "full splendor of her ochre-orange winter pelage." Her hunt ruined by Ranjitsinh's blunder, she walked off with what seemed to him withering contempt. Worse was the reaction of "that white man, whose piercing eyes stayed on me as I drove slowly past. If looks could kill, I would have died." Schaller had never seen a tiger kill a barasingha, and now this intruder had spoiled his chance.

Feeling guilty but also "curious who this chap was," Ranjitsinh sent a message asking if he might call on the Schallers that evening at their home in the park. He found Mrs. Schaller admirable in every way: wonderfully hospitable but keeping to the background like the "classic Indian woman, who lets men do their work." George was equally gracious, accepting Ranjitsinh's apology, particularly on learning that he knew the name of every Virunga volcano. "I'd never been to the gorillas, but he saw that wildlife was the passion of my life." For this hunter who would soon play a key role in India's rescue of the tiger and forever retire his guns, "those were the two defining hours of my life."

Kay and George had only recently arrived, under the auspices of the Johns Hopkins Center for Medical Research and Training. Newly focused on the ecology of disease, the university had established a campus in Calcutta* and asked George to study deer as possible vectors for anthrax and other pathogens. The best way to collect the needed blood samples, he decided, would be by scavenging tiger kills. So, with two-year-old Eric and one-year-old Mark, he and Kay settled in prime tiger territory—and the setting for Kipling's *Jungle Book*—Kanha National Park.

Though the tiger has been a creature of myth for millennia—as the mount for the Hindu goddess Durga, the guardian of the forest, a perfect killing machine—it had been little studied until Schaller arrived except, as he put it, through the sights of a rifle. So after years with his placid vegetarian gorillas, he would now have to learn to live with carnivores. He soon gleaned that a gathering of jungle crows could lead him to a kill, but that if the birds hung back the tiger was likely still there. "You sit there for an hour, quietly listening. . . . Nothing happens. So you very carefully go in.

* Now Kolkata

Usually the tiger has heard you and crouches down and waits, judging your actions. You don't even know the animal is there, until it suddenly rises, some fifteen feet away, and walks off."

George had chosen Kanha* after five weeks of reconnaissance by Land Rover, leaving Kay and the boys at Calcutta's New Kenilworth Hotel. With his young driver, Kumar, he traveled 4,500 miles—across the heavily cultivated Indo-Gangetic plain, north into the foothills of the Himalaya, west to the deserts of Rajasthan, and finally into central India. Inching down rutted dirt roads clogged with bullock carts, saving money by camping in gas stations, he caught his first glimpses of peaks above twenty-five thousand feet—instilling a yearning he would follow the rest of his life—and of the animals he had come to see: four-horned antelope, dancing sarus cranes, a buteo hawk swallowing a scavenged small Indian civet, only its banded tail still visible.

It was the human pageant, however, that first claimed his attention. Vendors lined the roads selling peanuts and mangoes, potato balls in crisp dough, biscuits in jars, and betel nuts. Farmers led oxen down a short slope, backed them up, and led them down again, to raise big leather bags of water from deep wells. Old women bent to cut grass with short sickles. Young women dressed for a puja festival strolled in purple saris, their foreheads painted red and studded with grains of rice. Some of it, George found challenging. No matter how far they drove nor how isolated a place seemed, "one never loses sight of cows"; under the hot sun, the bony white cattle had grazed much of the once-green land to dust. At the border with Nepal, on high alert since the 1962 Sino-Indian War, soldiers prowled everywhere "and I just hope no one asks me for a permit." The need for a toilet was "always a problem, since everyone simply squats in a convenient corner, something I find difficult to do in a crowded town." Cold nights ended in the

* He'd won assurances from the Calcutta center's associate director Charles Southwick that he could choose his study site by confessing his "problem"—a competing offer from Tanganyika's game department.

"loud retching, coughing and clearing of noses which seems to be part of the Indians getting up." The persistent boys' "shoeshine, yes, yes, yes" persuaded George that "I wouldn't have a shoeshine for anything."

Though much of what he recorded in his India field journals would never be published or read, George crafted vivid cameos, animal and human. A pig with spiky hair looked like an "unkempt cactus." A lone sambar deer against a pale red sky became "a black sparkling silhouette rising out of a low sheet of fog, breathing white cones of steam like some mythical creature. As the sun dries the grass, it whispers, a fine rustling sound like mice feet on straw." A puffed-up assistant wildlife warden ordered his underlings about but grew obsequious around his boss, the stout Mr. Chaturvedi, who sat at his desk in loose white linen being "fanned by a fellow standing outside his door, pulling a cord attached to a cloth suspended from the ceiling."

•••

After that long journey, George knew at first sight that Kanha would be their home. The other reserves he had visited were heavily logged, overrun by livestock, or hunted so relentlessly that the few wild animals who remained had grown exceedingly shy. Here, he found dappled glades of straight, slender sal trees, small villages with moderate numbers of cows and goats, and abundant wild predators and prey that did not bolt at first scent or sight of a human. The sand-colored stucco bungalow the family would live in had a veranda overlooking what he called a wildlife amphitheater stretching from the ravine (nala) across the meadow (maidan) to soft hills on the horizon. Out at dawn, he found a woman and three children waking in the rice field they had guarded all night from blackbuck antelope, fig trees heavy with fruit and screeching mynahs, and his first tiger, or maybe leopard, tracks. He didn't yet know the difference.

Returning to Calcutta, he and Kay took a brief, blissful, childless holiday in Darjeeling, flying on separate planes in case of a crash, then in December 1963 loaded the boys onto the Bombay Express for the six-hundred-mile trip to their home in the heart of the park.

The first weeks in Kanha were "hellish and hectic." There were the usual toddler realities: whining and temper tantrums, Kay and George irritated and bickering. When a couple visiting with their own toddlers departed, George cheered: "I'm glad to cut the crying by half anyway." Those ordinary stresses were heightened by their extraordinary circumstances. The bungalow had no running water, refrigerator, or electricity, except for a few dim hours of light on the rare days when a tourist came through. They lit the space with kerosene lamps and cooked on a small kerosene stove, bathed the boys by standing them over a drain and tipping tin cans full of well water over their heads, and sat on a toilet that emptied into a bucket that a servant later carried away.

Though complicated by kids, this was a domestic routine Kay was used to. Harder for her were the unfamiliar dangers outside their door. Taking daily weather readings for George required that she go into the meadow to swing a humidity meter. "I worried, what if a tiger comes and gets me while my babies are waiting inside?" More frightening was the possibility that the boys might climb off the veranda and down into the ravine where tigers and packs of dhole (wild dogs) passed; though tigers rarely eat humans, a little wobbly one might look like a fawn. The boys had already wandered once. In the chaos of the first days, Eric had taken Mark by the hand to go explore their new world. Park staff eventually built a bamboo fence to keep them in and predators out, though on more than one occasion, big cats broke into their shed to steal chickens or lambs. When after one such raid George found a wild pig killed by a tigress who had as yet eaten only the head, "we cut off the hindquarters to roast with mushrooms and onions foraged from the termite hill. This makes up for our last sheep."

Keeping the boys healthy was especially difficult. Villagers sometimes brought the family fresh buffalo milk or fruits they'd gathered in the forest;

a favorite was the juicy mowha (*Madhuca longifolia*), pale yellow, fleshy blossoms with a faintly resinous sweetness. But the closest food store was in Mandla, thirty-five miles away on a dirt track so bumpy it took several hours each way. Once there, they had to beg or bribe people to sell them food; after a poor grain harvest, the government had instituted rationing but issued no coupons to the Schallers. At home, George's efforts to rat-proof the powdered milk only half succeeded; the rodents enjoyed eating the containers as much as the food. Kay carefully sifted the weevils out of the flour before baking cookies, but only after they'd left India did George tell her that the odd bitter flavor that she could never quite mask was mouse urine. He knew the taste from his own childhood. Exploring everything, hands always in his mouth, Mark got two kinds of intestinal worms that, months after leaving India, Kay was still finding in his diapers. Visiting her family in Anchorage, "I had to get medicine without letting my mother know she had a wormy child in her house." For the rest of her life, she remained "raccoon-ish," forever reminding George and visitors to "wash your hands with soap."

Ever conscientious toward his sponsors—and aware that some at Johns Hopkins had tried to scuttle his study, not yet understanding the zoonotic origin of most human illness—George dove into his promised investigation of prey animals and the bacteria, viruses, fungi, and parasites that lived on or inside them.

The work, as always, had its miseries. The aptly named spear grass harpooned George's feet, leaving infected wounds. A contented hour in a tree ended with his legs suddenly on fire, the burn only spreading as he scrambled down, ants stinging his hands every time he touched the bark. More often it was simply tedious. After spending two hours watching chital deer graze, then two more hours crawling around weighing grasses to quantify what they'd eaten, "It's amazing," he wrote in his journal, "how much diddly work a study like this entails." He did experiment with a new methodology, getting permission to dart chital after first practicing with the syringe gun on the family goat. The gun shot erratically, but he finally hit one in the thigh; ten minutes later, when the tranquilized deer buckled, he bound its legs long enough to tag an ear, check for ectoparasites, and take blood. Even with the drug, "he usually has to wrestle them down," Kay wrote Bettina, "so ends up bruised, scraped, and sore." Finding the jugular often took fifteen tries. And the deer, he saw, didn't always survive. "They seem ready to get up but suddenly give up and die. I don't feel too badly since the bodies are taken by tigers and thus the animals are used as they would be anyhow."

It was by now standard for George to count, measure, and weigh everything: The chital masticated each bolus of cud "thirty to forty-five times at the rate of three chews in two seconds." As he had with the gorillas, he gathered such precise morsels whether or not he could find meaning in them—someday somebody might. And whatever the facts' ultimate value, the attention required to gather them served its own purposes. All his life, counting animals would be George's way of self-soothing. It would also be his practice, like yoga, for staying wholly present to all that was unfolding. He watched barasingha jab their antlers into tall grass and then walk around with tufts draped from their tines, their heads erect and steady as a debutante's to keep the adornment from falling off. He saw thick-chested sambar deer rearing in the behavior called "preaching," pushing their heads into high branches to scent mark the bark and leaves; and blackbuck antelope—the males strange and gorgeous with an arresting white ring around the eye and tall spiraling horns—spronking (as cartoonish as it sounds), bounding stiff-legged on all four feet in unison. One afternoon a gaur bull, a kind of wild cattle as glossy and bulked up as an extreme body builder, made a direct challenge. Accustomed to intimidating competing males by standing alongside them to prove its more impressive size, the six-foot-tall, two-thousand-pound bovid with its fearsome dorsal ridge lumbered up next to the Land Rover. When after a few minutes his metal rival hadn't budged, the animal turned aside, conceding defeat.

Unsurprisingly, ungulate sex provided many of the most vivid moments.

An aroused gaur, Schaller noted, slid the tip of his tongue in and out of his nostrils, saliva dribbling from his lips. The barasingha began its rut when its antlers hardened, neck swelled, and voice deepened. It sometimes displayed equine behaviors with German names: the curled lip (*flehmen*) or snort of air (*prusten*). A chital in rut became exceedingly restless, thrashing a bush with his antlers or, when the object of his desire urinated, thrusting his nose into the stream. Until consummation, he grazed, walked, and lay beside her, a behavior called tending, emitting at times a hoarse moan. After a few thrusts violent enough to move her several paces, he finished with "no postcoital display."

•••

At home, Kay's obligatory houseful of servants brought more stress than ease. "In the mornings our house is milling with people: a sweeper for inside the rooms and one for the toilet, a laundry man," George wrote in his journal. "By the time Kay has directed all these people, it would have been less wearisome to do it herself." Kumar the driver now had a debonair mustache, "thinks he is hot stuff and says he works only 9–5." It was a relief to George when a telegram arrived calling him home to be wed (an astrologer had determined that it was the most fortuitous time). Mary the ayah was also soon gone. Though efficient, she missed having a church nearby and "griped continuously about the cold and there being no running water or bath house and being too tired to get the children's food." George drove her to the bus, where "she made a final fuss of wanting to be taken to Gondia, a hundred miles away. I just drove off." They found the kids easier to manage without her. "There had been too many persons to satisfy their demands."

All did dearly love Jamil the bearer, whose job it was to run the household. He adored Eric and Mark, taking them each day to laugh at the *slap, slap* of elephants beating grass against their heads to knock off sand. And though he tried to follow Kay's instructions not to pick them up when they fussed—she was reading the experts and ignoring tantrums—he couldn't stop himself from squatting to give them a hug. "Indians seem exceptionally fond of and kind to children, and must think we are terribly mean," Kay wrote to Bettina. "George and I do tend to be impatient and expect too much. Maybe it's good they have a friend when we are short-tempered." George's own portrait of Jamil was equally tender. He is "thin and fragile-looking . . . makes clothes look as if everything is much too big for him . . . is not adept with his hands and seems always bemused. But he does not complain and pitches in with the children." He added a poignant coda. "He goes home once a year to U.P. [Uttar Pradesh] to visit his own four small children."

All in all, the move to Kanha had done the boys good. In Calcutta they had been "feverish or diarrhea-ish or something-ish" almost every day, Kay wrote home, with "the hollow-eyed look small white children seemed to have over and over again." Life in the forest brought renewed vigor and color to their cheeks. To Kay, too, who had learned to make comedy of the trials of motherhood. Having described Mark to Bettina as a "most homely" baby, his "huge red nose swollen and slantwise on his face," she now reported his looks improved but his character to be as stubborn as his father's ("he absolutely insists on crawling a certain direction, etc.") and as socially graceless (when he meets a little girl, "he stands right in front of her, sticks out his stomach, gives short loud screams and pulls her hair"). Eric had adopted a habit of thanking Kay for her every move. "Thank you for putting salt in, Nonnie" as she mixed dough; "thank you for putting it in the oven." When she found herself glamorized again as the intrepid Jane to George's Tarzan, she brushed it off. The "profuse admiration" of five charming male visitors (French, American, and Indian) left me feeling "terribly bushy," she wrote Bettina. "And when you describe me as kind and patient you make me blush and George laugh, for I am indeed the 'screaming idiot' you claim you would be!" Tiring of sibling spats, she wrote Chris,

"I begin to think your father's system of waiting ten years between babies has merit." Still nothing could mask that, in her jungle nest, Kay felt "especially content."

•••

For all the drama inherent in being prey, predators were clearly George's greatest fascination. And no place on Earth has more than India. Though covering just 2 percent of the terrestrial planet, it is home to a quarter of all carnivore species. That includes half of the world's last four thousand tigers, pressed into the remnants of a range that once stretched from Turkey to Indonesia.

That relative abundance did not make the world's biggest cat easy to find. Though with his tail a tiger can be ten feet long and weigh five hundred pounds, he survives by stealth. He doesn't run down his prey, as wolves and dhole do, but stalks to within about fifty feet, only then rushing to leap. The silence of a tiger's step and camouflage of its light-and-shadow coat meant George had to learn to spot flickering ears in dense vegetation or a sliding apparition that might materialize into a cat. Following the drag marks left by a cow killer felt "like gorilla tracking, this looking hard, straining to see a possible crouched tiger in the grass." He often spent days like Didi and Gogo, waiting for someone who never appeared. One day after lunch, he set up his camera overlooking a pool, hoping an animal would come drink. He waited until dusk. "No one came."

His eyes were often his least useful instrument. He began, like Olaus, to stick his hands and nose into everything, sniffing out scent marks that lingered for weeks, even dipping a finger into the everted preorbital gland of a displaying (captive) blackbuck to sample its perfume of acetic acid and musk. He learned the multilingual early warning system other species depend on: the flushing of a jungle fowl, sharp *ka-kao-ka* of an alarmed langur monkey, thumping hooves or belling of a sambar that—like trumpets along the parade route of a king—marked the path of a sauntering tiger. What he wrote of the chital, that it "has extended and complemented its perceptual world considerably by relying in part on other animals," was increasingly true of Schaller himself. And he found in sound not only information but also music. He loved the bugle of the barasingha, with its harmonic drone, and the song of the rutting gaur, "a strangely fine voice for a beast" that stepped down like someone practicing scales. He wrote it that way in his journal.

Even when he did spot a tiger, it was often fleeting. Sometimes it was George who retreated: detouring around a loud purring whose source he couldn't see or backing away from a tigress who—having appeared out of nowhere to crouch nearby at a stream, lapping water like a kitten—caught sight of him and laid back her ears. More often, it was the tiger who exited, even if it meant abandoning a kill. One tigress, surprised when she lifted her blood-spattered face from the body of a baby gaur to find Schaller just eighty feet away (he was surprised, too) snarled but backed away. "I see a tiger or two every few days," he wrote Chris two months after their arrival, "but it's difficult to get on friendly footing with them. They just walk off into the forest and disappear."

Slowly, man and beast began to get used to one another and hold their ground. When a passing tigress stopped to watch him from just ninety feet, Schaller stood for forty minutes watching back, though "quietly, and with a detached look." He was also getting better at finding them: looking for grass still flattened or brushed free of morning dew; calculating the cat's speed by the space between pugmarks; identifying individuals by the distinctive black markings above each eye—their "bar code," one of his protégés calls it. They in turn, he was certain, soon "knew me on sight."

More than a dozen times, he resorted to tying out a domestic animal as bait, a practice he later renounced. Sometimes the animal he tethered by its neck or foreleg was already dead. When one morning at 3:50 a.m. a servant awakened him to say that a tigress had broken into their shed, through a hole just twenty-two by thirteen inches, he found that she had killed their buffalo but could only exit with a sheep. Dragging the buffalo out himself, Schaller tied it to a tree, and was rewarded with hours of watching the tigress and her cubs feed. Occasionally, his conscience nagged. His original plan had been to tie out the buffalo alive, but leading him each morning to water he'd found him "getting quite friendly, rubbing against me." When a tigress killed a gaur three hundred feet from a tied-out goat, he knew that "the poor creature must have been scared, hearing the tiger kill and then eat all night." His gifts of livestock did accelerate habituation: When a *Life* photographer joined Schaller, the cats let the two men walk back and forth to their blind, and one came within forty-five feet to drag off a kill.

The real turning point came in March 1964, when late one afternoon George climbed a tree to watch his newest acquaintance, a notch-eared tigress he called Cut-Ear, tear at a gaur she'd brought down. To his delight he found her cubs gamboling nearby. The size of "setter dogs" at four months old, the three sisters and their brother raced and wrestled and stalked on little cat feet. One crept through the grass then leapt to its forepaws, vanquishing a bug. When, bellies full, the family went off to nap in a shady copse, George stayed in his tree, soon joined by excited vultures ruffling feathers and splattering feces. At night, as he'd anticipated, the tigress returned. "It was eerie to hear the bones crack loudly and see the shadowy creature jerk at the inert carcass." Though he was in shorts and shivering under a half-moon, he held still for sixteen hours watching the tigers eat everything down to viscera and soft bones. When in the morning a worried Kay drove out with the boys to check on him, he waved them away, knowing he was seeing as no one ever had these tigers' private lives.

•••

Fatherhood, for George, would always be a stop-and-go affair. In Kanha, he would spend twenty-four full nights and many partial ones out with the tigers. And, like his own father's, his absences sometimes stretched to weeks or even months at a time. That often was, or seemed, unavoidable. When he missed Eric's birth, Kay explained to Bettina, it was at her insistence. "Contacts and prestige are such an important part of his future" that he simply had to accept the invitation to present in Vienna. And when a gorilla was born soon after at the Basel Zoo, it only made sense for him to extend his stay to observe that baby. He'd gone off again, the summer before India, when Eric was turning two and Mark beginning to crawl: in June to watch seals on the Pribilof Islands, in July to Johns Hopkins, and in August back to Congo for *Life*.

If that was all more or less obligatory, the same cannot be said of his trip that May to Mexico. Despite all the work travel ahead, he spent two weeks in Veracruz with his friend Tom Lane, climbing the 18,406-foot Pico de Orizaba, then staying on to summit Popocatépetl, see Cuernavaca's Teopanzolco pyramid and the Palace of Cortés, and stop at beachside motels on the eight-day trip home. (Tom took a picture for Kay of George reading under a parasol.) By summer's end, Kay felt "a bit defeated with children and garden and missing George so dreadfully that I write him and ignore everyone else." In September, with George going from Congo directly to India, it fell to Kay to fly alone* with the boys from Alaska to Tokyo to Hong Kong to Bangkok to Calcutta, a nightmarish trip made more so by hours of delays.

George's field habits, in those early years, sometimes infiltrated his fathering, and vice versa. He charted Eric's first-year development as carefully as he had the gorillas' ("Aug 11 holds head up 1–2 seconds; moves eyes in direction of my hand. Oct 15 can put rattle in mouth to suck"). When they moved for the summer of '62 to Wyoming for George to study white pelicans, he continued his field notes on his little human subject ("May 23 pulled himself to kneeling") in parallel with his avian families. Though at

* This was against the counsel of Southwick, George's Johns Hopkins boss, who twice dared "meddle" in his personal affairs to strongly advise against so burdening Kay.

least on paper, the latter got more of his attention. He recorded the feathered parents' tireless labor—one foraging while the other tended eggs or hatchlings, then bowing with the *ho-ho-ho*s that marked the "nest relief ceremony"—and the babies' incessant demands. A fledgling might run to its mother, belly flop and beat its wings wildly to beg for a meal; wheedle open its father's beak and climb into his pouch to reach down his gullet for regurgitated food; or chase either so relentlessly the adult would flee into the water where the chick couldn't follow.

George himself left for eight of that summer's thirteen weeks to camp with the birds, leaving pregnant Kay and baby in the field station. But he also did things that he'd never done before. When a late May storm laid down several inches of snow, he kept away from the nursery island lest he step on the snow-colored eggs. When July's hard rains and winds broke waves over the backs of the young birds—exhausting them until they could no longer hold their necks above water—he took to a canoe with a visiting high school friend, spending hours rescuing all they could, hauling twenty heaving young to higher ground.

And if often absent, George was also extraordinarily present. In India he worked what Kay called the "usual 14-hour days that make him happy" but came home for lunch and tea and wrote up his notes after the kids had gone to bed. When Kay had to be away for a week in the hospital, he amended his day to include the boys, carrying them along on his rides. He told them marvelous stories, mining the A. A. Milne moments that were a regular feature of his life. In one, a sloth bear—comical with its long bare snout, bow legs, and "blue-black hair so long it was hard to tell front from back"—bumbled by, oblivious to four big tiger cubs. George, Kay, and Eric sometimes made up animal stories together, one starting and the others following in turn. Pictures find him playing with the boys in the little metal tub they swam in to escape the heat, or balanced atop a rock he is helping them to climb. "We took the children to a waterhole," Kay wrote Bettina, "and their father had a wonderful time. Eric isn't quite at the building stage but George certainly is. He built sand mountains and canals and even a slide for Eric's boat."

For the boys, whose stuffed toys had their real counterparts outside, exotic experience became normal life. On one 108-degree spring day, a huge cat walked past the car. "Eric was quite indifferent, saying it was a small tiger." Reversing his father's sympathies, the three-year-old's heart belonged to their two little lambs, which he led around and fed grass. When after a jackal attack they had to kill one sooner than planned (too small even to have chops, Kay complained), she had to call it simply "meat." Both boys scribbled notes in their "field journals" and played tiger and buffalo, clawing, biting, and devouring each other. "Sometimes I think their games will make them outcasts when we return to the states," Kay fretted. "Though they are fascinated and amazingly knowledgeable about George's work. Eric knows all the animals by name."

•••

India provided George his first experience watching killing over and over again. He was struck by how calmly the drama often unfolded; he watched one tiger kill a cow "with no ferocity, only methodical attention." Sometimes the attack began in stillness. When a buffalo turned to look directly at an approaching tigress, she froze mid-stalk, knee bent and paw in midair as if playing red light green light, only moving again when he turned away. By holding her gaze, the buffalo delayed his own death by forty-five minutes—and left George, camera ready, impatient: "a tedious time for me." More often the initial rush—massive forearms and grappling-hook claws wrapping a deer to pull it down, or a sharp bite and twist of a hind leg tumbling it to the ground—ended with a leap to the throat and a long silent finale, both animals waiting motionless till the one died of strangulation.

Alone for a few months (Kay had taken the boys to Calcutta), George discovered that a leopard had filched their lamb and, wanting to see the cat, decided to give him their goat too. He'd been sleeping on the porch to hear the jungle noises. Tying the kid forty feet away, he was awakened at midnight by a choked bleat and thump and found the "lovely" male leopard patiently holding the little goat's throat. There was no struggle. "All seemed almost abstract," the languid feeling only heightened by the unusual comfort he watched from. Lying in bed, holding the cat in the beam of his flashlight, "was like a dream, diffuse in the dimness and soundless, except for an occasional crunch of bone."

Sometimes the violence was more startling. When a jackal pup got in the way, the lovely leopard killed it with a single bite through the head. Vultures settled on an injured cow, reaching through the anus, udder, and neck wounds to eat it from the inside out (like Annie Dillard's famous frog), leaving just bones covered with skin. Spine-chilling groans heralded a successful midnight attack on a young deer. When Cut-Ear braced all four legs to jerk a buffalo's head, George and Kay heard the neck snap from eighty feet away. They were not the only fellow creatures affected. A domesticated elephant approached the buffalo corpse and began breaking branches and stripping leaves from a tree, dropping them as if to bury the dead.

George's relationship with Cut-Ear granted him entry to the lessons she gave her children, who by their first birthday nearly matched her in size and needed more meat than she could bring down on her own. Only the male, bigger than his sisters, had developed any aptitude for killing. George admired the wrestler's strategy he used to drop a gaur five times his size. Leaping to grab its nape, he then slid off to one side, levering the stolid animal down. The females, meanwhile, even as a trio, still faltered. When George tied out a buffalo, they ventured an attack, but the bull snorted and charged at any that got within ten feet. After more than an hour of this inept dance, their mother finally stepped in, biting one hind leg and swatting the other to tumble the buffalo, then retreating to let the young ones deliver the lethal blow. They failed, clawing and biting so randomly that the bull managed to get to his feet and shake them off, displaying a "pluck" George admired.

For four hours, nothing happened. The buffalo grazed and rested as if oblivious to his waiting executioners. Then the cubs tried again, darting in and out as the buffalo whirled with lowered horns to face each sally. Cut-Ear waited on the sidelines, determined that her daughters learn to fend for themselves. But when they made no progress, she again stepped in: throwing the bull down and this time pinning him with her paws while the cubs swarmed, lacerating him on his belly and back while he bellowed piteously. Still, when she let go, the remarkable bull struggled up again, sending the cubs falling to the ground. By now, however, his injuries were too great. He collapsed and—"a horrible spectacle"—the cubs feasted on him alive. When one young tigress grabbed his nose, the dying beast summoned his last bit of life force and nearly gored her.

As in Congo, George could witness such family dramas only by spending hours in the jungle alone, still unarmed and often staying all night. His encounters were sometimes closer than intended. One afternoon, as he scared vultures off a carcass with deep puncture wounds in the throat, his attention was diverted by a faint rustling. Turning toward the sound, he found a tigress standing just ten feet away. As startled as he, she quickly walked off and disappeared. Another day, he approached a boulder without noticing the large cub asleep on its far side. When, awakened, she looked over the edge, their faces were three feet apart. (A tiger's face, it's worth noting, can be fifteen inches across.) She roared. He leapt and backed away. She bounded toward him. He scrambled up a tree. Satisfied, the cub stretched out below, where ten minutes later she was joined by her brother, unaware anyone else was there. After a few minutes, George spoke up. "Hello there," he said. The male looked around, and finally up; when his eyes came into focus, he growled and fled. The female remained, soon joined by her sisters. All three lounged, watching their treed quarry, until after half an hour he tired of it and, shouting, drove them away.

His scariest moment came on a day he was following drag marks. Intently focused on the ground, he heard a low growl and raised his head to find a tigress bounding toward him with the coughing roar that signaled attack. She stopped at seventy feet, but for twenty minutes froze him in place. When she left, he saw what she'd been protecting: a tiny cub. Still, when Cut-Ear killed a gaur calf (its bereft mother walking "irresolutely back and forth for 5 minutes" before abandoning all hope) and went off to fetch her cubs, George hurried in to weigh the calf, collect its jaw, and examine the mauled skull.

Most surprising, in these intimate days and nights, were the tender moments he witnessed among animals said to be unremittingly solitary. Though Cut-Ear was rough with her kids—sating her own hunger first, cuffing or hissing at any who dared move in—she also tolerated their chewings on her ear or tail, reciprocated when they rubbed their cheeks or bodies against hers, and lolled on her back purring when one baby held her head to lick her face. That sociability extended beyond the family. On several occasions Schaller heard the beautiful roar that carries for a mile or more, the *aaau-uuu* of a tiger with food to share, and saw not only Cut-Ear's cubs respond but sometimes also a big male or even a tigress trailing a cub of her own. The female would stop only long enough to eat, sometimes with an exchange of swats and snarls. But the male often tarried. One night Schaller saw him nuzzle each cub, then wait hours for all to finish eating before taking his first bite.

The park's other predators also cast their spell. Late one night, waiting by a kill, Schaller heard stealthy steps and a hyena emerged, "a lovely, ghastly creature, part of the moon and the shadows, as beautiful a sight as I have ever seen. It ate and I could hear the ribs cracking." Finding a hundred screeching vultures churning up water with their wings as they fought for a place on a half-submerged carcass, he watched a pair of jackals rush in biting at the birds to clear a space; in a moment their little bodies had vanished, enveloped by vultures settling back in. He watched aerial attacks, as thrilling as those on the ground. After multiple strikes, a crested hawk-eagle killed a frantic peahen. A pond heron's evasive maneuvers—sharp veers and plunges like a desperate pilot in a dogfight—ended when a tawny eagle snatched it right out of the air.

These dramas were fast moving and multilayered, with casts of hundreds, often running for their lives. A single meadow could fill like the stage of a grand ballet: sambar neck-deep in silver water eating pond weeds, "their reflections bright"; fawns and piglets scampering, oblivious to the crocodile waiting to capitalize on their curiosity; monkeys sitting in rows as if in pews; shy, delicate muntjac slipping out of sight. How to watch without missing anything, but also preserve the details? George's solution was to scribble key words without looking at the page; then, as soon as he could, to write it all down more thoroughly. Nothing was too small for his notice. Opening a dead snake, he discovered a fully formed young one inside. For thirty minutes he watched a tiny scarlet mite spin and stroke the rump of another busy digging herself into the ground. When she vanished, George spent another ten minutes watching the stroker wander off.

•••

Most of the letters that survive from the Kanha years are from, or to, Bettina. That's in part thanks to her habit of saving everything. But Bettina also provided the most reliable connection to their faraway families, writing weekly letters that she numbered, reaching at least ninety-three. Kay wrote regularly in return, with a growing warmth and candor. "Mrs. Iwersen" became "Mother Iwersen." She still worked to charm ("you are the youngest and loveliest 'old lacer' and I'm pleased you aren't interested in the arsenic part") but also confided that a breast abscess had caused her to quit nursing ("more George's idea than mine; he wants to prevent future doctors' bills"). If she let some weeks go by without writing, it was only because, as one does, "knowing you better and sincerely loving you I seem to neglect you more."

George's letters to Bettina were rarer and still, sometimes, barbed. Bettina worried about the danger a restless China might pose to their safety. "It's probably more dangerous to live in the U.S.," he replied in December 1964, a year after the Birmingham church bombing and six months after the killing of three civil rights workers in Mississippi. "Here churches don't get bombed and one is not likely to be shot for saying that dark people are as good in every way as white people."

News from Kay's family in Alaska was more intermittent and often delayed. It was two weeks before she learned of the March 27, 1964, earthquake—the most powerful recorded in North American history—that opened a twenty-foot crevasse two blocks from her mother's home. The first news came in a cable from Bettina, which George tore open in alarm before seeing letters from Anchorage also in the pile. Kay's mother and sister Diane were unhurt, but neighbors' homes had split in two, the JCPenney was in flames, and no one had water, light, or heat. Still fearing a tidal wave, the family wore coats, mittens, and boots to bed (here Diane chimed in to say that their mother also put on their late father's flannel pj's, bed jacket, robe, sweater, scarf, and slippers). Her mother was an "unusually courageous and self-possessed person," Kay wrote Bettina, but was terrified by the powerful aftershocks. Thank goodness Alaskans were not prone to "panic or wailing" but instead set to work. When the grocer shoveled a path through his collapsed aisles so his neighbors could wade through to gather what they needed, Mrs. Morgan found hot dogs and hot chocolate, "food for the gods."

Had she not had her hands full, Kay would have rushed home to help. As it was, she and George were deep in preparations—stockpiling rice, firewood, chickens, and lambs—to sustain the family through India's own annual extreme event. Their local colleagues were surprised that the Schallers planned to spend the monsoon in Kanha, which would be completely cut off for months by flooded roads and swollen rivers. But both looked forward to the "peaceful isolation" and to the relief the rains would bring from 110-degree days and air so heavy they could scarcely breathe. With the forest bleached by the heat and the sand burning his feet through the soles of his shoes, George now began his day at 5 a.m. and was home by breakfast. Even his journal entries grew sluggish. "May 6, A tiger killed a cow but for once I did not sit by it all day. . . . I saw a skink carrying a cockroach."

As the night winds grew stronger, hammering the roof with twirling sal seeds heralding the approaching downpours, their resolve faltered. Kay, who never fared well in intense heat (she once collapsed at a Calcutta cocktail party) was losing weight and always tired. The boys, too, were showing signs of a vitamin or calcium deficiency. Mark's toenails were cracked and falling off, and Eric was becoming bandy-legged. While George hated to again separate the family and unsettle the boys, he and Kay decided that she should take them to Calcutta until September.

Though the monsoon was a bit less intense there—an average of fifty-four inches of rain versus seventy-one inches in Kanha—Kay was soon navigating flooded streets alongside "people carrying their shoes and walking with their saris and dhotis held high." She renewed friendships made the previous fall: with the "charming women" she taught English conversation at the Indo-American Society, and with other expat moms, who met to compare notes over tea. "I had my Dr. Spock and could reassure them." She dissolved in tears every three days, she wrote to her mom, at Eric's infected mosquito bites, and even more on learning that Cousin Ed had split with Arden. "I thank God for my own marriage despite my occasional resentment of George's complete absorption in his work. At least he is happy in it which creates happiness for us all."

Mark was still small enough to take the change in stride, and Eric adjusted quickly to the night braces and corrective shoes he wore to straighten his bowed legs. But he was terribly homesick for his dad. "Eric told me that he saw blackbuck, tigers and swamp deer this morning," Kay wrote George, "and talked about you all day. 'Daddy puts tigers in house in Kanha. Daddy come get Eric and Eric go back to Kanha.' I said you would take Mark and Mommy and Jamil too and he said 'Kanha too full of people.'"

George, meanwhile, was contending with molding books and clothes, dwindling food (their chickens had all mysteriously sickened and died), and walks in the forest that grew ever clumsier as vines sprang up, as in a fairy tale, to entangle him. None of that eclipsed the exhilaration of watching "all life burst forth as if waiting for this moment." Deer frisked, millipedes mated, termites "swarm[ed] from their subterranean chambers . . . a continuous stream on silvery wings." A small pool appeared and just as quickly filled with sulfur-colored bullfrogs; yellow and blue butterflies fluttered like confetti; blooming grass heads turned the meadow white as if covered with snow.

Missing Eric's third birthday, as he had the first two, George sent one of the cartoons he drew to mark every important occasion—in this one, tigers rode an elephant waving a "Happy Birthday Eric" flag—plus more tales from his storybook life.

> *Dear Eric, Last night I saw the tigress with the 4 big cubs. She'd killed another swamp deer; their bellies were big and round. I've seen three little snakes since the monsoon started, also a turtle and great big yellow frogs. Last night thousands of black ants waded into our living room, carrying their eggs. I hope you and Mark come back soon.*

Kay replied: "Do be careful walking; take anti-venom. The boys are watching vultures on a dead cow."

•••

More than three months passed before the rains let up enough for George to drive out of Kanha, reach a train, and retrieve the family. Navigating the still-saturated roads required its own kind of fearlessness: One ferry's approach was just a steep mud bank that George had to slide the Land Rover down, sunk to its axle, maintaining enough momentum and (he hoped) control to keep sliding onto the boat deck. A month with the family in Calcutta gave him time to deliver his collections of plants, reptiles, ticks, round and flatworms, suckling lice, and hippoboscid flies to the scientists who could identify them and the herbariums and museums that had requested them. He visited J. Juan Spillet, who had joined him briefly in Kanha before Johns Hopkins reassigned him to study lesser bandicoot rats, whose fleas carry plague. Spillett worked in huge sheds, which George found a fantastic sight. When the stored mountain of grain began to move, he realized it was carpeted with rats; when the wall rustled, that it was papered with roaches. He also visited journalist Desmond Doig, who had recently gone with Sir Edmund Hillary in search of the Yeti.

Even in October, the trip back to Kanha remained treacherous. First the ferry escaped from the men poling it and drifted far downriver; the Schallers waited hours for its return. Then the Land Rover had to climb over logs and through swollen creeks, including one that required hours of shoveling to slope its bank. Arriving in Kanha, they found the forest glistening with spider webs, some four feet across. When their battery died, they got the elephant to jump-start the car, pulling it fast enough for the engine to turn over.

With Eric now three and Mark turning two, George began taking them more regularly on his morning and evening rides. He could still lift both at once, a useful trick when he got stuck in the mud and had to carry them several miles home. And on the day when, walking home from a swim, they met a tiger strolling toward them in broad daylight.

Their first Christmas in Kanha, amid the chaos of their arrival, had been slapdash and rushed. This year, they decorated a branch from an Indian gooseberry tree (its leaves resemble pine needles) with candles, chestnuts George had gathered in the Himalayan foothills, and tin blackbuck, gaur, wild pig, and peacock he cut from discarded cans. They decorated the windowsills with red berries and lacy asparagus fern, and ate the treats they'd saved: canned ham and plum pudding flambéed with brandy.

Sometimes, leaving the boys with Jamil, George brought Kay into the field. Out one day to investigate a livestock kill—soon regretting his choice to ride an elephant bareback, jabbed by its protruding backbone and bristly hair—George found a calf bitten through the neck, gurgling blood but still kicking. He went to fetch Kay, who on a properly saddled elephant rode within thirty feet of the tigress, when she returned to claim her prize. Better still was a morning watching Cut-Ear and cubs devour a gaur. As lunchtime neared, Kay said she had to get home to the boys. "Well, go," George said.

> And I was petrified because I would have to circle around those tigers. I wasn't about to let him know how absolutely terror-stricken I was, so I stood up very straight and started to walk down a gully. And he called out in a whisper: "Duck down!" He was afraid I was going to disturb his tigers! He just had no conception that his wife was petrified she'd never get home to her babies!

Kay pretended this story meant that George "cared about his tigers more than me." But her savor in the telling gave it away: She knew this was his ultimate sign of respect—believing her as capable of this unafraid life among wild creatures as he.

•••

The Schallers were not alone with the animals. From their veranda, they could see the mud huts of the only other families who lived in the park, and had for thousands of years. Like the Gwich'in in Alaska, many of the Baiga were at first leery of George. "Our conversation went like this: 'How many people are in this village?' 'I don't know.' 'What wildlife is around here?' 'I never see any.'"

But the presence of a woman and small children eased their nervousness, and their curiosity soon won out. Kay sometimes found neighbors peering in the window "watching us like television," gathering in the doorway, even climbing on their roof or crowding inside. She hoped that "our novelty will soon wear off" but also appreciated the hospitality offered in turn. Whenever any of the five villages within the park had a festival, they invited the Schallers; since George's was the only car, twenty or more people often squeezed in with Jamil and the boys or rode on the hood, roof, fenders, or running boards. In November, members of the Ahir community celebrating Diwali came to the Schallers at 11:00 p.m. to do a stamping dance accompanied by flutes, tin cymbals, and bells. The village water bearer, Sukku, wearing the lattice apron made of cow-hair and cowrie shells whose shaking added to the music, "turned round and round while bent double. I gave him a box of candy. Sukku kissed my foot." When two men picked up Eric and Mark to dance with them, the boys grew shy.

Getting to know the villagers was crucial to George's understanding of this ecosystem. It was the very reason a medical institute had sent a zoologist to India: Animal and human ecology are inseparable, they were coming to understand; not two webs but one. Going house to house, George learned that the wild predators killed on average one in ten of a family's domestic herd. The villagers, in turn, competed for the tigers' wild prey, hunting sometimes even in the park's protected core. When George found a mile-long bamboo fence designed to funnel deer into snares, he cut it to pieces.

The far bigger impact on wildlife came from the growing numbers of livestock: George calculated that the five villages' 2,500 head of buffalo, cows, and goats weighed three times as much as all wild ungulates in the park combined.* The tigers did well in part because cows were easy to kill.

* The combined total in Kanha of wild and domestic grazers (13,000 pounds of biomass per square mile) was nowhere near the destructive overloading Schaller saw in places like Rajasthan's Keoladeo Ghana sanctuary (266,000 pounds per square mile). There, over-

But with barasingha numbers down from 3,023 in 1938 to just 94 in 1964 (82 by Schaller's count), and blackbucks from 168 to 16—in India's last best refuge—the ratio of domestic to wild seemed to George awry.

The locals sometimes came to him for help. On one of the nights when he was camped on his porch watching his lovely leopard, men came to say that another cat had broken into a hut and attacked a sleeping nine-year-old. Walking the five miles to the boy's village, George found him on the floor, his scalp lacerated and left eyeball on his cheek, swarming with flies. Though an old woman sat crying as George bandaged his head, the boy never uttered a sound. When that leopard broke into a second house and killed a man, George moved back inside and locked the doors.

Beyond the villagers, the Schallers knew few locals. In Kisli, the small settlement seven miles away where they got their mail, the postmaster found their visits so diverting he often prolonged transactions. One day, when it took more than an hour to pick up a registered letter, Kay realized he was having food prepared for them at his home and was stalling until it was ready. The man in charge of Kanha, Mr. Gahelot, also sought out their company. He frequently "just comes and stands for no apparent reason," wrote George, who got an early lesson in diplomacy when one day his driver refused a ride to Gahelot, whose official vehicle was a bicycle. Furious, Gahelot ordered all the villagers to stop working for George, until Kay smoothed his ruffled feathers.

Once in a rare while the Schallers socialized with the men still busy extracting the wealth of the subcontinent. An English engineer visited from a manganese mine forty miles away; three Indians working for Esso invited them over, pouring vodka until they became quite dizzy. George had nothing but loathing, however, for the "great white hunters" still about. A fat

grazing had turned fertile ground to desert, exposing soils to erosion and replacing palatable grasses with bunch grass and thorny shrubs neither domestic nor wild animals would eat. The livestock also transmitted illnesses like rinderpest to the wildlife.

Canadian brigadier, "jovial but not likable," shot a too-small tiger cub but stretched it to legal length. An Austrian named Fend posed in his Gurkha hat astride the leopard he had just killed. These types generally loathed George in return: While others might be impressed by their embellished tales of danger, he regularly got far closer to these animals without a gun.

Far more rewarding were the relationships George developed with the small coterie of dedicated amateur naturalists working to understand and protect India's fast-vanishing wildlife.

The few Brits in the group had lived much or all of their lives in India, arriving as colonials but staying on after independence. The self-taught E. P. Gee, a tea planter in Assam, had published the only field accounts of such threatened mammals as Kashmir stag. His landmark *Wild Life of India* (1964) had a foreword by Jawaharlal Nehru, with whose help he saved the greater one-horned rhinoceros in Kaziranga. Visiting Kanha, Gee watched George watch chital and left "amazed at how much information Schaller could extract from what appeared to be an uneventful hour."

Anne Wright, with whom George forged a tender and lasting friendship, had moved to India as a baby in 1930. Her family having "served the Empire" since 1803, she remembered watching her father, the deputy commissioner of Delhi, parade with Viceroy Lord Willingdon; at Mahatma Gandhi's funeral pyre, the family sat with Lord Mountbatten.

Like many daughters of the Raj, Anne grew up hunting, often near Kanha in what became the family's Kipling Camp. In those days of abundance, they sometimes waited hours for thousands of blackbuck to cross the road. As soon as she was big enough to clamber up his trunk, she was allowed to go out alone on her elephant and even sleep outside, though on her midnight kitchen raids she sometimes met a leopard. She tried to but never shot a tiger, and by the 1960s had been moved by the plight of the rhino to become a conservationist. Hearing of the newly arrived young scientist, she took her ten-year-old daughter Belinda to meet him in his

Calcutta hotel. Anne soon "absolutely adored George, worshipping his knowledge and patience." When Indira Gandhi appointed her to the Indian Board for Wildlife* and to the task force that created Project Tiger—the first national effort to save the Bengal tiger from extinction—she leaned on George as her "guru."

The Indian naturalists also came from elite circles. Most were "anglicized Brahmins," explained Ullas Karanth, a Schaller protégé who became India's leading tiger biologist, "who read English newspapers and went from hunter to gentleman-naturalist to conservationist." Learning of their work through the Bombay Natural History Society, George offered his help. Jivanayakam Cyril Daniel, director of the BNHS, wrote of how the "world-famous" ecologist walked with his team through 380 square miles of forest, teaching them the latest techniques for counting buffalo. Billy Arjan Singh, another royal who had renounced hunting and—following on Schaller's study—persuaded Prime Minister Gandhi to create Dudhwa Tiger Reserve, invited George to visit his farm and ride his elephants into the slough. George was watching the mahout steer their elephant by digging his toes into the soft spot behind its ears when all at once five hundred barasingha galloped onto a rise. With wetlands everywhere being converted to rice and sugarcane, he knew he was witnessing the last such sight anywhere in India.

Schaller published regularly in the BNHS journal and twice in *Cheetal: Journal of the Wildlife Preservation Society of India*, with local colleagues. In the US, his tiger findings appeared first as a cover story in *Life*, which paid him a much-needed $5,000. His matter-of-fact way in those pages of relating tooth-and-claw dramas was something new to the world. He begins on their veranda, enjoying with Kay the peaceful rustle of peafowl as the boys

---

* Tasked over her twenty-three years of service with monitoring the elephant market, which prohibited the trade of any calf under five feet, Wright had to go to sales with a measuring tape.

nap inside. A terrified chital doe and fawn burst onto the scene. In a few sentences, the fawn's life is over, the tiger's grip on the small throat "muffling its dying bleats," the doe returned to her calm foraging. He then swings our empathy to the predator, noting how many nights she has spent on fruitless hunts, her belly hollowing as she and her cubs make do with frogs, crabs, locusts, even berries. With at least twenty stalks to every kill, George often hears the moan of failure. Kipling called it the "dry, angry, snarly, singsong whine of a tiger who has caught nothing."

In an editor's note, photographer Stan Wayman described his eighteen-hour days with Schaller, sitting in complete silence in a flimsy ground-level grass blind. At last, one dawn, the tuft of grass Wayman had been staring at "took the shape of a tiger. I magnified his image in my 600-mm lens and realized why this beast commands respect . . . His direct gaze bores into the soul, leaving one weak with awe." Compounding his fear was the knowledge that a man-eating leopard was about. "Schaller doesn't 'think' there is any danger but until he convinces me he can talk leopard, and get the beast's personal assurance, I will continue to be nervous."

George had his own concerns. With both men tense after a week without sleep or sightings, tempers grew shorter still when Fend showed up, threatening to chase away the tigers if they didn't let him in the blind too. Things got worse when the tigers did show: The elephant George and Stan were riding was charged by a cub and took off, trumpeting madly. With thunderheads approaching, they had to take down their strobe lights, which meant chasing the tigers away from a fresh kill. In the storm-lit jungle, the trees pale green against a black wall of cloud, Wayman did get magnificent pictures. A tiger dissolves into a golden meadow, both animal and grass a vertical play of dark and warm light. Another stares at the camera with nose, cheeks, and chin dripping with vermillion blood.

A children's book on the tiger, his first for kids, extended Schaller's public reach. But it was his monograph, *The Deer and the Tiger*, dedicated to

Doc Emlen, that fundamentally changed field science in India and rewrote the tiger's fate. "Suddenly comes this book, a landmark," recalled Ranjitsinh, who became India's first director of wildlife preservation, helped establish two dozen national parks, and drafted India's Wild Life Act, which still guides conservation.

> Nothing like this had been done before. This field biologist was a new creature to India. He showed us how to study an animal; to look at habitat, food, prey; that the whole is interconnected. This was research that could be put to use; more than fifty years later, it is still valid and referred to. And it catalyzed a whole world of Indian wildlife ecology, stirring scientists but also citizens. It remains one of the most absorbing books I have ever read, a pathfinding piece of literature.

George began the monograph as he often would: with a picture of incalculable natural wealth, squandered. Encompassing almost every earthly biome—from tropical rainforest to 28,000-foot peaks, sand desert to mangrove swamp—India had until the arrival of the East India Company been home to a nearly unmatched diversity of wild creatures. In addition to all the species George studied, there had been urial sheep, ibex, markhor goats, takin, serow antelope, musk deer, gazelle, snow leopard, and Asiatic lion, in numbers that defied the imagination. Tigers alone had numbered in the tens of thousands, stalking hundreds of thousands, perhaps millions, of prey. Some species—cheetah, pygmy hog, hispid hare, pink-headed duck—were already extinct. Others had been reduced to a few hundred and confined to tiny, compromised islands of refuge. Every last Manipur brow-antlered deer lived in a single swamp.*

In tracing responsibility for this tragedy, Schaller let no one off the hook.

---

* In a 1968 piece in *Audubon* magazine, Schaller estimated that the tiger population had been reduced in the fifty years preceding his arrival from 40,000 to 4,000. Per Karanth, between 1875 and 1925, 100,000 tigers were killed.

For the Brits, killing animals had been a great perk of assignment to the subcontinent; George took sardonic note of a certain Colonel Nightingale who alone shot more than three hundred tigers in Hyderabad, including dozens of cubs. India's own royals also had abundant blood on their hands. Schaller reproduced a letter sent him by one precise maharaja tallying his tiger kills—"my total bag is 1150 (one thousand one hundred and fifty only)"—and recounted coming upon another hunting with the visiting president of Finland right from the prince's Rolls-Royce. The killing had only increased in the postindependence state, with its intent focus on industrialization and food production. By the time the Schallers arrived, three-fourths of India's forests had been cleared and the government was issuing guns to farmers to protect crops from "pests." The insecticides brought by the green revolution were being used to poison predators. When in 1970 actress Gina Lollobrigida made a sensation in her tiger skin coat, India awarded ten thousand export licenses to capitalize on the craze.

To staunch such losses—even to understand what was being lost, and why it mattered—required answering basic questions like: How many tigers were actually left, behind the official, highly inflated numbers? *The Deer and the Tiger* filled that vacuum. "For the first time on this subcontinent," E. P. Gee wrote in his review, "a dedicated scientist has remained for fourteen months in the finest remaining natural habitat in Asia. The book's wealth of factual data almost bewilders the reader."

From parasites to apex predators, Schaller traced the myriad interactions that held it all together. Deer waited below trees where langur monkeys fed to eat the fruit and half-chewed leaves they dropped. Mynah birds and jungle crows rode those deer to snack on mites and antler velvet. That it was crucial to maintain the vast diversity of these forests was underscored by his catalog of the plants the gaur eats—forty different grasses, forbs, and leaves, plus the bark of *Adina cordifolia*, the fruit of *Aegle marmelos*, and bamboo seeds.

Returning to the predator-prey interdependencies he had first seen in

Alaska, Schaller quantified the relationship as no one ever had. This is how much meat, he calculated, a tiger actually needs: 3.5 tons a year, the equivalent of seventy chital or thirty cows. That was the key to stabilizing and even restoring tiger numbers: The more animals tigers had to eat, the more they would reproduce. And where forests were empty of cats, this was not primarily because the tigers had been shot but because humans had taken their prey.

For all its density of data, the fact that *The Deer and the Tiger* was also a gripping read, often tapping the visceral present tense of his field notes (a tigress cuts the skin of a gaur calf "with her carnassial teeth," bolts down hunks of meat, then licks the rib cage to harvest the maggots swarming there) woke up a whole generation of tiger champions. Valmik Thapar, who would make a dozen tiger films for the BBC, was stunned by

> the complexity of what Schaller was doing when we had no laws protecting wildlife. Everything was up for slaughter: for meat, sport, trophies. And there was George, one of the most amazing characters the world has ever seen, an old-world scientist being very classical and particular, trying to understand tigers, probably the most challenging task ever undertaken. His book was my greatest source of inspiration. . . . For all of us who work in tiger conservation, it is the bible.

For Ullas Karanth, discovering George's work

> was like a person with poor eyesight putting on glasses. This is how you do tiger science. Not "how magnificent the tiger is sleeping under the banyan tree," but in this quantitative, analytical way. How far does it go? How often does it kill? With just binoculars, camera, and pencil he became a beacon for a whole generation, the face that launched a thousand biologists, a wildlife revolution in India.

A. J. T. Johnsingh, the first Indian to earn a PhD with a field study of a wild mammal (dhole), credited George with inspiring him to "make this my life." In a series of interviews historian Michael Lewis conducted across the country's premier natural history institutions, Schaller's name invariably came up as the pivotal early figure in Indian ecology.

That did stir jealousies. "His was a far more incisive book," said Ranjitsinh, "than those local chaps who'd been with tigers all their lives but knew less and felt upstaged." While many found in work with Americans a way out of their historic colonial scientific power relationships, others argued that the time for foreign experts was done. None was more adamant than Kailash Sankhala, who became the first head of Project Tiger. At the Tenth General Assembly of the IUCN, with George in attendance, he stood to dismiss *The Deer and the Tiger* as "absolute rubbish." Another rival accused Schaller of being CIA, a suspicion he understood, since "the Smithsonian had used the guise of natural history to spy on India." His own brusque habits didn't help. "I can imagine," said Karanth, "this young man, single-mindedly focused on his mission, not being obsequious enough to please self-important people. India has a habit of deference. George can be caustic. He is not by nature a stroker, and had not yet learned that it's necessary." Along with larger shifts in geopolitics, those rivalries and suspicions scuttled his hopes of working next in the Indian Himalaya.

Still, his impact far outlasted his tenure. It was Indian conservationists inspired by Schaller, and alarmed by a census E. P. Gee completed using his methods, who persuaded Indira Gandhi to ban tiger hunting and launch Project Tiger. Had she not, said Karanth, the tiger would have almost surely gone extinct in India.

In February 1965, Kay and the boys said goodbye to Kanha, posing in Madras plaids for a picture with Jamil, who wept to see them go. While George stayed on, Kay took Eric and Mark to Germany to meet their grandfather. Though the three spent several months there, they settled an hour away so as not to be "too much on his doorstep," and saw him just a few times.

George spent the spring working with local colleagues and entertaining job offers—to study lions in Tanzania, rhinos in Java, monkeys in Puerto Rico, or birds in the Canadian Arctic. He met again with Doig and Hope Cooke, the American-born princess of Sikkim, to discuss a renewed search for the Yeti, although he saw little chance given Chinese border troubles. In May, he joined the family in time for Eric's fourth birthday and a drive across Europe—to the Italian Riviera, where he found the "fat Germans on the beach getting red" ridiculous, and the Dordogne caves, where he loved the red-ochred bison, rhino, and mammoths. They finished with two months in Cornwall and London. The children "don't seem to mind this life," he wrote Bettina. "They've never known any other kind."

Returned to the US for a year at Johns Hopkins in Baltimore, he hung the pictures his mother had sent, of Dresden and Webster Groves, and settled in for an interlude of the ordinary. Kay's letters were now written at the hairdresser, head under a dryer, stationery propped on a limp magazine. Eric played one of the wise men in his nursery school Christmas pageant. The boys built their first snowmen, kept guinea pigs, caught American colds.

George had settled on lions, although he was worried about funding "with all that money burning in Vietnam." Before leaving for Tanzania, he and Kay bought a 135-acre Vermont farm with a rundown house to fix up on their return—a place, as George wrote Chris, "where the boys can run wild." If our plane crashes, he added, "or we all get killed some other way, I want you to have it." He suggested Chris keep the letter to substantiate a claim.

5.

A Peaceable Kingdom

Serengeti, 1966–69, Lions, Wild Dogs, Cheetah

Arriving in May 1966, the family's first days at Seronera in Tanzania's Serengeti National Park offered a taste of the joys to come. Out their window appeared a giraffe, his "creamy belly" so close they could have stroked it, his blue tongue plucking pods from acacia branches overhanging their roof. George, as the giraffe was also known, sometimes broke the clothesline, but George (the human) found "something spiritually satisfying in having a giraffe do so." Elephants rested in the shade of their yellow-barked fever trees; hyenas stole cushions from the veranda; vultures arrayed themselves on the roof like gargoyles, tearing at the heads George tossed up there until he could clean them. (Accommodating neighbors often left new heads on their porch.) Lions spent one night chewing up their neighbor's deck tennis net and the next killing and eating a zebra in the Schallers' own backyard.

The larger fairyland opened itself just as readily. Hans Kruuk, who had arrived two years earlier to study hyenas, took George out for his first evening ride, past a blasé leopard "flaked out" in the setting sun, onto plains black with migrating wildebeest. Seeing the vast herds Africans had kept alive stirred thoughts in George, as such sights often would, of the millions of bison that had once blanketed the American prairie, until they were exterminated by the US Army and traders in hides.

Out alone the next day, George watched the slow dance of two giraffes in ritual battle, each in turn swinging his neck down and then up to hit his rival's chest or throat. Spotting a sleeping wildebeest, he stopped to pet it. With small planes as common here as in Alaska, Tanzania National Parks director John Owen took Kay up to follow the herds. Hugh Lamprey, director of the new Serengeti Research Institute (SRI), flew George into the Ngorongoro Crater to visit Kruuk at his work.

Their wild neighbors were not all benign. When soon after their arrival a deadly puff adder took up residence in the boys' play area, Kay asked the teenager next door to shoot it. A few months later, she was stepping into a bath when she heard George—giving the boys dinner on the veranda—tell them to go inside and "don't ask questions now." He'd spotted a black mamba, the deadliest of all, Kay told Bettina, since they could not get anti-venom. Still, George never wasted an opportunity: When Kay found a six-foot cobra curled at her feet, he borrowed her sunglasses so he could take a picture without the snake spitting venom into his eyes.

Blood-sucking insects posed the other great danger. George and Kay both weathered bouts of malaria. Several of their neighbors got sleeping sickness from tsetse flies, which wouldn't die if slapped "but had to be popped with a satisfying snap." Those neighbors, airlifted to Arusha, faced a year in the hospital. Others just brought misery. Kay had to iron everything because warble flies laid their eggs in pant legs and sleeves; once next to a warm, moist body, the eggs hatched and the grubs burrowed into the skin. Digging them out only killed the larvae, causing serious infection. "You have to wait until they make an air hole," George explained, "then you squeeze the big, corrugated maggot out."

Despite the occasional snake in this garden, settling in was far easier than it had been in Congo or India. Their house, painted a cheerful yellow, had a gas-powered refrigerator, stove, and water heater, and electricity until 11 p.m. Fresh produce was plentiful and delicious meat was cheap (filet mignon for fifty cents a pound) or free (scavenged zebra). Living for the first—and last—time among expats, the Schallers' solitude gave way here to a social whirl. On day one, cocktails at the Kruuks'; on day two, at the home of Chief Park Warden Sandy Field. And on and on, in a round that George occasionally complained to his journal had grown tedious. He was more pleased by the many family visitors. It had been six years since he had seen Cousin Ed, who arrived with his brother, Bill; Bill's wife, Ani; and plans to climb Kilimanjaro. First, they all went camping, the boys cracking up at Ed's relentless singing of his new two-line ditty: "In Ngorongoro Crater, the old Land Rover broke down." (It had.)

More serious was George's own rare breakdown. Surprising the group one afternoon by announcing that he would stay behind in the rest house, he finally agreed to see a doctor, who palpated his liver and spleen and diagnosed dengue fever. He convalesced at a sanctuary for orphaned and wounded wildlife but after three days insisted on giving the mountain a try. He was the only one in the group to make the 19,300-foot summit, hiking another twenty miles the next day with Ed. His fever later returned, spiking to 104 and leaving a sharp pain in his back that would plague him all his life.

The visits continued, of family and colleagues. Kay's mother came, as did Chris and his new bride, Diane. George invited his half-sister, Renate, but "father and Annalise believe in the Dark Continent and its many dangers," he wrote Bettina, "so she can't come." Researchers Phyllis Jay (Indian langurs), Iain Douglas-Hamilton (elephants), and Roger Tory Peterson (of the Peterson Field Guides) passed through, as did Robert Hinde, the ethologist who had taken Goodall on at Cambridge to do her PhD. At times, their yard looked like a trailer camp. "I am running a short-order hash house," Kay wrote Chris. "I never know how many will show up for a meal."

All added up for Kay to "a perfect time in our lives." After years apart from female friends, she now had other wives to talk with about such "woman things" as the boys' homeschooling and her zinnias, which she fertilized with elephant and zebra dung. She had vanquished any residual squeamishness, as she wrote Chris.

> *George reminds me how finicky I was when we met and how much I've improved since(!) Our veranda is so full of horns and jaws that our whole house reeks of putrid meat, though George (and the lions) do remove as much as possible. The other night I was holding a flashlight while George cut off the jaw of a wildebeest. Maggots were crawling out of all the nasal passages but when he slurped the brains out, I couldn't help but yelp. I am so afraid Eric and Mark will be the same; they watch him collect maggots in cans with absolute calm as if all people do those things.*

Putrid flesh aside, pleasure permeated her long letters to her mother.

> *I wonder if I could be more content than I feel this evening, in spite of the smoking stove and the fridge not working. We have white-necked ravens such as I loved at Kabara. The sun is casting a streak of gold across running wildebeests. We went out to see the lions eating a giraffe, with their two six-week-old cubs. I am so glad not to be living an ordinary life.*

No longer afraid of any animal, Kay rarely worried when George stayed out for nights on end. She herself often sat alone on the veranda with a midnight snack, watching lions in the soft blue moonlight. When George was home, she and the other wives went to watch wild dogs ("we could always count on them making a kill") or to find her best-loved cheetahs, delighting when they leapt atop her car to scan the horizon. She realized that she was now rooting for the predators when, on seeing the fifteen-month-old cubs she'd known since birth growing gaunt—their mother having abruptly left them forever, as cheetahs do—she scooped up a dead gazelle fawn to carry to them.

Her heart did still also go out to prey, especially baby wildebeests, who in the chaos of calving season can get separated from their mother and not find her again. Some of the lost ones slowly starve; George found their mummified skin and skeletons littering the plains. Others were "rescued," as he put it, from that fate. He watched a hyena snatch a tiny wet calf taking its first stumbling steps and a confused orphan trail two lions before spotting another orphan, toward whom he rushed in joy. "It ran on in its escape from loneliness oblivious even to the lioness that rose as if out of the ground and pulled it down. As she held its throat until the new life faded from its eyes, hopefully the last thing it remembered was the other calf hurrying toward it." The first time he took Kay to see them, a bleating "orphan galloped up to our car as if this steel monster were its mother. Kay all but burst into tears," and, after seeing several more such forlorn babies, "refused to ever visit the calving herds again."

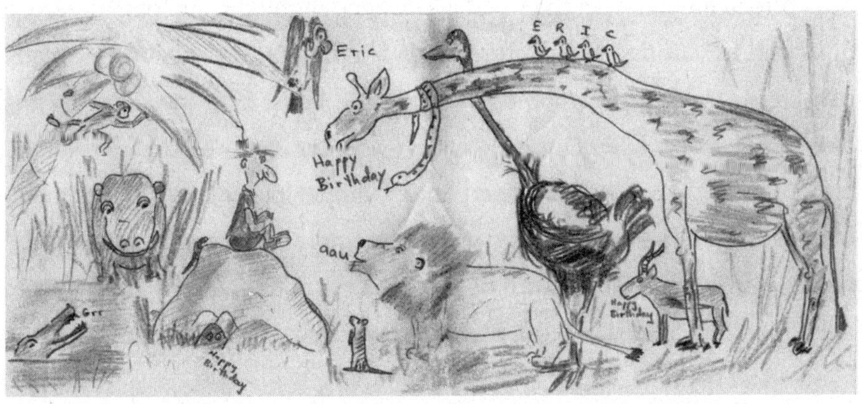

George's marvelous menageries celebrated every birthday, even if he was a world away.

The boys, now old enough to enjoy their wild childhood, were yet another source of joy. While Kay taught first grade to "serious" Eric, the "devilish" Mark, "with my quick temper and George's stubbornness," hung outside the window learning to read by listening in. Identifying animal crackers was a cinch. From their dad, they learned how to blow out an ostrich egg for a giant omelet and to sort the rows of skulls in the yard, each marked with the name of the carnivore that had killed it. Their budding field skills pleased George. Despite its perfect camouflage, Eric spotted a praying mantis. Mark identified the age, species, and gender of a tommy (Thomson's gazelle) chased through their front yard by a cheetah cub.

The grandest adventures were in the dark. Hearing water running one morning well before dawn, George found three lion cubs playing with the tap and hurried everyone into the car to follow them as they ambled off. When the engine died, he had to chase the lions away so the family could walk home. Headed to a friend's for dinner on a night so rainy he feared the rising river, George left the car and carried the boys one at a time across a rope bridge, swaying across reflected stars. Another night, out for a drive to show the boys shining eyes, they found a giraffe kill and listened for hours to a "seething mass" of hyenas "tear, slurp, crunch and lap." The boys often camped with their dad and were delighted when one night two lions padded around their tent roaring—although, if a new comic book had come, they easily ignored even fighting lions. Eric loved telling the *plink, plink* story: The night maggots crawled out of a skull on the roof and into a crack, then slowly dropped onto a dinner party below. He remembered a scrap of red felt snagged on a bush where Santa had passed, as well as the animals his dad carved for the crèche—lions standing over a zebra they'd just taken down. In his letter to Santa, he asked for a magnifying glass—and courage.

Of their many pets, George's favorite was a warthog.* Brought to them by a neighbor who had found her in a den with two dead siblings, the frightened three-week-old, tiny with bristling yellow hair, hid under the

---

* As Peter Matthiessen will later write, George's "shy warmth is most apparent when he speaks of crows and pigs" (*The Snow Leopard*, 42).

sofa until George's gentle grunting lured her out. She soon settled into his lap, suckling milk from a bottle "with much smacking." When she followed him outside and got frightened by a noise, George again had to play mama warthog, trotting alongside the galloping piglet until she came near enough for him to scoop her up. "It likes to be held tight in my arms," he wrote, still using the impersonal pronoun, though for a week his journal records little else. When she shivered, he took her into their bed. "Never had I held a body that was so solidly built and generated so much heat."

Giri, as they called her, shortening the Swahili word for her species, tried Kay's patience. In bed, she rooted all night with her hard little snout, demanding to be fed every few hours. They moved her into a box with a hot-water bottle, but if her blanket slipped, she fussed until George tucked her back in. When she began roaming around, "hooves clacking like castanets," they tried banishing her to the living room. But she butted the door with "such piercing wails and humanlike screams that Kay was afraid the neighbors might accuse us of having a marital squabble of epic proportions."

Daytime was little better. They tried walking Giri on a leash, but she tugged, lay down, and refused to budge, or spun wildly. If they left her in the car while they walked or picnicked, she hurled herself at the window. Or attracted attention: On a drive with Kay in Kenya's Masai Mara, George stopped to let Giri graze, bringing a passing jeep full of tourists to a halt. People popped from the hatches to photograph man and pig "as if we were wildlife."

An unleashed warthog was no easier. Out one hot day with Kay in the garden, the piglet startled and took off, running farther and farther as Kay chased her for more than an hour. ("I was frantic," Kay wrote to her mother, though George's notes focused on the pig's terror.) Twice, George came home to find the house in turmoil and Kay in tears. Giri had gone missing and despite hours of looking could not be found. Fortunately, his "daughter," as Kay called her, responded passionately to George's voice and step.

The first time, his arrival brought her out from behind the bookcase; the second time, she burst from a thicket where she had fallen asleep. She made a giant mess eating her favorite foods: pineapple beer, cooked rhubarb, and lettuce dressed with a vinaigrette. She also loved rice, if it was Uncle Ben's, refusing the cheaper local brand. She let Mark and Eric hold her but hid from strangers or, as she grew bolder, nipped them. George teased Kay about what lay ahead. "Our pig has grown," she wrote to Bettina, "and George just told me they can reach 180 pounds."

George and Giri, meanwhile, could not get enough of each other. If he lay on the floor, she flopped across him for belly rubs that sent her into an "ecstatic trance." If he sat, she covered him from ankle to thigh with a saliva foam or pressed in behind to be his pillow. Wherever he went, she pattered after, talking. He in turn found her tipped-up snout and tusks "enchanting." When Kay climbed out of the bath, he often plopped Giri in, then rubbed lotion on her dry skin until it was black and shiny. "I had to do my own," Kay recalled. George was charmed by the pig's play: tossing her head as if to impale him, or spinning about in midair, the move a warthog makes when chased to its hole so it can back in with tusks out. When he left, she cried. When he returned, we "greeted each other with such joy and prolonged affection that Kay complained a little peevishly that it sometimes took me half an hour to notice the rest of the family." One afternoon, George brought home a dozen tommies he'd found killed by hyenas, laying them in the yard to retrieve heads and fetuses (and hindquarters for dinner). Giri found one lactating and for thirty minutes suckled the dead gazelle.

The piglet had been with them just three months when George noticed something amiss. Feverish and dry nosed, she stumbled on legs gone stiff and uncoordinated. They gave her penicillin, but "she whimpered and bit the edge of the box," George wrote in his journal, "as if suffering." Two days after Christmas, returning home after dark, "I saw her box in the garage and knew she was dead." She'd grown so weak that Kay had asked Lamprey to shoot her. George then chronicled her death a second time, turning the story to data, his wall against pain. "12-26 Does not eat or drink. Lies all day on side. Diarrhea in evening. 12-27 Too weak to stand, breathing slow and labored. Can't raise head, refuses food. She was then killed. Weight with empty alimentary tract: 22.5 lbs." Mark's first grade essay was just as blunt. "She couldn't stand up so we had to shoot her dead."

Two months later, yearning for another pet, Kay suggested a banded mongoose, which, full grown, weigh just five pounds. They'd watched a troop move into the termite hill in their yard—evacuating for a time when, as Mark unhappily discovered, fleas moved in, too—and been plotting how they might capture a young one. One afternoon, when a marabou stork began probing the hill with its long beak, George saw his chance. As adults streamed out in panic, babies in mouths, George rushed after them, startling several into dropping their charges. Grabbing a pup, he held on even when it sank its needle teeth into his finger. Tucked into a box, the little creature made his high-pitched plaintive "I'm lost" call, lunging to hit George with his tiny nose at any approach. But within a day, he had settled in: taking milk, letting the boys hold him, and following at Kay's heels, cheeping. He liked oatmeal, honey, hamburger, and rubbing smooth, hard objects on his belly. Sometimes he fell asleep tucked along George's neck, purring.

Named Fungus for the way he glued himself to Kay, the baby rode around in her apron pocket like a joey, suckled a doll's bottle, stole cheese and chocolate from the boys until they learned to eat surreptitiously, and at 7:00 p.m. put himself to bed, emitting a musical *tu-tu-tu*. "By that time," wrote George, "we felt as we did after putting the boys to bed: relief."

Fungus soon began roaming outdoors, revealing a Rikki-Tikki-Tavi–sized courage. George saw him charge and flush fifty vultures; he stopped him from attempting the same with a stork, fearing he'd be killed. Whenever George the giraffe came by, the little mongoose reared up on his hind legs, to greet or perhaps confront him. One day he came in with the end of his tail almost severed. Mark was sorely offended when his dad accused him of experimenting, again, with scissors. George then caught Goosey, as he was sometimes called, confronting a raven with humped back and hair on end and deduced the more-likely perpetrator. When Eric's tortoise went missing, Goosey himself wound up falsely accused. The family had seen him hike an egg through his hind legs like a football to crack it. For a school lesson, Eric wrote of his worry that he'd done the same with Torty: "thrown him against a stone to break his shell." He hadn't; they soon found the tortoise hiding behind Mark's dump truck.

Twice, they nearly lost Goosey. One day, seeing his troop outside, Kay left him on the veranda to see how they would all react. When Fungus called to her, distressed at being alone, the troop responded. Several started up the steps, sending Kay rushing out to scoop her baby back up. One night, he took himself to the termite hill, his birthplace, to sleep. When he wouldn't come to his usual call, George sealed all but one hole to bring him out.

At six months, however, it was time for Fungus to go. The children didn't mind. Eric had a new tadpole to keep him content. And Goosey had become obnoxious, especially to unfamiliar women and children, whose toes he bit hard or rubbed with his musky anal secretions. When the family left him while traveling with photographer Alan Root, things had not gone well. Goosey devoured the white mice Root fed his pet caracal. But when the wild cat wanted to play chase, the mongoose nipped its feet and got swatted in the head. Finally given to a research assistant, he didn't last long. Left outside one night, he was eaten by a raptor. "We'll have no more pets for a while," George wrote. "Eric only wants a rabbit anyway, which one can't get here."

East Africa was in the late sixties a glamorous destination, made famous by writers and filmmakers including Isak Dinesen, Beryl Markham, Bernhard Grzimek—winner of the 1960 documentary Oscar for *Serengeti Shall Not Die*—and George and Joy Adamson, whose *Born Free* had become a global sensation. (Joy's 1960 bestselling book and the 1966 blockbuster film told the story of the lioness cub Elsa they raised and returned to the wild.) George and Kay had barely unpacked when Robert and Ethel Kennedy passed through, making a "very bad impression." Although Mrs. Kennedy chatted with the wives at an evening soiree, the senator was interested only in the several dozen journalists in his entourage. Another evening brought drinks with Russell Train, who would soon join the Nixon administration, helping to create and lead the Environmental Protection Agency.

Less congenial was a dinner with German zoologist Ernst Schäfer, whom George described as "a buddy of high-ranking Nazis who'd gone to Tibet and made the Nazi salute." The first part of that is unequivocal: In the 1930s Schäfer traveled under Himmler's patronage to Tibet—the "cradle of the Aryan race." The second part is stranger, involving a Yeti Schäfer ostensibly shot and stuffed with its arm raised as if heiling Hitler. Schäfer had read George's books and sought him out, but George was "thoroughly unimpressed . . . he is running to fat and very dogmatic." Robert Ardrey, his own fame just boosted with publication of *The Territorial Imperative*, joined the party and "as usual dominated the conversation."

The great explosion in those years of wildlife television lent at times a circus atmosphere. That first summer, ABC descended to do a four-hour documentary on "Africa" helmed by Eliot Elisofon, a brilliant photographer but a prima donna, in George's view. With Elisofon was British wildlife presenter Grahame Dangerfield; producer Helen Marcus, who had booked George on *To Tell the Truth*; and Ardrey, hired as "script man" and to conduct interviews, including with George on animal society. All were at the Schallers' one evening for eland roast when at 10 p.m. Louis Leakey's twenty-one-year-old son, Richard, roared up, driving over Kay's flowers, to announce that the crew wanted a dead buffalo Dangerfield had stumbled upon earlier that day. The party rushed to the scene, winched the thousand-pound animal up into a tree, and then dropped it into George's Land Rover. It was 2 a.m. when George finally got to bed, though "for obscure reasons" Leakey then chased a passing lion, got lost, and had to be found.

George thought it nonsensical to haul in the whole buffalo, especially since giving all that meat to the lions would leave them too gorged to hunt. But the young men "had to prove their strength, manhood, or what have you," and "my advice was not asked. I'll be glad when they're gone."

The circus continued with the arrival of the Baron Hugo van Lawick and his new wife, Jane Goodall, who was fast becoming a celebrity thanks to *National Geographic* magazine covers and films. The Schallers had them to dinner several times, sometimes joined by David Attenborough, who was in the park for the BBC. Attenborough never forgot a long interview with Schaller, the two stretched out on a kopje. When they wrapped, George "went into the bush not more than ten yards away and came right back out saying I should walk away slowly. 'There's a pride of lions there. They've been watching us for half an hour.'" Grzimek, whose son and co-director Michael had been killed when his small plane hit a vulture, came to dinner with his widowed daughter-in-law, whom he later married. Though he was clearly not entirely averse to the attention, in George's journals, again: "This social life is getting excessive."

Alan Root, who had been cinematographer on the Grzimek film and spent three months filming George's gorillas, was the one among the bunch the family grew close to. His antics endeared him to the boys; Eric remembered him mouthing gas to blow dragon fire and showing up one day in hippy hair and love beads insistent on meeting the famous George Schaller. The man who could find lions in the dark failed to see through the disguise. Root had been bitten so often by wild animals that writer George Plimpton profiled him in *The New Yorker* as "The Man Who Was Eaten Alive." Having lost part of his rump to a leopard when he foolishly jumped down to investigate a freshly killed jackal—game warden Myles Turner chided him for violating the rule against feeding park animals—Root landed in the hospital again after being struck on the knuckle by a puff adder at Joy Adamson's Meru camp. "He will lose his finger or perhaps his arm," Kay wrote to Bettina, "for the poison ruptures all the blood vessels, and unless they rebuild themselves, gangrene sets in. His flesh is already black and swollen more than twice its normal size." Root gave his amputated finger to Louis Leakey, calling it a first installment, the other bitten-off bits to follow in jars.

Though it would be hard to call George's life in the Serengeti conventional, his work here was as mainstream as it would ever be. For the first time, he relied on a vehicle and tracking technologies, had multiple big magazine spreads before the work was anywhere near complete, and even aligned in ways with the sixties zeitgeist, his observations enlisted in various debates on man's, and woman's, nature.

While in Congo and India he had been the sole outsider, here he worked almost entirely among whites born in colonial Africa or expats doing the kind of immersive field study he had pioneered. He did find kindred spirits. Fritz Walther, who had also lived in Dresden as a child, was "curt and loath to be distracted" until the subject turned to gazelle. Kruuk kept a pet hyena, camped with their wild clans, and was fighting a "moral battle" to rehabilitate them in the public mind after slander by such as Ernest Hemingway, who described them "obscenely loping" and the "mirth provoking" sight of them shot and "tumbl[ing] end over end" to wind up legs in the air.

When graduate student Tony Sinclair arrived to study buffalo, George took him under his wing, though not without razzing him about the ridiculous figure he'd cut stepping out of the plane in coat and tie, holding an umbrella. "George took me home to meet Kay, sat me down, and told me in his inimitable way how I should go about things. He was the person I strove to emulate. And lions eat buffalo, so that connected us. We would meet up in the bush, sit at the fire and tell stories, and next morning go our own way." Parks director John Owen later thanked George for the critical role he played, though still in his mid-thirties. "You were a sheet anchor. . . . Just what we needed when so many of the scientists were young and immature. And in that closed little society, you and Kay kept things on an even keel by your example. When you left, you left a hole we were unable to fill."

Though George always preferred to walk, to feel and smell the animals' world, the Serengeti demanded a car. At 5,700 square miles, it is sixteen times bigger than the core area of Kanha.* Its animals spooked to a man on foot but ignored a vehicle. And while a solitary tiger couldn't surround him, a pride of lions might.

Cars cost him the intimacy he craved. They also set fires, broke down, or hurtled into termite mounds, especially when driven forty miles an hour alongside galloping zebra—"it's exciting to go with hands not on the wheel but snapping pictures"—or in the dark after lions, without headlights that might disturb prey. Thorns pierced tires, grass seeds clogged radiators, mud grabbed like quicksand. One night, when George didn't return from the swampy upper reaches of the Mbalageti River, Kay called Lamprey, who fetched Turner and went up with him in the Cessna 180. Like Pyramus searching for Thisbe, they found George's empty car axle-deep in mud and, twenty feet away, three fat lions. Heartsick, they flew back to break the news to Kay. George opened the door. He'd walked eight miles in a hot wind to a road, then waited hours for someone to pass.

George's car itself sometimes tangled with the animals. During his first month, out in a pickup truck, he felt its shocks give and looked back to find a lion had jumped in. Another day, when cubs stalked his Land Rover, he found himself wishing he'd not removed its windows; starting the engine, he blared a tape of hyena calls, which puzzled the lions as well as twenty hyenas, who "appeared as if out of the earth." He had, fortunately, windows to slam shut the afternoon a young male lion crouched, snarled, and—though Schaller thought he was bluffing—charged. A picture from that day shows Mark admiring the two big holes pierced through the fender. Ten cubs once crawled into the cool shade beneath George's car. "I could hear their heads bumping the floor." Stopping one moonless night on see-

* Even the 1,500 square miles George focused on, sixty by twenty-five miles, was far too big to study on foot.

ing the weaving yellow eyes of wildebeest, he switched off his headlights, and—crash—a female ran headfirst into the hood, lay down, lowed, and died.

Schaller's turn to tracking technologies was in part impelled by the expanded scale: He had to monitor too many lions to recognize each one. He was also the ultimate beta tester as the tools improved. The anesthetizing darts he had tried on deer in India had been refined for bigger mammals by Toni Harthoorn and Sue Hart, who often hosted the Schallers at their home in Nairobi. Kay was fascinated by these "vegetarian, homeopathic, yoga-practicing vets." Sue reciprocated her regard. "I always thought," she later wrote, "that but for his stabilizing, gentle and very beautiful wife, George—tense and highly strung—would many times have gone up in smoke."

Armed with a Cap-Chur gun that fired Harthoorn's flying syringe into an animal's flank, Schaller began practicing tranquilizing lions with a low dose. Though he knocked out four on day one—long enough to clamp a numbered tag in an ear and sometimes to collect blood and ectoparasites—he worked in a nervous sweat, alert to both the drugged cat and the nearby pride. By the end of the first week, he'd "lost at least 10 pounds and by evening [was] beat, though do little but sit in the car." Ultimately injecting and handling 166 lions, sometimes more than once, he learned valuable lessons, including that the drug worked less well if the animal had been running or just eaten a big meal; that a lion, or his friend, sometimes pulled out or punctured the shaft; and that when the dart hit, its target occasionally lunged at whichever lion was closest, as if they'd caused the pain.

Though Schaller could rarely resist the "atavistic pleasure" of sliding his palm over a sleeping lion's slightly oily hide, he never relaxed. Tagging cubs, he wrote Bettina, "it's always interesting to wonder if the mother will suddenly return." An underdosed lion might wake too soon. Bringing the family along one evening, he saw Kay startle when a big male suddenly woke with a wet, guttural growl. (The boys were too busy squabbling to notice.) Sometimes George could buy time with a stirring lion by tossing a tarp over its legs for it to claw at while he finished, but the drugged lions were also vulnerable. Once when he covered an estrous lioness, her male "consort" dragged the tarp aside, nudged her, then lay alongside. Twice, other lions attacked the tranquilized animal and had to be driven off. George didn't weigh the cats or make casts of their teeth, unable to justify the additional disturbance. To avoid distraction, he began working alone, banishing even Kay after she emitted a small involuntary *oh* when an angry lioness reached through the open car door. "Gorillas, tigers, lions," she wrote her mother. "How I wish sometimes he would do a mild little dickey-bird."

Far worse, for George, than the peril to him of dosing too little was the danger to the animal of dosing too much. On three occasions, he had to resuscitate a lion by pumping a foreleg to compress its chest. His sigh of relief when one young lioness resumed breathing was drowned out by a growl; he looked up to find her mother charging, and just barely managed to dive into his car. A second lioness never did recover. "It breathed well for nine minutes," George wrote in his journal, reverting to impersonal language, "but then suddenly ceased. I gave artificial respiration for ten minutes, but the heart failed to respond. The pupils remained large; the eyes glassy." With the help of two "witnesses" he loaded the cat into the car. "Found 3 well-formed fetuses. I certainly did not give her an overdose. Perhaps her pregnancy had some effect. A horrible shock." Two days later, another young male's breathing almost stopped, and the day after that, "things really went to pieces. I shot high, low, the gun or syringe misfired, the dosage was too low. Nine shots and no animal down. We ended the long day without a blood sample. It's as though I have a psychological block, have lost confidence after that death." The next day, again: "I aimed at one male's flank and hit his jaw! I am most dejected and ready to give up." When he finally got one down and drew blood, "the rat we injected with it died so all the effort was wasted."* At last, two days later: "The hex is broken. Shot a male who climbed into a tree. Couldn't tag him but stood below and lanced his dangling tail for blood." The one death, he soothed himself, was a fatality rate of 0.6 percent.

George's experiments with another emerging technology were mostly just frustrating. Howard Baldwin from Tucson's Sensory Systems Laboratory supplied him with six radio collars to test their practicability on big cats. Since the primitive devices lasted just a few days and could not transmit beyond a mile, George had to continuously track the collared lions anyway; he followed three for the life of a single collar and a fourth for twenty-one days, replacing the collar twice. Baldwin also brought thermometers, to implant beneath the lion's hide. During one such surgery, the cat woke and swept the air with his enormous paw. Schaller ducked, grabbed the mane, and held the lion's head while Baldwin madly sutured the incision shut. But neither technology delivered any real insights.

George would come to detest all these intrusions into an animal's life, believing they inflict both physical and emotional harm, and ruing the growing reliance on remote sensing among wildlife scientists. But he was already, in some quarters, unpopular for "defacing the lions." Myles Turner was dead set against it; "for a while [his] lips tightened into a hard line when he met me." That may have reflected a deeper resentment, as the expert status long granted to hunters and game wardens was now being challenged by biologists. As Turner told Peter Matthiessen, "Scientists are in charge of the animals these days." And the ear tags and collars ruined the aura of danger for hunters, George realized, taking ironic note of the "bearded fellow with a waxed moustache" on whose car was emblazoned Professional Hunter, who on seeing him tagging a lion shouted that it was "the most disgusting thing" he'd ever seen.

---

* Lions contract various blood diseases, Schaller explained. Rather than repeatedly sample the lions to track their development, local vets hoped to study the parasites in white rats instead. It did not work out.

Schaller's interventions occasionally went further. Twice he tranquilized a courting male to see if a second male loitering about would step in. He did. On several occasions he used the drug (succinylcholine chloride) to euthanize emaciated lions. Once he drove a zebra so sick that he could pat it on the head toward a sleeping lioness, who he then woke up. He even meddled with two gray-headed sparrows who had stolen a nest that mosque swallows had spent a month building on the Schaller porch. When George blew smoke into the hole, one sparrow died, the other left, and the swallows came home. He did balk when hunters and Turner's wife campaigned to feed or shoot starving cubs. (Tourists were bursting into tears at sight of the listless creatures with jutting hip bones and sunken eyes.) The population, he argued, should regulate itself. And managing the prides as if in a giant zoo "would ruin my statistics." He promised to remove any he felt sure would die, but "with great luck, they died on their own."

Continuing to set the standard for field studies, George outdid even himself: Covering 93,000 miles he completed 2,900 hours of observation, stayed out 104 nights, and learned to recognize 60 lions. He was as painstaking as ever, as in these notes on a nomadic male: "1020 lay on side; 1235 drank at puddle then lay back down on belly; 15 minutes on side, 15 on belly, 15 on side, 30 on belly, 10 on side, 10 on belly, 220 on side; 1845 raised head and licked forepaw." Another male marked with "20 squirts of fluid from the caudally-directed penis at an angle of 30 degrees in a powerful jet that may propel 4 meters." Table 8 of the seventy-nine he would eventually publish in his monograph records "Frequency of head-rubbing by members of the Masai pride . . . based on a total of 780 interactions and 690 lion-hours (Note: 1 lion observed for 1 hour=1 lion-hour)." Female L, the most generous, initiated head rubbing 1.6 times per lion-hour, and got her head rubbed the most (2.6 times) in return. His drawings added both information and charm. With just four quick strokes, he sat a lioness on her haunches, her head vanished into a warthog hole.

This pileup was a bid for credibility, all the more important now that

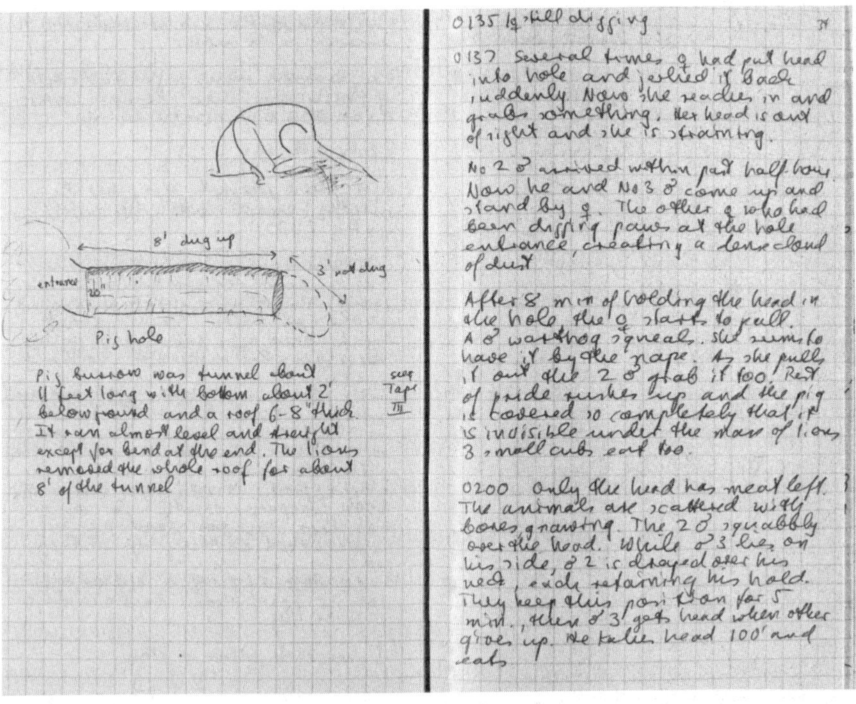

8' dug up
entrance
20"
3' not dug
Pig hole

Pig burrow was tunnel about 11 feet long with bottom about 2' below ground and a roof 6-8" thick. It ran almost level and straight except for bend at the end. The lions removed the whole roof for about 8' of the tunnel

see Tape III

0135 ♀ still digging

0137 Several times ♀ had put head into hole and jerked it back suddenly. Now she reaches in and grabs something. Her head is out of sight and she is straining.

No 2 ♂ arrived within past half-hour. Now he and No 3 ♂ come up and stand by ♀. The other ♀ who had been digging paws at the hole entrance, creating a dense cloud of dust

After 8 min of holding the head in the hole, the ♀ starts to pull. A ♂ warthog squeals. She seems to have it by the nape. As she pulls it out the 2 ♂ grab it too. Rest of pride rushes up and the pig is covered so completely that it is invisible under the mass of lions. 3 small cubs eat too.

0200 Only the head has meat left. The animals are scattered with bones gnawing. The 2 ♂ squabble over the head. While ♂ 3 lies on his side, ♂ 2 is draped over his neck, each retaining his hold. They keep this position for 5 min., then ♂ 3 gets head when other gives up. He takes head 100' and eats

A mortal game of hide-and-seek.

Schaller had taken his first real job, beginning a fifty-year affiliation with the New York Zoological Society. "In every project," he wrote in a history of the art of field notes, "there is pressure to amass quantitative data, thereby justifying financial support and enhancing one's scientific credentials." But the heap was also revelatory. Praising Schaller's work in *Science* for a "new level of resolution"—tracing individuals from birth to death and recording "in clinical detail" their idiosyncrasies and alliances—E. O. Wilson noted how often "behaviors become apparent only when the observation time devoted to a species passes the thousand-hour mark." Nobelist Nikolaas "Niko" Tinbergen, asserting a measure of parity between such repeated observations and the lab, saw in them a kind of "natural experiment." Watching all those head rubbings, George began to glean their function: to butter up another lion—in order to then, say, pilfer his wildebeest tail—or to make up after a fight.

George also made critical headway in illuminating the predator-prey relations John Owen had brought him here to study, although figuring out whom the cats were killing and eating required some sleuthing. The feces he had relied on to parse grizzly and gorilla diets was hard to come by for lions: Dung beetles liked to bury it, and vultures and jackals to eat it. Instead he had to catch the pride in the act, knowing he would miss "snacks" like gazelle fawns that were gone in minutes. Since lions often scavenged from other predators, he also had to figure out who had actually made the kill. Tooth marks on the throat pointed to the lions themselves; lacerations on the hind legs suggested they'd stolen the prey from hyenas. Complicating matters further, such thefts were often sequential: In one case a cheetah killed a gazelle, which was then appropriated by a hyena, which was then driven off by wild dogs. However the lions got their meat, Schaller found they could eat a startling amount—so much they'd have to sit or stand to make room for their distended belly—consuming in one go a quarter of their own body weight. As the masters of nose-to-tail eating, they wasted little: He admired how they used their tongue to reel the intestine past their teeth to squeeze it clean of vegetal matter, the only thing besides large bones and hooves they discarded.

Again, George demonstrated that predators are, in the aggregate, good for prey: helping to keep their numbers within the carrying capacity of the land, weeding out the diseased, and evolving their antipredator talents. Thomson's gazelles had become fast enough, able to run forty-five miles per hour, to dust lions, and so agile that Schaller once saw one jump right over a lioness's head. With their dodges and zigzags and twisting leaps, the tender little creatures won most of the time: For every eighty-five stalks, the lions got just fifteen. Zebras had evolved a different strategy. Close-knit families bunched tightly with stallions as rear guard; Schaller saw one stallion kick in a lioness's jaw, and a whole family run head-on at a group of crouched lionesses, who froze. "To grab a zebra would have been like jumping on a fast-moving train from a standstill."

This theme of the weak defeating the strong became a favorite of Schaller's. He saw predators cowed by a porcupine rattling its quills or a monitor lizard's lashing tail, and defeated by the carapace of a leopard tortoise. He saw a valiant mouse rise with arms raised to ward off four wild dogs. When one dog stretched a paw to give it a nudge, the mouse bit it, surviving a full five minutes, "longer than it usually took to kill a wildebeest." He watched a chameleon, blown up rigid, thwart a big green snake's attempt to swallow it, and a six-ounce crowned lapwing on her nest part charging wildebeest like the Red Sea by flashing her bright white angel wings. She stood in the dust as the split herd thundered by, her marbled-gray egg safe.

When the lions did manage to eat, it was usually by working in sync—fanning out and then rushing simultaneously from several directions. E. O. Wilson deemed these unprecedented observations to be Schaller's most important, revealing a "degree of cooperation . . . that is one of the most extreme recorded for mammal species other than man." His findings explained why lions evolved to live in prides: because "group hunting is a superior means of catching large herbivorous mammals in open terrain." And because without the group, things often went badly. George watched a lone lioness hang helplessly from the neck of a male buffalo who just walked until she let go, at which point prey chased predator into a tree. His meticulous sketches using arrows to delineate the strategic moves of lionesses working together—which reminded Ardrey, who followed Schaller around to write *The Social Contract*, of "nothing so much as plays in American football"—also scratched yet another trait off the list of those said to distinguish animals from men. Lions possessed theory of mind: the ability to inhabit the perspectives and intentions of others and anticipate their responses. That capacity could also enable anticooperative behavior. Schaller once saw a lioness kill a gazelle during a communal hunt, then sit nonchalantly with it hidden in the grass until everyone else wandered off, so that she could have it all to herself.

Some of Schaller's discoveries bore directly on how the park should manage the cats, and the men who hunted them. In a drama of regicide and infanticide that played out over days, he came to understand how male lions serve the pride even though females do most of the hunting and cub-rearing. It began early one morning, when he came upon Yellow Mane, from the Seronera pride, lying bloodied, a fist-sized hole in his chest, one eye closed by a deep gash above his brow, left thigh ripped to the bone, and shreds of his fur and mane littering the grass. He'd been torn apart trying to defend both a zebra kill and a lioness in heat from intruders from the neighboring Masai pride. The lioness now nosed his wounded body with a plaintive moan unlike any Schaller had ever heard. When she suddenly fled, he followed her gaze to the victorious Masai Black Mane, come to gloat. Schaller watched with respect as the yellow male raised his head with a "terrible effort" to growl a last note of defiance. Then the tattered beast's bladder emptied, a spasm rippled his thigh, his pupils glazed and went slack, and "he died before me, quietly." Waiting five minutes before going to touch him, George felt almost ashamed for intruding on his final moments.

The yellow lion's death upended both prides. The remaining Seronera male—whom Schaller thought a hapless beast, his only memorable adventure one with bees that left his eyes so swollen he had to go about with raised muzzle for days to see—could not stand alone. After watching him slink about in fear, Schaller saw him finally chased off forever by two males from yet another pride, who then returned to a fallen tree where they had killed three Seronera cubs. (As George was the first to document, new males often killed infants so their mothers would mate.)* One tore open a cub to eat its viscera; the second carried off another little corpse like a trophy. By the time it was all over, the Masai males had taken over the Seronera pride and nomadic lions had taken over the Masai. Two-thirds of all

* Jonah Western deems this one of Schaller's most significant findings. Though the idea was challenged at the time, infanticide was subsequently observed in many species and cited by sociobiologists as proof of the selfish gene. (Email to the author, Feb. 14, 2025.)

cubs were lost in the transition, and it took two years for the prides to regain the stability vital to raising young.

Schaller saw these effects of male protection, or its absence, across the landscape. In a pride with one male, just two of twenty-six cubs survived, but in a pride with three males, more than half grew to adulthood. The males shielded the babies not only from rival prides and the occasional hyena or leopard but sometimes from their own mothers. A lioness will nurse any cub but often cuff her own starving young away from meat. In the dry season, when the wildebeest and zebra migrate to wetter climes and the prides struggle to survive on fifty-pound gazelles, Schaller often saw adults remain in excellent condition while cubs died, to be left behind without a backward glance. At moments, his own past seemed to nourish his empathy. He imagined that for a cub to emerge from its neonatal hiding place and have to compete for meat in a world of snarls and mortal danger "must be a traumatic experience." Fortunately, the males—who typically claimed first right to a kill, before the lionesses who'd brought it down—sometimes let little ones join them at the meal. The common practice, therefore, of letting trophy hunters take an "extra" male in fact presaged disaster, tearing the social fabric and dooming cubs.*

Since Alaska, Schaller had known that no species can be understood in isolation. To make sense of lion-prey relations he needed also to study the other big carnivores in the ecosystem. None captivated him more than those most loathed by humans and feared by prey: the wild dogs who are "such proficient killers," as he wrote in *Life*, that vultures start "circling even before they bring their prey to ground." Dogs are coursing hunters, meaning they run their target to exhaustion; Schaller stayed with one pack as it ran twenty-five miles to get a zebra foal. When the little zebra finally

* Schaller had to "break the back of the knowledge white hunters still controlled," as Western put it, "even though most of what those hunters thought they knew was wrong. . . . His inside view of the lion had an impact far beyond biologists. It changed the way all kinds of people thought about lions in East Africa."

lagged, the pack swarmed. Tearing a hole through the foal's soft belly to feast on its intestine, the dogs ate it alive, a slow process as they have no killing bite. A picture in the *Life* piece captures a wildebeest paralyzed in a similar moment, an unforgettable terror in its eyes. The dogs' seeming lust for such killings, played out in daylight (unlike lions, who "strangle in the night"), stirred men's own desires; hunters bent on extermination shot puppies on sight. Even the remorseful gorilla-killer Carl Akeley had gunned down a whole pack, calling it the killing he most enjoyed.

George, by contrast, found much to love in the homely little creatures with their harlequin coat, round cartoon ears, and faithful and curious natures. Following a pack one afternoon, an ill-timed glance at the speedometer, which he thought read forty-three miles per hour, landed him in a warthog burrow. When he climbed out of the car to dig himself free, the dogs gathered amiably to watch. He saw that the males and females shared tasks, including care of the young. Pups too small to leave the den greeted the returning pack with joyful spins and chittering, pushing their noses into the mouths of the adults, who obliged with regurgitated chunks of meat. When big enough to join the kill, the little ones ate their fill while their elders waited, arrayed in a protective circle to keep other predators out. The babies were so playful and floppy that when a new litter arrived, it took several days for Schaller to get the count straight (there were fourteen) in the tumbling mass of black and white. When a few weeks later rains flooded their den, he watched their mother hesitate, then carry them one by one—by a head, leg, or tail—to dry ground. Five died of exposure.

The dogs cared just as selflessly for the lame or sick. In the spring of 1968, Schaller saw adults staggering, with gray mucus clouding their eyes. He found puppies dying. Catching one, so weak it just cowered, he laid it on his back seat to take to a vet studying wildlife diseases. It died in the car, of what the vet diagnosed as canine distemper, likely caught from domestic dogs. Yet even in a pack where sixteen of twenty-six dogs sickened, those who could still stand and hunt went on caring for the others. Knowing that such cooperation (dividing labor, sharing spoils) was considered essential to human evolution—but had not yet been observed in monkeys or apes—deepened George's appreciation for our diverse commonalities.

The dogs also, however, became an occasion for regret. A filmmaker bent on getting footage took to driving around looking for George's vehicle—which if parked meant that he'd found a den—and then "simply usurped the site for weeks." In a rare instance of letting anger sabotage his work, George abandoned the packs, failing "the dogs and myself." Though he never named the photographer, he didn't have to. Anyone there at the time knew that Baron van Lawick, his jeep emblazoned with his title, had a habit, as Kruuk put it, of "shoveling in on others' work, presenting it as his." In this case, he managed to get two books and a film out of his usurped observations.

The other predator that attracted significant attention from Schaller was the cheetah, though these observations were largely Kay's, which George published along with his own. For twenty-six consecutive days, the two took turns staying with a family. George found them elegant but aloof, without lions' "intense and uninhibited desire to touch"; one wouldn't deign to turn around even when he crept within feet of her. Kay's observations were warmer. She noted how a mother chirped to summon her four-month-old cubs, then dropped a still-kicking gazelle fawn at their feet. They failed the test. She recorded the strange submissive posture in which a male cheetah approached a female, moving on his elbows, and admired the topi who ran off a cheetah who had toppled her calf. Kay's bliss in these years clearly went beyond her friends and blossoming boys; her fullest engagement yet in the field also brought the deep satisfactions of work. In fall 1968, when George returned to India for six weeks, she carried on observing the lions and cheetahs on her own. When, as his study neared its end, he asked for a volunteer to take over, she raised her hand.

One of the surprises in George's work is his willingness to confess bafflement. Newborn cheetahs are black with a long blue-gray mantle of hair on their heads and neck, but "I am unable to explain the adaptive significance of this coat." After twenty hours observing leopards, he could say only that they were nearly always alone, sometimes ate cane rat or aardvark, and cached their prey in trees. He retained his love of the detail apropos of nothing. Twice he found a python that had swallowed a whole adult gazelle. In a bat-eared fox's stomach, he found one each of a lizard, dung beetle, grasshopper, and spider, plus five ants.

Being at least partly in the mainstream clarified all the ways Schaller persisted in swimming outside of it. Although many scientists had by now signed on to sustained observational studies, George still had more of everything—stamina, focus, patience, monomania, drive—than even his strongest peers. "What struck me," recalled Kruuk, "was his enormous energy. He only needed four hours of sleep. And he could sit with lions for days on end, which is hard because they spend masses of time doing absolutely nothing. He drops so fully into where he is, he leaves the past, everything, behind." Tony Sinclair remembered his "almost blinkered dedication, to the exclusion of everything else. He poured scorn on any scientist too involved in the parties, dinners, tennis. He thought they were wasting their time in the best place in the world." Schaller's lifelong habit of reaching the most remote places, often via extreme technical climbs, still awed Sinclair five decades later. "He stands above everyone else in the world. There is George, and then there is the rest of us."

William Conway, the NYZS director for whom almost every great field biologist has at some point worked, never forgot a bit of film shot by Alan Root. George, in shorts and sneakers with no socks, approaches a lion gnawing on a zebra and claps: a nonchalant *scat*, which the lion heeds. Moving in, he then just as casually rips free the zebra's lower jaw, carrying it off, wet flesh dangling, to assess its age. He once dragged home zebra bones with a yearling cub hanging on right to their front door, finally putting boot to nose to push him away.

These weren't acts of bravado, nor were they ever without purpose, as Matthiessen noted after an outing with Douglas-Hamilton, who insisted on driving their roofless vehicle back and forth beneath a lioness in a tree, although they were after only pictures and she was crouched to spring. A move "merely stupid," wrote Matthiessen, compared to Schaller's "disciplined courage." Journalist Martha Gellhorn, who visited for two weeks to write a piece on "Animals Running Free" for *The Atlantic Monthly*, marveled that George, "whom the Africans call Mr. Lion, goes off alone for three or four nights of full moon and lives silently inside his Land Rover." Once lions gathered around his car biting his tires. "He says they are playing. Was he not frightened? 'Less frightened,' he said."

His stature now beyond question, Schaller made little effort to mask his unorthodox ways. "As a scientist I am by tradition expected to control my experience rather than yield to it," he wrote in his book of Serengeti photographs. "In this I failed yet am not displeased to have done so." Knowing that two of three lion cubs die, he did not name them lest his notebooks fill with obituaries. Though he still at times "perpetuated the myth of the dispassionate scientist . . . when I saw a starving cub, whose development I had watched from woolly toddler to sleek youngster, whose vigorous play had brought me delight . . . I felt a deep anguish."

Olaus Murie had been his first model for this unarmored way of entering the world. Now Schaller had the opportunity to meet the other scientist he most emulated, on several occasions taking Nikolaas Tinbergen into the field. Reading *Curious Naturalists*, the 1958 memoir Schaller cites as a primary influence, the affinities are striking. Tinbergen spent formative time in the Arctic, often accompanied by his intrepid wife; his Kalaallit hosts shared not only their seal meat and blubber-washed crowberries but also their deep knowledge of wildlife and technologies for survival. Schaller's experiments at Wisconsin on fear in chicks built on Tinbergen's earlier work flying cardboard birds of prey like a child's kite over the heads of goslings and baby turkeys. George's analyses of the mix of species and total prey animals required by his carnivores were a more challenging version of work done by Tinbergen on the hobby falcon.

Beyond these concrete parallels ran deeper resonances. Tinbergen saw even the humblest creatures as individuals—he distinguished the digger wasps he studied with paint dots—whose biographies it was his job to write. As George had with the drowning pelicans, he sometimes saved a creature in distress: Upon finding eider ducklings stuck in crevices and abandoned, Niko hurled them out to sea in hopes they'd find their mothers. Like George, he often made fun of himself. He wrote of sitting in a hide he had built to watch gulls and realizing they had perched on top, the one place he couldn't see. Listening to a colleague cluck to the eider eggs he'd incubated in his eiderdown sleeping bag, Niko beamed when the "happy event" was complete and proud Olaf presented five charming ducklings. Perhaps most important, the elder scientist, who in 1973 would win the Nobel Prize,* was like Olaus in his open delight in the world and his belief in combining focused looking with "long periods of relaxed, unspecified, uncommitted interest," to see the "innumerable things that just happened although we had not set out to see them."

Tinbergen had been instrumental in toughening ethology into the quantitative science George pursued so assiduously. He had also resisted the persistent ranking in biology of "hard" lab science over "soft" observational studies, and increasingly of molecular over organismal biology. Only through naturalists, he wrote, could science understand "the entire behaviour, the phenomena in all their complexity . . . We must not lose them or we lose all contact with reality."

This tension was playing out right in George's backyard. He saw the labs at the Serengeti Research Institute attracting biologists "who are primarily

* Tinbergen shared the prize with Konrad Lorenz—although having been imprisoned by the Nazis for supporting his Jewish colleagues at Leiden University, he'd broken off with Lorenz for nearly a decade over the latter's work for Hitler on army psychology.

technicians with little inclination to go into the field," or who remained wedded to collecting huge numbers of specimens. "Killer ecologists," Gellhorn called them. In his journal, Schaller described watching colleagues shoot fifty buffalo, "the chop-chop of the rifle, buffalos keeling over." He understood the rationale: Some were investigating the distribution of ages; others the diseases that move between buffalo and livestock. The meat was distributed in local communities. But George hated these mass killings.

For all his affinities with Tinbergen, more telling still were the ways George departed from his example. For all his "relaxed" looking, Tinbergen was a theorist, with hypotheses to test. George had no such frames, only the confidence, as historian of science Lily Huang put it, "that if he just went and sat, a place would explain itself to him." Though at first he hurried past distractions—a warthog enlarging its burrow, a secretary bird in display—"the calm rhythm of the days dissipated some of my urgency and as time went on I spent many moments in trivial pursuits, my life the richer for it." He rejected the checklists ethologists increasingly relied upon, aware that "an important detail may be ignored . . . because it lacks a discrete category," and that the categories themselves often flattened the complexity and opacity of these beings. "When a lion sees gazelle and then lies down, is it hunting, waiting to hunt, or simply tired of walking?"

Schaller's wide-openness of senses and mind, joined to his training and experience that by now spanned three continents, enabled him to see things as no one ever had. Two lappet-faced vultures approached each other in "goose step," their leg feathers like wooly pants; the engine roar of their dive was the sound of air rushing through pinion feathers. The moon rose hot orange then turned white, as if "recapitulating its history" from molten to icy. Again and again, he caught life at its most elemental, a tale of brute force, cunning, ruined labor, narrow escape. A striped kingfisher slammed a skink on an acacia branch. A martial eagle spread its wings to hide a gazelle when another raptor flew overhead. An amethyst sunbird toiled alone to build a nest of spider webs and feathers and incubate two eggs; then, overnight, her eggs disappeared. His economy and rhythm, even in notes, make him hard to paraphrase. "There is a big croc, 12 feet long. Every day he lies in the sun. His left hindfoot is gone." Arresting in its specificity, his language can be like found poetry. On the back of one page: "Crematogaster ants raise young in galls of whistling thorn."

Where Schaller departed most from Tinbergen was in their deepest motivations. Reflecting on why he so enjoyed his experiments with digger wasp navigation—for which he cut off their antennae and painted their eyes black—Tinbergen concluded that they satisfied a universal human "desire for power." That's as far as can be from the yearning that first crystallized for Schaller in the Sheenjek: to secure not dominance of other beings but their release.

Schaller's quantifying rigor never flagged. Of all lion lickings, he reported, 44 percent were of forepaws, and "not until I had laboriously recorded who licked whom . . . did I discover that youngsters less than about six months old seldom initiated." But the *who cares* of that finding gave away another truth: that his scientific busyness was often a door to something else, a way for him to be receptive enough to achieve what Goethe and Wordsworth sought—the experience of oneness. For a twenty-one-day period, he lived as if changed into a lion, following a nomad night and day, avoiding anything that might "break the spell."

He was always drawn to these nomads, who lived outside of a pride and seemed as pulled as he was between untethered joys and the comforts of settled bonds. He both envied these loners—"I also liked to be on the plains at night. . . . I know of no solitude so secure"—and pitied them. They never knew whether a stranger would be friend or foe, had no one to turn to in misfortune. Menaced by a pack of hyenas, an elderly lion who had lost both territory and friends lowered head to paws in supplication. The sight of such lions, exiled like Lear to roam and die alone, "filled me with unutterable sorrow."

Visibly discomforted when asked to talk about himself, it makes sense that George would probe his own interior via four-legged stand-ins. It is an ancient habit, says Jonah Western: visible at Lascaux, where cave painters rendered caresses between elephants but no people; and in Masai Mara petroglyphs, precise in the length of an eland's tail but reducing humans to stick figures. George called his days of an "existence revolved around lions . . . saturated with them" a personal corrida, "a quest for understanding, not just of the predators but also of myself."

Mixed with George's empathy—an imaginative inhabiting of an animal's mind that allowed him to see in one lion's face "a bold and somewhat derisive way of looking at the car"—was his growing recognition of all that was unknowable about another being. He wondered if lions dream, what they thought of him, *if* they thought of him. Quoting Henry Beston's famous passage placing animals "in a world older and more complete than ours . . . gifted with extensions of the senses we have lost or never attained, living by voices we shall never hear," he described watching the Seronera lioness find the last of her three murdered cubs. "After sniffing the carcass and licking it briefly, she ate it." At the crunch of the small leg dangling from her mouth, "I realized that I had still not penetrated the mystery of a lion's mind." Other animals behaved just as inscrutably. He watched a buffalo standing belly-deep in a swamp abandon his refuge and walk toward the five lions waiting to kill him, serenely, "as if this had been preordained." He saw a giraffe rear up to pound her newborn with her forehooves. The calf was weak but alive, its side covered with hoofprints. "I don't know why."

What Schaller somehow managed was to be both anachronistic and cutting-edge at the same time: E. O. Wilson dubbed him "a Victorian who has successfully adapted to modern science." Gellhorn, too, picked up on this fertile hybridizing. "The science of wildlife ecology is so recent," she wrote, "that it is nearer to art than to science; techniques have to be invented, imagination and intuition are more useful than microscopes." She bowed to the dual gifts that enabled George to capture, as she could not, the intoxicating vision of impala and wildebeest racing over "their own" land. "Not being a poet or a scientist, I have no suitable language for these animals."

The most unexpected description of his iconoclasm came in a 1971 *Science* article by chemistry professor Thomas R. Blackburn, who deemed Schaller a paragon of "countercultural epistemology," uniting analytics with intuition and "direct sensuous" experience. George would quote the piece at some length, appreciating its recognition of all we can know, can *only* know, through our own animal bodies. We have "in our very selves 'instruments' capable of . . . understanding the blooming, buzzy, messy world," wrote Blackburn, "with staggering sophistication and sensitivity." Trusting his own visceral responses as information, Schaller felt his way into both predators and prey. The tension in a lioness's quivering flanks before her final explosive rush "infects me. I hold my breath; muscles flex as if for the attack." As the Masai pride moved at dusk "patiently, inexorably, heads low across the plains," it was impossible not to "participate emotionally, knowing that the death of a being hung in the balance." After then watching a feeding gory enough to raise his "rudimentary hackles . . . I dream of lions and toss and turn." And if his gut responses contained meaning, so did the animals' responses to him. He was the cardboard raptor overhead, the mirror held up to gauge their fascination or fear.

For all his ongoing transgressions against twentieth-century scientific norms, George had barely begun his other great turn, toward the human communities whose lands were expropriated and traditions deformed by colonialism and its "modernizations." That history had played out here in the Serengeti under German and then British rule, disrupting the lives of the Maasai as well as their stewardship of the grasslands and wildlife.

Jonah Western, who'd grown up in Tanganyika when people and wild animals shared the landscape, was then documenting those traditional practices for a PhD. When Owen asked him to present his work at SRI,

"George was one of the few to listen with interest and sympathy." Still, Schaller remained unconvinced. When Conway asked him to add Western to his team at NYZS,* "George said my 'United Nations' approach didn't fit with field biologists."

It was not that Schaller doubted who posed the greatest danger to wildlife. Millions of animals had survived in East Africa, he wrote, only "because Europeans . . . with their mania for killing that which is neither good to eat nor otherwise useful did not penetrate many areas until the turn of this century." He had seen how traditions could work to protective and humane effect. When hyenas attacked far more gazelles than they could eat, three African rangers slit the throats of those writhing in slow deaths so the Muslim guides could eat the meat. His relationships beyond the white expats were few, but warm. A truck driver, knowing Schaller's odd appetites, brought him a wildebeest calf dead of starvation; he, in turn, seeing an eland bull totter and die, cut off the filets for the Tanzanian staff.

He even found, briefly, a local collaborator. Learning that a young tourist guide named Stephen Makacha had been recording detailed observations of lions in Lake Manyara National Park, Schaller helped him get a job at SRI and worked with him on his first paper. Makacha's data deepened Schaller's understanding of the ecological specificity of behaviors. While in the Serengeti a pride needed a territory of 150 square miles, buffalo were so abundant in Manyara that Makacha's pride needed just eight square miles.

George also knew, however, that animals and people could harm each other if crushed together—as a return trip to India in fall 1968 served to remind him. Traveling first to Gir National Park to see the world's last Asiatic lions, he recorded sweet encounters reminiscent of his time in Congo: Curious children trailed him; a boy smoked out bees, then dribbled honey for George into a cup rolled from *Butea* leaves. But the eight thousand

* At the time (1977), Schaller, Roger Payne, and Thomas Struhsaker were the society's entire global science and conservation staff. Western did join but reported to Conway until in 1986 George asked Jonah to replace him as director.

people he found living inside the park were desperately poor, their only wealth the ghee they produced from the milk of a few cows, which the lions sometimes stole. The cats were just as impoverished. Too few in numbers to stay genetically robust, and starved for habitat, these abased kings of the jungle padded after a guard leading a buffalo, then waited for permission to attack. If they stalked too soon, a wave of the hand backed them off. When the buffalo wheeled to face a crouched lioness, robbing the cat of her nerve, a guard stepped in like a breeder on a horse farm, positioning the buffalo's rump for her attack.

Continuing on to Kashmir to see the last of another species, the hangul or Kashmir stag, George enjoyed gathering walnuts from beneath trees where bears had recently done the same, cracking the shells to eat the soft meat. Though a breakfast of hot water followed by six hours in a cold wind brought no sightings, a fine view of yellow rice fields glinting in the sun made for "a good day." But here, too, the animals brought mostly sorrow. Barely more than a hundred hangul survived, George estimated, and only in this sanctuary, yet pieces had been carved off for a sheep farm and cows had crisscrossed the rest with ruined trails and forage. Reading histories of cow veneration, George copied into his notebook charts showing total milk yields falling even as cow numbers grew.

The other great threat to wildlife was also here. Out with a forest officer named Quasim, George heard a shot and, glassing the hill, spotted four men with guns. The chase that followed was Keystone Cops all around. First, George and Quasim nearly caught two of the poachers when one paused "to dung" and the second to grab a stack of chapati, which flew from his arms as he ran. Then George stopped—to eat the chapati. That was a mistake. When the first came roaring back, they saw from his clothes, brash manner, and excellent English that he was an army officer. The still-sometimes-rash George nonetheless confronted him: "Dirty poacher." "What happens in Kashmir is none of your business," the officer spat back, shoving the frightened Quasim. George wound up convinced of a position that remains controversial: that in some circumstances, poaching can be stopped only if park staff are armed.

George's education in the business of illegal hunting continued on his return to the Serengeti, where after much pestering Myles Turner finally let him join a patrol. He could not keep pace with ranger Okech Onduto and his four men, who on discovering a half-butchered zebra encircled by human footprints ran more than a mile to apprehend three poachers. George only caught up when they stopped to roast a confiscated eland for lunch. These men were legendary trackers, said to be able to tell from a distant animal's breathing whether it was a buffalo or rhino. They were also always in danger; the bows the poachers carried, strung with impala tendons to launch poisoned arrows made from nails, were sometimes turned on them.

In just the two days George spent with them, Onduto's patrol found sixty-seven newly killed animals plus piles of gazelle horns, warthog tusks (sold as medicine), and the marijuana the poachers smoked to ease frayed nerves. They found dozens of snares; if the wire that held the trap to a tree had cut deep into the bark, they knew the animal had struggled. Following their noses, they found a male lion, its skin flayed and its feet cut off; though not as valuable as the hide, the claws were sold as trinkets. Poachers also killed predators so they wouldn't steal prey animals from their traps. Turner's reports detailed daily horrors: zebra cut to pieces by wires but still alive, a rhino walking with a wide "collar" of suppurating flesh where a snare was still embedded, dying by inches. In a decade as warden, he had collected twenty-one thousand traps, some set eight feet high to garrote giraffes, others simply a sharpened stake with meat hung above to impale leopards.

The abattoir in a nearby village was easy to find, marked by hanging impala legs and a rotting eland hide. Watching the strapping guards arrest three "spindly-legged fellows" in dirty loincloths, as two sad children with the swollen bellies signifying protein deficiency looked on, George understood that this burgeoning industry exploited humans too. These children were clearly not getting the meat their fathers hunted: This was a commercial enterprise, enriching people far away. Porters carried off hundreds of pounds of dried meat, tusks, and skins to markets that over the next two decades would explode, taking with them a quarter of Kenya's big mammals, half the continent's million elephants, and nearly all its black rhinos.

Corruption abetted the destruction. One jeep intercepted by the patrol carried twelve illegal animals, two senior government officials, and an Italian contractor. Most trophy seekers were also illegal: Licenses were granted annually to shoot six male lions, but professionals and their clients shot ten or fifteen times that many each year. George scorned any man who sought to "prove himself by obliterating a lion," especially in the hunters' preferred spot just outside the park, where nomads arrived having never learned to fear humans. "To cause unspeakable pain to another creature, to kill it without purpose, not for food, not for hides, not for protection, for nothing except enjoyment, is a form of sadism." That these men invariably mounted their trophy with bared teeth unwittingly revealed the truth: That grimace is an expression in lions not of ferocity but of abject fear.

George was particularly stricken by the death of nomad 57, as his ear tag named him, perhaps sensing echoes in their biographies: The lion, too, had spent his youth in perennial exile before forging a bond that lasted the rest of his life. For six months, he and another young male had been inseparable, until in January 1968 they wandered out of George's sight. That in itself was not unusual. But nine months later, when George opened a letter from an outfitter and tag number 57 dropped out, he knew that a man "may or may not have gotten out of his car. Lion 57 may have turned to look, or not" at the source of the bullet that ended his life. "I cupped the blood-encrusted silver tag in my palm with a terrible sadness. I would rather have retained my vision of [him] wandering through his kingdom. . . . Now I see him nailed to a wall, glassy-eyed, his teeth bared in supplication."

It fell to the Tanzanian guards to navigate these complexities. Gellhorn, who also shadowed a patrol, had been moved by the clear distinctions they made between the "Big Business" types who set hundreds of wire snares or poached from trucks, and the local Africans, treated less like thieves than "like men done out of their ancestral rights." (She was also taken with the guards' style: One wore a safety pin earring, another wound his long, hollowed earlobes around the tops of his ears.)

As for what George could do, he inaugurated here what would become an effective strategy, confronting those driving demand. Writing in the magazine of the American Museum of Natural History, he noted that for "the vanity of a few," 3,168 cheetah skins had been imported to the US in a single year. That was as many cheetahs as would be found in 98,525 square miles.

•••

Nearly a year had passed since they had exiled their mongoose when, in June 1968, Kay wrote to her mother. "George brought home a lion cub! Two weeks old and almost lifeless from starvation and exposure. He found him abandoned next to a wildebeest kill and couldn't bear waiting for the vultures to start on him." The lioness had insufficient milk, George guessed, so in a desperate maternal gesture had left the baby at the carcass despite his being too weak to even lick it. He watched a male lion pick the cub up, then drop him, thought about his agreement with Kay (after coming to know the *Born Free* Adamsons) to never adopt a lion, and told himself he shouldn't intervene. But watching the baby's breathing grow shallower, he scooped up the five-pound "mound of bones" to carry home. "I felt in a small way that I was atoning for all those permitted to die."

The cub was matted and filthy and so near death that even the inside of his mouth was icy cold. Kay and the boys bathed his limp body in hot water, vigorously rubbed him dry, and tucked him into a box with blankets and a hot-water bottle. One blue-gray eye was still closed and he was comatose most of the day, but Kay hovered, picking off ticks. By noon she could drip milk down his throat and "by evening he tried to lift his head and even stand." She filled pages with details: the feedings every two hours all night long, the diarrhea and Kaopectate, the first wobbly steps. "He snuggles against me and licks the children's feet, though he is so frantic at feeding time he scratches me, so I wrap his legs in a towel." George's journal also filled with the cub. "When I *uu* like its mother it perks up its ears, answers a harsh *iau* and toddles over. When everyone leaves it, it calls." Even Eric, for his homeschool homework, drew vultures lurking over the cub. "We can feel all the bones, and that's why Daddy brought him home. If he lives, we will keep him till Christmas."

All seemed well—the cat rubbing the family's legs in greeting and trying to trot—when suddenly both he and the boys collapsed. Mark and Eric, it was soon clear, had chicken pox. The cub was harder to diagnose. "He couldn't stand, his head drooped, I felt sure he was dying," George worried. "We hope to pull him through, though don't know what to do once he gets bigger."

By week two all three little ones were better. "The cub has accepted the household routine without fuss," George wrote. It "had been not only a resurrection but also a reincarnation. He seemed to have forgotten his lion past and accepted his human pride mates unconditionally." Named for the pharaoh who fought with a lion by his side, Ramses learned "come" and "no" but still treated food as something to fight over, growling, clawing, and pushing Kay while he suckled, so violently he sometimes tore the nipple off the bottle. He feared and hissed at strangers but with his family made "indescribably soft purrs" while "gently mauling" their feet, hands, and hair. George found "something tremendously satisfying about lying in the grass and having a lion exert gentle pressure of his cheek against yours." Kay was just as smitten, if less enthused about the sand boa given them by George Adamson, who had to be fed live mice. "I'm finally getting the cub to take milk," she wrote her mother, "and now have to get a hypodermic syringe to feed a tiny baby mouse."

By six weeks, Ramses was fully restored. He dogged Kay's steps, meowing for food and grabbing her ankles if she ignored him; dragged towels about as if they were a kill; wrestled with the boys and chased the toy they'd made him of a wad of plastic bags tied to a string. By nine weeks he was stalking and rushing, sometimes bowling one of the boys over with a flying tackle. Alan Root, filming a half-staged scene for *National Geographic*, captured Eric in the grass confidently grabbing the cub by his nape to flip him into his lap and rub his belly. When Kay—in slim light blue slacks, her hair set in neat waves—steps out with a bottle, the cub toddles toward her. After a feeding in which his sharp little claws are visible, she burps him over her shoulder, her freckles matching his baby speckles.

The parade of luminaries, meanwhile, had not let up. On July 3, 1968, two weeks after Ramses's arrival, George noted in his journal that he'd taken fourteen-year-old Robert F. Kennedy Jr. to look at lions, accompanied by a Mr. Kirk LeMoyne Billings. (Lem Billings, who had been at Choate preparatory school and Princeton with John F. Kennedy, was a lifelong companion to the family.) On his visit a year earlier, RFK Sr. had told John Owen how much his son would enjoy the Serengeti; Owen had invited him to send the boy the next year. Plans for the trip were complete when on June 6, the elder "Bobby" was assassinated. Mrs. Kennedy decided their son should go to Tanzania as planned, to escape the chaos at home and again show the world the brave Kennedy face. He wound up in George's care, on a visit covered at length by *Life*.

Still, Ramses claimed the lion's share of George's journal. He retained his baby sweetness, climbing into his box when his belly was full to sleep.

But he was also growing aggressive. When George scolded him for leaving his sandbox, the cub growled and bit his hand. "He's colicky at night; perhaps that accounts for his foul temper." George now had to scavenge meat for him daily. On July 7, he gave Ramses a tommy head, which he tore into. The next day, he took out Bobby Jr. and Lem again, "tough since they don't want to sit anywhere for more than a couple of minutes."

With a cub at home, it seemed time to visit the Adamsons, now living in separate camps in Kenya. The Schallers were leery of Joy,* whom they'd first met when the couple came to the Serengeti in search of the cubs they had released there after Elsa died. In a letter to Bettina, Kay described her as "a mental case and acknowledged b——. A talented painter and sculptor but terribly emotional and competitive." More than fifty years later, Kay's judgment had not softened. "The last thing we wanted," she recalled, "was for Joy Adamson to know the Schallers had a lion cub. Come emote over our lion? She was so outrageous, such a drama queen."

George did laud both Adamsons for overcoming "cross-cultural barrier[s] . . . to know lions as individuals better than anyone I have ever met . . . They told me of subtle gestures and sounds I had overlooked." But he had far greater respect for George A., whom Kay also found "quiet and dear and reserved." One of Schaller's most memorable meetings came walking out alone near George A.'s camp. Unable to tell if the big male lion coming toward him was wild or tame, he didn't move, turn his back, or stare. He was relieved and thrilled when the lion leaned against him in a typical greeting, then ambled on. But he also saw the high price paid by all for the Adamsons' breach of boundaries. The three cats George A. was then trying to rewild had been so deracinated by their experience starring in *Born Free* that he had to shoot zebra for them daily. And the gamble of living so intimately ended tragically. One lion badly mauled a child and

* Like Fossey in 1986, Joy Adamson was murdered in 1980, in both cases after many years of mistreating locals.

then killed a man. Adamson shot him, but the other two did their own damage: Released into the savannah, they killed the wild lions they found in their way.

By August, Ramses was unmanageable: He bared his teeth at strangers, often snapped when touched, and once charged Myles's ten-year-old daughter. In mid-September, they shipped him off to Nairobi. The last time they saw him, six months later, "Kay, who had loved and cared for him more than any of us, went alone to the cage and talked to him. But he gave no definite sign of recognition." He eventually landed at the Milwaukee County Zoo, where he fathered many cubs. "To keep a large wild animal as a pet is a selfish gesture, which I deprecate though I indulge in it," George wrote. Even if secure and contented such an animal can never be "complete." For their final Serengeti year, the Schallers settled for visits from a three-pound genet, who would follow Kay's trail of meat into the living room, then vanish into the night.

In January 1969, Schaller met Peter Matthiessen, who had arrived for a few months of reporting for *The Tree Where Man Was Born*. Their first outing was a picnic with Mellon heiress Cordelia Scaife May, their only animal encounter "the loud rustle of termites rattling their heads." But a few days later, George took Peter out in a slashing rainstorm to watch wild dogs hunt zebra. The pack singled out a foal and disemboweled it with the mother standing by, a killing both men later wrote about. Matthiessen devoted two pages to the "hounds of hell." George, for whom such events were by now commonplace, gave them just a few matter-of-fact sentences. That same contrast reappeared when a bull elephant, as George recounted, ambled up behind them, raised his ears, trumpeted, circled, then turned to face them when the wind shifted. "We retreated and got home in time for tea." In Matthiessen's version, "the mighty bull . . . bore down on us . . . looming higher and higher . . . A froggish voice said 'What do you think, George?' and got no answer." When the bull scented them, "the dark wings flared, filling the sky, and the air was split wide by that ultimate scream . . . that primordial warped horn note of oldest Africa." He did note George's understated response: "You don't want them any closer than that."

Schaller's austerity was sometimes more effective. Peter's book was a finalist for the National Book Award, but lost—to George.

Recent discoveries in nearby Olduvai Gorge had set George to thinking about another creature. His work on lions and wild dogs, he realized, might have light to shed on our own ancestors, who for two million years had also been "social carnivores"—more light, at any rate, than further study of "some vegetarian monkey." He'd briefly enlisted Peter to play early hominin with him, walking the savanna counting gazelle calves they could have done in with sticks or stones and predators they might have chased off a kill with primate shrieks and their frightening upright stance. Six months later, in his final Serengeti summer, George devoted a week to a more concerted reenactment of life as *Australopithecus*, walking a hundred miles with anthropologist Gordon Lowther. To warn off prides, he shook a rattle he had made of pebbles in a beer can.

He and Lowther gained two key insights. First, that in times of scarcity, hominids would have had to hunt, relying on such lion tactics as driving and encirclement. Though the wildebeest George charged wheeled on him, he saw that with a group he could have clubbed it to death. Twice, he briefly wrapped his arms around an animal—first an abandoned and sickly zebra foal, then a blind giraffe calf—to confirm that he could have taken the prey.

When times were good, however, man the hunter was more likely man the scavenger, trailing more-effective predators to get bits they left behind. Finding a zebra carcass, Lowther used a rough-edged stone to bash open a bone for the marrow, then handed the crude tool to George to smash the skull to get at the brain. George also made "our one lucky find, a buffalo dead of disease." Together, those animals would have supplied three hundred pounds of meat—though the pair didn't actually eat what they found, Sinclair recalled, but ate Spam in similar portions. This second finding was unfortunate, noted writer John Vaillant, for the (mostly male) scientists who had built entire careers theorizing that both language and gender roles had emerged from our ancestors' bloody and virile dramas.

•••

George published four books on the Serengeti, more than on any other place or animal. With his gorilla bestseller, the many stories in *Life* and *National Geographic*, his star turn in *Sports Illustrated* and *To Tell the Truth*, he was by now a minor celebrity. One day three cars pulled up in the yard: a tour of old ladies determined to meet George.

First came his monograph, *The Serengeti Lion*, for which he won the National Book Award. He began with a paradox: It was the very worship of this cat that had often presaged its doom. Paleolithic cave painters gave them pride of place, but they were gone from France long before recorded history. Rome emptied Greece, Turkey, Iraq, and North Africa of thousands for the Colosseum; Caesar consecrated his forum by killing four hundred. Barring the sad few George had visited in India, sub-Saharan Africa was *Panthera leo*'s last home.

Schaller's second Serengeti book showcased his stunning Ektachromes; the lioness hanging by her teeth from the throat of a zebra, its head twisted and teeth bared in screams, was his most hard-won. The text accompanying the photos is acute and pure; "a little masterpiece," editor Angus Cameron promised Knopf editor-in-chief Robert Gottlieb. Describing lionesses in a stalk, Schaller matches their pace, conjuring excruciating suspense without a single flourish. Their bodies strain toward the herd, as if willing it to approach. As the delicate gazelle draw near, the cats fan out, "mere wisps of wind among the tall stalks." Only when surrounded do the prey catch the fateful scent and scatter; one, realizing too late that she is rushing toward death, leaps. "But with exquisite timing the lioness reached up and with gleaming agate claws plucked the animal out of the air."

Schaller has a poet's ear and gift for metaphor. Scrambling up a kopje, he finds "in moist clefts sometimes a gloriosa lily." Late at night a galaxy of stars flashes across the horizon "as if the heavens have retreated into chaos"—the eyes of a fleeing herd of gazelle. Heat waves turn plump gazelles into "lean viscous creatures" and zebra stream by, "their black-and-white columns in the flat light making one so dizzy . . . that only the zebra appear stable in a moving world." He is just as attentive to sound and scent: the unearthly boom of the kori bustard inflating and snapping shut his throat sac; the slurp of lions on ostrich eggs; the haunting braying of zebra "filling the void," their hooves trampling herbs into a minty perfume.

As Schaller moved for the first time from an academic to a trade press, his correspondence with his new publisher reveals a prickly, exacting author. Negotiating with him in fall 1970 over the two books they would publish, Cameron warned Gottlieb that "Schaller is more knowing about contract than I realized." When the picture book came out, George wrote them that he was pleased, except "four sunsets in the first pages is too many, on p. 83 they should have cropped out the right lioness and the printing sometimes loses highlights—the giraffe are muddy, cheetah flat instead of sparkling."

At work meanwhile on the manuscript that would become *Golden Shadows, Flying Hooves*, he rejected Cameron's urgings for a more dramatic opener. "It would not do to have drug darts flying all over the opening pages." Cameron had recently published his own book about owls and was clearly starstruck, gushing over George's letters sent from a monastery at thirteen thousand feet "with a couple of ancient lamas and their fleas" (Schaller had moved on to Nepal), or with "endangered bears in Baluchistan." The manuscript, he told the copy editor, could be trusted without checking. "As his wife says he has a 'Germanic exactitude.' He does not make mistakes."

George was particularly rankled at Knopf's repeated delays in releasing *Golden Shadows*. He'd heard rumors of a Van Lawick book on big cats: "It would seriously hurt our sales to have theirs come out first, especially since they've been collecting everyone else's data for years." "Still no books," he wrote in fall 1972. "All I see are ads for Matthiessen and Porter's book" (Dutton had bundled *The Tree Where Man Was Born* with photographs by Eliot Porter). He was frustrated that copy and images had not reached him in time for revision. "You've known my schedule for months. I asked for a layout last year." Cameron's assistant often had the unenviable position of go-between. Having laid into her—"As usual it is pleasant to get material too late to make changes. Hopefully the errors of grammar were caught. And tell the [jacket copy] writers to go easy on the ad agency adjectives"—George immediately apologized, asking her to pass his complaints on "to the proper culprits."

When an exceptionally intense monsoon trapped him in the Himalaya, Kay took over, further toning down the flap copy: "George would not like its exaggerated style." At last home and kept there by a "well-timed bout of malaria," George got books in November. Ever scrupulous about credit, he was unhappy that the flap copy claimed he'd "added importantly to the concept of the hyena." That honor belonged to Hans Kruuk. And he loathed the cover. "Could you not have gotten an artist, someone with a bit of imagination, rather than an illustrator from a second-rate men's magazine? I don't like seeing myself there. It needed mystery," not harsh realism. And was "factually wrong: lions would never lie there indifferently while hunting dogs chased zebra behind them."

For all his grumbling, these books' impact was again far-reaching. For the third time in a decade, Schaller had produced what all recognized as the foundational text on a charismatic mammal. "I cannot see," John Owen wrote, that *The Serengeti Lion* "will ever lose its place as the major scientific study of the species." (It indeed has not.) And like his earlier works, the book inspired a generation. In 1974, Delia and Mark Owens, newlywed students at the University of Georgia, wrote Schaller asking for advice on how to get started in work on Africa's wildlife. "Just go, and hurry," he answered. Their first book, *Cry of the Kalahari*, used Schaller's data as their baseline.*

As his gorilla work had been enlisted in late-fifties debates over the "natural" family structure, Schaller's study of predators now became fodder for arguments inflamed by the war in Vietnam. Anthropologist Loren Eiseley saw in his lions "a world of swift death not one fraction as ruthless, cruel or destructive as our own civilized kind have proven themselves to be in the century's wars." E. O. Wilson saw the opposite. Schaller's accounts of lion-on-lion violence and the hyenas' blood frenzy that left a hundred tommies dead and many more hunched in agony, bones jutting from bloodied legs, had crumbled a "cherished notion of our wickedness . . . that man alone kills more than he needs to eat." As measured by murders per individual, "we are among the more pacific animals."

Wilson was then at work on his own magnum opus. When *The New York Times* took the rare step of previewing it on page one, the article named Schaller among the few scientists whose observations were critical to its making. In his own review in *Science* of Schaller's monograph, Wilson underscored that debt. "The emergence of authentic theory," he wrote—meaning his own—"has created a large demand for information" of the sort Schaller was generating. "Though not a theoretician, Schaller has a strong intuitive feel for the important questions of sociobiology."

Robert Ardrey, in his more facile and wildly bestselling distillations of the science in what his critics derided as *Just So Stories*, also leaned heavily on Schaller. On why wild dogs ate nothing until their puppies were sated, for instance: "We discussed what course of natural selection could have produced such inhibition favoring the young. Only a very high adult death rate could place such selective value on the successful raising of young to

* Delia Owens's multimillion bestseller, *Where the Crawdads Sing*, features a heroine much like George in her taste for solitude, affinity with wild creatures, and exceptionally acute eye.

maturity . . . 'I think disease,' he said . . . And six months later Schaller wrote me that distemper had hit the pack, leaving only nine survivors."

Ardrey's overly clever voice ("A lioness came into heat and Schaller enlisted for the duration. The two alpha males reported likewise for duty") was not a particularly good influence on George. Though Cyril Connolly rightly praised *The Serengeti Lion* as "the antidote to Elsa"—rigorous rather than sentimental, focused on the animals rather than the author—in his popular writing George occasionally edged into shallow analogues. Females fared best under the vigorous "jurisdiction" of their "overlords"; a lion harassed by hyenas was a "king humbled by a mob, something not unknown in human affairs." And though his Knopf editor deemed *Golden Shadows*'s last chapter "the most notable piece of writing I have ever read on man and animals," it holds up less well than Schaller's other work, which stood adamantly apart from intellectual currents of the day. Closing with a banal quote from Ardrey on the failure of man's "new brain" to communicate with his "old brain . . . the animal within us," George renamed that inner animal the "savage that lurks in our depths"—a jangling phrase from the person who had unmade all the conventional claims for the savagery of nonhumans.

The extrapolation to humans was of course catnip to the critics. In his *New York Times* review of the monograph, Columbia professor George Stade, who later wrote a novel about a serial killer targeting feminists, found grist for his own obsessions. "Lions . . . remind one of American society as described by Women's Liberation. Lionesses do 90 percent of the hunting but their lordly masters get most of their meat. . . . Females commonly have to initiate copulation." Hyenas, by contrast, "remind one of American society as described by a timorous misogynist. . . . Female hyenas are larger than males and dominant," with external genitals including "a remarkable clitoris [that] seems to have evolved in imitation of the penis, perhaps out of envy."

No critic seemed able to ignore the stunning stamina of Schaller's lions.

Connolly marveled that "the king of beasts can pack 300 orgasms into a weekend." A reviewer in *The Washington Post* reacted with a mix of lasciviousness and horror to Schaller's account of one lioness, still not satisfied after 157 copulations in 55 hours (once every 21 minutes), who turned to a second male after the first "not surprisingly" disappeared into the bush. By the end, her rump was worn shiny.

In 1972 Matthiessen and Schaller met again in New York City and agreed to walk together the following year across the Himalaya. The mix of rivalry and respect in the lifelong friendship thus launched had already been evident at their Serengeti dinner table, as Kay wrote to Peter's widow Maria following his death in 2014. George had been "under the spell" of Matthiessen's book on the Kurelu tribe of New Guinea. But his comment (over zebra curry) that Peter was lucky that he could always write a novel when he had "nothing else" to do had inadvertently landed as an insult. "Peter firmly replied," Kay wrote, "that he was 'very proud of his fiction.'"

Before following Kay and the boys back to the US, George stopped again in India. An article in the *Journal of the Bombay Natural History Society* had piqued his interest in Nilgiri tahr, a wild goat endemic to the Western Ghats. The article's author was one of India's hunters turned passionate amateur naturalist, a lawyer named E. R. C. (Reggie) Davidar who lived in Tamil Nadu with his wife (a tribal doctor), and three children. Davidar nearly swooned when one day "the world-renowned George Schaller" showed up at his door. Not realizing his guest would stay for two months, Reggie hastened to arrange everything—from an old taxi to a bandy-legged guide who even when going out for a week did not bring a blanket but shivered and coughed until George crept alongside to share his. Exploring his first shola—the stunted, remnant forest found at altitude in South India—George discovered familiars from other beloved places, including Virunga's yellow *Hypericum* flower and leeches as ripe for intimacy as Sarawak's, who here left his underpants spongy with blood.

Davidar's daughter Priya, then seventeen, remembered the thirty-six-year-old Schaller as "lanky and solitary," eating mostly chocolates. When not searching for tahr, "he read his way through my father's library of books and journals, from volume one on, making notes." Though the local swells invited him for drinks at the Wellington Gymkhana officers club before going off with horses and hounds, and Priya noticed his eye for beauty and flirtation with any lovely young visitor, "there was little wasting of time." Accustomed to florid storytellers, she appreciated how low-key he could be. Returning from a jaunt in the decrepit taxi, he told of looking out the window to see a wheel roll past, realizing only when the car lurched and dropped that it was their own. The driver fell out the door; they stopped only by crashing into an embankment. Out with her father, George found dhole on a young chital, its eye mangled and flanks already half eaten. "Then my father, who loved deer meat, cut off a piece to bring home. My mom was horrified. 'How can you eat meat eaten by dogs?' Schaller reassured her. 'Oh, this isn't the part the dogs had eaten.'"

George's visit drew attention to the imperiled goat, including via an ad for Canon cameras featuring his photo of five astonished tahr staring down at him from a ridge. "Understanding," the copy read, "is perhaps the single most important factor in saving the Nilgiri tahr and all of wildlife." More lasting was the inspiration he provided both father and daughter. Reggie asked him to read a paper he thought "would establish my credentials as a scientist." George read it and handed it back with an attempt at tact: "So, you enjoyed yourself." That response did not demoralize Reggie but rather the opposite. The desire to produce something that George Schaller "would recognize as meaningful" became a powerful motivator. He apprenticed himself, taking careful notes as George showed him his journals and how he classified animals. Over the next five years, Davidar used that training to census the tahr across its range, work relied upon by the IUCN when listing the animal as endangered.

Priya was also forever changed by that visit. Her father had often taken her, as the eldest child, camping, fishing, and trekking, but as a "traditional

Indian man, he saw his sons as the inheritor of his legacy." George saw only her keen interest in animals. "When he said to me, 'You must take up wildlife study,' that opened that door for me." After completing a PhD with ornithologist Salim Ali and a postdoc with E. O. Wilson, she returned to a professorship in ecology at Pondicherry University and decades of work conserving Asian elephants and tahr. As both a woman and a member of a low caste, her career broke two barriers in India's scientific establishment.

6.

Into the High and Magical Realms

Pakistan and Nepal, 1970–75, Snow Leopard, Punjab Urial, Blue and Marco Polo Sheep, Markhor and Wild Sindh Goats

Since returning from the Serengeti, the family had settled at their Vermont farm for their first-ever long stretch in the United States. Supported by a Guggenheim Fellowship, George spent the better part of two years there, writing his lion books. He still dreamed of the grand mountains he had glimpsed from India, with their equally grand wild goats and sheep. But he needed a friendlier home base from which to explore them, and so in 1970 made his first reconnaissance trip into Pakistan. Taking George to the airport, Kay left the boys, nine and seven, alone for the first time. "If I'm not home by 6:30, please call Mrs. Clark," she instructed in a note. "To call you must: pick up receiver, dial 1917, hang up. Listen for the phone to ring several times. As soon as it stops, pick up the receiver and say hello and tell them you are calling because mummy has not come home." Beneath the note was tucked a "present to you for being such good boys": a George cartoon adventure, featuring one of them about to crash on his bicycle while his brother and an impassive raccoon look on. More drawings would come over the next three months, George's favorite way to bridge distances. In one, cartoon George climbs a pinnacle, oblivious of what the goat perched on top is plopping down on his head.

On this first survey, George took along an eager local apprentice. Well, not eager at first: Zahid Beg Mirza was a taxidermist in Punjab University's zoology museum but had never heard of George. "At that time," he explained, "it was difficult to get up-to-date scientific literature." It was Chris Savage, a hydropower engineer and founder of WWF-Pakistan, who told him of their esteemed visitor. "He asked, 'Would you like to go into the field with him for several months?' I said, 'When?' He said, 'Tomorrow, or the day after; you'll meet him and then go.'" When the university refused Mirza's request for an immediate vacation, he took three months' leave without pay. "When I told Schaller, 'I want to go with you to learn how to study wildlife,' he asked, 'What is your take-home pay?' I said, '530 rupees.' He said, 'OK, that I will give you.'"

Traveling from Lahore by night train, past camel-drawn plows tilling the desert, the men first went into the Salt Range, to the private reserve of the Nawab of Kalabagh. Except for the occasional raid by outlaws, this was the best place to see Punjab urial, a species of wild sheep. Assigned a bodyguard by the Nawab to discourage kidnappers, George found it distracting to watch sheep "while someone behind me played restlessly with the bolt of his rifle." But he soon grew fond of the ostensibly fierce Usman Khan, with his white turban and handlebar mustache. Catching lambs for George to weigh, Usman cradled the gaunt bundles, crooning "*Shabash, shabash*" (well done, well done) into their moist, dark eyes. Setting them back down on legs sprawling like Bambi's, the men sometimes had to run away lest the babies imprint on them and follow.

To understand any animal, George needed to see it in every season of the year and of life. He had timed this initial trip to observe the rut—mating season—when the rams are "handsomest . . . their pelage . . . like burnished copper," and one can see "the marvelous subtlety of sheep society." Going out before dawn, when the males "seek their goal with special urgency," he watched them strut stiffly, displaying immense horns that arc like a 1950s bouffant flip. He marveled that their skulls had evolved into battering rams, withstanding head-on crashes so intense their hind legs often lifted into the air. Though a siege of sandflies made sleep impossible, every day "we walked and walked and walked," Mirza recalled. (In his journal, Schaller commended his paunchy companion's fortitude.) Together they found four hundred sheep, including six dozen newborns, a good fraction of the estimated two thousand total in all the world.

Outside of the reserve, the sheep didn't stand a chance; their own guide would have shot two pregnant females had George not laid an imploring hand on his arm. But even inside, he found warning signs. In well-managed Iranian reserves, urial rams reached two hundred pounds; here they weighed half that, stunted by competition with livestock. There, well-fed ewes regularly birthed twins; here only half had even one young at heel, and just half of those survived. The villagers' closeness to the animals was also, paradoxically, disruptive; seeing George catch babies, they jumped in to bring him more to weigh, but then caged them to suckle on domestic goats. Beyond frustration that it "will throw off all my data," George worried at the babies' fate.

Over the next four years, Kalabagh would offer the nearest thing to the kind of intimacy with a landscape and its animals George had grown accustomed to. He would make the six-hour drive from the family home in Lahore many times; once Kay would join him for four days, her only time in the field in all their years here. He would make findings important to the animals' survival: disproving, for instance, the hunters' lore that a ram travels solo only if past breeding age, and so was fine to shoot. (Like those taking "extra" male lions, trophy seekers everywhere seem adept at contriving rationales for killing the big beauties they want on their wall.) In fact, George would establish, males are alone when looking for estrous ewes.

From Kalabagh, he and Mirza headed southwest to Baluchistan, another province George would return to until political instability shut him out. Here his aim was markhor, a Dr. Seussian goat that climbs high into trees and leaps branch to branch despite five-foot-tall horns that twist like Baroque church spires. Though he found "something ludicrous in looking for a hoofed animal in a tree," George sometimes spotted the telltale swaying branch or tip of a horn poking through leaves. More often, he depended on local guides, one of whom could pick out animals with 8x binoculars that George struggled to see with his 20x scope. Most memorable was a silver male with towering corkscrew horns stretched out on a ledge, foreleg rakishly dangling over the edge, grand beard blowing and "his" females all around.

Though most of the world's markhors live in Pakistan, George found that they, too, were on tenuous ground. Hills once covered with juniper had been stripped, leaving the goats without food or places to hide. Without trees to hold the scant rains, villagers also suffered; many of their wells ran dry. But with no other source of fuel, they had little choice. With the added pressure of hunting, soon "all our wildlife problems will be solved," a forest officer assured him. "There won't be any."

Their final destination was in the far north: a remnant holding of another royal, the twenty-year-old Mehtar of Chitral. George was excited to reach the flanks of Tirich Mir, his biggest mountain yet at 25,300 feet. He had again timed their visit to the rut, when markhor males court with behaviors he hoped to see, like dousing their own faces with urine. But nothing had prepared him for winter in the Hindu Kush, against which the broken windows in the hut he shared with Mirza offered little protection. Sleeping out one night in the machan (an observation platform), George was so frozen by the time he spotted goats, he could barely hold his pen and half hoped they'd move out of sight.

Mirza was surprised at how this American luminary traveled, in third-class compartments, eating street food, never bathing. "He'd just remove his fur-lined trousers to get in the sleeping bag, then in the morning put them back on and say, 'I'm ready.' Only in town would he finally take a bath and shave, and look very beautiful." Seeing George's willingness to challenge the forest guards, who "sat in their huts making up reports of a wildlife paradise," left Mirza forever changed, as did his mentor's undistractable teaching. "At night, by lantern, we recorded our observations and he answered my questions. If I asked him anything unrelated to our study, he would not answer, just come back to our business. We would have looked like two gloomy persons, barely talking at all."

It was on December 14, 1970, that Schaller met the animal that would become his totem, seeding an obsession he would never shake. Hearing that big cats might be about, he and Mirza had moved to a speck of a village called Merin, George fretting in his journal at the ten-rupee (two-dollar) cost per night. As blizzards began to close passes, he knew they couldn't stay long, so he spent a few more rupees to buy domestic goats to tie out. Each morning, taking them a breakfast of oak leaves, he hoped to find spilled blood. Each time, "three lively goats greeted me with cheerful bleats." When one did die—strangled by his tether—George cut it open to attract raptors, but too late: The gas accumulated in its gut burst, splattering him with viscera.

They had seen no other action when, out one morning watching markhor, George spotted their guide Pukhan running his way, gesturing toward the ridge, the sleeves of his tweed overcoat flapping. (Global clothing donations had eroded local ways.) Scoping the crest, George saw them, backlit by the sun. A snow leopard and her tiny cub. Here was a beauty like none he had ever seen; not the taut, rippling gleam of a tiger, but a thing almost without boundaries—her gray and cream coat soft as a fog, her luxuriant long tail thick as a Wonderland caterpillar. Mesmerized, he moved slowly, remaining (as with the gorillas) always in view. She watched his every step with her intense pale eyes. But she did not move. Several hours passed before he could pull himself away, racing to his hut to grab more film and ask shikaris to bring up the strangled goat. He was surprised to see that the hunters feared the snow leopard, which never attacks humans.

Finding her in the same spot on his return, George climbed three hundred feet to get close for pictures. He had (unaccountably) worn stiff new boots that gave him such painful blisters that he had to switch to sneakers and balance on frozen feet. He'd also brought no gloves, so had to clear snow from handholds with bare fingers.

The snow leopard watched, wisps of mist encircling her, a cap of snow on her head. Given her unmatched skill at vanishing, George saw her choice to remain as a kind of consent, and took the first pictures anyone ever had of this "ghost of the mountains." *National Geographic* published them the next year. In one, she sits on a snowy rock as if riding a cloud; in another, she rests her chin to watch him, like Junior on his belly watching Kay TV.

For a whole week, the snow leopard let George stay, sleeping just 150 feet away—he, soaked every night by wet snowflakes; she, dry under a rock overhang. By day three she lay in the sun, eyes closed, heedless to his presence. Then, miraculously, she called her cub out of the rock cleft in which she'd hidden it. Bounding to touch its forehead to her cheek, the baby gnawed awhile on George's goat, returned to rub its fuzzy face against its mom's, licked the top of her head, and vanished again. As two young lammergeiers tumbled in the sky above, each with talons sunk in the other's chest, the mother leopard let George witness a kill. Running straight down at tethered goat number five, she reared to swipe the air when it wheeled about, then lunged, holding its throat for eight minutes until it fell still. George thought it "a good finale to my goats." Then the cats slipped away.

Only one other Westerner had ever seen a snow leopard, and few anywhere had met on such intimate terms. Being a part of her world for those few days filled George with quiet ecstasy, the rare, fleeting state "beyond the subjective," when one seems almost "to become what one beholds." For the rest of his life, this cat would be his grail: an animal but also his portal to the dissolution of self, the reunion he would forever seek. He would sometimes lament the obsession for diverting him from more earthbound work. Returning here two years hence, he would spend a month waiting in vain for a leopard that never came. "He tries not to let the leopard interfere," Matthiessen will write, "but the great cats have a strong hold on him."

Trapped in Chitral for five days by heavy snows, George missed Christmas. At least, Kay wrote Chris, everyone at the post office was impressed by the thank-you notes they mailed off to highnesses and princes.

•••

Schaller had been pushed from India to Pakistan by politics; now war between those two countries closed Pakistan's northern mountains. So, while still planning to move the family to Lahore, he began 1972 in northern Nepal, in the Kang Chu (now Lapchi) valley halfway between Kathmandu and Everest.

The family had managed a festive holiday, Kay wrote Chris, "in spite of George being grumbly" at Bhutan, which had also denied him entry. Gathering a crowd at the farm, she had finished up in the kitchen to find the living room littered with bodies and "the spirit of Christmas buried in grunts and snores."

The trip began as they now invariably would: with deserts of lost time. First came the quest for permits. For two weeks George went from office to office in Kathmandu; in his journal he lists twenty-eight visits, to the home ministry, forest section, immigration, foreign ministry, and round again. Then Mahendra, the king of Nepal, abruptly died, of his second heart attack, while hunting swamp deer. (The first had felled him while hunting tigers.) That shut the whole country down.

George would get help from the "colonial types" he had met in East

Africa who'd now washed up here. Frank Poppleton, who had sheltered Kay at Queen Elizabeth Park while George drove back into Congo, would soon arrive with the UN as advisor to Chitwan National Park. John Blower, a former game warden in Uganda, was already here. Blower had lived a swashbuckling life. Deep behind Japanese lines in Burma* in World War II, he would have been killed but for an assignation with a woman that kept him off a plane that crashed. In Ethiopia as wildlife advisor to Haile Selassie, he made the first descent down the upper reaches of the Blue Nile, first smoothing the way with the gift of a puppy to the chihuahua-loving emperor. Schaller's own rules for living echoed in his praise for Blower's "quiet, efficient, dedicated" reserve. "He seldom talks about his family or all the exciting things he's done."

With permits at last in hand, the next step was to find porters to move two months' supplies to Lamabagar. This was the first time since Congo that George needed porters; he always preferred the freedom of carrying everything essential on his own back. But in these formidable mountains, support would prove vital to every expedition.

Finally setting out, George walked ahead of the line, to move faster and for "less noise; less wind-breaking." But already on day two, that plan went awry. Realizing as night fell that he was hours ahead of the rest, who were carrying his gear, he was settling in to sleep—feet stuffed into plastic bags and then into his pack—when a porter arrived with chapati and the news that the others had all quit. Since they were waiting to be paid, George must hike back down by flashlight. His irritation turned to gratitude when the porter who had stayed spent the whole next day making him snowshoes out of fresh-cut bamboo.

George had also hired three Sherpas, the world's most expert mountaineers, who not only carry but also act as trip leaders, guides, and cooks. With them, too, he got off to a rocky start. When Mingma, left to guard

* Now Myanmar

their equipment, fled at the approach of strange men, George snapped at him. He grew just as impatient with the head Sherpa, Kancha, who before delivering bad news always "stood silently, screwing up his courage."

He would soon apologize to Mingma, belatedly realizing that he was just a teen; and he would find many occasions to appreciate Kancha's seriousness and initiative. But he would grow closest to the young cook, who always packed him extra chapati whenever George went to camp alone. Though little interested in his animal watching, Phu-Tsering often went along, for the companionship both enjoyed. And though the Sherpa's open expression of emotion could not have been further from George's own way of meeting the world, more than once the scientist saw its power. When an officious villager barred their way into Lapchi monastery—even though the lamas had invited them—Phu-Tsering pushed the man aside with a light laugh that ended the matter. More unexpected was his response to a letter from his wife saying that she was leaving him for another man. With tears streaming, Phu-Tsering climbed atop a rock cairn in the police-station courtyard to read the letter out loud. Constables, shopkeepers, his fellow Sherpas, village children—all formed a circle around him sobbing loudly in sympathy.

The first weeks seemed to promise equally warm relations with the Tibetans who inhabited these border villages. George was touched at their hospitality. Seeing him, in sneakers yet again, breaking through the crusted snow, a cobbler made him a pair of the locals' high woolen shoes, cutting special pieces for his big feet. A family invited him to a wedding, lovely in the glow of yak-butter lamps. He brought gifts of biscuits and soap, a self-conscious stork above the rest. When he visited a hut where Phu-Tsering said people had heard a snow leopard cry, he found five men sitting around a fire with long knives, stabbing chunks of yak meat out of a pot. They shared their meal.

The lamas were most gracious, hosting him for ten days. He took careful notes: on their walking meditations and turnings of the prayer wheel, the purpose of the stones piled near entryways (to paralyze thieves), and the chants, cymbals, and dances that accompanied the dead as they were delivered, dismembered, into the river. The meditative rhythms soon became his own: He spent hours doing little beyond an occasional stir of the fire. "There's no incentive to read or write or wash; gradually one is happy to eliminate the nonessentials and just sit."

This peaceful interlude did not last. Police had been interrogating villagers on his doings; now he was given a full-time minder, all the more irritating because the man had to be fed ("and his appetite is great.") George rebelled by refusing to learn his shadow's name, calling him simply *the policeman.* Matters got worse when he and Phu-Tsering made the ninety-minute walk to the Tibetan border. Two days later, foreign ministry officers arrived to investigate a boundary marker they said had gone missing. When they told George he had two days to leave the monastery, he sent Kancha to town to plead his case, but soon received a plaintive note. "Please come back today. If you do not then soon will catch policeman to me." George did return, only to wait weeks for permission to leave for Kathmandu and home. He finally begged to be arrested so he could be taken out as a prisoner.

The villagers, meanwhile, had grown less welcoming. Were they worried, he wondered, that he might be a thief? Did they fear he'd stop their hunt for protected musk deer, or poach them himself? They began ignoring his requests to move their livestock away from where he was observing wild sheep because—why would they? He knew he hadn't explained himself in a way that made sense inside a Tibetan Buddhist world, where nature is not a thing apart to count and dissect, and his doings seemed at best eccentric. Yet when he was interrupted taking his first pictures of blue sheep, he fumed that it took several minutes of "violent gestures" to get the shouting man "to shut up and stay away." He tried paying for help, with little more success: When he offered thirty rupees to herders to show him the snow leopard they'd seen, they kept him waiting hours while they cut grass, then waved vaguely in the direction the cat had gone. Things got downright hostile when a yak died and local men hung the stinking meat in his room, although an empty room sat nearby.

Most distressing was that even at these elevations—the monastery is at sixteen thousand feet—where villages seemed far too small and scattered to make any mark, scant space had been left for wildlife. As in northern Pakistan, every possible inch of land had been terraced for wheat and every tree carried off—the latter at the urging of USAID, then pressing Nepal to develop a timber industry. In such high and fragile lands, a wheat field or livestock can disrupt ecologies that take centuries to heal. It took George two months to see his first Himalayan tahr, long-haired cousins to the Nilgiri tahr he'd studied in India; these scaled cliffs in a singular manner, gripping with chest and knees.

With few wild animals to watch, George resorted to observing domestic yaks and hens, finding companions where he could. "A dismal morning," he wrote in his journal on March 5, 1972. "Then saw a bird with a brilliant blue cap and yellow belly flitting about. The sight cheered me and the morning seemed better after that."

Still, for the first time, his notebooks filled with loneliness. Knowing no local language was isolating. "I long for the chance to share an idea." On the back of a journal page, he jotted "some thoughts: you feel an intense form of loneliness; you try to find consolation in external beauty" (though lower on the same page, he quoted Bertrand Russell on the peace of walking "with nobody to be careful of.") When camping alone, he slowed tasks to make them last: gathering wood stick by stick; brewing and sipping malted Bournvita, then beginning again with tomato soup; listening hour after hour for the gunshot sound of ice tumbling off cliffs; even rigging his camera for self-portraits.

Kay's own loneliness leaked into the letters she wrote every few days, starting just hours after he departed. "I don't know why it is that when you leave for a long period, I simply can't settle down to read or do anything. Long spells such as the one that began this afternoon just defeat me."

Having allowed herself that moment's gloom, she rallied her own spirits. Returning from the airport, she'd found tracks in the snow. First a canine's: "Dog or fox, I hope the latter." Then a boy's roundabout zigzags, which led her to Mark, who

> *obviously had a nice walk in the quiet by himself. . . . Oh darling, I do miss you so tonight. But I shall get better and things will go swimmingly, I know. When I told the oil delivery man where you are, he asked if you're in the service. I said "No, animal behavior." He said you didn't look like that. I said, "It doesn't really show."*

A week later, Kay was again low. "I'm so miserably lonely for you this afternoon that I will write about nothing just to talk to you a little." Those nothings included that she was waiting for her egg to boil, would then wash her hair and watch David Frost, and that mice now clumped around the attic even in daytime. Mark set traps "so I hope to collect bodies soon." A few days later, hearing her "world-renowned scientist" quoted on TV (on the twilight of Indian wildlife) stirred her yearnings anew. As did falling on ice posting the mail, gouging a knee and spraining a wrist "all just to mail a letter to you. In true GBS tradition I gritted my teeth and sawed down the lilac and now my back is miserable. I miss you terribly. At least the bed seems too big and empty just for me. Cold weather needs company."

She did remind him of all she was shouldering: typing on "your" *Golden Shadows*, reviewing page proofs for the lion picture book, indexing and numbering pages. "A damn nuisance having you gone at such a time I must say I must say I must say!!" (George meanwhile confessed to Chris that "Kay is typing for me and proofreading and other things I'm just as happy to escape.") Two months on, winter was wearing on the boys. They remained resourceful: Keyed up after a Celtics game, they spent an afternoon tossing an empty oatmeal carton into a basket made of cardboard boxes.

But after India and the Serengeti "they don't seem to be snow children." When Saturday cartoons ended and the fighting began, Kay sent them out into a snowstorm "to work it out of their systems."

George had been home from Nepal just three weeks when on May 14, 1972, he headed back to Pakistan, though going the long way round. At thirty-eight, he had just taken on vastly expanded responsibilities, agreeing to head up the new Center for Field Biology and Conservation created by Conway to expand the NYZS reach. Though Schaller would lead a global team, Conway had no intention of tethering him to the Bronx. To the contrary, he considered it his job to protect George's freedom to roam. And anyway, "telling George Schaller what to do would have been like trying to be the boss of Albert Einstein."

Beginning in London, where he saw Tinbergen and Kruuk, George and his climbing partner Tom Lane spent a month driving a new Land Rover to Pakistan. In Germany, they stopped two nights in a guesthouse near Georg Sr. The family walked together at Walchensee and dined on sausage and bread, then George played badminton with Renate while their father checked his mole traps. (He'd caught one.) In Turkey, George and Tom climbed Mount Ararat, finding no animals even where Noah had landed them after the Flood. Summiting on their own after their guides gave out, they sledded down two thousand feet on their nylon pants. Their embarrassed guides kissed their hands and invited them to dance, which George did. In Tehran, he planted the seeds of future work on Asiatic cheetah.* In Afghanistan, where he would also later work, he shuddered at the hordes of "dirty barefoot hippies walking as if asleep" and the shops selling snow leopard skins—but enjoyed their breakfasts of cucumbers and Coke. After almost seven thousand miles, he and Tom crossed Khyber Pass and reached Lahore.

With the family due in June, George found them a home in the Swedish

* He met with Fred Harrington, an American invited by the Iranians to oversee their game department.

expat community, swayed by the mango trees and pool, neighborly volleyball games, and good Chinese restaurant for Sunday dinners. After three years of scant social life in Vermont, Kay looked forward to "the fun of a big city" and "being among people who travel, understand our work and have interesting work of their own." A fine school awaited the boys, offering their first opportunity, as she wrote Bettina, "to have such contact with other children." (Though George credited their isolated life so far with keeping them "pleasant.") With schoolmates from twenty countries, they would soon be playing board games in Swedish and baseball in Japanese. Both would find the "buzzy chaos" exciting, especially when a meeting of Islamic heads of state filled the neighborhood with machine-gun-toting men.

Even before the family arrived, however, the synchrony between Kay and George began to slip. On June 20, Kay called to say that she was delaying their arrival to visit Tom Lane's wife, who was worried about her husband's depression. "That made me angry as I must get into the field," George wrote in his journal, "and surely she has local friends." He killed time with walks to the market for samosas and cake. But after a week he could wait no longer and cabled Vermont. "Imperative Kay arrive Lahore July 4 or not meet until September."

"It had better be imperative," Kay wrote Chris. "Though it will be nice if he just misses us too."

When July 4 came, George met all three flights into Lahore. No Kay. The next day, the same. Finally, on July 6, after a "miserable overfull flight with bundled-down Greeks and six Chinese with bad colds behind us" the family arrived, tearful and exhausted. They were soon revived by a birthday party with dancing: "Fox trot so even George and I could participate." But five days later George left for a week in Kalabagh, and then after just three days home left again, for a month in Chitral. The family planned to join him there for a packhorse trip, but when George went to meet that plane, he was handed a letter from Kay saying Mark was sick and the next flights were all booked. George sent back a message telling her to go home. "Damn! For once I'd hoped the family could take a trip together and now things are screwed up again." Since school was starting, Kay wrote Chris, "we'll remain in town and hope George visits us once in a while. Probably three days a month but better than if we were in Vermont and did not see him at all."

George had returned to Chitral to trek west, toward Afghanistan, in search of markhor and ibex. This time he was joined by Major Amanullah Khan, a wildlife enthusiast who would become his most frequent travel companion. A dashing character, Aman had served as a Tochi Scout in Waziristan, chasing an "infamous brigand," the Faqir of Ipi. A devoted samba dancer and Dean Martin fan, he was called Colt for the .45-caliber pistols he wore low on his hip and handled with skill.

With friends everywhere, Aman frequently charmed their way into permits and out of trouble. No matter the quandary, "I have a plan." His fluency in languages local to the regions they traveled did help overcome what George saw as a natural distrust toward outsiders among people living in the remotest places. And though by his own description George was "Teutonically punctual," up before dawn and crabby if he had to wait for others, he somehow forgave Aman for being "mindless of time"—enjoying his leisurely cup of tea in bed and methodical morning toilet. "If I displayed impatience, the Major grinned in his fetching way and tense moments passed." By trip's end, despite heat and dust, breakdowns and bureaucratic dead ends, and the challenges of working in "unsettled" tribal areas, the duo had managed to survey markhor across its range, publishing their findings in *Biological Conservation.* As with Mirza, collaborating with George solidified Aman's engagement in conservation. Six years later, he would draft the hunting regulations for Pakistan's Northern Areas,* built on a principle then just beginning to take hold: that villagers should be the ones

* Now Gilgit-Baltistan

both to patrol against poaching and to collect foreigners' trophy fees, to spend locally on irrigation, health services, and schools.

•••

Even in summer, reaching Chitral was a challenge; of the nine flights George tried to take into or out of the remote valley over eighteen months, only three left on the appointed day. Without reliable phone lines, he could never know in advance. Each time, he would have to lug hundreds of pounds of food and equipment to the airport, join the pushing crowd at check-in, then push again to retrieve his luggage if the flight was canceled.

The Chitral royal family again hosted George, this time long enough for him to devote several pages to describing the men. (The women were kept out of sight.) He began with Shahzada Asad ur-Rahman, the young Mehtar's uncle. "Short, paunchy, pop-eyed, slow-moving and dreamy," Asad had been regent for twelve years following his brother's death in a plane crash, until the rightful heir was invested with full powers at age sixteen. Now twenty-two, that heir—His Highness Muhammad Saif ul-Mulk Nasir—seemed to George pampered and bored, proud of the embryonic mustache he stroked at every opportunity but without the force of character needed to keep the royal legacy intact. Watching His Highness play in a village polo match—and perhaps recalling his own daredevil approach to Alaskan ice hockey—George was unimpressed. "He dodged contact and did not scramble." Saif's great uncle, Prince Burhan-ud-Din, was a grander presence, "like a boisterous bear," but full of contradictions. He agreed with George that there was too much shooting. "I worry about the ibex in my reserve. You must go there to see how many there are. Take my good rifle and shoot as many as you need. Have you eaten ibex? No? They are very good."

George again needed pack animals, though he felt guilty at the rough treatment of horses ill-suited to this terrain. To make them climb, the men bent their tails past the point of pain; going down the sliding scree, they dragged on them like a brake. He was no happier to be reliant again on porters, who dawdled and kept reopening disputes over matters (pay, load weights, daily distances) he considered settled. "I've never seen such people for arguing. . . . All one had to do was voice disagreement" and in no time thirty men had gathered. "One steps back quietly and they all just shout at each other about matters that don't concern them." The fights were often theatrical, with men hurling his rupees to the ground. Though he usually let others sort it out, on this trip George joined the fray, bellowing and cursing until the men took his money and stormed off.

It is hard to know what to make of these fights, or of George's "private tirades" in his journal against "rapacious" porters. His stinginess was never about his own greed or appetites. All his life, he inflicted his parsimony most harshly on himself. The disputes seem more a matter of honor: both his own to his backers and that of the men with whom he had shaken hands. Still, to fight over a few rupees, which then fluctuated around fifteen US cents, can seem both miserly and foolish.

At the same time, George's journals filled with gratitude for Pakistani hospitality, which was warm even where people had nothing or had not seen foreigners in the three decades since the British withdrawal. (Over several years, the only entries in the Merin guest registry were his.) Wherever he arrived, people bore a charpoy—the traditional laced-rope bed—into the shade so that he might rest, bringing him tea and fresh apricots. His sweet tooth never sated, he ate and ate the sun-warmed, dripping fruit, cracking the pit to get the "almond" within. He admired the flat-roofed mud houses, which seemed to him the opposite of intrusive; "they belonged here." And he could scarcely believe the lengths people went to on his behalf. One merchant closed up shop so that he could bring George to his father, a craggy, big-chested army man who had taken British officers on hunts and had a wall of trophies. In these barren mountains, measuring heads—in royal residences and military installations, atop graves or the roofs of mosques—was often as close as George got to animals. He seems sometimes like an archaeologist, sifting the bones of a civilization already lost.

In these difficult circumstances, George developed profound respect for men like Mirza Hassan, a game warden who lived alone in a cold and barren hut because he could not afford to bring his family but worked tirelessly to enforce hunting bans, even prosecuting princes. Together, the men surveyed markhor in two key reserves: the Mehtar's Chitral Gol and Prince Burhan's Tushi. Hassan remembered when hundreds of the goats roamed every valley. Now they found scarcely a hundred in all, confirming a decline so severe the species would by 1994 be classified as endangered.

It was by now depressingly easy to locate the cause. Scrambling up scree fields, the sun on the white limestone hurting his eyes, George found livestock trails up to fifteen thousand feet. Domestic goats, he knew, were the mainstay of local livelihoods. For generations, families had used or sold their milk, hair, meat, and dung. And from Extra Assistant Commissioner Wazir Ali Shah, he'd learned of the new burdens these families faced: Having given up a lucrative levy when it banned hashish, the government had begun taking 10 percent of villagers' annual production.

But this overgrazing was hurting everyone. The only plants left were those with repellent chemicals or spines; domestic or wild, a goat would find no forage here. Looking through the royal family's photo albums from the 1930s, he saw how much forest had also disappeared. Much of it had been burned by officials, collectively allocated a million pounds of wood a year; most saved their ration to warm the small rooms they slept in, shivering in ice-cold offices all day long. Herders added to the damage, lopping branches off oak trees so their goats could reach the leaves, ruining them for the wild tree-climbers. Whatever didn't starve was shot—by border police using their official rifles, or by men like Safari Club International founder C. J. McElroy, who had vowed to shoot one of every grand species on Earth, and had been here hunting protected markhor. George would later meet him shooting Punjab urial, to complete "his private mortuary."

In September, after ten days in Lahore, George headed off again. "It was good to have him home, even for such a short time," Kay wrote Bettina, "but it does make it hard when he leaves again." Two weeks later, she wrote again, trusting perhaps that Bettina would remember what it was like to have two sons at home and a husband away. She was enduring a

> *dreary, lonely Sunday, usually my favorite day as my servant is off and I can putter without disturbing his routine. But every so often I feel particularly left out with George gone and all the other husbands home from work. Neighbors try to include me but it can be difficult, for a couple of the men have wives back in Sweden and are a bit "on the loose" here and I have to angle out of situations. I told George if his field stays got any longer, I might just join their bachelors' club! His last trip was five weeks, this one for six, the next for two months. So it certainly does not get better!*

This time, George had gone south, to Sindh province, to search in the Kirthar Range for the endemic wild goat coveted by hunters for its scimitar horns. Joining him was Andrew Laurie, who as a teen had spent a summer in the Serengeti assisting George and the other scientists. The two had met again in May when George passed through London. Having recommended Laurie to study one-horned rhinos in Chitwan, George invited him down from his studies at Cambridge to meet and make plans. When he declined—"I can't break away from exams"—George asked just how serious he was. Andrew came.

Andrew was in Rwanda assisting Dian Fossey when the note came via Kay asking if he could join George in Sindh. He saved Fossey's reply giving her blessing, "certain that you will gain immensely from working with Schaller." Three weeks later, he arrived by Aeroflot, landing in Karachi at 5:00 a.m. George had been waiting since 4:00.

The twenty-two-year-old proved a fine collaborator. He more than passed the fortitude test, regularly hiking farther and staying out longer than George, now nearing forty. He also evinced some of his mentor's acute attunement to life's entanglements. Modeling his note-keeping on George's, Andrew recorded how parched bees landed to sip their sweat—declining to sting unless inadvertently trapped, as in an armpit—and were then devoured in turn by the hedgehog George had caught one night beneath his bed.

Nothing lived easily in this bleached, desiccated landscape. Every morning and evening, a merciless wind hammered the men and shook their scopes so violently they could scarcely see through them. When at midday it eased, the heat grew so fierce that even the goats lay down beneath the white-hot sky, becoming hard to find, let alone to count or classify. George gave up and joined their "quiet torpor," sipping syrupy tea that never cooled, rereading murder mysteries and Jack London's *Martin Eden*. He described these hills with words like "bitter," "abandoned," "cruel." The sun-blistered cliffs are "hostile," the cairns built by Brits when defending this frontier each "just big enough for a sniper."

Nights—passed in the bodily intimacy George inescapably shared with all his fellow travelers—offered little respite. The men braced for the slam against their hut, the wind howling and tearing at its roof of reeds. One evening guards shook them with the news that fourteen armed bandits were headed their way. Packing quickly, they moved to the nearest police station, where they struggled to sleep with a camel ruminating loudly nearby. Most unnerving was the night George reached in the dark for pants he'd flung on a nearby chair and heard a heavy panting. He paused, wondering if "young Andrew might be responsible." Then, recognizing a hiss, he let out a shout. "Andrew, the flashlight, quick." On his predawn walks, George had often seen a pale saw-scaled viper. He'd always looked for it—"acquaintances are difficult to come by here"—and, when it vanished, hoped no one had killed it. Now, finding the venomous creature nearly in his bed and poised to strike, he grabbed a rock. Though the throw broke its back, the snake could still move, so he pinned it with the chair and beat it to death.

As everywhere, the people George met in Sindh presented a complex picture. Dr. Rizvi, the ophthalmologist who had invited him, devoted most of his time to building and staffing free eye clinics in rural villages that had no other access to care. But Rizvi also had a direct line to President and soon-to-be Prime Minister Zulfikar Ali Bhutto, which he tapped to get Kirthar National Park established here in 1974 and later to support George's similar effort in the north. Their local guide Bachhal was supremely gifted at both finding wild goats and baking bread; George hovered awaiting his *tok*,* dough wrapped around a hot stone, set in embers till crusty, then cracked off "like a skull." But the zoologists at the University of Karachi reminded him of the old-school types he'd known in Alaska, assuring him that their study of the elusive fishing cat was going well: "We've shot five already." And conversations with his "expert" counterpart were surreal enough to merit transcribing.

> **MR. WASIM:** "Goats must drink every day."
>
> **GBS:** "There are so few water holes they probably go many days without a drink."
>
> **MR. WASIM:** "*Of course*, goats can go weeks without water."

Mr. Wasim's imagination remained lively even in sleep. One night, dreaming that George wanted tea, he woke the cook, who in turn woke George at 3:00 a.m. to say that breakfast was ready.

These contradictions began to seep into George. One moment, he slipped into misanthropy: The villagers "scuttle like desert rodents . . . as if

* Better known as "kaak"

fearful of the brilliant light." The next, into mortification at the meal laid out for him by the US consul, of chilled Chablis, caviar, and oysters. He bristled at the shots contrived by a *National Geographic* TV crew that began following him about: Schaller stopping to drink, Schaller building a fire, Schaller silhouetted against the rising sun.

Home and family drop out of sight for long stretches in these years. None of George's letters to Kay, and few of hers to him, have been preserved, or at any rate shared. So it's hard to know what to make of artifacts like the bit of a Pashto song George scribbled on the back of a journal page. "Your eyes are two loaded revolvers / and your narrow smile has destroyed me." These verso jottings are often where his interior life spills free. In one, lying flat on his back seeking respite from the wind, he travels fifty million years into the ancient sea that left these sands behind, to float amidst the urchins and snails whose fossils he has gathered. Desert wanderings have always been fertile ground for visions.

Over the next two years, Schaller will return to Kirthar four times, though typical of the wheel-spinning here, his timing will often be off. In March 1973, he will come to observe newborn kids but find none; delaying his return home, he will finally see two, just two, and as small as bunnies. Returning that fall after a "sterile" summer, he will find eighty goats but lose them just as quickly. "I am always in the wrong place." Exhausted by the heat, he will accede to having animals driven his way, even though that distorts behaviors. Saddest will be his encounter with the mouse who nibbles on his plant press and peanuts until George resolves to end the incursions. Hearing the trap snap, he will find the lovely little thing—limestone gray with huge whiskers—with the bar across its back, gasping. "I felt sorry and put the trap away. Here where life is sparse and living hard, the mouse seems especially precious." He would later learn that it was a Cairo spiny mouse, only the second recorded specimen from Pakistan. "Something to expiate." He watched so many deaths impassively and then—one pierces.

George's return trip to Chitral that November 1972 again began badly. He got as far as Peshawar, then three mornings in a row dragged his stuff from Jan's Hotel to the airport and back again; one of those days was Thanksgiving. On the fourth day, the plane took off but was turned back by a storm. He could have gone home but didn't, telling himself that he didn't want to give the family his cold. Instead, he and a similarly stranded mine manager joined forces to drive as far as they could. Stopped by foot-deep snow, they continued on foot, making Lowari Pass at 4:15 p.m.

The sheer-ice path down caught George again unprepared: "I did not realize my sneakers were worn smooth." As the better-equipped porters and mine manager went on ahead, he fell every few feet, his hands frozen without gloves, keeping himself going as darkness fell by "cursing the stupidity of making such a trail." He finally put his socks on outside his shoes, emulating porters he'd seen wrap their feet in goatskin to leap through unstable boulder fields, though now his feet froze. It was 10:30 p.m. when he at last saw lights and crawled into a hut with road construction workers to sleep. "I picked up a flea but was grateful to stretch out." The flight, he learned the next morning, had made it on a second try. All the trouble had been for nothing.

Things got little better from there. Returned to Merin, site of his enchanted meeting, he found old snow leopard tracks and spent ten hours retracing the cat's path. Though he could decipher the cat's doings—here scattering black-naped hares, there sniffing fox scat—he wished he'd waited to come until snow had pushed the animals lower. "This whole autumn has been poorly planned. The trouble is snow leopard. It would have been better not to worry about it." He tied out another goat, who bleated whenever he passed. Each time, he stopped to pat it. The only mail from home was magazines. "I used to enjoy being off alone. Now I don't. Perhaps I've had too much of it."

He did make it home in time to decorate for Christmas, as he liked to do. The boys had enjoyed Santa's arrival in a one-horse tonga, Kay wrote

Chris, and her school board volunteering was keeping her occupied "with more than coffee and bridge." But George had stayed just five days. "I hope that when he returns [in a month] he will be home at least a week so we can get reacquainted. It seems awful to end 1972 without him, but reflects the whole blasted year. I just hope his being away to start 1973 won't set the tone for it—but of course it will, for it will go on as this one has."

Joining George this time was colleague Brad House, curator of mammals at the Bronx Zoo. A quick visit to Kalabagh had not boded well: After a day's walking, House said he had a sore foot and better stay home. George suspected "he is loath to hike because he's overweight and out of shape," but continued anyway to the far more demanding mountains of Chitral, getting halfway up Lowari Pass in a salvaged UN jeep. Such "taxis" were makeshift affairs, typically jump-started with a running push by a dozen or more passengers, who then crammed inside or climbed atop teetering rooftop loads. Drivers saved on petrol by switching off the engine going downhill. With bald tires and all that weight, that meant careening down cliff-edge switchbacks with little control. Schaller once kept a "morbid tally" on this famously dangerous road; "in one month five jeeps fell off the mountains, killing twenty-three persons." This particular vehicle had its own special miseries: the smell of leaking gas mixed with House's cigarettes, a cracked wheel, and an engine so punky they had to push it up steep hills and through mud. When policemen waved them down to say bandits had blocked the road ahead, they waited for an escort, then continued sandwiched in a convoy of lorries packed with armed men.

The taxi made it to Dir, but from there they would have to walk the nineteen miles and four thousand vertical feet through one, then two, then three feet of snow. Getting miles ahead of the rest, George waited three hours for House to catch up. Exhausted and "complaining of fainting spells," he cursed George for not warning him about the trail. "In fact," George wrote in his journal, rather uncharitably, "the trail could not have been easier. Obviously, he'll be useless in Chitral and a health worry. Nor has he been a pleasant companion. I'd hoped he would break my solitude but he makes no effort at conversation. I could not ask him to go back but made the trail ahead sound worse than it is." That worked. After a night on rope beds, House turned back. George continued over the pass to where more jeep-taxis waited. Offered one with a huge flap of loose rubber on a front wheel, he rode ready to jump if a blowout sent them over the edge.

That Herculean effort brought the usual scant reward: The markhor rut was over. "I missed it by going home for Christmas." With little else to do, he began taking pictures of domestic goats. "I loathe them in principle for the damage they do, yet cannot help liking them when they look at me quizzically, head cocked."

Then a note arrived from Prince Burhan that was so promising he pasted it into his journal: "On your account, I did not shoot a snow leopard that has been driving away all the markhor at my preserve." George raced to Tushi, making it in two days. "Where have you been?" the prince greeted him. "You're too late. My keeper shot the leopard." (The cat had surprised and frightened him, the man said.) George nearly wept. "Burhan had even seen it stalk markhor. If I'd gone to him instead of Merin I could have had photos and saved the cat's life." Invited to lunch with the prince's elite circle for Eid al-Adha, the holiday marking Ibrahim's near sacrifice of Ismail, he found the dignitaries sanguine about the killing. All went to admire the leopard hanging in a shed, skinned and "grotesquely" stuffed with straw. "I feel very depressed. So far this winter nothing has gone well."

For all his labors, George was learning little about the leopard except how hard its life had become. Even the mother cat who had let him sleep nearby, he learned from villagers, had begun with two cubs. Malnourished herself, she had not been able to sustain the twin. When George managed to find scat, it often contained only plants or dirt or, in a few cases, domestic goat hair; though depredations were rare, the cats did sometimes take a single goat or sheep. Unlike wolves, who "wolf" their meal and run, leopards eat slowly. When caught, they were usually beaten to death with axes, shovels, or staves.

George accepted the prince's invitation to stay on, though he had little hope that any snow leopards might still be there. At least seven had been shot at Tushi over the past two winters, he knew, two of them by the prince's brother. But with six flights in a row canceled, he had nowhere else to go. And he got his wish for someone to talk to when several shikaris arrived, including the one who had killed the leopard. "I feel like hating him, but he is efficient and helpful. He shows me markhor kills, and I collect stomach samples. Collecting a specimen adds something solid to one's data, and this is satisfying." Undeterred by the dodgy conditions—crusted snow over unstable shale—George and the leopard-hunter found ninety-two markhor, half a dozen in a single tree. For dinner, they roasted a leg of lamb sent by the prince. Knowing the mine manager always craved company, George dropped in on him the next day for a lunch of curried goat brain.

George is not much one for exclamation points, but January 22, 1973, merited two. First, "a goat has been killed!" Then, glimpsing a passing shadow, "a Snow Leopard! No other cat dissolves into its surroundings in quite that way." Still yearning for a reprise of his communion two years earlier, he rolled out his sleeping bag. When Himalayan griffons picked the first carcass clean, he fetched a live goat, then sat as the cold hours ticked past.

At last, at dusk, a lammergeier flushed. And there she was, body flattened, creeping down the slope. When she stopped, they waited together for darkness to fall; by the time she made the kill, George could only see two bodies struggling. Unable to sleep with her electric presence nearby, he lay on his back counting shooting stars, then rose at first light. He looked and looked, and had just accepted that she had gone when something moved. She had been lying in the open but, with her coat "melting into the rocks," had materialized only when she lunged at scavenging magpies. (The birds made a game of it, waiting till the last second to jump aside.) Finally, uneasy in full daylight, she vanished. "It was the last time I saw her. My first encounter had left me exhilarated, but this one left me regretful and melancholy, as if seeing something beautiful fade away forever." Writing about this later, he would quote Virgil: "All things doomed to die touch the heart."

At home, Kay's excitement at life in a city had given way to chafing at having neighbors so close "they know what I'm doing all the time," and whose children were not raised "as strictly as my own." (She thought one German couple particularly lax, letting their two-year-old swing out over an empty pool. Not everyone would find that riskier than raising your toddlers with tigers.) She had also joined a group exploring the "new women's studies."

The spring of 1973 was George's most restless yet, his stops home rarely longer than a few days, the purpose of his roamings around the western edge of the country often hard to discern. Over many thousands of pages, he recorded data of clear use—bird lists, routes then on no map, weather records—but also every chapati and every rest house: mat or bed, stove or none.

He never failed to find beauty: in the dappled groves of pistachio and tangerine trees, or the strings of camels wearing beaded chokers, their tall loads often topped by a small human child, a newborn camel, or a crèche full of baby goats—tiny black and white heads peering out like hatchlings in a nest. But the drives were long and dusty, and again and again he arrived too late: too late in the morning to get ahead of the village grass cutters from whom the markhor fled; too late in the century to see the hundred goats a guide promised to show him on the Surghund massif. He was not surprised to find the tracks of just one. Two days earlier, a magnanimous extra assistant commissioner had invited him to shoot the protected animal. "I can give you a license."

In contrast to his methodical surveys across East Africa and India,

George sometimes seemed in these months to be killing time. He drove to Khyber Pass to visit a famous smugglers' bazaar, toured a town dedicated to manufacturing guns, hunted partridge with a local farmer. Aman was often along, or meant to be: "No word, naturally, from Aman" became another journal refrain. The search for permits was ever more tiresome. Going from Quetta to Peshawar and back again, into sweltering offices and shoving crowds, this official was too busy, that one had gone to Islamabad, a third couldn't help. "It takes a year before officials cease to be suspicious," George wrote to Chris, and even then, the "bureaucratic nonsense takes days instead of hours." Worst of all, what officials called "tribal uprisings" were putting off-limits many of the best wildlife areas. George soon gave up on Baluchistan, although not before a meeting that left a lasting mark. A forest guard, an "honest man" living with his wife and four children in a traditional *gedan* (tent), explained that he could not support his family on the hundred rupees a month the government paid him. So he cut the trees he was supposed to protect and guided illegal hunters. He then offered to kill one of his three chickens to serve his visitor a proper meal.

Finally, with the school year ending, George's focus returned to the family. He began planning a trip with the boys into the heavily glaciated Karakoram range, to look for wild sheep and cats in the Hunza Valley.

"I am being sent off on vacation for six weeks," Kay wrote Chris in May 1973.

> *George suggested I use the $1500 I made as a consultant on a National Geographic book last year, and now that I've decided to blow it all and go to Anchorage, I think he is shocked but can't back down. He's been wanting to get off alone with the boys and we both agree they do not get enough Daddy time while they have too much of me. I'm so inclined to over-direct them and they need some more relaxed atmosphere where no one is complaining about the heat and dirt and worrying about disease. Having been home only once in ten years, I do look forward to seeing my mother. But it's the first time I've ever*

> *blown money in that amount so I feel a bit guilty. . . . I'll consider it a bit less to leave in my "estate" which by then will have greatly decreased in value anyway. Can you help me think of any more justifications? . . . George will then go to Nepal . . . for three months. The boys and I will fly there with him and stay a week, then come home to wait in loneliness until after Christmas, I fear.*

On May 28, with Kay gone to Alaska and Eric and Mark home in Lahore awaiting their father's return and June camping trip, George had the strangest, saddest misadventure of his life. Returned to the North-West Frontier Province* to again search for markhor, he had been waved past a police barrier but was then stopped, told he needed a permit, and sent back thirteen miles to get one in Daraban. There, an officious extra assistant commissioner (EAC) told him he had broken the law and was probably a spy. George pointed out that in the spring dust storms he could not see more than a hundred yards, but the EAC said he knew about special cameras. Going on about his love for Americans, and calling Schaller "brother," he ordered him to drive, with two policemen riding along, to Dera Ismail Khan to be interrogated.

They'd gone just a few miles toward D. I. Khan when, seeing a bicycle ahead, George tapped the horn to warn of their approach. The cyclist looked back but then, just as they were passing, careened toward them, perhaps unbalanced by the big hookah slung across his back. It all happened too fast for George to brake. He did swerve away, but his fender caught the front of the bike and he heard a terrible thump. Jumping out, he found the man unconscious but breathing, with a big gash across his forehead. Lifting him into the Land Rover, they took him to the Daraban infirmary, where the EAC soon arrived to tell George that he was under arrest.

Confiscating his money and belongings, the police locked Schaller into

* Now Khyber Pakhtunkhwa

an "indescribably filthy" cell. That evening, word came that the cyclist had died. The next morning, George was handcuffed to a five-foot chain held by two policemen as they took him by bus to D. I. Khan. (True to routine, the bus hit a bridge, had a flat, and broke down along the way.) George could make no calls but was allowed to see a Finnish missionary named Paavo Norkko before being taken to the central jail, where the iron gates slammed shut behind him. Put into one of twenty cells around a courtyard, he was met even here with hospitality: One prisoner passed him a mat, another a sheet, a third a boiled egg and tea. The next morning, he learned that while most of the 1,200 inmates were packed into big cages, each with a hundred others, he was in the section for educated and land-owning prisoners. One twenty-year-old, Mohd Ismail Khan, loaned George a razor and invited him into his cell to be cooled by his fan. Mohd had killed an army officer in a knife fight; the officer's family had refused the customary monetary compensation, so he had been here for three years. Another English speaker, Asmat Ullah Khan, had wild hair, a scar across his cheek, a passion for politics, and abundant intellectual curiosity. Also "involved in" a murder, he'd been here nine months without bail and was equally generous—lending George a lungi, cooler than pants. In the evening, they included him in their wrestling match.

Norkko soon came round with a well-connected politician who said he would get George a lawyer. The police had booked him for unbailable murder, but after three nights they released him into the missionary's custody. He had a swim in Norkko's pool and joined a volleyball game but felt "emotionally quite worn and dejected." The young bicyclist's parents did agree to the customary resolution: George would pay them five thousand rupees ($500). Meeting the frail father, he added a thousand more, feeling the utter helplessness of the gesture.

It all took a week to sort out. The central intelligence chief confiscated his store-bought Pakistani road map, leaving untouched his diary and flight maps. The US consulate had to clear up its own confusion when Schaller's name got mixed up with that of a compatriot who had been arrested with seven pounds of hashish. His June 7 journal entry strikes a typically sardonic note: "Free at Last!" He also pasted in a letter from Mohd, thanking him for the letter and magazines he had sent, "giving salaam to you and my kiss to your children." Finally home, he was relieved to find his own boys fine, although it was 122 degrees.

Five days later, the three set off at 5:00 a.m., up a narrow winding road with precipitous drop-offs made more treacherous by the need to dodge livestock being driven to summer pastures. They spent their first nights camped in the Naran Valley, catching trout for dinner and riding horses to Saif-ul-Maluk Lake, their noses blistered because George had forgotten sunscreen. At night they had hot cocoa with marshmallows. Hearing a cuckoo, George shared the memories it stirred—of his own father holding his hand as they walked through a German forest, singing a silly *Kuku-Kuku* song.

Heading on to Abbottabad, they waited for Aman—"June 17, Major Kahn did not come," reads Mark's journal. "June 18, Major Kahn did not come"—passing four contented days in a cheap hotel, reading comics, playing dominoes, and living on plums, samosas, and soda pop. Both boys retained vivid memories of their meals whenever Kay was away. Eric recalled a bag lunch his dad once packed for him, of chocolate chips and pickled herring wrapped in leaky foil. Mark remembered long hikes, eighteen miles or more, with no lunch. "He'd say, 'We'll take a piece of bread and be fine.' He wanted nothing but what was absolutely needed. The rest of us plan for the worst case. He thought there wouldn't be one or, if there was, he'd figure it out." With their father's Zelig-like gift for crashing into notable people, they met the "gnomelike" Ardito Desio, who led the 1954 first ascent of nearby K2. He taught Mark that tomato soup is better with olive oil.

When Aman finally showed, they drove five hours north, found the road closed by a landslide, and drove five hours back to go the long way around, the boys in the back seat discussing what seemed to George their only topics: what they'd do when they returned to the US and the Incredible Hulk. Now in middle school, both had fashionably long hair. Eric remembered his father bursting out laughing, then turning around to explain that Aman had just complimented him on his beautiful daughters. "I did part it on the side, so had this Veronica Lake swirl in front."

Encountering boys their own age intrigued both sides. Two young shepherds watched intently as Eric and Mark pitched their tents. Eric wrote in his journal about the crowds of boys selling apricots "who came running toward us" and a polo match. "The red team won." Mark loved the terror of rope bridges. All three took awed note of Nanga Parbat, the world's ninth-tallest mountain, and Rakaposhi, a tower rising twenty thousand feet from where they stood. Eric sometimes sketched the landscape. George drew pictures reminiscent of his mother's, of the women in their elegant pillbox hats draped with gauzy cloth.

Approaching the Chinese border, they entered the realm of the grandest and rarest sheep of all, named for the explorer who in 1273 first brought news of them to outsiders. The horns of Marco Polo sheep *(Ovis ammon polii)* coil twice and grow as long as six feet. When crashed together by rutting males, the crack can be heard a mile away.

George had heard that row upon row of these coveted trophies—"the head of heads," Kipling called them—hung on the walls of the seven-hundred-year-old Baltit Fort. So upon reaching its high perch, George and the boys climbed the massive stone plinth to knock at the door. Yes, they were told. They could measure the heads. And the Mir of Hunza would like to meet them. Dressed all in white, the Mir served cherries as he spoke of his own son, a drummer in a rock band, and the history of *polii* killings that had emptied his lands. An American hunter "lost his head and shot and shot." The son of president Ayub Khan blasted five from a helicopter. A German industrialist and his two sons killed nine; an army officer fifty to feed his men.

Continuing to their northernmost destination, the vast Passu Glacier,

George led the boys over the moraine and onto the ragged, twisted ice to peer into the unearthly blue crevasses. Now two weeks out, he sometimes let Eric and Mark (almost twelve and ten) fend for themselves. One night, he left them to camp alone, not far from a rest house. "We read, played ping-pong, watched people play ping-pong, read, and played with [local boys] Pervez and Asef," Mark wrote in his journal. "I drew a super-guy by posing in front of a mirror with muscles flexed." George and Aman, meanwhile, drove many hours to Skardu, picking up a hitchhiking Swede out hunting psychedelic mushrooms. In every town, the three searched for horns and magic fungi. Another day, when Mark didn't feel well, George left him in the tent while he and Eric climbed a twelve-thousand-foot crest. Pushing hard on the long drive home, the boys lived as their father often did: skipping dinner, sleeping on the jeep's roof.

Eric had become feverish toward journey's end; by the time Kay got home six days later, George was too. "He so rarely admits that he's ill, it disturbs me when he does," she wrote Bettina. "As he wouldn't go to a doctor, I started treating him for malaria." Two days later, he was up and off to Kalabagh to check on lamb survival, though to no end: It rained too hard for him to find the tagged ones. Returning home, he had planned to leave the next day for a five-week trek with Aman's nephew Pervez Khan, but "decided to stay home to spend time with Kay and the boys." By "time" he meant one more night; he left early the following day.

With Eric and Mark, George had explored northeastern Pakistan. With Pervez, he walked a 140-mile horseshoe through the northwest: up the Yarkhun River to its headwaters in the Hindu Kush, east along the edge of Afghanistan's Wakhan Corridor and the Pamir of Soviet Tajikistan, then back south across the Hindu Raj to Swat.

The trip offered variations on his usual travails: bedbugs in their six-rupee-a-night hotel, a jeep that had to be hand cranked to start and its brakes madly pumped to stop, a road so pitted it took six hours to go twenty miles. At one particularly bad washout, only a promised tip and a bit of taunting ("You're scared") persuaded the driver to cross. The trek on foot brought more ordeals. Weakened by a rare bout of diarrhea, George fell behind and had to shelter in his aluminum emergency blanket when sleet began to fall. Climbing to 16,900 feet to peer into Afghanistan, he felt the frustration of political borders. "We are a day's walk from Marco Polo sheep but cannot go there." Sure enough, that afternoon, an intelligence officer arrived on horseback with a stern warning.

Finding porters was harder than ever. In Buni, where the road ended and the walking began, villagers told them, "We are contented; we don't need work." The few who did sign on rarely wanted to travel beyond a day, so every night they had to search anew. George's dependence on others again turned him sour. He upbraided Sher Panah, the cook he had brought from Chitral, for forgetting to heat his metal plate before putting food on it. He scolded Sher's brother, Said, for failing to stay with the donkey carrying his cameras. The animals wore no halter and were controlled by what George called "literal kicks to the head"—all intimidating to Said. As he often did, after such bouts of ill temper, George immediately repented. It did not escape his notice that the porters climbed these passes sustained only by bread, salt, and tea. Nor was there judgment in his description of one of the (toothless, shriveled) herdsmen and his equally thin companion lying together for an hour beneath a blanket, smoking opium.

He was glad for the companionship of Pervez, an experienced mountaineer who, though slower than George, tortoised his way through this most demanding trek yet: with two passes above fifteen thousand feet and one at nearly seventeen thousand, glacial crevasses to leap, and unstable moraines to dance through. George did grow tired of doing all the morning chores while Pervez took an extra half hour in bed, his habits as languid as his uncle's. If nudged, he merely grunted "like a torpid bear"; the porters learned to collapse the tent behind him as Pervez crawled out the front. Watching him try clumsily to make a fire and heat soup, George realized he'd never had to learn how. When they crossed churning rivers, with rolling rocks that could smash a foot, Pervez had himself carried.

The villages they passed through were as small and isolated as any George had ever seen. Nine stone huts made up the whole of Pechus. Zhuil, at 12,300 feet, had just four, with doors barely three feet high.

His journals from this trek occasionally lapse into generalizations of the sort he rejected on principle when applied to any being, human or otherwise. Chitralis were hospitable but gullible, easy marks for the canny Pashtuns. Kohistanis were hostile to outsiders. "Men stopped us to demand money, cigarettes, or medicine. Many are armed, an affectation. When a boy threw a rock at Pervez, the men jeered."

But when he lingered with individuals, especially those living in the fluid borderlands still linked by the ancient silk roads, his portraits captured the rich texture of these cosmopolitan lives. Over refreshing bowls of yak curd, their host Meerza Rafi recounted how his uncle, a Tajik from Sinkiang,* had settled his family in this high valley in 1933, after discovering it on his way to Mecca. In 1942 Kyrgyz raiders—thinking that since Rafi's family had come from China, they must be rich—had killed his father and sister. The family had stayed nonetheless, any thoughts of returning home sunk for good by China's 1949 revolution. Now forty and married to several Wakhi women, Rafi spoke his wives' language as well as Urdu, Persian, and Chitrali. The women, uncommonly relaxed about being seen, brought the yaks in to milk while adorned for their visitor in brightly colored clothes and turquoise rings, with silver buttons in their tightly braided hair.

Rafi's knowledge of wildlife was comprehensive and grim: Ibex were scarce here, snow leopard and bear scarcer, Marco Polo sheep entirely gone. Continuing south into the Gilgit Valley, George heard a similar

* Now Xinjiang

report from the seventy-year-old Raja of Gupis who "with precision outlined the past and present status of each species." Two local populations of Ladakh urial were now extinct. The once-abundant markhor had all been shot. No musk deer had been seen since 1947, and no snow leopard since 1952. A few wolves survived. But while the wolf scat George found on fourteen-thousand-foot Karambar Pass had been dark with soft ibex wool and the monkey-like paw of a marmot, here they had only livestock to kill. Last winter, Gupis told him, the villagers had poisoned the carcass of a depredated donkey, killing several wolves and also foxes and crows.

These local reports confirmed what George had already found, or rather, not found. In five weeks, he had seen a lizard, a snake, a bat with huge translucent ears, and a vole that turned out to be the first of its kind recorded in the subcontinent. He spent hours watching a horsefly bite his hand, fascinated to see smaller flies circling "like jackals [around] a lion kill," scavenging his blood from the horsefly's proboscis. But the only animal he found in abundance were marmots, because (he guessed) Muslims do not eat rodents.

He did see one bear—but wished he hadn't. A few local entrepreneurs had discovered that a captured cub could fetch as much as $200; they needed only to shoot the mother so she wouldn't interfere. Itinerant entertainers then trained the babies to "dance": rising on their hind legs at the sound of a flute to shuffle and sway. George had seen bears on the scorching pavements of Rawalpindi and Peshawar; his photo of one—standing with arms flung wide—suggests a fate worse than death. In his journal he sketched this first he'd seen in "the wild," its teeth smashed to blunt stumps and a leash strung through its mutilated nose. "A desecration of a magnificent being."

The only other mammals he saw were four remnant ibex. The "survey had been a failure. . . . [I am] thirty years too late." Having pushed ever farther in hopes he might find hidden treasures, he at last fully grasped the scope of the great dying he was witnessing, the sequel to the erasures of their native fauna that Europe and North America had concluded a century earlier. He now saw that he had been living in a dreamworld. Alaska, the Virunga, Kanha, the Serengeti: All were aberrations, last oases of abundance. Most of the world was already like this: largely emptied of wild life. Animals that had roamed the planet since the Pleistocene were being shot, poisoned, captured, or starved out of existence.

For the first time ever, mountains became for George a place to visit "as local people do . . . out of need, not pleasure." Without animal intimacies, he felt "imprisoned by this waste of stone," lost all desire to climb the barren slopes "for ahead lay only more barren rock." In what had always been his truest home, he felt as stranded as the yak that accompanied him across Chiantar Glacier and fell into a crevasse, legs dangling in thin air.

The emptiness pushed him to despair but also, as voids sometimes do, to raptures. He was "lifted by the sight of pitiless walls of ice," of remote summits "bridging heaven and earth." Out before the sun cleared the ragged ridges, he felt himself dissolved in a world rendered, as in a Chinese ink painting, in black and white, a vision of "clean, stark simplicity . . . our caravan dark silhouettes in mist and, above, the shining peaks."

Light and shadow, illusion and substance, the underworld and paradise: Dualisms began to pervade his notebooks. A labyrinth of crevasses was "filled with sinister darkness" but also watered Elysian meadows of yellow cinquefoil, lavender aster, and sky-blue forget-me-nots, all scented with mint and carrying him back to Alaska. "Ethereal cloud [met] impregnable stone." The mountains changed from "mere translucent vibrations in the noon haze to lucid and stark in the westering sun."

That same duality infiltrated his feelings toward people, tossing him between contradictory yearnings: for solitude and companionship, extreme exposure and shelter. Being always the stranger, uncertain how to act, had become exhausting. Though he had loved it when gorillas made him the object of their scrutiny, he felt less easy when it was humans who reversed the gaze, instead feeling "trapped when, as usual, a dozen kids and men sit in a row watching every move we make." Descending six thousand feet from "lonely, cold alien peaks," he felt just as estranged from the huts and fields and people "who groped their whole lives away." He knew that rumors circulated: that he could hypnotize and hold a leopard or was scouting a missile base.

Yet he also tarried in the tidy villages, enjoying seeing the families napping on their flat roofs or spreading bright orange apricots and purple mulberries to dry, and how the ripening barley shone gold against fields of deep green cannabis and white and pink poppies. For any failed human connections, he faulted himself. In the village of Bang, when a group of men hunkered around him staring mutely, their necks swollen with the telltale goiters of thyroid disease, he realized they were just as uneasy as he was, and that he needed to make the first overture. He couldn't, and hid in his journal until a boy broke the ice by bringing him a dead *Plecotus* bat to examine.

•••

On August 21, 1973, George arrived home with a Swat chest carved of cedar (a sixteenth-anniversary gift for Kay), and his usual gaunt frame and insatiable appetite. After an uncommonly long stay of nearly a month, minus a week in Sindh checking on goats, he packed and left for Kathmandu to rendezvous with Peter Matthiessen.

The two had kept up their correspondence since first meeting in the Serengeti. George ("GS" in Peter's writing) wrote of the hot winds in Sindh—"in these desert mountains one must really suffer for one's data"—and of the "bleary-eyed" hippies invading even Chitral. His Christmas greetings were particularly glum. "An awful amount of time is wasted just getting some place, and as often as not there are no animals left to watch. Just spent a month in the mountains to get data on snow leopard. Trouble is, there were none."

Still, he had high hopes for the trip he was inviting Peter to join, to the Crystal Mountain in northwest Nepal. John Blower had shown him pictures of the blue sheep (also *bharal* or *na*) thriving there under the protection of the lama at Shey Gompa. By observing the rut, meaning again going to high altitudes dangerously late in the year, he hoped to solve a taxonomic puzzle: whether this sheep was in fact a goat. And in this rare place where prey was plentiful, he promised Peter, the siren snow leopard was also bound to appear.

Peter's wife, Deborah Love, had died of cancer a year earlier. Making sense of that loss would become a central focus of his journey. But it also left open the question of who would care for his son, Alex, during the several months he'd be gone. Twice, George wrote to say that Kay would be happy to take the nine-year-old into their Lahore home, "if he's the kind of fellow who adapts reasonably well. She would of course not like a child who would feel abandoned and unhappy." Peter declined—"My boy is in fine shape; friends will stay in my house"—but was relieved to know George would get them home for Christmas, that being "one of Kay's stipulations." (The Schallers had considered moving the family to Kathmandu, Kay wrote Bettina. "But George will be away there too and here in Lahore we have friends and a good school.")

Over several letters, George dispensed with logistics. For the sake of the permit, he'd had to make Peter his assistant. "It might complicate things to say you are a writer. Dolpo is sensitive because anti-Chinese 'bandits' seek refuge there from Tibet. Detente about visiting Peking does not apply to Tibet." He sent packing lists ("bring clothing to 15F . . . and any treats you can't live without"); estimated they would not spend more than a few hundred dollars each; and asked, "Have you climbed? We'll cross a 19,000-foot pass."

Peter tried to joke away his nervousness. "I must confess that 19,000 is an awful lot higher than I've ever climbed, and that I have no experience of climbing, really, except an intense dislike of cliff edges and narrow windswept ledges. . . . I'll probably make it alright. . . . If not, I can always stay at Shey and eat ghee for 2 months." He waited until later to tell George of the note Richard Van Gelder from the American Museum of Natural History had included with the special mousetraps George had asked Peter to ferry over. "I should warn you the last friend I had who went walking with George in Asia," Van Gelder had written, enshrining the Brad House story in legend, "turned back when his boots were full of blood."

George was reassuring. "If you've been at 15,000 you should have no trouble with 19, especially since there is no difficult climbing involved."

The two-and-half-month, 250-mile trek across the Himalaya the two began on September 28, 1973, would be chronicled four times. First in the travelers' journals, all preserved, though George left his unmarred while Peter's became a first draft, with scribbled revisions and redactions often past the point of readability. Each then wrote it again, as a book or, in George's case, a long chapter. It's hard to think of another historic journey recorded simultaneously by two such peerless observers. These separate accounts—of dizzying, luminous landscapes, wolves the color of gold, and porters who abandoned them far beyond the reach of help—open a *Rashomon*-like window on both the adventure and each man's character.

Typical of his first weeks anywhere, George began the journey more as anthropologist than zoologist. He found a pleasing calm in the villages they walked through, all absorbed in the autumn's work. People swept yards and spread pulses on mats to dry, the *thump-thump* of rice huskers an ostinato under the cicadas' metallic whine. He noted the bent knife (kukri) a man used to kill a buffalo, nearly severing its head, and how the villagers collected the blood in tightly woven winnowing pans, drinking it fresh and warm. He heard sweet bleats coming from overturned baskets, where women put baby goats to stop their wandering, and pondered the problem of human feces when no pigs were around to eat it. He made detailed drawings: of the wood woman that guarded a bridge, her vagina stretched open; the roofs arrayed in serene geometries of drying pumpkins and seeds;

women's hairstyles and earrings; the clay relief of two interlocking cobras, "strange for people who will never see one." As everywhere, he found some villages ordered and lovely, others degraded. In one "miserable clutter of huts," "squat women" scurry and screech and spindly-legged children "play among refuse heaps . . . enshrouded with veils of dust and swarms of flies. . . . Seldom have I been . . . in an atmosphere so Stygian." Arrived a month later in Ringmo, its white, shining "pagoda-shaped stupas" and stone houses against "cliffs rising into everlasting snows" seem "terribly vulnerable, a medieval vision that might vanish with the shrug of a mountain."

Four Sherpas joined them for the journey, repeatedly demonstrating what George had assured Peter, that "when the going gets rough, Sherpas take care of you first." George was glad to be reunited with Phu-Tsering, now with a roguish red cap and shining gold tooth. The other three—also in their twenties—soon won his equal regard. Though "slender, soft-featured, diffident and reserved," head Sherpa Jang-bu proved unwaveringly loyal and determined to make the expedition a success. Restless whenever they held still, he often walked a day, or several, to surrounding villages, to ask about wildlife and bring back good cheer: delicious disks of heavy green buckwheat bread or a bottle of the local white spirits called Arak or Raksi. "Shy Dawa"—that was the Homeric epithet Matthiessen gave the most bashful Sherpa—often ate outside their circle, his back to the rest, but was a powerful and willing worker. Gyaltsen, the youngest of the bunch, brought along schoolbooks and made the whole trip in sneakers. (George had given each Sherpa a hundred rupees to buy boots, but Gyaltsen had spent his on something else.) Despite that lapse, "which can endanger a whole expedition," George admired the boy's toughness. No matter how rough the day, he remained smiling and willing to help, not once complaining of frozen toes. When at dawn George heard twigs snap, "someone, probably Gyaltsen, is building a fire."

Their first bunch of porters were equally simpatico and reliable. Five

Tamang men stayed with them all the way from Kathmandu to Ringmo, though they walked barefoot, saving the shoes George had bought them to sell later, and had little experience at high altitudes. Bimbahadur, a stocky Magar who had served in the British army, saluted them each morning but kept to himself. "At dusk, with the drizzle now an icy downpour," wrote George, "I see his earth-colored figure hunched alone by his solitary flame." Walking far up-valley with Peter one morning, George spotted "a kind of gnome" in a burlap bag. In Peter's version, they took Bimbahadur for a Yeti, or a "mountain lunatic," howling and brandishing a stick. So that he might sleep alone in a cave, "this oldest and slowest of the porters has given himself an additional six miles of stony walking." Most complicated was a Sherpa named Tukten who had signed on as a porter. After eight years in the army in Malaysia and Hong Kong, he had a worldliness the others lacked and spoke excellent English. Peter became quite mesmerized by him, seeing in him magical powers.

Despite this sound team, they immediately met obstacles. The airstrip they'd planned to fly into could not handle a plane big enough to carry them and months' supplies, so they would have to walk from Pokhara to Dhorpatan, adding nine extra days. Two weeks of rain stalled them further and promised deep snow on the high passes ahead. George worried they would arrive too late for the rut and—since he had to pay porters even on idled days—run out of money. Checkposts added more uncertainty: Though their permits were in order, "the farther an official is from the constraints of Kathmandu, the more he may become a local despot." Once, climbing a notched tree trunk to the police station roof, Peter joked about the dour official presiding there on his metal folding chair: "I bet he hasn't cracked a smile since he was weaned." George replied quietly: "Some officials speak English." Fortunately, this one did not, and waved them on, though they got a shock when a shot rang out at their backs. A constable climbing the ladder had accidentally discharged his rifle.

George's parsimony made everything tougher. Already at Pokhara, Peter lamented, "we could not afford an extra porter to carry more paraffin for the lamps, or more canned food to vary our diet, or even a single bottle of strong drink." Though he admired George's ethic of "rigorous responsibility" to his sponsors, he sadly tolled the end of their sausage, crackers, and coffee, then their sugar, chocolate, and sardines. "We shall soon be down to a pallid regimen of bitter rice, coarse flour, lentils, onions, and a few potatoes, without butter. [With] dwindling lamp fuel . . . life promises to be stringent." When late in the trip they were offered yak meat and butter, they were out of money. "As to finance and frugality," Peter wrote in his journal, George "may be a bit fanatic."

As always, finding and keeping porters posed the biggest challenge. After much searching, Jang-bu enlisted seven from the Kami tribe, though he did not trust them so followed closely "like a police escort." The "dirty Kamis," as Peter punned, got them over the first pass (13,400 feet), but then went home. Next, they hired a man with five ponies, who after a day of pouring rain ditched them too. Peter recorded George's profane response. "Goodbye, you sonofabitch: I don't like fair-weather people who back out on an agreement, leaving others stranded!" (Though it likely didn't cross his mind, the abandonment recalled the farmer who left him and his father halfway across a muddy field on their return from Dresden.)

"Tibetans say that obstacles in a hard journey . . . are the work of demons, anxious to test the sincerity of the pilgrims and eliminate the fainthearted," Peter later wrote. "GS has certainly been tested . . . but perhaps because [he] is not fainthearted, there falls almost immediately a stroke of luck." Travelers heading home to Tarakot agreed to carry if paid three times the usual fee, a demand George met, saying, "They know we are stranded, and they have us against the wall, it's perfectly natural!" Although they were also barefoot, the Tarakot travelers got them over the second high pass (15,300 feet). Descending with giant steps through the sun-softened snow, George found their pink trail of blood spots marking the way.

That pass brought their first real suffering. George had urged everyone to protect their eyes from the dazzling white light of the reflected sun. Phu-Tsering had goggles, and George gave his extra pairs to Peter and Jang-bu, advising the rest to tie a strip of cloth around their eyes with slits to peer through. Except for Tukten and one porter, no one heeded him. Dawn revealed "a scene of utter desolation. Some porters kneel on the frozen ground, moaning and holding their hands over their faces, while others stumble about, their red eyes draining fluids and swollen almost shut." The stricken men tried to help one another by blowing on their eyes. But snow blindness is cured only by time, so George sent the Tarakot men home. Peter described him as "disgusted" at the poor adaptation to their environment of these "childish" people. George had perhaps forgotten his own failure to bring sunglasses when last in Nepal. Though he had tried looking at his own shadow to avoid the brilliant reflections, he had learned the pain of sunburned eyes.

Anguished to see old Bimbahadur suffer, George made him a poultice of tea leaves, stopped a night to let him heal, then left him with Phu-Tsering to rest another day. After a breakfast of frozen rice soaked in cold coffee—they were out of fuel—he and Peter would carry the porter's and Sherpa's loads over 14,880-foot Jang La. Peter, head pounding and legs weak, set off with "a faint heart." George, though also dizzy from the glare, felt a "wild, inner joy" at their self-sufficiency. Bimbahadur caught up with them in Tarakot but turned back for home, accepting George's parting gift of new white sneakers and white ski socks. Socks "pulled up high on his short rootlike legs," Matthiessen wrote, "the old Gurkha salutes us, crying, 'Sahib!'"

George and Peter's relationship was as complicated as each of the men in it. They sometimes sought one another's company, especially in the early days: sharing an immense cucumber for lunch or—when the rains finally ended—baking side by side in the sun. George gave Peter a haircut, "'short like a pelt,' according to instruction," and even lingered in bed one morning while Peter made the fire and fetched water. Both exceptional naturalists, each shared what they'd seen. PM: "There's a tree back there with lavender, orchidlike blossoms . . . Do you know its name?" GS: "A *Bauhinia*, I think. I saw an orange-bellied squirrel a while ago." Occasionally, they reminisced about the Serengeti. They wondered how Asian vultures would react to the egg of an ostrich, a bird once common here. After watching a wolf hunt, George asked Peter if he remembered the wild dogs killing the zebra foal in the strange storm light.

Mostly, however, these private men kept to themselves. George often hiked far ahead of the rest. And even when together, they rarely spoke more than a few words. If depressed by rains or delays, they ate their rice or tinned cheese in silence, or retreated to their solitary tents.

Each was awed by the other. As Kay would later tell Peter, "George admires your writing so much, and being shy, has never wanted to impose. . . . He was so appreciative of your teaching." Peter in turn marveled at George's utter ease in these mountains. The "human chamois," he called him in his notes (appropriately invoking a goat-antelope native to the Alps), watching him nonchalantly cross ledges so narrow that Peter had to bear the humiliation of crawling on hands and knees. Until now Matthiessen had thought himself an "inspired" walker but he deemed George "formidable." If not for the slow pace of the porters, "he would run me into the ground." Descents were no easier. Coming off a fifteen-thousand-foot ridge, Peter discovered that "GS's idea of a safe route is not the same as mine." From this "true mountaineer," he learned to hold his stave short on the uphill side, the better to punch it through the crust should his boots slip. When George took off a sneaker to find it full of (leech-spilled) blood, Peter was relieved to see "that he is mortal, prey to the afflictions of the common pilgrim."

The many delays had Peter worrying that they might get trapped at Shey for the winter; he feared running out of food but even more that he might fail to get home to Alex. George understood—"he felt guilty leaving his young son behind, indulging himself"—but found the repeated discussion of how Peter could leave early tiresome and distracting. "Rarely does one think ahead weeks or days, in the moment choosing the trail, scanning the sky for a change in weather." Peter felt shame at fretting but also vented in his journal at George's shrug at the dangers. If snows blocked one route, he'd said, they'd just go out another, without permits, passing the police at night.

The climactic adventure came, conveniently enough for both writers' narratives, crossing the last and highest pass. Arrived in Ringmo, they had been greeted by the handsome headman; astride a pony in a wine-red *chuba* with a silver dagger at his hip, he'd put his hand to his neck to indicate the depth of snow they would find on Kang La. Their US Army map said the pass was nearly twenty thousand feet (it is in fact 17,450), an elevation for which the Tamang porters were not prepared. Resigned to their decision to return to Kathmandu, George bought a goat for a thank-you feast. Phu-Tsering slit its throat, poured boiling water over it to make the hair easy to pluck, smeared the skin with mustard oil, and put the head and legs onto the fire. The stomach and fat went into a pot, and as the meat sizzled and intestines bubbled, Peter and George wrote letters home. Gyaltsen and Tukten would carry the letters to Jumla ten days away, pick up the mail the men hoped was waiting there, and replenish supplies.

All of this unsettled George, as he told his journal. "Writing home, thinking of home, causes restlessness of the mind. The even tenor of packing, hiking, finding the next camping place—the immediacy of the trip, is broken." Though the letter he wrote has not been found, Kay captured its essence for Bettina. With porters abandoning them and record monsoons turning easy passes into great obstacles, she wrote, George was in some ways "having an awful trip."

The appearance of twenty-nine blue sheep ("Males and females!" George exulted) promised a turn in their fortunes, as did his first sighting of ravens. "A good omen. My most treasured wilderness memories include these birds." But with Shey Gompa still a three-day walk away, they would need help from the people of Ringmo, and those negotiations went quickly off the rails. Though George thought the loads light and pay generous, the prospective porters protested for two hours, egged on by spectators on nearby roofs. George finally agreed to take two extras, for a total of fifteen, but then went mad at their slow pace. He went madder still when twenty minutes after starting out, they stopped for a long meal, and four hours later insisted on stopping to camp.

Though the first half mile of the trail was harrowing—a vertical rock wall traversed on stepping stones laid atop wooden pegs driven into the cliff, between which the blue lake was visible far below—George as usual walked ahead. Browsing barberries and rose hips, he was rewarded with the best omen yet: the pugmarks of a snow leopard.

But the second morning brought more rancor with the porters, who prepared with "deliberate inertia." When one began complaining loudly, goading the others into revolt, George told him to "shut up or get the hell out." The man "sullenly subsided." Next came a dismal gorge, where they had to wade through freezing water because the rocks they might have hopscotched across were wrapped in ice. Again, George got out ahead, this time vanishing. "He told me later that he sat near the trail and watched us pass," Peter wrote; "he just needed to get away from human company." ("I've been snappish today," George wrote in his journal, "a surfeit of people.") It fell to Peter to push the porters, who wanted to quit at 1:30 p.m. Though he got them to walk one more hour, they stopped at 13,600 feet, a spot they would call Cave Camp, leaving a four-thousand-foot climb in brutally thin air for the following day.

The next two days, October 27 and 28, were so exhausting that for the first time ever, George wrote his notes after the fact. He felt strong and full of anticipation when he and Peter set out at 7:00 a.m. from Cave Camp bound for Shey; they climbed almost two thousand feet to the head of the valley in less than two hours. Then George's judgment failed him. Though Peter had a 1956 book by a scholar of Tibetan Buddhism showing the climb beginning a mile farther, George's map showed the trail striking off to the left, up a creek. Certain he was right, he left Peter to wait for the porters while he forged ahead, seized by a force he later acknowledged as irrational.

> *The snow is hard and I travel fast, determinedly escaping from the terrible solitude of people crowded together . . . away from the pressure of the constant demands, decisions, and worries imposed by others. . . . To reach the pass, to force my will against the gods and devils that seem to make the mountains retreat before me, becomes my supreme purpose. In the shattering light of the snow my mind shuttles between illusion and reality, and though I now feel that this may not be our route, the blue notch of the pass remains a shining beacon, beckoning me on until the world vanishes at my feet.*

By 1:15 p.m. he was at 17,600 feet, but with no sign of porters behind and a "pathless chaos of peaks beyond," he knew he had gone wrong. Hurrying back to where he'd abandoned Peter, he found tracks heading up the right-hand creek and began to follow.

Peter, meanwhile, had deferred to the Ringmo porters, two of whom had been to Shey and agreed with his Buddhist text on where the path lay. He again kept them going until, at 2:00 p.m. and 16,500 feet, they balked, cached the loads at what they would dub Camp 2, and turned back to Cave Camp. With clouds gathering, Peter and the Sherpas worried about George. Was he trapped in a crevasse, waiting for help? They'd barely begun to backtrack when he came into view, clearly "not happy to see us here on this side of the pass." Though in his journal, George laid the failure to his own error, Peter felt blamed: "'Looks like this day has really been bollixed up,' says he accusingly . . . I do not point out that both yesterday and today he was elsewhere when the porters quit." Attempting to make peace, Jang-bu offered to take Peter's pack. Better take George's, Peter said. "And GS accepts this offer without thanks to either Jang-bu or myself and stalks off down the mountain without a word."

That evening, at what George called a "war council," Peter argued for giving up on Shey. They'd seen blue sheep and snow leopard tracks in Ringmo and would have access there to the outside world. No, George said. The animals at Ringmo were harassed and skittish. And even if the Sherpas had voiced doubts about making the pass, they would follow his lead. He dismissed Peter's concerns: "This is his first journey into high mountains and he has no basis for judging the limits of the possible." They were just a mile from the pass, just a day from Shey, and no real try had been made. In the morning, traveling light, George would go with Jang-bu, Phu-Tsering, and two local guides to reach Shey. There, they'd recruit new porters, since the Ringmo bunch were not likely to last. Dawa would carry up firewood in case they got caught in snow or darkness, then rejoin Peter.

As anticipated, come morning, the Ringmo men quit. They had demanded a raise, which George refused, certain that even if he paid, they'd bolt short of the pass. Peter agreed they were "robbers, but since the total increased expense . . . comes to about twenty-five dollars, our decision seems to me a false economy."

With his four companions, George set off early for Shey. The "supposedly impossible route [was] not difficult" at all. Although for the second day in a row he climbed four thousand feet to an elevation of almost eighteen thousand feet, sinking to his calves while the lighter men stayed atop the crust, they made the crest by noon. Sitting, or rather lying, on the black scree, smashed down by an icy gale, George was again "filled with a savage joy. Shey is ours. We have passed through the Himalaya . . . and now . . . the Tibetan Plateau lies ahead." With numbed fingers, he scribbled a note for Phu-Tsering to carry back to Peter, instructing him to move their fourteen loads to the pass while George and Jang-bu went on to scout Shey.

George then began the descent with a move that even Phu-Tsering, who had recently been on the world's fifth and eighth highest mountains, thought crazy; Peter was not wrong that George sometimes undercounted risk. The first hundred vertical feet were a sheer wall, overhung by a cornice. George chopped through the cornice, had the Sherpas belay him over the lip, then began cutting steps into the wall, bent against the wind. At the bottom, where the snow turned powdery, he strapped on his snowshoes—a piece of equipment he'd neglected to suggest Peter and the others bring—walking rapidly while Jang-bu floundered to his knees. At a frozen waterfall, they had to detour a quarter mile through deep snow, then cross the stream on iced rocks. George paused long enough to take note of a white-throated dipper.

It was dusk by the time they arrived at Shey to find no people, no dogs, no sign of life. Only after much calling did a woman finally appear and—once convinced they were harmless—offer shelter, potatoes, and tea. Namu, as she liked to be called, told them that she and her two children plus an old woman were the only ones remaining. The rest had all gone over the next pass to Saldang, to escape the early snows.

For himself, George was unconcerned by the deserted village. He had spotted sheep for his study, and would simply count on light snows through early December—for, as the women said, once it got deep there would be no way out until March. But he had to figure out how to get their loads here without porters. And "Peter is another problem. With the lama gone there's not much for him here. And he wants to leave by November 18. Who will guide him back to Kathmandu? Will Tukten and Gyaltsen get here with our mail and food?" With a goat snuggling against him for warmth, George lay awake worrying. "Tomorrow, we'll have a day of rest. Then I will have to go back over the pass to talk to Peter."

With the Sherpas and George all off fetching mail or climbing the pass, Peter had spent the day alone at Cave Camp. Then Phu-Tsering arrived with George's note, the "famous note," as Peter will call it. Its "peremptory tone" so dampened his joy that in *The Snow Leopard*, he reproduced it in full.

1200 hrs Top of Pass

Peter:

The route yesterday was obvious—sorry I was not there. Straight up the ridge you talked about and in two hours you're at the pass.

Move all gear to Camp 2. Establish Camp 3 on pass. Two trips a day can easily be made. With your trusty knife cut some thin 2–3' long willow wands and mark the trail every few hundred feet, in case it drifts over.

Pay porters 80R's each

GBS

Again, Peter heard an accusation: "his deluded intimation that the porters would have carried on to Shey if only he'd been there," as he wrote in his journal. And the orders made no sense: Why camp at eighteen thousand feet "when we must keep returning anyway to 'Camp 2' . . . a thousand feet below?" Slashing at willows, grudgingly following instructions, he again cursed George's fanatical thrift. "For twenty-five dollars' worth of porterage . . . we could have been across this pass."

The next day, powered by his fury, Peter, Phu-Tsering, and Dawa began moving loads. With sixty pounds of lentils on his back in a broken basket that shifted about and stabbed at him, Peter climbed first to Camp 2 and from there the last thousand feet to the pass. "Why isn't George here to help . . . instead of firing off those silly orders?" he fumed. "If he'd packed a sack of lentils 1000 feet up knee deep in slush, he would send no more damned messages about two easy trips a day." Or maybe it was a good thing he hadn't come. "I might have been unable to contain my temper."

Soon, however (at least in retrospect), "I come to my senses. . . . I know this man, and if he has stayed in Shey . . . he has good reason." In fact, had he known what George was up to, he may well have been more furious. While Peter was "floundering through soft snow," George was enjoying "a wonderful morning [sitting] with a wide view of dozens of sheep" and then a "lazy afternoon in the sun," ruffled only by "a bit of desultory sheep-watching and exploring of the prayer wall."

October 30 was brutal for both men. Peter and the two Sherpas moved two more loads apiece from Camp 2 to the pass. Arriving at the top the second time, they saw George in his bright blue parka making the three-thousand-foot climb back up from Shey, clearly slogging through heavy snow. Exhausted, they did not wait but returned to camp. At the pass by 1:00 p.m., George paused to appreciate how hard the others had worked to get nine loads there, then slid and ran down the 1100 feet to camp in twenty minutes. He spilled the bad news: There were no porters at Shey, and no food, and the monastery was locked shut. The five of them would have to finish moving the fourteen loads. He and Jang-bu had left a tent and firewood at the lake along the way should they want to stop a night before going the last four hours to Shey.

As he talked, George felt Peter seething and guessed why. While in his journal he was unrepentant—"He thought my note was a scolding . . . when in fact it apologized. . . . The hard labor and high altitude are having an effect on him"—with Peter he was more diplomatic. "It was I who failed the expedition by being truant at a crucial time," George told him, suggesting that *both* were suffering the ill temper of insufficient oxygen. Peter appreciated that George brought up his "curt note" of his own accord, and accepted his explanation that

> *it was so cold and windy at the Kang Pass he could not write more than bare essentials . . . [and] meant no criticism but recognition that my instincts had been correct, whereas he'd held us up by 'waiting so long on the wrong mountain.' That he wrote it as he did, and I took it so amiss, he ascribes to the pressure of the days preceding and high-altitude irritability that has ruined so many mountain expeditions. . . . Quickly and happily we drop the entire business.*

Though again the parallel didn't seem to occur to George, the exchange recalls his young peacemaking with Bettina, some of whose account-keeping Peter shared. Whether in penance, or to drive home that it *was* an easy climb, George rested briefly from his six-hour slog back from Shey and then spent two more hours carrying a load of sugar to the pass. He and Peter shared a tent that night. Though George was in pain from frostbitten toes, he was as ever "a remorseless sleeper."

On October 31, Peter and George climbed to Kang La for the last time, each carrying a heavy pack; the Sherpas broke camp and followed with the last loads. George coached Peter down the treacherous first wall: "Never move more than one thing at a time!" Phu-Tsering then began lowering their packs and food by rope. One basket burst, scattering Jell-O and spaghetti across the snow. Without looking, Phu-Tsering then dropped Schaller's big case. Alerted by a shout, Peter looked up to see it bouncing erratically toward him. At the last minute, he guessed right, jumping and flattening himself on the snow just as the case hit the very spot where he'd been standing. Chewed out by George, Phu-Tsering again dispelled the tension (his own and the others') with his merry laugh. Then all carried on, bent under one pack and dragging another, returning twice to do it again. Sapped by the "murderous work," George and Peter stayed the night at the lake, sharing the last tin of sardines.

Finally, after five weeks of walking, they reached Shey. George set up his own tent and helped Peter with his. The Sherpas moved into an empty hut, where for the next five weeks they kept a cozy fire.

As George had been discovering for the past three years, no place was remote enough to escape the impacts of tireless human enterprise. Here as elsewhere, stripped forests, overgrazed meadows, and eroding soils had spiked lamb mortality: Only half survived, auguring a grim future. Already, Shey had too few blue sheep, George calculated, to support even one snow leopard full-time.

Still, for the first time in all his treks across these mountains, George was able to establish a research routine. Recording such previously undescribed behaviors as male-on-male "rump rubbing" was deeply satisfying. As were two wolf sightings with Peter, including a thrilling chase and near kill, foiled when the sheep escaped onto a precipice. George was happy to have this canine predator about: scent marking prayer walls, lacerating the jugular of bait goats, delivering him the chance to assay the rumen contents of a wild kill. A snow leopard was about, too, but toyed with him. Camping in a cave where he'd found fresh tracks, George slept through her visit, only waking when in the morning Tukten brought him chapati and jam.

Having animals to watch put George at ease, as did discovering that they were not as cut off as feared. Saldang, where they could get food, was just a day's walk away, over a pass free of snow. The two routes heading south from there, the wandering Jang-bu reported, were also open. And George readily settled into local ways. When the owner of their hut showed up, wanting a rupee-a-day rent plus tea and George's extra boots, George threw in a biscuit tin and got twenty pounds of potatoes and a fossil too. In exchange for a plastic container, Namu gave them yak butter, delicious on the potatoes. When travelers came through (locals of all ages crossed these passes as a matter of course), George happily talked and traded with them as well. At day's end, he joined the Sherpas by the fire, feeling "doubly cozy . . . when the peaks give a sense of intense and remote coldness."

Peter, meanwhile, seemed increasingly withdrawn, reading his books and seldom joining the evening chats by the fire. Though he still sometimes went with George to watch sheep, "after half an hour [he] became restless and wandered off 'to meditate.' All to the good because with each doing our own thing, we did not clash." The question of who would guide him out of Shey early was answered when Gyaltsen and Tukten returned—though their long trip had been a waste of time. The most recent letter from Kay was already a month old, and the "outside intrusion" again made George edgy, breaking the peace and routine. Peter did not read his mail, not wanting bad news.

In *The Snow Leopard*, Peter devotes many pages to George. The resulting portrait is vivid and, in many ways, astute. Though Mark recalled his mom wishing it had been more purely heroic, he found it "pretty damn accurate and funny. . . . I can easily see my dad doing the stuff Matthiessen found irritating, and also what comes through later, his gentler, more meditative side."

That Peter often found George "blunt and abrasive" is no surprise; he was not the first ever tempted to throttle him. " 'When Kay is typing up my notes, and I don't hear the typewriter,' " he quotes George saying, " 'I go and ask her what's the matter: she gets wild at me.' He often says this—'Kay gets wild at me'—as if to remind himself that his wife may have good reason." When Peter indexed his journal, he included an entry under GS for "querulousness."

If George's reputation for irascibility was already secure, Peter was among the few to wonder at its source. "There is anger in the man (as there is in me)" he wrote in his journal. "I know that much, at least some strong emotions that he's unable to express. Though he talks little of himself, that little is instructive." Alluding to what George had shared of his "near-orphan childhood," he noted that thirty years later, "he is in touch with but emotionally out of touch with his parents. . . . Has seen his father on 2 occasions since war?" (It had in fact been five, though never for more than a day or two.) "An aloof and lonely man . . . he simply does not know how to read people."

Peter also captured George's nobler qualities. Though he pokes fun at his exultations over data, he recognizes that they express the kind of awakened state the *Be Here Now* generation yearned for. Watching the wolf hunt, "he was yelping right alongside them." Peter forgave the "frowns and occasional moans and sourness [as] more a bad habit than a quality; he is remarkably patient and resilient in trying circumstances [and] has fine, old-fashioned qualities in abundance . . . how many of one's friends could be entrusted with one's life?"

Above all, Peter saw that for all his briars, a "gentleness shines through. The Sherpas see it; they are fond of him and respect him very much. Phu-Tsering, hunkered on his heels, is often to be seen humming contentedly at 'Chorch's' knee." When George scolded Tukten for the porters' slowness, or Phu-Tsering for letting the Sherpas eat a whole can of jam for breakfast (he and Peter had made theirs last two weeks), they "accept his reprimands in good spirit, since GS is faithfully considerate of their feelings and concerned about their welfare." Having seen his anguish in the Serengeti at the prolonged death of a lioness, Peter had learned to recognize how George veiled pain. When they see a child with useless legs pulling herself along "like a broken cricket," George said "stiffly" that in Bengal beggars "break their children's knees to achieve this pitiable effect for business purposes: this is his way of expressing distress."

Though both had been happy to blame the altitude for their clash of wills on Kang La, Peter also saw these high landscapes softening George.

> *In the lowlands, GS was a formal man who could not quite communicate his feelings; in the freedom of the snow mountain he is opening out in true, warm colors. On two occasions, he has managed to say how glad he is of my company on this trip, and he astonished me in Ring-mo when he spoke of an impulse to "cuddle" a child (the child being snot-nosed and filthy, the impulse was stifled).*

Where Peter's portrait falters is when it becomes mostly about himself—as a force thawing George, or prying him open. In the first instance, Matthiessen casts himself as teacher in how to be a social human. He tallies George's lacunae, beginning with his commandeering of the good tent—twice the size of Peter's and dry—"without so much as a mention of taking turns. This is his expedition brought about by his hard work so he deserves it, but all the same." Peter does not voice that protest but speaks up the next day when George tosses used cans into a schoolyard. Even if locals wanted them for containers, "why not set the cans upon the wall instead of . . . making the people pick them up out of the mud?" He describes George actively seeking his guidance. "Perhaps you can teach me how to write about people," he had entreated. "I don't know how to go about it."

It is easy to imagine George saying such a thing; he is the weaker writer on humans.

When it came to dealing with actual men and women, however, George thought Peter flattered himself. He is "not as good as he thinks he is in learning about people," he wrote in his journal the evening of Peter's departure.

> *In fact, I think he is . . . not very interested in doing so. He seldom asked a probing question [of me], and when Dolpo people came to visit or trade, seldom came and sat and asked questions. I was struck by how much time he spent writing while he was in the field. The sherpas asked me if Peter doesn't like them because he didn't interact with them much.*

The wartime history that had left George an isolate had also provided some bridges. Both he and Peter were born with a silver spoon, but only George had seen his snatched away, replaced by such relevant skills as how to survive in a barter economy. Peter's unbroken life of privilege (even if rebelled against) had preserved his social graces but also created its own kind of remove. Early in *The Snow Leopard*, he worries how he and George will fare with "no company but each other"—as if the Sherpas and porters were not there. "Basically," concluded George, "he was on a sentimental journey."

More tenuous still is a second arc Matthiessen develops, in which he plays George's spiritual teacher. He starts this thread with a callback to his description of Schaller in Tanzania, as "a stern pragmatist, unable to muster up much grace in the face of unscientific attitudes." He takes comic note of a night when a fearful Phu-Tsering chanted *Om mani padme hum* so incessantly that George longed to throw him off the cliff. And he finds him generally skeptical of Buddhism.

> *GS refuses to believe that the Western mind can truly absorb nonlinear Eastern perceptions; he [thinks] Eastern thought evades "reality" and therefore lacks the courage of existence. . . . I remind GS of the Christian mystics . . . [he] counters by saying that all these people lived before the scientific revolution had changed the very nature of Western thought.*

This depiction slots neatly into an understanding of modern science that would necessarily see the men as opposites, split halves of a whole: the no-nonsense rationalist impatiently harrying the porters while the seeker listens as "a dove calls from the secrets of the mountain." This understanding has a fine lineage. Keats despaired at investigators like Isaac Newton who "unweave a rainbow"; Emerson warned that scientists are "apt to cloud the sight" and "freeze their subject under [a] wintry light"; Max Weber saw a disenchanted world.

Peter then traces the blossoming in George, under his guidance, of a kind of Zen mind. After a month together, he seems "less dogmatic, more open to the unexplained." At Peter's urging, he begins writing haiku, several of which, Peter concedes, are much better than his own.

> *On cloud trails I go / Alone, with the chatting porters /*
> *There is a crow*
>
> *Cloud men beneath loads / A dark line of tracks in snow /*
> *Suddenly nothing*

When the snow leopard had still not appeared by the time of Peter's departure, he attributed to George a comment anticipating Matthiessen's own famous rendering of that nonappearance as a kind of Zen koan ("Have you seen the snow leopard? No! Isn't that wonderful?"). "We've seen so much," he quotes GS saying, "maybe it's better if there are some things that we *don't* see."

"When I say, 'That was the haiku-writer speaking' . . . we both laugh."

Peter was George's teacher in important ways. The haiku practice further refined his already spare voice, as if the seventeen-syllable experiments George scribbled on the back of journal pages seeped through to the field notes on front. "I follow the trail of the others. Bare feet in the snow." It was thanks to Peter that he read core texts, gaining insights that became ever more valuable as he committed his life to historically Buddhist regions. Peter had sent Shunryu Suzuki's *Zen Mind, Beginner's Mind* before they set out. On the trail, after "compulsive" George had plowed through all his own books (and chocolate), Peter shared others he'd brought—The Bardo Thodol (Tibetan Book of the Dead), John Blofeld's *The Tantric Mysticism of Tibet*, Alexandra David-Neel's *Magic and Mystery in Tibet*—all of which George read, taking notes all the while. Peter introduced him to Milarepa, the eleventh-century Tibetan poet whose verses on the sacredness of all life George would later print on cards to offer people he met across the region.* With "Peter as teacher," he explored ancient shrines of the pre-Buddhist Bon religion and, at Shey, visited the lama, Karma Tupjuk, at his hermitage. (The monk had been there, it turned out, all the time, his privacy protected by Namu.) They waited two hours for an audience, the lama's helper Takla serving tsampa (the barley staple), bits of sun-dried green yak cheese, and salted buttered tea.

But George was unmoved by the lama, in whom he felt no aura of sagacity or mystic vocation. He would later distress Peter by calling him "a real scrounger" when, having asked for and gotten Phu-Tsering's pants, Karma Tupjuk wanted Jang-bu's boots too. Since the lama could barely walk on his arthritic legs, and Jang-bu needed his only pair to cross three seventeen-thousand-foot passes on the way out, George said no.

---

* Peter's discussion of Milarepa may shed light on George's ease with viscera and maggots. The monk followed his guru's guidance to embrace all he finds repugnant, the better to realize that everything in the universe, being inseparably related, is holy. Returning to his village to find his mother's decaying corpse, he made a headrest of it which he lay upon for seven days (Matthiessen, *The Snow Leopard*, 92).

George also tired of Peter's raptures, a feeling he let loose in his journal on their parting day:

> *Peter can be hard to take when he is a "writer." He thinks . . . that only he feels and sees the beauty of a place. . . . Only he notices the snow slopes seemingly moving in midday heat waves. Every little thing is verbalized, unique. Because I don't comment, he assumes I've not seen.*

When George does observe the "mountains move," Peter makes much of it, wondering in his journal "if he would have acknowledged this a month ago" and giving it a whole scene in *The Snow Leopard.* He points out that George has commented on this seeming motion many times. Still the scientist "resists the implications. . . . 'Well,' he mutters, 'from a certain point of view, I mean, geologically, of course, the Himalaya is still rising, and then there is a downward movement due to erosion—' I interrupt him. 'That's not what you meant,' I say."

Yet in a way, it was. Schaller's most spiritual reflections did often take the form of imaginative journeys across geologic time. For all that it served Peter's narrative to imagine George climbing under his tutelage to a higher consciousness, this scientist's quest beyond dry reason had begun decades before: in his struggle to express the completeness he felt sitting with the Muries overlooking the Brooks Range, his intimation peering into a Virunga volcano of a transitory planet in an eternal universe, the use of his own animal body and intuitions to make sense of the Serengeti ecology, and especially that moment in the Sindh desert with a pocketful of fossils when he imagined himself afloat in the primordial sea. Even lying prone, the wind and sand had lashed so brutally he feared he'd be blown off the narrow rock shelf. Clutching the ground, he'd inadvertently grabbed the fossils he had piled in a mound. "The wind roars over me like the waves of the sea, and for a moment I relive an ancient reality, floating among swaying stems of crinoids, water pressing me down, sands from the ocean floor covering me slowly . . ." Sleeping out at Shey in hopes of seeing the snow leopard, he returned to that vision. Stretched out among fossilized creatures of the ancient Tethys Sea, "I lie buried beneath the sediments of the ocean floor, and perhaps my skull will become a *salegrami* stone."

Even as Peter crafted his self-flattering narrative, he himself saw through it. That George was usually ahead of him on the physical path became symbolic: It was this solitary figure leading *him* into a higher kingdom. "In the next hour," he wrote in *The Snow Leopard*, "GS moves steadily ahead, until his blue parka turns to black. . . . The emblematic figure rounds a pyramid of white; then there is nothing." And in his journal: "GS is the Buddha; Tukten is the Buddha . . . pilgrimage to find Snow Leopard, find Deborah Love Matthiessen, find self; find death; find the Yeti—all the same."

Most telling is Peter's description of the Bonpo porters gliding along the terrifying ledge that had put him on his hands and knees. Despite heavy, awkward loads that at any moment could nudge them over the precipice, the men move "with an easy, ethereal lightness, as if some sort of inner concentration was lifting them just off the surface of the ground . . . GS appears, moving as steadily as the rest." On the next page, Matthiessen wonders if the porters might have been practicing "a primitive form of the Tantric discipline called *lung-gom*," a focus so intense it becomes almost a trance, mind transcending matter, "matter returning to energy . . . so that it flows." Schaller had achieved this same flow. As one scholar put it, "this scientist has [what] Matthiessen is seeking." Emerson counted us "strangers in nature. . . . The fox and the deer run away from us; the bear and tiger rend us." All of us, except George.

Paradoxically, this trip with Matthiessen was Schaller's most conventionally scientific, focused on the kind of "tyrannizing" taxonomic question (Emerson again) "which evermore separates and classifies things." When he appeared on *The Dick Cavett Show* in May 1980—the men flanked by ferns; George in a black turtleneck and sideburns—Cavett introduced him with a tease, mocking the pedantry of George's quest. "You've all heard, I think we all have, all our lives, about the importance of separating the sheep from the goats."

Aptly, the blue sheep resisted such categorizing. Though sheeplike in their preferred habitat and lack of knee callouses and beard, George found them goatlike in the way they fought and, when it came to the most telling trait, to be halfway in-between. Goats, but not sheep, take their own penis into their mouth, urinate, then spritz their face and beard to funk up their smell. While male blue sheep do not go quite that far, they do lick and nibble their own genitals in an unsheepish manner. As George noted in one of his dry comments in which he pretends no analogues are intended or will be heard, "an extended, unsheathed, erect penis has a number of useful functions." Neither *Ovis* nor *Capra*, exactly, blue sheep (*Pseudois*) likely resembled the ancestor of both.

Beyond that, taxonomy can be the opposite of reductive, an act of sacralization. W. H. Auden called "proper names . . . poetry in the raw." Nabokov's fictionalized father in *The Gift*, who like George searched for "animals in Tibet, the Pamirs and other blue lands," was happiest "in that incompletely named world in which at every step he named the nameless." Photographer Vincent Munier, who in 2012 reprised Schaller's search, didn't care whether a yak "lord" had "ruminated twelve times this morning" but insisted on calling Tibetan antelope *Pantholops hodgsonii*, "accustomed to speaking Latin in the presence of animals."

George's spiritual practice may have deepened in Dolpo, but it always consisted of watching animals. Predators in particular demand an acolyte's utter focus and stillness: Jonah Western calls tracking them "meditation but outward." Similarly demanding was George's effort—impossible but fruitful—to inhabit each animal's *Umwelt*, its distinctive way of perceiving the world. He pasted into his journal fragments of D. T. Suzuki's *Mysticism: Christian and Buddhist*. If a man "aspires to paint a plant or an animal, there must be in him something which corresponds to it. . . . The secret is to become the plant . . . the mind thoroughly purified of its subjective, self-centered contents." And on the Buddha: His "whole philosophy" came from "seeing things in the state of suchness (*tathatā*) or is-ness." In *Stones of Silence*, he will quote *Zen Mind, Beginner's Mind*: "When you understand a frog through and through, you attain enlightenment."

On November 18, George and four of the Sherpas saw Peter off, accompanying him to 16,900-foot Saldang Pass. Dawa and Tukten would continue with him to Kathmandu; Jang-bu and Gyaltsen would go as far as Saldang village to buy a goat. George would return to Shey, where Phu-Tsering waited alone. "Peter and I shake hands, a silent affirmation of gratitude. . . . It has been a good trip, a marvelous trip, and we wish each other luck in the days ahead. I watch Peter's tiny caravan retreat and as I raise my arm in farewell the figures dissolve into the immensity." In Peter's version, George blurts out, "I'm sorry as hell to see you go." When Peter turns back, a silhouette raises its arm. When, on seeing wolf tracks, he turns back again, the sky is empty.

George spent eighteen more days at Shey, altogether the longest he had stayed anywhere for some time. Their first night without the others felt quiet and lonely; when George sat to write notes, Phu-Tsering came to sit alongside, rocking and chanting. Both were happy when Gyaltsen and Jang-bu returned, bearing turnips for stew. Phu-Tsering soon adapted to cooking for this reduced crew. When George asked why the tea and coffee tasted the same, he answered, "Half and half, sah. . . . Sherpas want coffee and Sahib wants tea."

The few locals still in residence or passing through took to gathering by the visitors' evening fire. Namu brought her young relative Sani, who laughed and chattered and when the crowded bodies got too hot, opened her *chuba* to let cool air flow over her bare breasts. A woman George called "Tundu's wife" came to ask for empty tins with lids, a coveted item, "in such a sweet smiling way one likes to give." The lama's helper Takla stumbled in, pleading to sleep with them because he'd bungled the buckwheat bread and been threatened with the lama's stick.

Provisions dwindled: first salt, then soap, though George couldn't imagine where the latter had gone since a "glutinous black soot . . . penetrates every crease and every pore of our skin [and a] strange gray fuzz . . . clogs our hair." On his first trip to Nepal, he had wanted to wipe clean the children's dirt-crusted faces, inflamed and scabby from seeping noses. Invited by an elderly man for tea, he had watched with chagrin as his host's wife cleaned a greasy porcelain cup first with a grimy rag and then with her thumb and spit before filling it for George with chang (sour barley beer) and a pat of rancid butter. After he drank, she rubbed the butter into her hair to deter lice. By now, having taken just two sponge baths in the months since leaving Kathmandu, he'd "discovered as Tibetans had long ago, that the luxury of warmth far outweighs that of cleanliness."

George liked Shey emptied for winter. "I suspect that in summer with people and livestock everywhere, its feeling of remoteness is lost." Every day he walked to the hermitage to watch sheep. An old woman and a black cow lived there now. When a vulture signaled the presence of a dead blue sheep, he regretted that the village was at that moment "full of people; at least ten or twelve," all of whom ran to the carcass with baskets and knives. Jang-bu got a hind leg for their household. George arrived in time to get only a stomach sample, though he did get a look at the glands by visiting families who had taken home bigger pieces. For dinner they ate the roasted leg, "excellent and very tender."

Still hoping to see the snow leopard, George bought a goat out of a herd of fifty driven through by three traders from Tibet, whose red pants, multicolored boots, and long wrapped braids he greatly admired. He tied out the little animal where he'd seen the cat tracks. Every morning he crested the ridge with anticipation, but every time found the goat still alive. (On this use of live bait, Cavett asked: "Do humanitarians, uh, animalatarians, become upset?" Schaller answered with a logic that most conservationists, if not animal rights advocates, would embrace. "The animal does not really suffer. It has been watered and fed. And for every domestic goat the snow leopard eats, a wild goat's life is saved.")

When December 1 brought the first storm clouds in weeks, George knew that though the cat had still not appeared and the rut was only half over, they had to go. The passes would soon close, and he had not forgotten his promise to be with the family for Christmas. Putting aside seven days' worth of food for the trip out, he resolved that when the rest was eaten, it would be time. He traded away what they wouldn't need: a scrap of mosquito net for milk, a frayed towel for yak butter. To mark his final day, he opened a last tin of fruit, his special treat since boyhood.

Since no predator had killed the goat, Phu-Tsering did. The Sherpas curried the lungs and kidneys; George fried the liver and heart; and they sent a leg to the lama, who sent gifts in return, including a prayer flag he had printed especially for George. Meat brought visitors, even the wary woman with her black cow. The wolves made a last appearance, "as if to see us off." Phu-Tsering sprinkled incense in the fire, chanting to bring good luck. They would soon need it.

When George left Shey on December 6, 1973, he did not look back or stop until he reached Saldang Pass. Namu and her husband had helped carry that far but parted abruptly, with "no word or gesture acknowledging that our lives had touched across the depths of our cultures." Suddenly George longed for home, "needing the solace of my family, its warmth and love." After descending an ice river, he, Phu-Tsering, Gyaltsen, and Jang-bu found shelter with a woman who, like everyone they met, kept at her work while she talked. As she spun thread, she told them how her neighbors searched out wolf dens to kill the pups.

The next three days would be so tough that George would once again write notes only at the end. He and the Sherpas and three new porters set out early, to get ahead of the midday meltwaters that would flood the canyon they would soon enter. Passing a caravan of yaks in brilliant red tassels and bells carrying soap and butter churns to trade for buckwheat and rice, George imagined the forlorn picture his own small band made in this immensity, "little figures moving over the grey and tan terrain, bent under loads against the wind." By dusk, snow was falling heavily on the peaks. Would they have another crossing like Kang La? Camped at fifteen thousand feet, George soothed himself with the (mistaken) hope that the pass was just 1,500 feet higher. By dark, the storm had enveloped their camp. All night, George listened to snowflakes tapping on the tent roof, knowing the porters would not go on.

Sure enough, waking to a blizzard, the porters refused to continue even for doubled pay, lest they not make it home again. George knew he and the Sherpas had to get over the pass that day, for just a few more inches of snow would close it for the winter. But with all landmarks obliterated, they could not possibly find the route themselves. Their only hope was a young man who, with a boy and a dog, had attached himself to their group the previous day. The man said he would guide them in exchange for whatever gear they would discard so that their original seven loads could be carried by four. George handed over cooking equipment, all but three days' worth of food, and even some specimens, which the youth cached. To the Sherpas, he gave every bit of clothing he did not have on: to Jang-bu a sweater and gloves; to Phu-Tsering socks for mittens; to Gyaltsen, still in "knickers . . . like a schoolboy on a picnic," more socks, a thick shirt, and an ear band. Still, each was left with a seventy-pound load. And after just a few hundred feet, their guide deserted them, the blizzard by then so intense they were immediately lost in a featureless expanse of swirling and deepening snow.

That would have been the end, but for the apparition that saved them. First, they heard bells ringing and hollow cries—*Yo, Yo*—as if someone was "dispersing a gathering of mountain spirits." Then, out of the milk-white nothingness, descending from the clouds, came a sinuous line of fifty black yaks and equally "wild-looking" men, their long hair wind-tangled and sticks held high. The spectacular picture George managed to take would remain close to his writing desk ever after. Though the travelers and their tracks soon vanished, a trail of yak droppings led the way to the pass.

Still, their trials were not ended. The yaks' sliding step had turned the trail to ice and the men fell heavily with their loads, each struggle to stand up again exhausting them more. With screaming winds whipping up the snow, they had to stay within a few feet to see each other, turning their backs to huddle like penguins against the most battering gusts. Losing heart, they prepared to bivouac, hoping to survive the night.

Then, again, they heard bells. As if in the allegory of the four ages of man, an elderly herder appeared, accompanied by a young man, a teen, and a little boy, driving six yaks up the pass. Mercifully, they agreed to carry George's suitcase, lightening each man's load by fifteen pounds. With the yaks breaking trail, they reached the 17,500-foot pass, where despite gale winds the young man unwrapped a prayer flag, climbed the cairn, and tied it on a pole. George quietly added his thanks to the mountain gods. The descent down to camp at 15,200 feet was brutal; George fell thirty times. But the yak men gave Phu-Tsering wood to build a small fire, and George shared his party's last cheese and biscuits.

Sadly, by morning this reciprocal trust had crumbled. The "Dolpo men" said they would stop for a day to let the yak feed. George asked if one might continue on as their guide. The young man and teen said they would, then said they wouldn't, then changed their minds again. When they suggested that George leave the gear for the yaks to bring tomorrow, he grew suspicious, "especially since the kid kept ducking under the blanket to laugh." Finally, the young man agreed to guide if paid sixty rupees in advance. "We were in no bargaining position, so I paid."

They had walked only an hour when suddenly the Dolpo man fell to the ground, rose to limp dramatically, and said he could not go on. When he said he could not give back the money because he had left it with the others, Jang-bu's reaction startled George. Exploding in rage, the Sherpa kicked and punched the man until he lay sobbing in the snow. His leg "cured," he picked up his pack. But he walked so slowly, George knew they would never make the next pass.

Still whimpering, the man began to insist there was an easier way, down a canyon. George knew canyons to be treacherous. But with Gyaltsen still in sneakers, he agreed—though he added that if this route were so easy, then surely the "Dolpo chap wouldn't mind coming along a ways." The man had no choice; Jang-bu threatened to beat him again and Phu-Tsering stayed right behind him.

Soon, as George had feared, the walls closed in. With the turbulent river twisting to touch first one and then the other sheer wall, they had to cross and recross ice bridges, struggling to maneuver with their heavy packs, water rushing beneath. George went first, to test the ice with his own greater weight. Stomping to sound out hollow spots, he crawled across the most dubious stretches and sometimes leapt desperately when a bridge gave way. Slipping on a rock, he got soaked from the chest down, his clothes immediately freezing. When the canyon got too narrow for them to remain at the bottom, they climbed up to a faint trail across the cliff face, a six-inch-wide iced ledge sloping outward. Worrying about the Sherpas with their more unwieldy loads, George again went first, heels hanging over the abyss and bare fingers dug into frozen cracks, shuffling sideways to scrape platforms for them with his toes. Worst was climbing down a twenty-foot frozen waterfall, like the shining glass mountain that tests a fairy tale prince. "I am sure we all wished for the pass."

At 4:00 p.m., well ahead of the others, George stopped and built a fire to dry his clothes. As darkness fell, the birches turned opalescent under a full moon, but still the Sherpas did not come. Without pad or tent, the cold stabbing deep into his hips and knees, George spent the night planning how, if one was hurt, he would get them out.

The next morning, after rekindling the fire to thaw his rock-hard boots, George began hiking back to find the Sherpas. He'd walked only a few minutes when he heard them coming. They had slept in a cave, without the Dolpo chap, who at a difficult spot had dropped his load and bolted. George's relief that all were safe was doubled when the canyon opened onto a sunny plain. Again walking ahead of the rest, he had scarcely stepped into the light when something bounded off. A gray and buff rump, a long flowing tail—snow leopard! In months of searching, he had not been allowed even a glimpse. Now, on his last day in her realm, the cat had granted his wish. Continuing downriver, he found fresh tracks and scat of three more. With that, and his return to the very spot at Phoksundo Lake where they had camped six weeks earlier, he felt the journey was complete. A lone diving duck floated on the water. Jang-bu tried to stone it for supper, but George stopped him.

The trek from Phoksundo to the airstrip in Jumla normally took ten days. But if George was to make the flight Peter had reserved for him to Kathmandu, where the family would join him for Christmas, he would have to make it in six. He and Phu-Tsering decided to go ahead, carrying just notebooks, film, cameras, and sleeping bags, leaving Jang-bu and Gyaltsen to follow with the rest. Even half running, George noticed the scarcity of tracks; he grieved especially for the "persecuted" musk deer, poached for their scent glands, long prized for perfume. Hiking late into the night, Phu-Tsering chanted to keep dark spirits at bay. Once a day, he stopped at a house to buy them a meal. On the third morning, they got eighteen eggs and ate all but two at one sitting. Fed a sumptuous feast in Ringmo—green buckwheat bread, chang, boiled potatoes, and omelet—George almost forgave the porters who had abandoned them.

Arrived in Jumla December 16, George felt jangled by the first radios and motors he had heard in ten weeks. He wrote a sardonic ode to his newest companion ("Oh flea / you too / are one / with me"), collected pine branches for the family Christmas, then flew on a tiny plane back over the route they had walked in on. Landing in Kathmandu, he rushed to the Hotel Shanker.

But the family was not there. Warned by the US Consulate not to come to Kathmandu without hotel reservations, which despite multiple cables

Kay had failed to secure, she had canceled. "It's one thing to travel by myself without reservations," she wrote Bettina, "but with the boys as well . . ."

George, bitterly disappointed, had to make new plans. Flat broke, but stuck waiting for Jang-bu and the gear, he borrowed $500 from John Blower at the UN. Instead of family time, he had meetings: with Andrew Laurie on the rhino work George was supervising, and with John Seidensticker, who under the auspices of the Smithsonian Institution had just become the first to radio-collar a tiger. (In his journal, George complained that Seidensticker did not ask him a single question. Since Seidensticker has called George his idol and inspiration, the elder scientist's feeling of being ignored may have distilled a larger anxiety, about being left behind by a more tech-savvy generation.) Though George made it to Delhi December 23, fog canceled his flight to Lahore. Again, he filled days: seeing Ranjitsinh, searching shops for illegal furs, spending "a cheerful Christmas in a damned hotel, alone." On December 28, he at last made it home.

While pacing about in Kathmandu, George had written a long letter to Peter. It began with a startling moment of disloyalty. "As usual things got screwed up. No Kay." The consulate had told her there was "no hotel space—what nonsense—so she canceled the trip. Certainly one of the most stupid things she has ever done. She wrote you to look for reservations, but the letter was with my mail, so you never got it. . . . I'll miss Christmas with the family and might as well have stayed at Shey." He guessed that Peter's hunger was finally sated ("it usually lasts 10–14 days after a long tough trip") and passed along bird identifications from the Flemings, the father-son duo who would soon publish *Birds of Nepal*: "The whitish falcon is probably a merlin; the bird I thought a black eagle was an immature golden." He recounted "the intellectual excitement" of his last weeks at Dolpo: a ewe who though not in heat let herself be "gang-raped"; the rams who ran other females ragged, sniffing and mounting; another sighting of "our wolves."

His account of the journey out was brief: just a few low-key sentences on being abandoned in the blizzard, the rescue by the yaks, the canyon, and the sighting they had both yearned for. "A snow leopard jumped up in front of me and ran off. There had been two but I did not see the other." The understatement was true to form. Though George had just come off the trek when they met in Kathmandu, Laurie recalled, he said almost nothing of it, instead devoting all their time to discussing Andrew's work—and to eating.

Intentionally or not, the letter salted Peter's sensitivities. Though his reply has not been found, it is clear from George's response that Peter again felt blamed. "I seem to have a knack of writing notes and letters that leave themselves open to misinterpretation," George began. "Of course, you did everything possible, and more, in Kathmandu. Things got fouled up because of Shanker ineptness and Kay's timidity." He tried to make amends for calling the lama a scrounger. "I don't consider him a 'low character.' I suspect he was more complex than we gave him credit for . . . and lamas have to make at least part of their living that way." To Peter's apparent claim that his own route out had been harder, George agreed: Going down the canyon was in retrospect "preferable . . . since we got out without mishap." He then spent a paragraph undoing that affirmation. Had anything gone wrong, there would have been no help; a German architect told him the two-pass route was easier, et cetera. Neither man ever seemed quite able to yield to the competition.

About the same time (February 26, 1974), Kay also wrote to Peter, sliding from an almost abject tone into something else.

> *Dear Peter,*
>
> *This letter is long overdue as I meant to write in late December to thank you for making the hotel arrangements etc. of which I did not take advantage. Neither your fault nor mine. . . . I would have been better off to forget about reservations and just take a chance, [but] I lost my pioneering spirit and decided it would be impossible to go,*

> *much to George's and my disgust. It certainly made for a miserable Christmas for us all. And I doubt I'll ever quite forgive myself.*
>
> *As if that was not bad enough, we got a copy of a letter Bill Conway wrote [NYZS primatologist Thomas] Struhsaker in which he says: "Most disturbing call from George's cousin Bill Barnes last Friday. Kay had written to report that there had been no news from George since the report that he was out of food with Peter high on a Pakistan mountain, it was snowing and the porters had deserted! Several weeks had passed. Then, Monday, Peter telephones to say that all was well, he was back, George was okay and that it had just been adventuresome."*
>
> *George is beginning to doubt my word that "I never wrote you were out of food, just that the porters deserted" and I'm wondering which Bill (Barnes or Conway) colored up my letter? Did Bill Barnes sound disturbed when he called you? I only gave him your number (which George says you probably didn't appreciate, so forgive me) in case he was as eager to hear about the trip as I was. He must have misread my letter or the tone of it or something. I really had no worries about the two of you up there. But I think George is beginning to think he'll have to censor my outgoing mail. Which he can't do because he is* NEVER HERE *anyway. I'm beginning to threaten to get a male concubine, number 2 husband, friend, etc. and leaving around books such as* Open Marriage, *but unfortunately he knows he can trust my Victorian attitudes. It has been a bad year for Schaller togetherness and the coming one doesn't show much chance for improvement.*

At least, she finished, they'd have the summer together in New Zealand. "I don't know if it appeals so much because I'm eager to see that part of the world, to be going home, or just to be traveling with George again. We've not had a trip together since we went out to the Serengeti in 1966."

As they readied their respective manuscripts for publication—both by Viking, Peter's in 1978 after first serialization in *The New Yorker*; George's in 1980—each shared a copy with the other. Peter appended a list of questions, ranging from how to convert Fahrenheit to centigrade (George provided the formula) to the earliest appearance of the sheep-goat ancestor. (GS: "The subfamily Caprinae appeared in the late Miocene, 20M years ago.") Beyond these answers, George offered just one substantive suggestion. "I feel Tukten is overdone; you seem to try too hard to make something out of an enigmatic smile." Peter was apparently unpersuaded. Tukten remains in *The Snow Leopard* an almost-supernatural character: "wanderer or evil monk, or saint or sorcerer" with "yellow Mongol eyes" and a "Bodhisatva smile that would shine impartially on rape or resurrection . . . the gaze he shares with wild animals."* He did add that "GS feels that I make too much of Tukten's spiritual propensities and . . . warns me to beware of him."

Kay offered Peter one additional note, defending George's character in a way he himself never did. (She attributed his scowl, for instance, to hours of squinting through binoculars, and when asked what demons George struggled with, said she didn't think he had any.) "Peter would be murmuring Buddhist prayers and George wouldn't want to interrupt him, so if he had something to say he'd write out a little note and toss it in through the flap. This irritated Peter; he didn't realize George was respecting his privacy. And I explained that to him and it made sense to him."

That change Peter did take, twice. When George "hurls goggles through my tent flap," he wrote, it is "more in exuberance than rudeness." And after fuming over the infamous note, he added this:

> *Surely he does not mean this note the way I take it. What seems abrasive in GS's behavior is often merely abrupt, as when he flings something that he wishes me to look at through my tent flap; once I flung something right back out again, as a hint I didn't care for these manners. But as I learn more about this man, I see that such acts are not bad manners but the intense respect of a private soul for the privacy of*

* Some have criticized what *Boston Review* (May 7, 2014) dubbed "Matthiessen's Orientalism," particularly in this aura of "dark magic" he attributes to Tukten.

> *others; for all he knew, I was taking notes, or meditating, and might not welcome an exchange of any kind. On a hard journey, with no respite from each other, such consideration (extended also to the Sherpas) is far more valuable than mere "good manners," which sometimes hide a mean spirit beneath, and may evaporate when things get rough.*

Peter's notes on George's manuscript were far more extensive, filling three mostly single-spaced pages. He began with one "quibble" on the writing, acknowledging that "this impertinence comes from somebody who not only much admires your writing but had an also-ran in the same category in which *The Serengeti Lion* won the NBA." What he thought George could improve was "a tendency at times to weaken strong writing by overstatement (usually with excess adverbs or rhetorical abstraction) . . . of points you have made elsewhere more subtly or directly." Kay wrote soon after to thank him for that note, which "(perhaps to George's dismay, though he says not) released my inhibitions about over-editing. I went through again with a very determined blue pencil, as did George. We hope you will note the difference in the final result." A letter to Bettina testified further to Kay's key role in refining George's work: "I will be meaner than ever so as to strengthen and tighten it."

The rest of Peter's long note challenged how he came off in George's telling. He found George's one-sentence account of a tense encounter with a Tibetan Mastiff—to which Peter had devoted two highly dramatic paragraphs—to be too "casual," and objected to his calling the climb to Kang La easy. The contrast recalls their divergent accounts in East Africa of the zebra kill and elephant charge. And, two decades later, Matthiessen would himself dampen the drama. Having made much in *The Snow Leopard* about the dangers—how far they were from help, for instance, should an appendix burst—he told *The Paris Review* that the "physical risks were minor."

Peter's critique continued. Tukten and Gyaltsen were not sent on the

German Vice-Consul Georg Schaller Sr. and his eighteen-year-old fiancée, Bettina Byrd Beals, in Chicago.

CHICAGO HERALD AND EXAMINER---A PAPE

Prepare for Exciting Onwentsia

BRIDE OF TUESDAY

Married August 30, 1932, the newlyweds left that same day by train and ship for Berlin, where George was born in May 1933.

He "moves all the time like an eel," Bettina wrote to her best friend about seven-month-old George, "and goes off in wild explosions of laughter over nothing."

Schaller was just nineteen when he made his first field survey, paddling the Colville River to the Arctic Ocean. The raven he befriended returned with him to campus and introduced him to the love of his life.

A week after his twenty-first birthday, Schaller made the first ascent of Alaska's Mount Drum with Austrian alpinist Heinrich Harrer, recently returned from his famous seven years in Tibet.

George (second from left) at twenty-three, on the 1956 Sheenjek expedition that spurred the creation of the Alaska National Wildlife Refuge and set him on his life's course. Fellow graduate student Robert Krear took this photo of him with professor Brina Kessell, visitor "Doc" MacLeod, and expedition leaders Mardy and Olaus Murie.

In camp, among friends.

At the University of Wisconsin, beginning what he thought would be an ornithology PhD, George was delighted to find his ducklings imprinting. He and Kay Morgan were married in August 1957.

Belgian Congo, spring 1960: A blackback Schaller called Kicker was open and curious toward young apes from outside his own troop, including twenty-six-year-old George.

At last overcoming her terror, Kay often climbed into the gorillas' abandoned nests to watch—and be watched. Like us, she said, gorillas are "more afraid of the unseen."

In nine months in the Congo jungle, the Schallers went to town just three times, making the five-hour, four-thousand-foot climb back to the cabin each time.

When the Schallers arrived with their toddler boys in 1963, the Kanha tiger reserve was a last refuge for the world's most fearsome predator, whom George opted to meet on foot.

The Schallers all adored Jamil, who ran the household but was too tenderhearted to follow Kay's "modern" instructions not to indulge the boys when they fussed. He, Kay, and the boys all wept when they departed for Europe, below, in 1965.

George's field journal from Tanzania, November 11, 1967: "At mid-day, I got the photo I have been trying to get for 1½ years. One of the Masai pride lionesses grabbed a zebra, [which] just stood there with the lioness hanging on . . . The main impression was the bared teeth of the zebra's twisted head and her screams."

Sunbathing on a Serengeti kopje.

Knowing of George's love of pigs, a neighbor brought him a three-week-old warthog found in a den with her dead siblings. Giri demanded to be fed every few hours and cried when George went off to work.

Summer 1968: Eric age seven, Mark age five, Ramses about eight weeks. "I wonder if I could be more content," Kay wrote to her mother. "I am so glad not to be living an ordinary life."

In July 1966, out in the Maasai Mara with warden Myles Turner, "we found an ostrich egg, laid at random and abandoned." The delicious omelet fed four Schallers plus three guests.

Nepal, late 1973: Schaller lingered most evenings with the Sherpas; here, with Gyaltsen, Jang-bu, and Phu-Tsering.

Pervez Khan joined George on long treks through the high ranges that collide in northern Pakistan, crossing the Hindu Kush, Hindu Raj, and famously forbidding Karakoram.

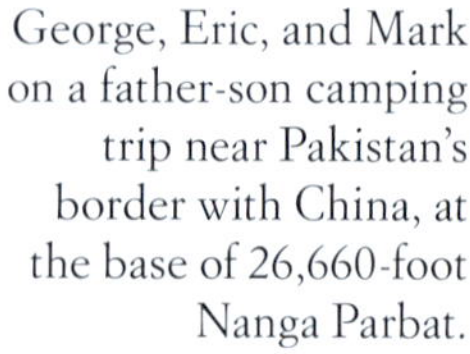

George, Eric, and Mark on a father-son camping trip near Pakistan's border with China, at the base of 26,660-foot Nanga Parbat.

\fter his two-month-long search or the snow leopard with Peter Matthiessen, George tempted fate by staying in Nepal's mountains or three more weeks. Utterly ost in a blizzard, he and his four Sherpa companions would have been finished but for the sudden emergence of fifty yaks driven by a handful of men. Though the passersby vanished just as quickly, the five were able to follow a line of yak droppings to the pass.

To track Brazil's elusive jaguars, Schaller turned reluctantly to radio collars, still in the beta stage in 1978.

With most of Brazil's Pantanal in private hands, George made his jaguar study on a ranch called Acurizal—until his presence became an obstacle to its sale. To drive him off, the foreman ordered the ranch hands to kill Schaller's study animals, then sell their hides. It shattered George to know that had he never come, the cats might still be alive.

Visiting Brazil for the summer, the Schaller boys' job was to catch fish for dinner and for baiting jaguar traps. Pictured here is Eric with a golden dorado.

A rescued baby white-lipped peccary became part of the family until she grew too unruly and George had to release her to an unknown fate.

As he had with tigers, George liked to study wetland predators at close range.

To tranquilize a national treasure was beyond nerve-racking. But by mentoring three generations in the field, as one panda scientist put it, Schaller "established the foundations of Chinese wildlife ecology."

Sichuan Province, early eighties: "It is serene and I prefer to be cold and with George than in the busy life in Roxbury," Kay wrote to Bettina.

One of Schaller's most famous pictures concealed an infinitely sad story of wild pandas reduced to objects for our entertainment.

Of Tibetan antelope, or chiru, Schaller wrote: "I am in a dream landscape of unicorns . . . the center of this consecrated space."

Autumn 1985, Tibetan Plateau: After a rare windless blizzard sealed all the grass away under a blanket of snow, starving the nomads' animals, George joined this People's Liberation Army rescue team.

Chiru, adapted to Earth's most extreme terrain, grow the finest of all wool, which must be plucked from a carcass. Schaller cracked a global smuggling ring that reached all the way from Tibet to New York's Park Avenue.

Tibetan visions sometimes stirred echoes of bygone places. These wild asses, wheeling about like zebra, cast George and Kay back to their bliss in the Serengeti.

long journey to Jumla "only because Peter insisted" on getting mail at Shey, nor was Jang-bu sent to inquire about alternate routes out "only to appease Peter." (George changed it to "partly to appease Peter.")

> *Finally, I wonder if "constant talk of leaving" is a fair description of my state of mind at Shey; if it is then I am truly regretful as well as astonished, and wish to apologize. It's true that I was restless about Tukten and Gyaltsen, since if they failed to show up, I couldn't leave Shey without disrupting your expedition and I was already feeling guilty about Alex. . . . But it is ALSO true that I was immensely happy and content at Shey . . . thus the instinct not to let news from home intrude on the last precious days, on the clarity of a way of seeing that had been so hard-won. Certainly, I felt anything but "taut," though my bouts of restlessness no doubt made me appear so, and I'm sad to learn that Jang-bu wondered if I didn't like them; I do wish you'd alerted me to that (those abrupt departures from the cook-hut were brought on by smarting eyes, to which I am very susceptible) as I was fond of all those guys, and very grateful.*
>
> *This is the only section . . . that really disturbs me, and I hate to speak about it, all the more so since you were so stoic about the references to yourself in my own ms. (I hope that you—or Kay, at least—have seen the revised references in the published version.) But by leaving out the reasons for the described behavior, this passage seems incomplete, therefore inaccurate, like a half truth, and I wince at the unattractive portrait of what I remember as a stark but beautiful exalted time, one of the most memorable and precious of my life. However, if you remember it this way then you have every right to leave it just as it is. So much for my protest, I won't bring it up again. Do let me know when you are back. I look forward to hearing your news.*

George changed it to "frequent talk of leaving," and took out all mention of the Sherpas' doubts about Peter's feelings for them.

The reciprocal influences went beyond editing, beginning with the titles, which might have been switched at birth. Matthiessen's recalls several of Schaller's (*The Mountain Gorilla*, *The Serengeti Lion*). Schaller's *Stones of Silence* is for the first time both symbolic and elegiac.

Peter also catches some of George's obsessions. He is thrilled to provide "the first datum" on the blue sheep's "harsh high-pitched" voice and delights at George's own excited observations. "Oh, there's a penis-lick! . . . A beauty!"

The reverse influence is more pervasive. With Matthiessen at his elbow, *Stones of Silence* became Schaller's most literary work. He quotes Robinson Jeffers ("Soon, perhaps, whoever wants to live harmlessly") and Ezra Pound ("Pull down thy vanity"). The long chapter about their shared journey has a saga's stately beginning: "We assembled beneath the spreading boughs of a pipal tree." Inspired by Peter's passing observation that villagers on their roofs are arranged "like the stage set of a play," he crafts a scene as detailed as a Brueghel: "One woman combs her hair, then ties it in a bun at the side of her head. . . . [Another] rhythmically sifts maize flour, while a third empties a basket of beans to dry." He evokes the painter more directly still when he sends the snow-blinded porters home in a pitiful procession, "the blind clutching the cloaks of the others for guidance."

And though he never was the disenchanted scientist, his writing is at its most mystical here. Under a dawn sky black as outer space, "only a rosy band high above reveals that order has not become chaos, that the earth will still lean toward the sun." His descriptions of animals fill with mystery. Bar-headed geese fly at thirty thousand feet, "where men gasp and die." The great white wolf appears and then abruptly vanishes, as if through a briefly opened crack in the mountain. The snow leopard, in particular—who had twice before come to him as an emissary, stepping forth like "stone turned to life"—now affirms her magic by declining to materialize; the appearance of spirits "cannot be willed." Here, at the limits of existence,

Schaller walks the knife's edge of the sublime, both exalted and terrified. One moment, a twenty-six-thousand-foot peak is a celestial pyramid of light; the next its nakedness "revealed my terrible loneliness, my utter insignificance." As always, he found release from that loneliness in another species. He spotted two black and white cranes, perhaps the rare Tibetan black-necks long revered by Buddhists. "They still fly on in my dreams."

This cosmic writing sometimes goes overboard. As he'd shown with Robert Ardrey, Schaller can be overly permeable to the influence of the more-famous writer who follows him around. "Beyond the facts, beyond science, is a domain of cloud, the universe of the mind, ever expanding," he writes, sounding very 1970s. "The heights glimmer, the sky sways; and, above, Crystal Mountain, ice-veined, a mysticism of stone." Saying goodbye to Peter on Saldang Pass, he reflects that "we shall meet again, some centuries hence, in New York."

Soon enough, however, he returns to the concrete writing that is his most transcendent. On his last day, he spots a Tibetan hare "drawn into itself with ears flattened . . . its life devoted to remaining cryptic." Finches sway on "naked boughs like autumn leaves refusing to fall" or roll like storms up canyons, "swerv[ing] as with one mind"; George marvels at the "internal fires that fuel these wisps of feather and bone." Footprints of a vole appear like a secret Tibetan script. "*Alticola stoliczkanus* probably left these signs."

George and Peter never traveled together again, although they did reunite for one strange, sad coda. They had discovered their shared openness to the existence of the Yeti over dinner in Kathmandu when a young biologist named J. A. McNeely showed up with a cast of a big footprint. Noting its similarity to the mountain gorilla, George said it could not have been faked except by someone who had spent years studying primates. When at last let into the monastery in his final days at Shey, George made detailed drawings for Peter of all he found there, including a painting on cloth of a wolf chasing what looked like a female Yeti, hairy and naked. Finally, in February 1976, Peter asked George to accompany him on a "top-secret mission" to New Jersey. A Boer carny had a creature he claimed was either a juvenile Bigfoot or a chimp crossed with a "pygmy" for which, with a letter from an NYU researcher attesting to its abnormal chromosomes, he had a buyer willing to pay $10,000. His claim that "Oliver" had come from Congo piqued Schaller's interest; he had not forgotten his friend Charles Cordier's own search for the Yeti there.

Peter treated the whole business as a lark ("determined to pursue *l'histoire d'O* to its inevitable end . . . we were borne away unblindfolded"). But George reported soberly on three unusual traits: a gentleness of temper reflecting either that the creature had been well treated or was drugged; a mouth that was not as prognathic (protruding) as expected in a chimp, soon explained by George's discovery that nearly all its teeth had been pulled; and an upright stance, with straight knees and spine—the result either of a congenital deformity, a childhood accident, or manipulation in infancy of his pliant bones. Oliver was just a chimp, he concluded. Though, regardless of how he came to stand upright, George thought him worth studying to understand what changes enabled the shift from an ape stance to bipedal locomotion.

•••

George had been home from Dolpo just a week when radiotelemetry expert Mel Sunquist arrived in Lahore. The two men had never met. But frustrated that his "old naturalistic techniques" were failing to locate snow leopards, George had sought Sunquist's help.

It was not a happy collaboration, on either side. Sunquist shared his memories in an email, agreeing to be interviewed and then changing his mind because "there is really not much to tell . . . hardly worth a phone call."

> *In late 1973, out of the blue, I received a message from George Schaller. He wanted to know if I was interested in joining him in North West Pakistan to try to catch and radio collar a snow leopard. This was during the early days of radio telemetry, and scientists were just getting started with this new way of studying secretive solitary animals, [but] I had serendipitously acquired several years of expertise. . . . I had just come off an arduous two-year pioneering study in the canopy of Panama's forests where we sweated off ten pounds a day climbing 100-foot trees, capturing sloths and using telemetry to follow their complicated movements through the tree tops. Snow and cold weather sounded good. George told me to gather whatever equipment I needed and get myself to Pakistan as soon as possible so . . . in January 1974 I flew to Lahore where I met George for the first time.*

Though he couldn't have known it, Sunquist walked into a situation already strained—by the Christmas disaster and constant travels George knew were taking a toll on Kay. "I've been gone too much," he wrote in his journal. "She does not like being a 'widow' in a big town . . . feels we ought to invite over neighbors but finds good excuses not to. And she gets tense just before I am due to leave." (Mark remembered that "it was shortly before he would leave that they'd have arguments. And when he returned. A disruption of the rhythm either way.") George's acknowledgment of Kay in *Stones* will be his most contrite. "The fact that she not only waited but encouraged me in my work in spite of the pain my absences caused, that she raised our children through critical years of development, creating persons of whom I am immensely proud, and that, in the end, she edited and typed this book, a book about journeys she wanted to take with me but could not, can only elicit my admiration, devotion, and love."

Now, as he and Mel waited for the airline to find the leopard traps it had lost, "this not knowing plans, and a virtual stranger in the house, did not make for a relaxed week."

At last, having collected the traps plus Aman, a jeep, and porters, the men headed for Chitral. Almost immediately, George's journal becomes a litany of frustrations. After waiting hours for his younger colleague, George took Mel's pack from him to carry over the pass. Abandoning Merin after a fruitless week for a second village that claimed a cat was killing its livestock, he grumbled at having to do all the cleaning and organizing since "Mel had little inclination to do such things, or for that matter to help look for snow leopard, judging by the fact that he stayed around home when he could have been looking for tracks." That Sunquist had studied sloths had a different meaning for George than it had for Mel. "He is not fond of climbing, so I searched the valley alone for spoor."

George knew there were no predators here. He could feel in his body some "missing vitality," a lack of "natural tension." His hosts confirmed that not a single cat had come through Chitral Gol or Tushi all winter, though the previous year George had seen leopards or tracks in both. "All this effort, time, money, and we'll not succeed. Every year there are fewer and soon there will be none." He and Mel baited traps anyway, hoping for something to test the gear. But even though every night a fox walked or dug near the trap, it never went in. "Two university grads ought to be able to outfox a fox!" In the end they caught just one animal: a raven that when released played dead, lying with wings pressed to its body until George stepped back and off it flew.

Meanwhile, local support for George's work was eroding. A wildlife warden said they didn't need him; no matter the animal, numbers were "more than sufficient." Some in Islamabad, the district commissioner told him, opposed his presence, saying Pakistanis ought to do this work—a view George shared, which is why he had sought out Mirza and Aman. The rumors that he was scouting a military base, the DC added, had been planted in the paper to embarrass his royal hosts. Now a new rumor circulated: A man shot by a jealous rival had crawled into George's bed to die in his arms. The first part was true. "The blood is still there, 200 feet from our hut, splattered all over the snow."

George never saw only darkness. He was grateful when a member of the Mehtar's household brought him sweet apples, fresh paratha, and permission to measure markhor heads. Climbing two thousand feet to tie out a goat, he rested "alone in a cloud world," the damaged land made clean and new by fresh snow, all silent but for a nutcracker's cry. He watched two daring wolves follow a markhor onto a ledge and, alerted by a villager to pugmarks, found that a big snow leopard had indeed come downvalley and crossed the stream—leaping seven feet from one glazed rock to the next.

After two hours of tracking, alas, he lost its trail. And his joy when his guide Subedar Afzal found twelve ibex—persisting in the search after George had quit—turned quickly to sorrow. Watching the females dig laboriously for each mouthful on slopes stripped by livestock, he knew that the "energy needed to keep warm, plow though snow, partake in the rut [and] nourish a fetus, must be so great that the balance between survival and death from malnutrition is tenuous indeed."

In mid-February, Schaller went with the Subedar to explore further valleys, leaving Sunquist with Sher Panah and instructions

> *for me to contact him if I caught and collared a snow leopard. That was to be the last I saw of him. I stayed on at the site for three months until the weather began to warm up and snow leopards and their prey moved to higher altitudes. . . .*
>
> *George was not the most talkative of field companions, he preferred his own company and was difficult to get to know. We only ever had brief conversations over the evening meal . . . a quick description of the day's events. My impression was that he preferred studying animals like lions and gorillas that were easy to find and watch. . . . He was not happy with the fact that snow leopards were going to require technology . . . they were the very definition of invisible.*

In the 1988 book *Tiger Moon*, Sunquist (who would go on to radio-collar more than thirty species of wild cat) added one additional detail. He had been losing weight rapidly to an intestinal parasite when left to fend for himself by what another biologist calls the "George disappearing act."

By month's end, having found evidence of at most five snow leopards in 1,200 square miles, George wanted only to go home. In four years, he had watched the animal of his dreams go from "tenuously secure" to seriously threatened. No snow leopard visited Chitral Gol the remainder of that winter nor the following year. "After that I did not seek further news . . ."

Schaller's last months in Pakistan failed to lift the gloom. After weeks of waiting, he got permission to return to the Hunza Valley just in time for a glacial surge to block the road. A final trip to Kalabagh found few newborn urial: the Nawab had clear-cut the reserve's trees to sell as firewood, leaving the sheep without lifesaving shade or watering pools. Traveling home with the family via Bangkok, Australia, and New Zealand, George had a last fight with the airport hotel over the seven and a half rupees they charged him for two small pots of tea, "which I refused to drink at that price." Leaving the South Pacific in mid-June, Kay took Eric to Anchorage while George took Mark to New York. "They've been bickering so we decided to separate them a while."

By fall 1974, with the family reunited in Vermont, Bettina's letters resumed. She appreciated George's brief visit ("it had been much too long . . . a stepfather can be a disturbing factor where sons are concerned") but wondered at the unfamiliar ways they were all raising her grandchildren. "My friends and I never played with our parents. The relationship was really rather formal. With our servants and nurses our parents did not feel they had to bother." Kay replied with thanks for the recipe calendar, from which she had already made the cottage cheese and ham pudding, and with news of their latest pet: a "darling" baby yak the Bronx Zoo hadn't wanted. Bottle-fed by Kay and often joined overnight in the barn by George so he wouldn't get lonesome, Tsampa romped happily with their black lab puppy, Tika. When George went off to South America to scout his next project, Mark wrote of his efforts to build a fence the yak could neither jump nor climb under. Defeated, they gave him to a local farmer, but the Vermont summers proved too warm and Tsampa soon died.

George did finally get back to Hunza, with Pervez in November 1974, retracing his trip with the boys but continuing on the new Karakoram Highway all the way to Khunjerab Pass, the crossing into China. In the Wakhi language, *Khunjerab* means "valley of blood," a reference to its murderous terrain, or perhaps to the mayhem that long defined this meeting place of empires: the bandits who raided caravans for slaves, the sons who killed their fathers for the crown.

Securing permission to travel to this still-tense frontier took more than the usual weeks of pleading. Even with a letter from Major General Imtiaz Ali, they were stopped after seven hours of driving by a captain who said they needed a permit from Northern Frontier Force headquarters in Islamabad. Left with no other option, they drove back seven hours, only to be told that no foreigner might enter headquarters for any purpose; that even if George were allowed, permission must come from the Ministry of Defence by appointment only; and that anyway the man in charge was out of town.

Sorted out with the help of connections (Uncle Aman's and Dr. Rizvi's), George and Pervez got past the nervous captain but still made only inching progress. Every few miles, the Chinese laborers hired to widen the road would wave them to a halt with a warning shout: "Boom-pow." The blue-suited laborers would then spend hours shoveling the dynamited rubble, so slowly that George sometimes pitched in—always met with a smile, thanks to the panda emblem on his borrowed WWF vehicle. They could finally see the pass when the driver stopped to relieve himself and forgot to leave the engine running. At ten degrees Fahrenheit, with no chance it would restart, George and Pervez unloaded the five hundred pounds of wood they had brought along (contractors had already burned every tree) and sent driver and Land Rover rolling back down the road with instructions to return in a week.

They did not stay even that long. Chasing his last hope of finding Marco

Polo sheep in Pakistan, George climbed alone to seventeen thousand feet. He saw no animals or sign, only Chinese military trucks carrying cabbage and turnips to road crews. After just a day, he gave up.

He did linger long enough to learn what had happened. Forty years ago, villagers told him, there had been many wild sheep; protected by the Mir, they had hidden here from the hunters in Sinkiang. But China's revolution and the Partition of India had flipped things: Now the Chinese government protected them while Hunza allowed uncontrolled hunting. Until the 1950s, several hundred still came in summer. But that had ended with the arrival of highway crews, who shot them for meat. If they came at all now, it was only briefly in spring for the new grass.

The Marco Polo sheep had quit crossing nearby Kilik Pass, too, outmatched there by the four thousand domestic sheep the local herders brought up every summer, who cropped to the ground plants already miniaturized by altitude. George did see a few ibex at Kilik, though they, too, were heavily hunted, often by the military scouts who patrolled these borders. In the intense cold, he could scribble only shorthand notes: "0815, five ♀ 3 yg 1 ♂ 2.5 yrs, w-facing slope, feeding on avalanche path, upper edge of scree." He even saw pugmarks of two snow leopards. But he did not follow them.

The Land Rover never returned. Told it was broken down eleven miles away, he and Pervez walked there, eating turnips that had fallen off trucks along the way. The mechanic said it needed a fan belt, so with the help of Mr. Beg (the man assigned to travel with them), they pushed an old junker across the road to force the next Chinese truck to stop. That driver gave them a fan belt and agreed to jump-start them with a push. But when after two miles of pushing the Land Rover still hadn't kicked over, he said he had to go. They begged for one last try, so he backed up 150 feet and came at full speed. Now their vehicle was not only dead, it also had smashed windows and a caved-in rear end. Their only option was to push it on foot another twelve miles, which they did in five hours with the aid of seven men.

("The help of the locals has been tremendous," George wrote. "Beg's friend came all the way to check on us; a shikari took us into his home.") Once they reached Beg's house in Baltit, Pervez went to find another jeep while George passed the hours with *Reader's Digest* and *Peking Review.*

Off again. More roadblocks. At the first, a jeep had come from the other side, so they just swapped loads and passengers. At the second, George took a walk and was taunted by military police. "Come here, hippy." Just the sight of "strutting uniforms" made his neck bristle. At last in Gilgit, but awaiting a plane that had been commandeered by the military for an Iranian general, George found happy distraction in a book found on a dusty shelf: a Swedish prince's account of gorilla hunting in the Virungas. "In my mind I was back at Kabara with Kay." Giving up on flights, they continued by truck, waiting at one dynamited site two full days as the clean-up crew dawdled. They set to work only when a Chinese VIP arrived, clearing it in just ninety minutes.

Out of that frustrating journey, unexpectedly, a national park was born, the first George had helped to create since the Arctic National Wildlife Refuge. In a report delivered by Rizvi directly to the prime minister, George summarized the sad state of the nation's iconic wild creatures and proposed a modest aim: to secure the survival of each threatened species in at least one place in Pakistan. A park on the Chinese border could protect five: Marco Polo and blue sheep, Tibetan wild ass (kiang), snow leopard, and brown bear. For the first three, this was the only place in the country they were found. He sketched rough boundaries that would permit the animals to make their seasonal migrations or hunt sufficient territory. On that recommendation, Prime Minister Bhutto established Khunjerab National Park.

Schaller has been criticized for his role in that park's creation, which some saw as perpetuating a colonialist fantasy of an ideal Nature "uncontaminated by people." With his line on the map, wrote novelist Amitav Ghosh, Schaller "set in motion . . . the exclusion of all human activity from this area. At one stroke, the way of life of the people of the valley was criminalized."

Such dispossession had in fact been the norm since the creation of Yellowstone National Park in 1872.

However, this particular story was not one of exclusion but its opposite: Schaller's first attempt to sort out how everyone—human and nonhuman—could carry on with their lives. Olaus Murie had seeded that vision two decades earlier, in his call for an Alaskan reserve that would also protect everything the Gwich'in needed to carry on "the ways of their forefathers." What George now similarly recommended—crucially influencing IUCN efforts then underway to create a globally recognized definition for a national park*—was that traditional pastoralism should continue everywhere within the 877-square-mile park. The sole exception would be four square miles of scree, right at the pass—less than half a percent of the park's total area.

That no-grazing zone was not a grasp at "pristine" nature but reflected the fragile realities of this land and these animals. Humans are integral to ecosystems, but not everything is compatible everywhere. Gorillas survived in the Virungas only because cattle were kept out. If Marco Polo sheep were to survive in Pakistan, they needed a few miles of safe passage from China.

Schaller further proposed that, since it was local people who would bear the burden of the small no-grazing zone, they should be the ones to get the new park jobs and entrance fees. The only bans he proposed were on firewood cutting by military contractors and possession of dancing bears.

This essential lesson, says Mirza—to protect whole ecosystems, people included—was what settled him on a career in conservation. Exhausting the land would doom humans too. "George saw that before the rest of us, urged us to act before we turned our home to desert."

* Schaller's vision also crucially influenced the design of Nepal's Annapurna Conservation Area. It was on his advice that forestry officer Karna Sakya recommended that no people be displaced; instead the ACA protects both hundreds of wild species and one hundred thousand human residents and their traditional livelihoods (Interview with the author in Sakya's Kathmandu home, October 26, 2025).

In the spring of 1975, Schaller made his last trip to Pakistan, still hoping to get to Khunjerab when the *polii* were visiting from China. Asked to give a wildlife talk at an event in Karachi honoring Prince Bernhard of the Netherlands, he took the opportunity to ask Mumtaz Bhutto, chief minister of Sindh, to reach out to his cousin on George's behalf. Bhutto did introduce George to the prime minister's private secretary. But after two weeks of calls and visits, he still had no permit.

Feeling once again "low and claustrophobic," George decided to go instead to Skardu, where he had been asked to evaluate a possible park near K2, the world's second-highest mountain. He and Pervez made it as far as Rawalpindi, then spent days condemned by bad weather to the familiar back-and-forth, lugging their stuff each morning to the airport, then back again to Flashman's Hotel.

It was a case of mistaken identity that turned their fortunes. Loitering about the hotel, George had grown curious about the other trapped guests. He soon gleaned that they were the K2 expedition led by Jim Whittaker, the first American to summit Everest. Among the eleven team members was Galen Rowell, a legend for his first ascents and photographs in Yosemite. Rowell later recalled the day a tall man interrupted a meeting to ask if they might have a Dr. Schaller in their group. "I've been getting calls from a young lady in the US. . . . She's on the phone now." The expedition doctor, Robert Schaller, shot out of the room, leaving Rowell to contemplate this "new" Dr. Schaller. "Something clicked. . . . But I thought, 'No, this man is too young to have done all that.'"

The same idea, fortunately, had occurred to team member Leif Patterson. Many elite climbers were awed by Schaller, aware that he did what they did, but incidentally—on route to his real purpose. As Rowell later wrote: "He has spent more time in remote Asian mountains than any mountaineer I know." Are you the man, Patterson asked, who studied gorillas and tigers? "Modestly, he answered 'Yes,' then ducked out the door."

That evening, Galen visited George in his room. They would travel the same route, they discovered, covering two hundred miles, including the nearly forty-mile-long Baltoro Glacier. Rowell's team had six hundred porters to move their tons of supplies; to get them from Rawalpindi to Skardu, they had rented a C-130 cargo plane from the Pakistan Air Force. And where were Schaller's provisions? George pointed to a small box holding some twenty pounds of nuts and meats, explaining that he, Pervez, and a single porter, who had been on the 1958 first ascent of Gasherbrum IV, would carry all they needed.

Over the next days, the two spent many hours together. "At first his 'thought world' seemed closed to me," Rowell wrote. "He answered questions like a computer." But that impression soon dissolved into its near opposite, of a man "shaped from clay that even in middle age had yet to harden. His interests are as diverse as a child's; his sense of wonder totally alive." Those who know him best will often describe George this way. "He was so gentle and attentive with our little daughters, ages four and seven," one Laos colleague recalled. "He wanted to enter their world, wandered into the garden asking them questions. Being one of them in a way."

George was still waiting for a seat on a plane to Skardu when at midnight on April 28 a note was delivered to his room. It was in memo form, handwritten on lined paper, from Rowell. "Our expedition is offering you a place on our flight," it began. Pervez had secured a seat on a military plane, so at 5 a.m., as per instructions, George showed up ready to fly.

With hundreds of boxes stacked in the hold and the men perched on canvas seats around the sides, the cargo door closed and the C-130 took off. To avoid the Kashmir cease-fire line they flew west of Nanga Parbat, George peering through the porthole windows at the massive block of ice. They then circled K2 to give the team a look at their never-before-attempted route, though they were allowed no photos of the north face, which sits in China. From the air, K2 stood alone, towering above the rest. Looking at the merciless "pyramid of rock too steep even for ice or snow," Schaller wondered what "spiritual terrors" assailed the climbers. (K2 is the deadliest mountain on Earth; one of four climbers attempting its summit have died.) He thought back on his own ascents, of Orizaba, Kilimanjaro, Ararat. "The complexities of life vanish . . . the objective is wholly clear . . . unifying one's existence. . . . [One feels] part of the natural world where death is meaningless and random." He understood the particular yearning to go where no man had gone before: For as long as Mount Drum "persists on the face of the earth," his own first ascent "cannot be surpassed."

Arrived in Skardu, the climbers resumed waiting—for yet more gear—while George began his work. He met up with a wildlife warden and forestry officer to search, he thought, for ibex; their fishing poles told him they had other plans. Rowell joined him instead, on a seven-hour hike for which the AWOL climber got scolded by the team.

George and Pervez then set off on their trek, which brought a few exhilarating moments: George saw his first ever Ladakh urial, or shapu, and a fox unlike any he knew. He watched a sleepy male ibex nod off, his heavy head jerking up every time his sinking nose touched the snow—the kind of intimate moment so commonplace in his other homes, and so rare here.

But the hiking was awful: over "damned" moraines and the ugly dark rubble that covered the glacier. George listened fearfully for tumbling boulders and navigated the dangerous bergschrund (in this land of mountaineers, he uses their lexicon): the valleys that form between the moving and static ice. "For ten hours I must take every step with precision." Even for ibex, this was treacherous terrain; at the base of an avalanche chute, he found two males that had been first swept to their death and then eaten by a bear. He was glad to finally be in a place where livestock could not go, and to find, consequently, a rare sight: deep grass. Leaving Pervez happily in bed reading his book of famous quotations, George went off for five days alone, spending his evenings cleaning (his fingernails, the binoculars) and thinking, awakened each dawn by a raven's gurgling bell.

Still, the internal fires that had propelled him through decades of danger and misery remained dampened. Just a day away from gaining a full view of K2, he turned back. He told himself that a storm was approaching and that Pervez had no tent—though George had rigged his companion a snug lean-to, and was not himself generally dissuaded by snows. On the way out, running into Rowell headed toward the climb, he shared his despair at the forsaken state of even this most inaccessible of places. Where the forest department had promised 4,500 ibex and 450 musk deer, he had found fewer than one hundred ibex and not one sign of deer.

Continuing down, George walked for more than an hour past the Americans' string of porters, then another hour past a French team just as vast, reflecting on the damage done by these expeditions: the trees mutilated for firewood, the garbage left behind. He found a moment's camaraderie with some shepherd boys, teaching them to count to five in English. When his flight was canceled, he whiled away a day running down sand dunes with a similarly stranded Canadian journalist. But on learning that he'd at last been granted a permit for Hunza, he found he'd lost all interest in going. He pasted another scrap of poetry in his journal: "if the great gift returns to me / if my heart's love again is mine / My love will increase but more for thee / As winds sing of my love till the end of time."

The scientific monograph Schaller published out of these trips is uncharacteristically heavy going. The prologue is effectively a seven-page apology for its deficiencies, beginning with the lack of historical information: Unlike Africa's gazelles, zebra, and wildebeest, not one of these high-altitude hoofed species had been studied. Of the few scientists who had made even one journey here, only Ernst Schäfer, who came twice under Nazi auspices and once after the war, left useful information. Explorers and spies had passed through, but only Nikolaj Przewalski and Sven Hedin had paid any attention to wildlife. And for the first time, Schaller had only meager data to add. Was his intimate way of studying animals obsolete, he wondered, in a mostly emptied world? He did make his strongest case yet for two advances that would become central to conservation. Building on E. O. Wilson's insight that extinction rates are higher in small and disconnected habitats, he urged a shift to the kind of "landscape-scale" protections, incorporating human uses, that he had championed in Khunjerab. And he pressed again to vest authority with the people who live within these ecosystems. With their traditional knowledge and sustained presence, they were the only ones who could consistently monitor, and protect, these wild creatures.

The record for the rest of the family remains thin in this period. Bettina wrote to Kay in April 1975 about her plan to visit them in Vermont after her Daughters of the American Revolution outing; it had been ten years since she had seen the boys. Kay replied with a family picture "to help you recognize us, though I usually do not simper on George's shoulder. He left two days ago and as usual I am depressed and desolate, which I always am to an extreme the week before and week after he goes." Kay wrote again, after Bettina's three-day visit, on how she wished "I had half your strength of character; I certainly see where George gets his." Bettina's own warm thank-you for the visit included an ad for Naturally Blonde, "the hair-do and color would be terrific for you." And her insights about men: "I know the waiting seems endless, but I'm sure you give him the courage and strength knowing you are home caring for your boys. For a man, that is the only thing that enables him to work and have peace of mind. Too bad more women don't realize that fact. It is rare these days for a man to have the kind of satisfaction George has in his life."

7.

New-World Jungles

Brazil, 1975–79, Jaguars, Caiman, Capybara

In 1975, with the boys now teens, George and Kay made two decisions. First, for better schools and a shorter trip to George's desk at the Bronx Zoo—where his responsibilities leading global field science for the Wildlife Conservation Society had been growing in tandem with the organization—they moved that summer from Vermont to Roxbury, Connecticut. Settling in a converted barn in a forest of maples and pines, they soon filled every corner with "memento[s] of exploration and desire"—a scrimshawed walrus tusk from Alaska, masks from Congo and Nepal, a Maasai shield from Tanzania and Dayak headhunting knife from Sarawak—to which they continued to add: from Brazil, a stone adze; from Mongolia, a dinosaur bone.

Their second decision was to accept still more separation. One last wild continent called to George, as did the only new-world great cat. He began scouting South America for possible sites to study jaguar, as always navigating politically treacherous landscapes. When a misunderstanding with the Argentine junta landed him face down in the bed of a truck with a gun to his head, he was reminded that—as with nonhuman predators—remaining calm was key. Kay was glad he made it home in time for a Christmas dinner of stuffed goose and buttermilk sherbet. "Though it takes so much energy," she wrote Bettina, "to be or look busy all the time!"

Having set his sights on the Pantanal, the world's largest tropical wetland, George wrote to Brazil's director of national parks. He heard nothing for months, and likely would have been ignored forever had not José Manuel Carvalho de Vasconcelos ("Vasco"), an advisor to the federal agency overseeing forests and parks (IBDF), recognized his name and invited him to come. In July 1976, the two traveled together into the Pantanal. Since this swamp—larger than England—is nearly all in private hands, they would visit fazendas, the immense cattle ranches that define the landscape, in search of both jaguars and an amenable landowner. Small planes were the quickest way to travel, so they went to the town of Poconé to find Inácio Tolentino de Barros, the pilot who knew the fazenda families best. The basis for those relationships became apparent when a guest at Fazenda Santa Isabel—the son of a general, there with friends—offered the pilot 10,000 cruzeiros ($400) if he and his fifteen dogs would take them hunting. A dozen jaguar skulls already ornamented a backyard fig tree; clearly the 1967 law protecting the cats was little honored here. But in deference to his present clients, Inácio declined. They continued on in his rickety plane, alerted to his worldview by the name he had painted on the side: *Universo em Desencanto* (Universe in Disenchantment). Two years later, just months after another flight with Schaller, Inácio would die in a crash of that plane.

George followed this aerial survey with more-meandering explorations. Floating the São Lourenço and Paraguay Rivers, he found jaguar sign at every stop: tracks, clawed trees, the occasional otter head or capuchin monkey tail. Driving the new Transpantaneira highway, with its more than one hundred bridges, he became entranced with caiman. "Ancient, dark and armor-plated," the crocodilians flowed through the muddy water "like black syrup." To lure mates, males played their own bodies like a guitar, vibrating to produce a humming resonance. Though usually leery of animals whose faces he could not read, Schaller admired the sophisticated fishing tactics of these "expressionless beasts." One lay open-mouthed under a waterfall, waiting for fish to drop in. Pairs fashioned themselves into fish traps, curving tail to tail across narrows. Watching a big caiman eat a little caiman, he wondered at a brain at once nimble enough for collaborative fishing and "brutally primitive" enough for cannibalism. Both reptile and man seemed to welcome proximity: On one occasion, eighty-eight caiman swam near and lay in a semicircle "as if waiting for me to give a speech." On another, George folded his long body into a tiny, tippy rubber raft, floating into their midst to lasso one.

In this watery world, lovely with its floating hyacinths, many of the strangest creatures swam mostly hidden below. A silvery fish George could not name had inch-long teeth jutting from its lower jaw. One catfish had eyes at the bottom of its head like headlights pointing down; another "monstrous" one, with "sandpaper" gums, squeaked by grinding the stiff bones in its anterior fins. "I don't like fish to talk," he wrote in his journal. "It makes it seem terribly callous to haul them out, indifferent to their suffering. If all fish cried, I'm sure fishermen would not dump them on the ground and let them slowly die for lack of air." The predators, of course, charmed him most. Reeling in a fish, he didn't mind when a six-foot caiman grabbed it and his lure. Out in a canoe, he saw a sungrebe dip its black-and-white head below water, then quiver and fall still. He picked it up to find a neck but no head, neatly bitten off. Its heart was still beating.

On land, it was armadillos that first won his heart. Found in the Pantanal in small, medium, and large sizes (a.k.a. three-, six-, and nine-banded), they seemed to resurrect the boy in him. More than once, seeing a puff of dust from a hole, George reached in to grab a tail, though he could rarely budge the animal, its curved foreclaws dug in. The nine-banded had the best defense: jumping straight up in the air when startled, startling George in return. The adorable three-banded, with its rounded shell, "pink nose, sleepy eyes," and Spock ears, just rolled itself into a coconut-sized ball; if picked up, it peered at him with an agitated chuckle and huff. In general, George messed with animals more here, whether because he could—many were small or slow—or out of some pent-up hunger after his barren years in Pakistan. His ostensible purpose with the armadillos, as with chital in India, was to estimate the total biomass of the species jaguar preyed on. And even when he couldn't lift them by their tail, he could sex them.

Like all the rest, this ecosystem was undergoing worrying changes. As in North America's great wetland—the Mississippi Delta—levees built to turn swamp into arable land were causing many kinds of ruin. Deprived of the mud and nutrients delivered by seasonal floods, the fertility of new-made pastures soon played out. The *rancheiros* then had to add fertilizer; when it ran off into waterways, blooms of plant growth choked out fish. The diverted floodwaters also inundated fazendas downstream, condemning hundreds of thousands of cattle to drown or starve or, pressed together on bits of high ground, spread disease. Bodies littered those islands. George saw one cow in water so deep she couldn't lower her weary head without drowning. By the next day, she had.

The most promising place for a study seemed to be Fazenda Acurizal, fifty-five square miles stretching from the Paraguay River to the Serra do Amolar. Its absentee owner, Horácio Coimbra, said George could not only do the study but also live in his ranch house. He might even sell it all to IBDF, to serve as headquarters for their Caracará Biological Reserve just across the river.

As George prepared to move to Acurizal, he got a letter from a biology student named Peter Crawshaw, offering his help. Like so many before and since, Peter was stunned when this "world icon" wrote back, asking if he could come to Brasília on a day George would be passing through. "For months, I could think of nothing else. When the time came, I missed a test and spent all my savings on a bus ticket from Porto Alegre. The trip was 1,300 miles and took 33 hours. I dreamt of jaguars all the way."

Meeting George at IBDF headquarters, Peter could see that he was unhappy. The parks director was telling him to forget about a study at Acurizal—better to go back to the US and leave things as they were. George made clear that he would forge ahead, though he refused when Peter offered to start right away, insisting he must first finish his degree. "You will be much more useful to the project and to your country." Heading out on his own, in a heavily laden truck pulling a trailer with three boats, George drove one thousand miles to Corumbá, then boarded a cattle barge to travel up the flooded river. Humans were corralled in the rear. At Acurizal, his very first walk seemed to promise a return to abundance: He found the tracks of two jaguars traveling together and saw caiman, capybara, and coati. Alas, he later wrote, "I was still naïve."

The odd doings began almost immediately. Staff locked George out of some rooms, came into his without knocking, listened in when he spoke on the radio. When he protested, they said they were acting on orders of the ranch manager Geraldo, a near cartoon of machismo in high boots, knife and gun on his hip, big mustache, and burning stare. Geraldo went out of his way to antagonize George, answering his request for a second bedroom to be readied for his sons' visit by having the whole house cleared of everything but a few moldy mattresses. George had to fly to Corumbá to buy beds, chairs, tables, dressers—even a refrigerator and stove.

Geraldo's hostility was part of a larger jockeying George did not at first understand. He was perplexed by a string of visitors, many of them the kind of scheming expat he'd met on every continent. A German ex–military man presented a letter "of dubious origin" revoking Schaller's permission to use the house. Arne Sucksdorff, a Swedish filmmaker who had been in the Pantanal for fifteen years, told him there were no jaguars here. (When Sucksdorff later pressed George for his opinion of the IBDF, "all the while I could hear a hidden tape recorder. Why is he playing Nixon?")

It was another fazenda owner who gave the game away. Mistaking Vasco for the Acurizal manager, he gossiped about the commission the owner had promised to anyone who found a buyer for the ranch. George's presence could only hinder such a transaction. And his aversion to schmoozing foreclosed any counterscheming; IBDF held weekly meetings by radio, but he declined to join.

That the parks director had his eye on a prize that had nothing to do with jaguars was evident in the coworker he assigned to George. Young Marcio worked eagerly only on his tan; out for a ten-day river survey, he wore "the briefest underpants and lotions himself." He refused tasks he deemed menial, like washing dishes, and was so slipshod about the rest that George had to redo most of his work. Coming home one afternoon in a downpour, George found Marcio enjoying a long shower with all the windows open and their sacks of food soaked. George also worried about him: Deaf in one ear, Marcio might not hear a growl.

After years in extreme places where so few species could survive that George could name nearly every one, he found his curiosity revived by this wet, warm world full of unknown creatures. He saw his first giant jabiru stork, its neck inflated as if it had swallowed a gourd. When his arrival upset a vast rookery of egrets and spoonbills, he watched a clever caracara falcon capitalize on the disturbance, stealing two eggs and a newly hatched chick. As dazzling as the variety—of birds alone, more than six hundred species—were their numbers. In seven minutes, he counted 228 passing kites, from a flock he guessed exceeded one thousand. At first baffled by the absence of song, it dawned on him that a bird could be heard in this crowded world only by screeching. A colony of cormorants and anhinga sounded to him like a pond full of bullfrogs, each returning bird croaking a greeting to its partner.

Of the mammals, he was most taken with the capybara: "a 150-pound mouse that barks like a dog," has a nose so blunt it "looks like it was shaped by running full tilt into a brick wall," and wallows with as many as twenty squeezed together like "some bemuddled encounter group." (Even George could not entirely ignore the Me Decade.) He watched a yellow-bellied bird land on one to pick bugs off its scraggly "bargain-basement coat." When the rodent waddled into the swamp, submerging itself like a hippo, the bird climbed from rump to back to head, riding across without getting wet. Reclining on the grass while capybara puttered nearby, "I felt like a strayed Lilliputian among field mice."

The insects—a hundred thousand species or more—fascinated him with their ornate constructions and grand parades. Termite hills rose out of the swamp like volcanic islands. Columns of army ants streamed to their bivouac bearing in their jaws the heads and thoraxes of the defeated. Watching a dredge sift a river bottom for diamonds, Schaller counted in its lights at least fifty kinds of moth. The channel of light cut by the road through the dark forest drew in still more butterflies, forty-four varieties of which he sketched with a lepidopterist's precision.

The interdependencies were as profound as ever. He found orange dryas butterflies clustered on ocelot scat, harvesting salts they would later transfer in their sperm to females; put into a kind of trance by the salts, they let George touch them. The canopy was yet another universe, transited by "bats on velvet wings swooping . . . softly sowing seeds."

Having grown accustomed to bare expanses of sand, rock, and ice in which everything from the tiniest mammal to the stars above was exposed, George did find it strange to see neither horizon nor sky, nor many of the most abundant creatures, except for the damage they left behind. Black ants cut his tarps into pieces to carry off; red ants plundered his rice; termites ate the labels off his drugs. Firewood crumbled in his hands, riddled with wood eaters who swarmed to bite. Worst were the carrapatos, tiny pepper ticks in clusters of fifty or more that left itchy marks with every bite. On one of Kay's visits, it took them an hour to pluck the invaders from her legs. George resigned himself. "I am a man of red spots now."

Bigger animals bit him nearly as often. Finding little brown bats clustered like bark on a tree, he caught three with a mist net; one sank fangs into his thumb. Out on a swampy trail, he discovered a gelatinous mass of eggs in a depression in the sand; the big-headed green fish guarding them lunged, snapping at him with wicked teeth. While cleaning piranha, the severed head clamped down, lacerating his finger. The festering sores that soon covered his hands throbbed at night, the infections sometimes washing him out for days. When a local brought him a small (four-foot) anaconda, oddly humped where its back had been broken and healed, it bit him, too, hooking him with teeth that angle inward to drag prey into its mouth. "Our first pet," he sighed in his journal. Though the snake seemed to get meaner rather than tamer, he gave her a fat rat (at thirteen ounces, a quarter of the snake's own weight) and watched as she slid over it and began to swallow. It took eighty minutes for the rat's tail to vanish. She later threw it up three-fourths digested, "a stinking mess" of an offering.

For all the bites, George continued to grab and poke things. When the ranch hand José reported a ten-foot anaconda near his house, the two conspired to catch it: José grabbing its tail and racing backward, George running after with the capture stick. Though the snake kept trying to strike—quick, hard thrusts with mouth open—George managed to lasso it. Holding its head in his hands, he stroked its smooth, cool skin before stuffing it into a sack to weigh. The next morning, he released it into a stream, amazed at how it vanished, just a few bubbles to mark its path. Still more remarkable was a snake that bulged like Saint-Exupéry's hat with what George guessed was a capybara; it weighed eighty-eight pounds. Prodding another that looked like it had swallowed a hanger, he felt the hard beak of a heron. And in a rare sighting that might serve as metaphor for many political (and intimate) relationships, he saw a giant anaconda that had wrapped itself around a caiman and been trapped in turn, its tail held fast in the caiman's mouth.

It was a little anteater, however, who finally got the best of him. He had been harassing one he'd spotted eating termites in a tree, shoving it back up into the branches so he could get a picture. Now, chasing another, he "did something stupid." When the little creature wheeled on his pursuer with arms spread wide, George nudged it with his boot. In a flash, anteater claws ripped through the rubber. When George used his other foot to try to push it off, that boot got slashed too. When he complained that everything here seemed to "scratch, bite, sting or otherwise lacerate you," it's hard not to think he didn't sometimes deserve it.

Venturing into the forests here stirred memories of Congo, his last real jungle home: the trees so wrapped by vines they looked like nodding goblins, the thorny palms and wild pineapple so thick he sometimes had to go on all fours—harder now that he was forty-four, not twenty-six. Hearing and smelling unseen wild peccaries all around him, snorting and squishing in the mud, he was reminded of the grunts and crunching of invisible gorillas.

But the Virungas had been a high, cold jungle; this one was low, hot, and far more impenetrable. When a big peccary boar came into view, raising his hackles and giving a hard warning clack with his teeth, George was acutely aware how hemmed in he was by spiked tree trunks. Unlike Darwin and Humboldt, who found their greatest inspiration in South America, this landscape was for George at best half loved. "Life . . . smothers one like a crowd on a city street . . . thousands of species . . . each defending itself . . . with thorns, spines, claws, toxins, and poisonous stings. The human intruder . . . is forced to enter this struggle, constantly combating encroaching vegetation and insects. . . . Such an existence promotes little reflection."

His prime quarry, meanwhile, remained elusive. Unlike lions and tigers, jaguars were "restrained about advertising their presence." George did find tracks and carcasses, reigniting the pleasure he had hungered for in Pakistan: the chance to read stories. In the skeleton of a marsh deer, he found the tiny clavicle of the fetus she had been carrying; in a capybara corpse, a record of the distinctive way a jaguar kills. Opening her mouth wide as if to swallow the rodent's whole head, the cat had placed a canine tooth in each ear and, with the most powerful jaws for its size of any feline, pierced its skull to the brain. Next to this primal force, he reflected, the tiger's grab of the throat and quiet strangulation seemed downright fastidious.

Drag marks led George to the dark hollows where the cat had taken the giant rodent, but growls warned him away. Returning at dawn, he wondered why she had eaten only the brisket, heart, liver, and shoulder. Was food plentiful enough to waste? Or did meat rot too quickly here to finish? As he dragged the carcass from the thicket, the veil of flies that had shrouded it resettled itself on George.

Despite his frustrations with telemetry in the Serengeti and Pakistan, George decided that to study these secretive cats, he would have to give it another go. Repurposed from Cold War surveillance tools and not yet two decades old, the technology remained crude and controversial. Frank and John Craighead had been collaring grizzlies in the Rocky Mountains since 1959, but as with most beta-stage technologies, lots had gone wrong. In the first two years, six bears had died.

One challenge, as George had seen with lions, was dosing: A yearling grizzly earned the nickname the Sucostrin Kid for his high tolerance to that immobilizing drug. A second was the bulkiness and glitchiness of these early devices. In 1970 Monique the Space Elk became a celebrity when she was fitted with a $25,000, twenty-three-pound collar designed to communicate with a weather satellite. The collar immediately malfunctioned and then, just six days later, Monique died, raising public outcry. Hitting closer to home for George were protests lodged by Adolph Murie and seconded by Olaus against this "laboratoryzing and demystifing" of wild creatures—turning them into data points or a kind of cyborg, subject to control. George had run into this debate in the Serengeti but in that heavily stage-managed landscape had found it easy to shrug off. It was trickier in the Pantanal, so far free from such artifice.

Most worrying was that he would have to trap the jaguars; only lions were blasé enough to let themselves be darted while free. And after the washout with Sunquist, George still had little experience. His first experiments with foot snares—aimed at learning how to avoid injury to the animal—had mixed success. He caught a fox, who was unharmed but terrified: chewing frantically at nearby plants and bolting the second George was inattentive. He caught a six-banded armadillo, who in the struggle ripped off several claws, though not, George was relieved to see, its main digging one. Then another fox, who bit at his captor, urinating in fear. Finding in one trap the feathers of a curassow and "his foot too," George began walking the trapline with apprehension. Though his goal was still to "chronicle the jaguar's private life," a new dissonance was emerging between language and method.

The first cat he trapped was not a jaguar but a puma. Dart one went into the dirt, but he landed the second in the thigh; when the cat got merely groggy, he realized that the drug, mixed a month earlier, had lost its potency. He managed to slip the noose of the capture stick over the nodding head and, while Marcio pulled at the thrashing animal, jabbed him again with a syringe. This time the puma slept, though his ear proved too tough to tag and the cable of the snare had so twisted they struggled to get it off. When, as they put on the collar, the cat reared his head, George gave it a third dose. The puma took many hours to recover, but George counted it "good practice."

Alas, the signal from the collar died almost immediately, the first of a string of equipment failures that each time required Vasco or another colleague to fly to Brasília or Campo Grande for repairs—losing in the meantime all visibility into the tracked animal. They never did hear from the puma. When by chance they recaptured him eleven months later, he was still wearing his dead collar.

Beyond the beta curse, George turned out to lack aptitude for communication technologies, or perhaps the will to learn. (In 2022, he still found the computer "so complicated that I've not been able to figure out how to send a document. I should just stay with the hill people in Vietnam, join again in their feast of termite larvae and not get tense over technology.")

When he later abandoned trapping and tranquilizing, it was because he could no longer sanction the stress it inflicted on the animal. And because he saw zoology growing too infatuated with remote sensing: "We don't need to get the home range of two more snow leopards." But he also loathed telemetry because it presented the rare challenge in the field that rendered him helpless, incapable of improvising a solution.

Nor were the jaguars making it easy. Though their broad head and shoulders and powerful jaws suggested "a triumph of brawn over brains," the cats thwarted George's efforts to trap or even see them so often and so cleverly that he began to think they were taunting him. Light-footed dancers, they stepped into his snares long enough to press a track, then out again before the loop closed. One set off a trap by pulling a carcass across it; another stole the bait by reaching over the side; a third stepped on the wire but not the trigger. He tried sleeping outside: hanging his hammock and a few enticing rotten cow lungs near where he had heard a growl, settling in to read *Kim.* He woke to find a trail of prints just sixty feet away, "as if to rub in my ineptness." Twice, capybara with crushed skulls were dropped in camp, a jaguar's graffiti tag. *Kilroy was here.*

After weeks of this, George acceded to the locals' way: using dogs to chase the cats into a tree. Not only was that the surest way to catch one, but a cat's bottom, perched on a branch, was an easy target for a syringe. An opportunity soon presented itself at Fazenda Santa Isabel, where he had seen those twenty skulls hung in a tree. The manager had applied to IBDF for permission to shoot a cattle killer, Inácio told him. Maybe they would let George capture and collar the jaguar instead?

They did. But "that hunt," he later wrote, "I do not like to remember."

On the morning of June 21, 1977, Schaller slogged on foot through waist-deep water. Ahead of him were the ranch owner, Inácio, and four other men, all on horseback with knives in their belts and .38s in their holsters. George was already tense ("too many people and dogs to suit me") when, coming upon huge tracks and a fresh-killed tapir, the hunters released the dogs. They soon erupted in anguished yelps. The coati they had cornered had used its sharp raccoon claws before escaping into a tree; several dogs were slashed across the throat and legs.

When soon after the dogs again burst into an uproar, Joselito (one of Inácio's men) told George to wait. Though George soon lost patience and pushed past, Inácio was furious, saying he had waited too long and the jaguar was killing his dogs. George was now fully alarmed—at the harassing of the cat and the chaos of men all shouting and heedlessly waving pistols in the thicket. When, as he feared, Inácio fired two rounds, the trapped animal attacked, chasing a dog directly toward him. "I see his 200-pound bulk coming and his face in a snarl," George wrote in his notes, "hear the growling cough as he slashed with forepaws." Missing the dog, the cat wheeled and ran. He had come within ten feet, but in the melee, George had no chance to shoot the syringe. In retrospect, he was glad for that; the dogs would have killed the debilitated cat before the men could call them off.

The dogs went after the fleeing cat and again Joselito told George to wait. This time he ignored him and rushed toward the howling. "Inácio, the professor is there," Joselito cried, betraying his own fear. "I have no gun, only a knife." As George circled the clump of brush where the jaguar hid, refusing to be treed, the cat broke away. A sudden yowling led them to a dog biting desperately at a branch trying to regain his feet, paralyzed by a bite through the back. Inácio shot the dog, vowing to avenge his death by killing the jaguar. But "the hunt seemed to be over, the men tired, the dogs bloody. Luckily, they were lousy shots and the jaguar escaped. 'A happy ending,' as Vasco said." Of the thirteen dogs, seven were injured and one was dead. Relations between George and Inácio were never again easy; attempts at reconciliation invariably ended with both turning heel to walk away.

George went home that awful day to his newest friend and surest balm against sorrow and loneliness: a baby wild pig called Fuxica. He'd bought the month-old white-lipped peccary for a hundred cruzeiros ($8) from the children of Felinho, a ranch hand who had rescued her from his dogs. Kept with their two domestic piglets, which were five times her size, she'd had to scramble for food and weighed just six pounds. While carrying the "endearing, coarse-haired creature" home in his arms, Schaller had met a big anaconda. Lassoed by the capture stick, the thirty-nine-pounder had wrapped itself around George's leg. "And so, we proceed to the house, me, a snake and a pig."

As he had with his African warthog, Giri, George let the piglet upend his life. At night he got up every three hours to let her out. Whenever her hair stood on end in terror—because "new feet" had entered the house, or a clap of thunder had sounded—he stopped everything to stroke her. Like Giri, she hated being alone. She slept on George's feet or atop his chest in the hammock, and followed him room to room, often spraddling on the slick floor. As she outgrew her timidity, she followed him outside. Trotting after him in the heat, she panted and, "when we reached a mud puddle [a "pud muddle," he wrote, in a rare but apt typo] she flopped into it." One morning, when he went off by boat upriver, Fuxica jumped into the water and swam after him. Worried about the piranha, George scooped her into the boat. The engine noise and rocking set her shaking with fear, but she soon adapted. In several pictures, Schaller holds the rudder in one hand and the pig in the other; at first, entire baby forequarters in his lap, soon just her snout in his arm.

His journal, of course, filled with pages and pages on little Fuxica. (For anyone committed to reading them all, he can be as tiresome as a new parent.) He compiled a dictionary of the five distinctive grunts that signaled her mood. He took her into the forest to learn what peccaries most like to eat; her favorites were his omelet and noodles, but she did scavenge beetles. He drew her from every angle: In several that focus on her three-inch-long eyelashes, she looks like Miss Piggy.

Most of George's acts of tenderness toward the pig would be famil-

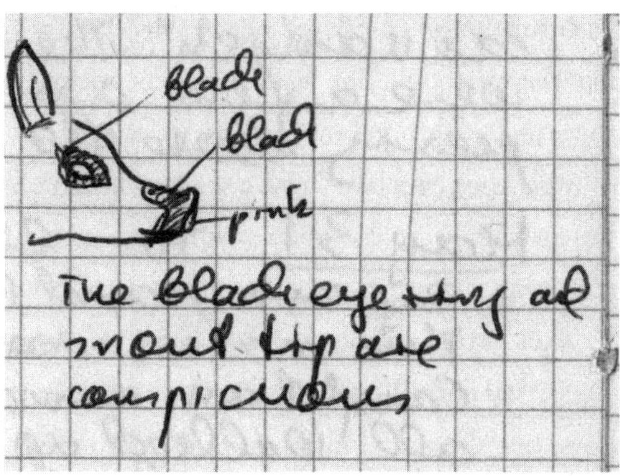

George loved pigs, especially baby Fuxica.

iar to anyone who has loved a house cat or dog: He scratched her behind the ears until she grunted with contentment; one famous picture finds him lifting her, lips puckered like hers, to give her a reassuring kiss. Other of their intimacies may be fathomable only to a zoologist. George showered with the pig (though she liked to escape, running covered in soap all around the house). And "today while she rubbed against the Cutter's insect spray on my feet, I rubbed her gland. She rubbed harder and clacked her teeth. If squeezed, her gland sometimes squirts its musky fluid an inch."

With the family's arrival imminent, George did begin weaning her: closing doors, cuddling less, and no longer feeding her at the table. "It is difficult for us both," he wrote. Though of the two, he seemed the more dismayed when, after a three-day absence to fetch Kay and the boys from the airport, he found the pig "stand-offish."

Fuxica soon resumed her attachment, and George his chronicles. He noticed that as she shed her baby coat, the pig also "changed psychologically. She is less intent on following me. Whether this is due to my absences, or to her . . . smaller need for 'motherly' protection, I don't know." He was relieved at how quickly she accepted the family. "She plays with us, gently biting, though unlike our lion cub never hard enough to hurt." Out in the yard, George watched her challenge the chickens, fleeing the second that one turned to face her. When a hen with chicks pecked at her, she turned in profile, showing her white lips, a submissive gesture. He wondered if her new habit of dust bathing was an imitation of her feathered friends and, with Kay, often compared her to Giri. Fuxica's nose was dry, not slobbery like the warthog's. She was less independent, but also less vociferous when thwarted, with none of Giri's "wild screaming." And the peccary was the only pet they ever had "with a sense of humor. She would wait till our backs were turned, sneak to open a cupboard, grab a bag of flour or rice and run scattering it, then stop to see our reaction."

Those antics grew harder to enjoy as the maturing pig began to stink. They tried corralling her in the courtyard, in a spot where she could see the boys, but she paced all night retching in misery, wearing her hooves sore and growing so frantic when three domestic pigs came near that she cut her nose on the wire fence. They brought her back in the house but put boxes in front of the bedroom doors so she couldn't rattle them all night with her prehensile snout. She fell ill for a week, "psychologically as well as physically." But she soon felt and "smells good again . . . even leaping all the way around. She has a most determined walk, nose raised, when her mind is made up, and is as cuddly as ever."

If George had too much to say about the pig, he still lacked the language to describe his feelings for his human family, resorting even in his journal to banalities. "It's nice to have someone to talk to and to help. All gathered in the evening. . . . It is pleasant to be a family again." Still he was clearly contented in a way he had not been for some time. Eric, nearly sixteen and like his father happiest when investigating the world, "is suddenly quite grown up, self-assured, more approachable and affectionate." Mark, fourteen, "is sweet and helpful, more people-oriented. He chats a lot, especially about the past, like our trip to the Karakoram."

Four years after that last trip together, George was glad to offer again what only he would describe as "small adventures that please the boys." Together they collected giant ants to see which species armadillo liked best, found thousand-year-old petroglyphs and a far older stone axe, and walked the beach on cool mornings looking for footprints in the sand. George let a ten-foot anaconda wrap itself around Mark and taught the boys how to cast jaguar tracks in plaster, taking pictures to sell to a children's magazine. In the evening they fished for pacu to use as bait, roasting a few over the fire to eat off palm-bark plates, or poled the boat silently toward the glittering eyes of congregating caiman. Most fled but some hung in the water, allowing them to slip the capture noose over their head, then sit on the muddy bank to weigh them. They caught the babies with bare hands.

On overnights, George took the boys one at a time. "It is more companionable that way. They talk more." With Eric, he found a two-foot-long mustard-yellow monitor lizard, prehistoric with its serrated back and slashing tail, and a second in a trap, shedding its skin into a new black and pearly coat. The two shared a fascination for this "Cretaceous scene of swamp and palms," recognizing ghosts from the dinosaurs' world in the three-toed tracks of a tapir, an agouti "humping along like the first mammal in Paleozoic mud," and the clacking of tortoises mating. George pasted into his journal a passage from *Moby Dick*, about the "wondrous period . . . when wedged bastions of ice pressed hard upon what are now the Tropics . . . [and] the whole world was the whale's . . . Ahab's harpoon had shed older blood than the Pharoah's." To Ishmael's question "Who can show a pedigree like Leviathan?" he proposed an answer in the margins: "Armadillo?" He caught a three-banded one to take home and "everyone enjoyed it. Tied to a string, it is like a mechanical toy."

Eric's observations were nearly as thorough as his dad's. He illustrated letters to his grandmother with detailed renderings: of the palm tree that gave Acurizal its name and the toothy piranha that menaced the pig as she swam. He described the double-decker cattle boats and his six-inch pet tarantula, whom he named Shelob after the spider-demon in *The Lord of the Rings*; she liked the frogs he fed her, sucking them dry.

His adult memories preserved that Schaller acuity. The piranha "which it was my job to catch and filet for lunch" not only took the beef he used for bait but often bit through the metal leader with a loud crunch. For the fruit-eating pacu, he baited his hook with the berries the fish stretched to pluck from overhanging trees. "When I caught one, Dad said, 'Cut open its stomach. You'll get a nice surprise.' Turns out they're full of little worms. When I got chiggers in my toe, he showed me how to clean them out with a knife."

Mark's most vivid memories were of the overnights with his dad. The low-flying bats they netted had a sweet smell and screeched as they untangled them. Finding a dead capybara, "My dad, based on whatever clues he was attending to, knew a jaguar had killed it within the past day, so decided it was OK to carve off some meat to cook for dinner. It was like eating a giant hamster. Obviously, it wasn't part of the plan, but he could improvise anything while camping." (George found it "strong-flavored but tender enough." He left its head in the river, to rot into ideal bait.) At night, Mark mistook the thousands of mayflies crashing into their hammock tents for rain.

Kay's joy at being all together in the field bubbles through a letter to Bettina. "With George's chickens, ducks, and pig, it seems like Old McDonald's Farm. . . . The boys and I take turns doing the 5:30 a.m. trapline walk with him. He does it again in the afternoon. Sixteen miles a day—no wonder he's slim!" When it was her turn to snare caiman, George left her holding a six-footer on shore while he ran home for his scale. "I was barefoot, and waiting in the dark felt very vulnerable indeed. He always has a new experience for me! The boys are quite proud to be really working for him. He pays them 40 cents a fish as he does the local people. I fish for free." For their twentieth anniversary, they walked to see the giant water lilies in bloom, serenaded from the canopy by the "mooing" of a tiger bittern.

Still, George couldn't entirely shake the desultory feeling induced by this jungle and his failure so far with jaguars. "I feel lazy and without interest," he wrote on one particularly hot, thorny, tick-infested, enervating afternoon. "A horrible day," he wrote on another, after a cow got caught in a snare, destroyed it, then lunged at George as he tried to free her. "I took a club and hit her head until she lay down and I could remove it." A week in a blind waiting for caiman eggs to hatch produced nothing, he wrote Chris. "There can be few things less exciting than watching a pile of leaves all day." He grew briefly elated when he shot a peculiar wailing monkey that he thought might be a new species. But as he stuffed it, "I rather lost interest."

Marcio's presence still rankled. When in August Peter Crawshaw visited Acurizal for the first time, George was darkly amused to see his indolent assistant suddenly officious and proprietary.

> *He whisked him off to show the traps, talking loudly and constantly. Things are coming to a head with him. I can't stomach his sullen behavior, especially his rudeness to Kay. And his childish bellowing, "Hey George!" At dinner he sits silently, ignoring our attempts in Portuguese. I spend more time now avoiding his company than thinking about jaguars. And I'm sure he dislikes me equally.*

When Crawshaw said that he could start sooner than expected, George celebrated. "A good day. Marcio is leaving! I will get more work done without his grating presence."

As it often did on the eve of change, family harmony frayed as summer drew to a close. Mark was growing bored and Kay tense, as problems arose with their plane reservations home. "The Marcio and ticket business," wrote George, "is rather ruining the tranquility of our first six weeks." He was glad when rains gave him an excuse to stay home for some last days of play. But watching the boys pack, he dreaded "the end of a fine two months, the best we have spent as a family, [and] the three-month wait before we can be together again." Eric and Mark went fishing a last time, to leave him a full larder, but had no luck. "And what will I do with the pig when I go off? She needs company. Her future looks tragic." Returning from the send-off, "the house seems empty. The pig wanders into the empty boys' room and lies down."

For more than a week after that parting, George did not write in his journal. When he resumed, he described "the lonesome thinking one does while walking—one begins to accept limitations." Two years into this project, he still felt disregarded. He waited in vain for an invitation to an upcoming IBDF meeting on Pantanal conservation. He tired of the words *espere* and *amanhã.* "Tomorrow" would come the plane, the mail, permissions, an expected colleague. "I need to watch animals," he wrote in his journal. For eight days, he went off to camp alone.

•••

In late October 1977, George left Acurizal for his first trip into the Amazon. Hugging Fuxica, he wished he could explain—that though he'd be gone several months, he would return to her.

His companions for this trip were a mixed bag. He was glad to have Vasco, as well as lepidopterist Keith Brown and Robin Best, a polar bear expert who was now hand-rearing Amazonian manatees, which felt to George like overinflated inner tubes. He was less happy to have along two of the kind of "scientist" he could never abide. Neither the "slow, potbellied" botanist nor the complaining geologist could muster the energy even to look out the window of their small plane, sitting up only at the prospect of a visit to the Xingu Indigenous reserve, where they hoped to see bare-breasted women. George celebrated when after a week both went home.

Ten years later, he would satirize these two in his foreword to a reissue of *White Waters and Black*, Gordon MacCreagh's hilarious 1926 account of his Amazon expedition with eight "Eminent Scientificos" from Harvard. "In temperament they were, to put it mildly, unsuited for enforced close association," wrote Schaller of that earlier band. Though undefeated by flesh-eating maggots and boat-crushing rapids, "they could not function at a human level; they were unable to master themselves enough to achieve unity as a team. It was perhaps a not unusual expedition." Borrowing MacCreagh's device of using only titles, he continued.

> Reading this account again brought to mind a trip in which I participated, though the word "expedition" does not apply to our effete effort. In the mid-70s, I found myself with a team in the same region. . . . By plane we spanned in hours distances that once took months. . . . Yet ease of travel did not lead to conviviality of spirit. . . . The Geologist . . . placed in the bow of a boat to spot submerged logs . . . dozed off. The Botanist . . . remained mute on all subjects save Manaus' women . . . being living proof that still waters can run shallow.

The changes George had seen in the Pantanal paled next to those just getting going in this greatest of tropical rainforests, with devastating consequences for the world no one could yet foresee. The road they set out on was brand new; one of many the junta was ripping through the Amazon to speed development and extend control. Winding their way along the Venezuelan border, they spent their first night at a new military camp and their second as guests of the other claimant on the land and souls here: Nossa Senhora de Lourdes, a Salesian mission.

Chugging up on a tractor to greet them, Padre Pedro gave them a tour. He showed them the frame houses the padres had built for the Yanomami to replace their traditional communal dwellings, and the new mission store. He took them to Sunday services, where George noticed that the women struggled to suckle their babies in the dresses they now wore. His Native guides would later tell him how their village had splintered, with half its original two hundred inhabitants moving several days' walk to the east, beyond the mission's reach. Even some who remained had refused conversion. George saw a chief lying naked in a hammock, turning arrows in coals to harden the shafts. Another man, transported by snuff, danced and chanted, naked but for feathers on his biceps, a necklace down his back, and a string tied around his waist from which he had suspended his penis.

The discordances were only deepened by his conversations, in German, with the elderly Padre Franz Knobloch, recently immigrated from Sudetenland. The priest's questions unnerved George. ("The Rockefellers have Jewish blood, don't they?") But he had also published a collection of Yanomami stories, worried for the curious young men who now paddled their canoes to the road construction camps, and voiced grief at the cultural erasure he knew the mission was hastening.

Five Yanomami guides then took George out to explore the forest; when he could not keep up, they bent twigs to mark the trail, occasionally backtracking with creatures they hoped might interest him. (The yard-long earthworms did: When you stepped on their castings, "you hear a gurgling as if the worm is withdrawing.") At day's end, he watched them cut poles of the imbine tree (*Annonaceae*) and build a frame for their hammocks, which they hung stacked like bunk beds.

Two of the men, Saba and Alexandre, carried 16-gauge shotguns they'd bought in the mission store. Walking ahead, they shot fifteen big birds including a curassow, two tinamou, three guans, and nine trumpeters. From the guans, they carefully removed the alula wings, explaining that they used them to paint themselves, and from the curassow, the curly crests for bicep decorations. Though George found the taste of the trumpeters strong, he was impressed by Alexandre's marksmanship: It took him just nine cartridges to get nine birds. When George took his own shot at a macaw, he missed.

George was most interested in mammals, but in a week, they saw not one: no otter, no armadillo, no deer—not even tracks. The guides said monkeys used to be here and taught George their local names. But roads, guns, soldiers, and miners had brought their end.

With no animals, George kept busy drawing pictures of giant dung-eating scarab beetles, which stayed close by in hopes of an offering from one of these rare upright mammals. He sketched or pressed botanical beauties, including the leaves of a *Macrolobium* dotted with a Modernist's circles of green and gold. His companions showed him the epena tree (*Virola elongata*), the source of a hallucinogenic snuff. Back when they still hunted with arrows, they told him, they dipped the tip in the snuff to relax monkeys' muscles so the animals would fall.

Near the military camp, an old couple had shown George two tortoises—flipped on their backs with legs tied together, saved to eat later—and an uakari monkey they had "plucked" for dinner. But a more fundamental cause for the absence of animals was given away in a radio conversation George overhead on returning to the mission. (He had by now learned a fair bit of Portuguese.)

> **BISHOP (CALLING FROM SÃO GABRIEL):** Padre Pedro, do you hear me? I need a jaguar skin to present to the colonel.
>
> **PADRE:** Jaguar is not easy, Dom Miguel. I can get you a maracajá [jaguarundi] cat. Several skins would be easy.
>
> **BISHOP:** Maracajá would be fine. Get me several.

The flight back over the seemingly endless trees revealed more pressures bearing down on this forest and the people who until now had kept it standing. Landing in Rondônia to refuel, George saw the fast-advancing frontier: towns sprung from nowhere to serve truckers and fortune seekers, forest reserves checkered with clearings for coffee plantations and cattle, mines grinding and thumping all night, the miners floating "like condemned shades" through the dark.

By the time he got back to Acurizal, he was in a dark mood. A fazenda owner was "an inconsiderate SOB" for forgetting a rendezvous. With Laurie, who had promised to join him, again delayed, "I feel like telling him to go to hell." And though Fuxica soon went from indifferent to effusive, the cook Leontina told George that she'd become aggressive: chasing dogs, goats, and sometimes children, and vigorously humping the leg of any man who visited. (George drew this, the pig's forelegs stiff, mouth open.) He said that if the behavior got worse, she should have "it" killed. "Leontina looked sad, as she genuinely likes it."

He was harsher still when he found Marcio there on holiday with a friend—in George's room; he had gotten past the lock with a screwdriver. Though George was leaving for home the next day, he threw Marcio's belongings out. "Later he came strutting up, expecting trouble. He got it. I told him to get out of the house. He rather disintegrated, stood in the yard and cried and screamed and threatened for an hour. To return to Marcio after the similar Eduardo [the complaining geologist] did not get me in a happy mood." George last saw the young man on the runway, shaking his fists at the departing plane.

Returning to Acurizal four months later, in March 1978, George was pleased to find Peter well settled with his pregnant wife, Mara; their four-year-old, Danielle; and a new arrival named Tommy. Peter had liberated the puma from a display cage at a Corumbá nightclub. The cub was so starved and weak that the pilot had agreed to bring his crate into the plane, strapped to the back seat. But the shaking and noise of takeoff had been too much and—perhaps fortified by the feral kittens he had snatched along the way—Tommy had burst free. Leaning forward to get as far from the cat as he could, the pilot had nearly plunged them into the bay. Peter did at last manage to pull the puma into his lap, calming him with coos and pets long enough to land at Acurizal. By the time George arrived, the cub had been properly fattened on discarded bits of slaughtered cows, and the whole menagerie had a sweet family feeling. Danielle played with Leontina's daughter, Zenil, for hours, often with Fuxica; their moms grew close; the pig even played with the puma.

The scientists remained, however, as unwelcome as ever. That became clear when, again under orders, the ranch hands cut off Tommy's slaughterhouse treats. For a time, Peter caught piranha for the cub. But with George worrying for the kids' safety, he sadly gave the puma away.

Peter was in any case by now fully absorbed in his apprenticeship. He learned first from George how to distinguish tracks: A puma's are long with oblong toe pads; a jaguar's are wide with rounder toes; and an ocelot's are like a jaguar's but much smaller. If the feet are smooth bottomed, the cat is young. Older ones show wrinkles and scars. From tracks alone, George knew that they had four jaguars in residence—an adult male, a mother with a fifteen-month-old daughter, and a second adult female—and that the male roamed both the ranch and the mountains, overlapping the females' smaller ranges. Coming upon the fresh carcass of a collared anteater, he taught Peter how to reconstruct the crime. The anteater had deep puncture holes above its nape and shoulders and small scratches all over but had been left uneaten. Surrounding the corpse were the tracks of two jaguars and matted grass, indicating a fight. The jaguars had most likely "played" with the anteater to ease precoital tensions, George explained, a behavior common among large cats.

Always open to young scientists, especially those with experience he lacked, George invited into the project John Weaver, fresh from collaring coyotes in Yellowstone. With tensions rising at Acurizal, he also added a second study site at Bela Vista, a ranch on an island just a few kilometers away. A military post there afforded Bela Vista's three resident jaguars some protection, as did the keen interest of the post's commander, Sergeant Jair, who gave them permission to trap one on condition that they show it to him. The ranch manager was also hospitable, though afraid of the pig, who he warded off with a chair "like a lion tamer."

A week and a half of journal entries from Bela Vista capture the strange texture of George's days. A jaguar moved a trap without springing it. Finding a dead dog, a favored prey, George took the rotting head home to examine the killing bite. When he wished out loud that he had a live dog for bait, Sergeant Jair gave him his. George tied the little black-and-white mutt to a tree with two ropes, but came back from fetching water to find the pup had chewed through both. He tied him up again while he went to build a doghouse. This time, he returned to find the dog strangled by his collar. So he put him into the trap dead. In spare moments, he worked on the chapter recounting his trip with Matthiessen to Crystal Mountain, "a little each day, though it's hard to place myself in the barren ice." In heat so intense his clothes fermented, the dog was soon "grotesquely bloated, eyes bulging and fur slipping." But no takers.

The jaguars were not far from where he and Weaver were camped. One set off two traps, leaving only toe hairs; another left huge tracks on the beach. A neighbor gave George another dog for bait, but when, after howling all night, he broke free, George began taking a gun on his evening walk to hunt for bait. The first peccary he met he could not bring himself to shoot; the second he did, then felt terrible. "At least he made no sound." On March 28, the bait dog came back, wagging its tail but with a cut, maggoty foot. George pulled out the maggots and gave him penicillin and the peccary's viscera, which the dog gulped down. (George had eaten the hindquarters.) He tied it out again but it just shivered all night. On April 4, a jaguar killed a heifer; George found the first vertebra bitten all the way through. At the peccary bait site, both traps held a caracara.

Finally, the next morning, George heard the clank of the rubber-padded leghold trap. His first jaguar! Through a screen of branches, he saw her leap against the palm tree to which the trap was tied, then lie still, her face calm. Filling a syringe, he snuck toward her. She glared when he snapped a twig underfoot, but not seeing him, turned away. Even the *pfut* of the dart did not excite her, and in seven minutes, she was out. The trap had caught her by one toe; beyond a small cut, she had no injury. As John ran to fetch the sergeant, George carried her into the light; at 130 pounds, she was not much heavier than Kay. Covering her eyes with a cloth, he affixed the yellow radio collar, then lay beside her, an arm slung over her body to monitor her breathing. Soon both were asleep. The cat stirred once, but under the warm weight of George's arm soon dozed again. After ninety minutes, Weaver returned with Jair and nine noisy men. They backed off when the waking cat growled and reared to her feet. George cut her free, then waited until, after a few woozy rests, she walked away. In honor of the donor paying Weaver's salary, they named her Ella.

Ella's signal immediately vanished, though they searched the island for days. Then Vasco flew in, and with antennae mounted on the plane's wing struts, they found her. Encouraged, George decided to try again to trap one of the jaguars he had tracked at Acurizal, this time camping with Peter. They experimented: setting "jumping-traps" that when triggered spring up around the paw; keeping cows out with an arbor of lianas; building a corral with a trap at the entrance and, inside, a small pig. Every day, when checking traps, they took food and water to the pig.

But then, calamity. On April 14, Peter caught up with George on the trapline. (He'd been twenty minutes late to their 5:30 a.m. rendezvous, so George had—naturally—not waited.) In the very first trap was a young crab-eating fox with a foreleg so badly broken George had to kill her. They measured her, collected her stomach, and checked her reproductive tract; she'd not yet had any young. Peter reminded himself that she was the only animal they had so injured, and that trapping was the means to a bigger purpose: The knowledge they gained would save many animals. But the awful feeling remained. A few days later, they found a coati with his forepaw almost severed. George cut the skin from which the paw dangled and released it. It was a clean break and he was hopeful it would heal. But "I hate the steel traps," he wrote in his journal.

Occasionally, the fiascos were more comedy than tragedy. Approaching trap number five on April 19, Peter and George heard the rattling of a chain and found an adult male ocelot with his foot in the snare. Before releasing it, they put the Ketch-All lasso over its head, though wrongly, as they would soon see. (Their accounts differ as to which of them made the mistake.) Holding onto the pole while George readied the syringe, Peter saw the cat slipping free and urged him to hurry. Too late. The cat leapt, and George right after it, managing to grab the tail. When the ocelot twisted around on him, George tried to hold the cat off with his boot but felt its teeth pierce first the leather and then his foot. Still, he held the tail while Peter tried to slip the noose back over the ocelot's head. Blocked by George, he couldn't. "Take your foot out of his mouth." "No. If I do, he'll just bite me somewhere else." Peter finally managed to loop the cat's other end, and George got free. The day before, a yearling fox had nipped his hand. "I worked with gorillas, tigers, and lions without a single injury," he told Peter, "only to be bitten by these little beasts."

In a week out, the men caught four foxes, one armadillo, and two birds: a tinamou and a three-foot-tall seriema. A ranch hand brought them the mandible of a porcupine he'd killed; the first record of the species in the Pantanal. But no jaguar. Even when the new Acurizal foreman Aníbal reported a jaguar-killed calf, they came up empty. Schaller's quick forensics found bleeding in the throat but no tooth punctures on the head or neck. The cause of death was strangulation; the perp, a puma.

Ella remained their one consistent link to the jaguars' world, and to her human neighbors. One day Peter was on Bela Vista listening through headphones when the ranch foreman, Francisco, approached. "*A onça está aí*?" (Is the jaguar there?) Peter nodded. "*Posso escutar*?" (Can I listen?) "Sure," Peter said, handing over the headphones. For a few seconds, Francisco listened intently, frowning. "*Eu não estou escutando nada.*" (I can't hear anything.) "*É esse sinal de bip.*" (It is that beeping sound.) Oh, said Francisco, his interest vanished in an instant, "*eu achei ela ia estar esturrando!*" (I thought she would be roaring!)

Not long after, Ella gave Francisco's neighbors the thrill he had hoped for. The ranch cook, Cida, had just walked by with a friend when Ella's signal grew so strong that Peter could hear it without the antenna. A dog yelped. A woman shouted "*Olha a onça!*" (Look at the jaguar!) and all rushed away.

Over dinner that night, Cida told Peter that they had been on their way home from collecting *bocaiúvas* (a local coconut) when the small dog running ahead of them had been snapped up by the collared jaguar. Still Ella's favorite meal.

Relations at Acurizal, meanwhile, continued to deteriorate. "The IBDF is trying its best to evict us," George wrote to Mark, now speaking man-to-man with the fifteen-year-old. "They're hassling us in small ways, like taking away our radio. But one carries on and to hell with others whose only aim in life is to make money."

George's edges didn't help. When he was "tired and feeling the limits of his Portuguese," Peter was "glad the local people couldn't understand what George was saying, so I could smooth his words in translation." He soon had help from the consummate smoother, when Kay arrived in May. "Your mother gives much moral support, and is someone I can gripe to, let off steam," George wrote Mark. He hoped he'd come too. "I need you and Eric. If you're interested, be sure to let me know."

Though she made little headway with the ranch hands, still ordered by Geraldo not to help or even socialize with the outsiders, Kay remained her stoic self. "We are running out of many foods but can get by on rice, beans and an occasional egg," she wrote the boys. "I may have to go fishing yet!" She had forged a bond with their neighbor Felix, a squatter allowed to "scratch out a living" for his daughter and seven grandchildren by raising chickens and manioc. "Felix is a truly good man. Your father employed him to look for sign as a way of giving him money. 100 Cr a day. He was so pleased that he gave us a dozen eggs and a sack of manioc. Your father accepted the eggs but paid for the manioc." When Kay visited with her Polaroid camera, the little ones posed excitedly with their grandpa, his long hair flowing from beneath the black plastic hard hat he always wore. All then watched with delight as the instant photo emerged.

Though Ella gave George and Kay the kind of nights they loved best, camping near her on the island, George could not stop worrying about the Acurizal jaguars. That he'd not trapped them was one thing, but why did he not even see sign? Then a telegram brought new cause for worry: Kay's mother was ill and she must return to the US. When the air taxi came, they spent their final half hour together flying over Bela Vista in hopes of a last contact with Ella.

Determined to find the missing cats, George again resorted to dogs. This time he hired Tony de Almeida, a guide famous among American and European hunters hungering to kill the world's biggest jaguars. Almeida's book on those hunts, which included the trophies' sex, size, stomach contents, and behavior, had provided George with his most useful data so far. With Almeida was Richard Mason, a former mercenary in Angola, and houndman Manoel Dantas. Dantas had hunted in the Pantanal since 1952 and shared valuable observations, including that on several occasions he'd seen this cat, reputed like tigers to be solitary, mingle with friends.

The men had come once already with their legendary dogs, but in days of searching had not found the jaguars. Peter had seen the worry then in George's face, as he turned away from the boat to walk the twelve kilometers home alone. Now Mason and Dantas were back with their top dog Gigante, a "castrated yellow mongrel" they conversed with: When he barked a query, the men answered *hup-briii* to signal their position. On June 21, out early enough for dewdrops to still hold molecules of cat scent, they quickly found one. But it was Freddie, the puma George had collared the prior August, resting calmly in a tree. George sketched what happened when he shook a nearby sapling. Dotted lines first show the fifteen-foot leap Freddie made to another limb, then his climb down—on feeling the dart's sting—to a low fork, his rump sagging and tail limp. Gathering palm fronds into a soft landing, George gave the cat a gentle push. This time the radio collar functioned for two months, allowing them to track Freddie over his extensive range.

Good fortune, alas, never lasted here. Out three days later with Gigante roaming ahead, they heard first excited barks and then awful yips and whines. "That dog is being hurt," Dantas shouted, releasing the others who raced baying toward their leader, the men running behind. They caught up with the pack leaping against a tree in which a jaguar lay, watching impassively. Finding Gigante badly cut, Dantas bristled at the restraint imposed by this client, who stood preparing the syringe. "On any other hunt that cat would be dead. I would be skinning it now." Hit with the dart, the jaguar raced down the tree and up a tangled steep slope. With one dog (Jagunço, "Henchman") tracking, Dantas cut a path with his machete and soon found her, deeply asleep. It was a struggle to carry the limp body to a flat spot where George could attach the collar. To honor Gigante, Mason led him to the jaguar whose claws had just torn into him. The dog merely looked at her. Now noticing the fresh small cuts on the cat's flanks and neck, Dantas surmised that she'd had a male companion, who had attacked Gigante while she ran.

With the rest gone home, George and Peter released the jaguar, who was still unsteady. Testing the receiver signal and finding it low, George darted her again to change the collar. They named her Feminha (Little Female). Returned to the house, George treated the deep wound in Gigante's abdomen and gave him a shot of antibiotics. But eleven days later, the dog died. Since Dantas could not take the jaguar in compensation, George paid him the $500 he'd agreed to for any pack member lost. Still, he understood what it meant to these men to lose Gigante, who had accompanied them for many years, and felt little satisfaction despite two captures in five days. That he hadn't recognized this jaguar also gnawed terribly at George. She was neither the mother nor the daughter he knew. Where were they?

The sad revelations now came quickly. First, Geraldo told Schaller that they would clear more forest to expand the cattle herd. He quoted a Brazilian proverb—"You can't whistle and chew sugarcane at the same time"—which meant, George understood, the jaguars had to go. Three days later, Peter brought worse news. Leontina had overheard a ranch hand's wife gossiping about the killing of two jaguars weeks earlier. The vain search for the two resident females suddenly made sense. "We were both," Peter recalled, "swept with a sense of helplessness."

Though with little hope of a response, they followed protocol, notifying IBDF and the ranch owner that protected animals had been killed. They then set out to learn more, calling on the men who had remained friendly even under their enforced ostracism. Cláudio, who had often brought George small mammals and reports of tracks, sat on his mule, looking over their heads. "I don't know anything. I wasn't with the group." João da Mata also avoided their eyes: "In some things, I have to obey orders." That night, at 2:00 a.m., Peter woke to see a light under George's door. He knocked and found him storing his belongings. "I could feel that he was shattered. He said he felt like just packing up and returning to the US but would finish what had to be done first."

The next morning, they visited Felinho (from whom George had bought Fuxica); he confirmed that he had killed the mother on Geraldo's orders. He had surprised her on a calf kill, then sold her skin to a *mascate*, one of the traders that ply the river. George could understand why. Acurizal was a lousy place to work: The men were paid half what hands earned elsewhere and got meat (given freely on other ranches) only by killing wildlife—or by stealing cattle and blaming it on jaguars. George spent the afternoon in his vegetation plot, weighing flowers and leaves.

Late that night, he awakened to find João in their house. Pretending at first to be after rubbing alcohol, which some locals mixed with milk and sugar to drink, the hand soon confessed that he was afraid to be seen with George by Aníbal, who had been there when João killed the younger jaguar. The foreman had cut out the cat's tongue to eat ("like a vulture," George later wrote, though he usually admired vultures) and hidden her skin in his house. As George had suspected, Little Female was a recent settler. And it had been a cruel joke, he now understood, when at a time Aníbal would have known the search was futile, he'd ridden up to ask, "Have you found the jaguars? They are perhaps on the other side of the hills."

The killing of jaguars was hardly uncommon. Cowboys counted it a badge of machismo; one old rancher in Poconé boasted to George that he had personally shot eighty-three. Retribution for cattle killing, they called it, though George estimated that for every cow taken by a jaguar, one hundred died of disease, drowning, or starvation. (Those seeking excuses to poach regularly pinned those deaths on jaguars too.) The poaching had climbed wildly since 1962, when Jackie Kennedy's appearance in an Oleg Cassini leopard coat set off a global craze for spotted furs and sent prices soaring: A local could get 3000 cruzeiros ($120) for a jaguar and 1500 for an ocelot. In just two years (1968–69), the US imported 23,347 of the golden hides. Officials did little to stop the trade. As one put it, after apprehending and then releasing a boatload of skins, "I don't want to be a dead hero."

But these particular jaguars, George knew, had been killed for another reason: to drive him off. Had he never come, they would almost surely still be alive. In their grief and anger, he and Peter tried again to rouse the ranch owner, traveling to his sumptuous São Paulo office. (Coimbra would not come to the ranch, which he visited just three times in twenty years.) He told them he "regretted the unfortunate situation, but there was nothing he could do." They persisted until an IBDF investigator came to Acurizal, but he threw up his hands, too, despite it being an IBDF-sponsored study. The most he would do is confiscate the skin Aníbal had taken. "I finally met her, the young female who in life had been so adept at eluding me," George wrote. "The hide with its sorrowing beauty, its hollow eyes, its bullet hole—I did not want that memory."

In the days that followed, Peter could feel that "George had changed. He was a mix of hopelessness and urgency. He rarely spoke unless spoken to. It was almost as if he wasn't there." Though he stayed another month to monitor the three collared cats, George did so "dutifully, with little joy."

Tracking, at least, had become easier, thanks to Marcos Rolim, a pilot who had begun ferrying fresh local fish to urban markets, using the airstrip belonging to his friend David Cardoso (whom George described as an "erotic" film star). Mounting antennae on Rolim's wing struts, George and Peter soon found Freddie the puma. They knew Ella was about when Sergeant Jair told them that a jaguar had taken a dog: her signature. When they then picked up Little Female's signal, they went to find her on the ground. They never saw her, but George was soothed just to hear her collar beeping nearby. "I listened to her at rest, her signal calm and constant, until the call of a tinamou announced the coming night." Deciding to camp, he and Peter built nests of the tall grass called *rabo-de-lobo* (wolf's-tail), transporting George for a sweet moment to his nights with gorillas.

But how far this beep was from that first silverback roar. Did George love his time with gorillas best because they were the only animals he met with nothing but binoculars and a notebook? No bait, no darts, no tags, no traps, no collars. Of Little Female, he wrote: "We stayed on together, the three of us, in the darkness. I slept, as did she, until dawn, when at the hoarse chorus of howler monkeys, she moved up the valley into the forest."

Before going home for a month, George visited sites to which they might move now that Acurizal was a graveyard. At Fazenda São João, he and Peter slept where Teddy Roosevelt and his son Kermit had stayed on their ill-fated River of Doubt expedition. After using caiman for target practice, Kermit had shot a treed jaguar, which "fell like a sack of sand." The former president celebrated his boy's "vigorous manliness." George thought it "about as sporting as executing one in a zoo."

They did find a new home: a site near Poconé used for agricultural expositions. Though they would see no jaguar here, they could study its prey, especially caiman and capybara. On his way there in September, George stopped at Acurizal to retrieve his things. No one greeted him. A dry period had thinned the canopy so George went out with his .22, taking a marmoset, a rattlesnake, a coati (pregnant, it turned out, with two big fetuses), and a squirrel he had heard chattering to its friend, "a lovely animal but one I needed as a specimen." He saw a few howlers and a marsh deer skeleton, "but without enthusiasm. I do not care for this place anymore."

Their new home was not without sorrows. George found a caiman washed up on shore, abandoned with a bullet in its brain, and a capybara struggling to stand on useless hindquarters. To end its suffering, George bashed its head with a shovel. When it "collapsed with a sigh," blood pouring from its ears, he cut it open to find a shell in its spine. Beyond carrying in more guns, the Transpantaneira highway was also taking a direct toll. Among the "litter" of road-killed bodies, George found thirteen smashed capybara; some drivers ran them over for sport. "I can feel the broken bones of their skulls." Still marshalling the equanimity demanded of anyone who works with bodies in extremis, he often seemed unmoved by the pain and death he witnessed, and sometimes caused. But occasionally, unpredictably, the dam breached. He suffered watching a capybara die, racked by spasms, and still more when, to diagnose its lung disease, he had to kill three healthy ones—to collect glandular exudate and heart tissue. There were moments, even, of horror. When one of the dying crawled into a pool, he had to wade in, feeling with his feet till he found the body.

As the road builders laid gravel, cars drove faster, bringing more carnage. George and Peter complained to Firmo, a guard posted by IBDF to enforce wildlife protections, then a day later saw him barrel by, killing a capybara right in front of them. Watching it "twitch in its blood," the scientists were furious. But they knew that Firmo got no support from headquarters, not even gas to patrol. And by now, the metamorphosis from animal to meat was swift. They measured the capybara, then cut off a leg and fried it. When their rescue of another of the rodents from a dog proved too late—it died from deep bites on its rump—they took the forelegs for lunch.

Despite all that, George found here—between their base at Poconé and Porto Jofre ninety miles to the south—his best hours with animals since the final weeks at Dolpo. After the ocelot incident, he and Peter sorted out how to work together to capture carnivores. Peter would lasso a caiman and let it struggle a bit till it tired; then George would tie its mouth shut long enough to measure and tag its tail. Not all yielded. A battle-scarred ninety-pounder, half his tail lost to a fight, flipped and flipped, quieting only when George used his shirt to blindfold him. One female whipped sideways and snapped her long jaws, nearly taking George's legs. "Sooner or later, one will grab one of us."

The capybara were of course more amiable, letting George watch them play, mate, and fight. "Finally, to be observing animals again with binoculars and scope, like old times. I haven't taken so many notes since I started in Brazil." On his "most interesting bird day ever" he watched a phalanx of wood storks "in pink stockings" move shoulder to shoulder like a medieval army, poking their bills into the mud. "Nothing can escape such a search." He saw young limpkins rush to their mother, recognizing her voice; as she cracked open a snail, using her bill as a hammer, crowbar, and finally tweezers to extract the meat, her children stood "watching, learning, anticipating." Another limpkin demonstrated an alternate technique: crunching and then dipping a snail in a puddle to wash away bits of broken shell, repeating until only flesh remained. The fish, too, remained wonderfully bizarre. At one pool thousands lined up just below the surface, faces to the sun "like worshippers." Even the gaudy flowers had secrets, to which George was introduced by visiting botanist Ghillean Prance. Insects drawn to the delicate pink flowers of floating bladderwort fell through them into the plant's underwater dungeon. Once the trapdoor closed, digestion was swift.

George had not lost his gift for capturing in a phrase a being with a life story. A disheveled little capybara sat alone, stunned by a bite through its cheek. A big male, harried by one female, turned for comfort to another, and was attacked by her too. "After all this hard luck, he withdrew into a bush." A few seem almost fairy tales, absent the happy endings. Brought a big snake by a ranch hand, George "popped it into formalin until dead," then cut it open. Like Red Riding Hood's grandmother, a toad emerged alive and unhurt and set out walking across the grass. "I would have liked to let it go," wrote this woodsman, "but it was a peculiar toad such as I've not seen before, ballooned like a rubber toy with a tiny head and mouth." In one

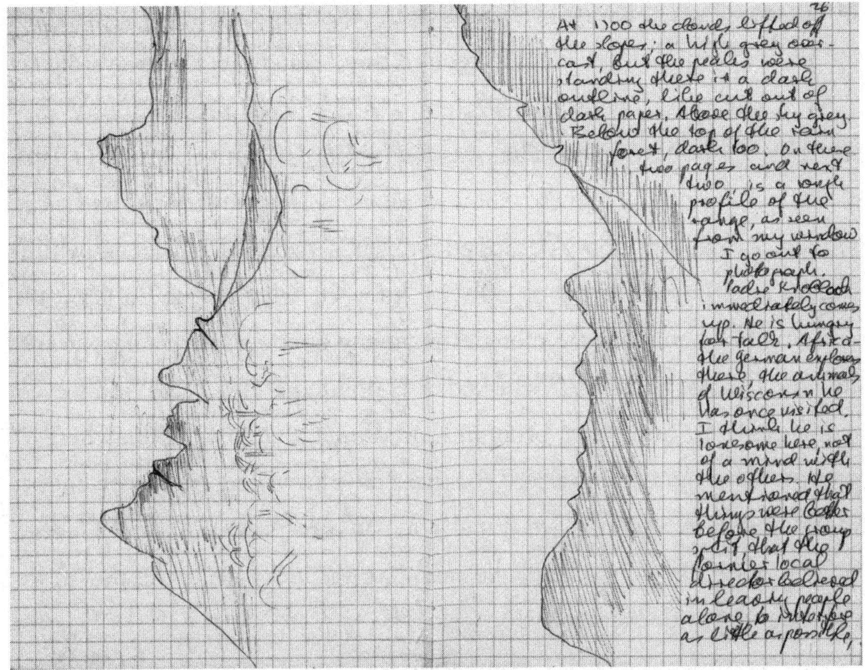

Turning field notes into concrete poetry.

of his rattraps, he found a possum caught when she snatched up a cockroach eating the bait. The roach was still alive in her mouth, as were the five pink young in her pouch, attached to her nipples though her body had gone stiff.

Some of George's visual play also returned. In his Amazon journal, he had run a continuous pen-and-ink drawing of the mountain skyline and palm trees down the left side of several pages, the text following its contours down the right, like concrete poetry. Now he resumed his quick portraits. A snake who had somehow caught an eellike fish by both head and tail looks flummoxed by the unswallowable horseshoe in his mouth. A small black catfish seems to grin as it stiffens its pectoral fins to make itself too wide for a bird to swallow. A fleeing capybara comes to life in three quick calligraphic lines: a brown ball atop two hurrying legs.

Andrew Laurie had finally bailed on Brazil for good when another young admirer wrote to George. Fresh out of college, Kent Redford was in the Chaco region of Paraguay; he was trying to study armadillos and anteaters but had found none. George replied immediately, as he always did to eager young ones, saying that if Kent was ever near the Parque de Exposições, he should stop in. That was invite enough. Kent flew into Cuiabá and, speaking no Portuguese, showed a taxi driver the letter. The driver nodded, drove two hours, dropped him at a deserted cluster of whitewashed mud houses, and took off. Not knowing what else to do, Kent sat on a porch. Then sat some more. After many hours, a lady came out and smiled. "George Schaller?" he asked. She pointed at her watch and waved her hand mysteriously, then brought Kent dinner and showed him where to put up his mosquito net to sleep. For three days he waited, ever more doubtful he was in the right place. Finally a VW van pulled up, and George stepped out. "Oh, you're here," he said. "You're lucky we broke down; we weren't going to be back for two weeks."

The very next morning, George took Kent out to capture caiman. The first one bit Kent's pole in half. But George wanted a picture, so he instructed the twenty-one-year-old to drag a second one into the water, holding it while it twisted and lashed. George then suggested they both wade into the murk where the caiman hid beneath floating carpets of hyacinths, Kent probing with the pole for the hollow thud of an armored back. Only when George stepped on one did he call Kent back to dry land.

So far, Kent had passed what seemed a kind of test. But when George suggested he stay out two nights alone on the trapline ("if I'm paying for his food, I need work from him"), he thought he detected reluctance in the young man's reply. He wondered in his journal if Kent was "driven" enough; though helpful, he seemed disinterested in the "hot, foot-slogging" work. Schaller rarely failed to discern which young scientists would make the field their life; while Redford would become a central figure in tropical conservation, it would not be for his work on armadillos. Kent himself could think of no one whose commitment matched George's, as fervent as any religious vocation.

Returned to Poconé at the end of February 1979, George found a world in flood. Pastures, forests, roads: All were submerged. He sometimes didn't know if he was on water or land. Nights in the hammock tent he'd carried since 1952, his first summer in Fairbanks, were like sleeping in a puddle. And the only other animals about were the enterprising rats who had found dry land in his quarters, nesting in his book box and using his toilet bowl to store seeds.

Fortunately, Peter had been equally enterprising, paying locals to find the nests caiman build as summer ends. An expectant mother piles leaves into a mound, then excavates a chamber inside of which she lays a few dozen sandpapery eggs. She then closes up the nest, pyramid-style, for seventy days of incubation. The effort most often ends in heartbreak. After hours of pushing through chest-deep water and jungles of aquatic weeds, George and Peter usually found only broken eggshells, left by plundering foxes or coatis. Of forty-eight nests, just six produced young.

Using an egg lifted from each of those surviving nests, George aged the embryos to predict when their nestmates would hatch. He then made the rounds listening for the telltale yipping of babies summoning Mom to come dig them out. He was on his knees near one such noisy nest when he heard a hiss behind; he turned to find mother caiman just five feet away. He moved, but only to dig from the other side. Finding a half-cracked egg, he helped it open; out popped a snapping baby, belly still bulging with the yolk sac. One other egg cheeped. But the rest had rotted in the flood. Another nest, raided by red ants, offered a still more grotesque sight: Skulls eaten clean peered from shells, inside of which the bodies remained intact.

At home, Eric didn't mind, at least in retrospect, that his dad missed his high school graduation and valedictory address. He hadn't made "a major mess of it," he wrote to him in Brazil; and in the hot room "we were all glad when we got to clap as that cooled off your hands."

Kay was having a harder time, though in place of the frustration she had felt alone with middle school boys in Lahore, her letters now were pure yearning.

> *Dearest, it was so good to hear your voice, though faintly. I want very much to be with you and even if it means deserting Mark and Eric and the garden harvest, I can come to you. I get so torn between being with you as I want and my duty to the boys that it gets emotionally difficult. But that is not as important as you doing what you wish. You are the most important to me. I want you to truly tell me what you want us to do.*

George had just been making his own anguished family decisions. Though he had been "rather ignoring" Fuxica, he dreaded their inevitable parting; the pig had been a great comfort when things turned dark at Acurizal. But she had also killed one of José's prized fighting cocks. And having grown too big and unruly to go with George into the field, she had to pass her days locked up alone.

The only option was to release her and hope a herd might take her in. He stapled an orange plastic collar around her neck so people would know she wasn't wild, then set her loose near Jofre camp. Like his heron Siegfried, she would not go. When George drove off, she ran after him for two miles. She followed again the next morning but he outstripped her, then regretted it when he couldn't find her on the way back. "Where is she? I fear she will get shot and eaten." When days passed with no sign, he concluded that she'd turned off and never regained the road. It was the last time he saw her.

That was all on George's mind when he sent Mark condolences for the loss of his dog Tika, who had become unmanageable, thieving and killing neighbors' kittens. Rather than keep her forever chained, Mark and Kay had agreed it was better to put her down. "You know the reasons, but I fully understand if you feel resentful and hurt. I can only apologize for being a cause of it. . . . love, Daddy."

Six weeks later, George did get news of the pig. She had made her way to a fazenda ten miles from where he had last seen her. There, a girl "entranced by her friendliness" had named her Duquesa and given her a home.

It didn't last. The Duchess continued to kill baby chickens so was banished again, now without her orange collar. Whenever George saw a herd of peccaries, he called, *Fuxica, Fuxica*, "hoping so much that one would race grunting to rub herself against my legs." None ever did.

Though the work on prey species had been useful, George felt guilty about "abandoning" the jaguar and found in two zoology graduate students a way to revive that work. Both then fish out of water at the University of Tennessee, Howard Quigley and Alan Rabinowitz met up most Sundays for burgers and beer. On one such afternoon, Quigley shared something he had read: China had asked George Schaller to come save the giant panda. Since Quigley had just immobilized and collared two dozen black bears, he joked that "Schaller needs me for that project." The bolder Rabinowitz replied, "Yes, he does, so you'd better write him," then pestered Quigley until he did. A month went by, then two. Then one night the phone rang. "Hello, this is Kay Schaller. George has been away but will be in town to change his suitcase and would like to talk to you."

Quigley's first thought was that Rabinowitz had put someone up to it, as a prank.

> It's hard to convey what that call meant for a young wildlife biologist, the stature that George had for my generation. It was like Jacques Cousteau calling. I said, "Of course I'll be available," then called Alan and said, "You won't believe what just happened!" The next day, George called. "I'm interested in what you do," he said. "I don't know what will happen in China. But if I have to give up jaguars, would you come for a look?"

Three weeks later, Howard was on his first trip to the tropics, flying twenty-five hours to reach Miranda Estância, the new base Peter and George had found for the work. At more than eight hundred square miles, Miranda was the largest fazenda in the Pantanal and popular with jaguars, as evidenced by hunters' success. A cowboy named Celestino—who became a vital source of information—had killed sixty-eight in seven years, until the owner said he wanted the killing to stop.

They again enlisted dogs, a "sorry pack" that in George's telling has a hint of self-portrait, with inexperienced pups surrounding the disgruntled, wounded Solito (Loner), known to be prone to fights. By trip's end, the human solito felt "superfluous . . . Howard can handle this. And my mind is on China." He did publish several papers: with Vasconcelos on jaguar predation and marsh deer; with Crawshaw on caiman nesting behaviors, jaguar movement patterns, and the social life of capybara; with Prance on vegetation. But for the first time, he wrote neither a scientific nor a popular book. Peter asked if he ever would. No, he answered, "it would be too sad." The blow of the jaguars' death had just been "too hard," said Peter. "I think he never recovered."

Schaller would leave a lasting legacy in Brazil: The young scientists he helped train, and their students, have reversed the cats' decline. In four years at Miranda, Crawshaw and Quigley gathered the deep data that had eluded George, tracking two males and four females across sixty square miles. Though reliant on radio collars, they developed a kind of understanding with the animals. At regular intervals, when the collar batteries ran out, they had to recapture each of the six. But rather than becoming more skittish or fierce, the cats seemed to learn that they wouldn't be hurt. "As soon as we'd put the dogs on the trail, the jaguar would tree," Crawshaw recalled. "Every capture got easier. The animals seemed to know what was going to happen and collaborated, cooperated to a degree."

The work at Miranda became the basis for Quigley's PhD, and for a life devoted to protecting wild felines, including as head of the Teton Cougar Project and of Panthera's Jaguar Corridor Initiative, working to restore the cat's historic range from Argentina to Arizona. Crawshaw, too, wrote a dissertation on jaguars and ocelots, then led the development of Brazil's National Research Center for the Conservation of Natural Predators (CENAP), addressing rancher-carnivore conflict and creating jobs and incentives via wildlife tourism for keeping the cats alive. "That's what's most important in George's work," Crawshaw said. "Wherever he goes, he leaves roots."

In 1980 it became Rabinowitz's turn to meet Schaller. After an initial trip to the Wolong panda reserve, George had come to Tennessee to consult with black-bear expert Mike Pelton. Since Rabinowitz was studying raccoons and bears—both posited as possible panda relatives—Pelton asked him to take Schaller for a hike. The twenty-seven-year-old found the forty-seven-year-old to be "stern . . . strongly opinionated" and intensely competitive, racing him up Mount Le Conte. But "what I took to be rigid views," he soon saw, "came from an understanding of nature and wildlife that few humans ever attain." Just days after the visit, Pelton tracked down Alan. George had called, asking if he might like to study jaguars in Belize. Alan answered as quickly as George had on gorillas. "I said, 'Tell him yes.' My professor said, 'Think about it.' I said, "I don't need to. Just tell him yes.'"

George welcomed Alan to his Bronx Zoo office by opening a map. A single dirt road crossed Belize, traversed by an occasional bus. "George said, 'Put on a backpack, get on the bus, have it stop, and just go survey.' He gave me $500 and six weeks, and off I went walking into a jungle full of fer-de-lances [pit vipers] and jaguars. George was always a minimalist. And everything I am and have accomplished began with that initial trust he put in me." Schaller would later explain to *The New York Times* why, after one day, he had asked Alan to take on jaguars. "He had a vision for himself that he hadn't realized. When you meet someone like that you have to give him a try." Rabinowitz would go on to win protections for wild cats in Myanmar, Thailand, and across the Americas.

8.

Rescued to Death

Sichuan Province, China, 1980–85, Giant Pandas

George's first weeks in China, preparing for life in the panda forests, stirred new varieties of doubt. After the heartbreak of Pakistan's emptied landscapes and the revenge killings in Brazil, would he have the strength not only for the physical travails in these steep, soggy mountains of dense bamboo but also for his most sustained immersion since childhood in treacherous social and political conditions?

Just months earlier, he had nearly been lured back to the place he and Kay would always remember as their lost Eden. He now oversaw twenty-five WCS projects on four continents; it was in that capacity that in February 1980 he returned to Africa. Jonah Western had proposed a project on white-eared kob, an antelope with a migration as grand as the wildebeest's, which for unknown reasons was dying in droves. For two weeks, the two men flew in Jonah's single-engine Cessna across the great Sudd swamp of southern Sudan, watching the herds flow through the liquid landscape. When that river gave way to a Pompeiian expanse of dead animals, Jonah proposed they set down to have a look. If there was a landing strip, George couldn't see it; in decades together, this was the only moment Jonah ever saw him hesitate. The two stayed at Jebel Rajaf with Peter and Ann McClinton, there running a conservation program for the Frankfurt Zoological Society. George's notes on the isolated couple have a lost, doomed feeling. Ann, ill and depressed, seemed to subsist on sherry; Peter, consumed with anxieties, kept postponing flying into the reserve George wanted to see, then did not want to land.

Jonah had not forgotten George's dismissal a decade earlier of his "UN approach" to conservation. So he was surprised that—though they found the kob starving, cut off from the Nile by human settlements—what George proposed was focused on those humans: a wildlife tourism program to be run by locals, who would keep the proceeds. He asked to meet with the Nuer and Dinka communities, to understand their traditions and subsistence hunting, and how they were managing the famine they had seen filling Juba with refugees. "I saw this whole other side," Jonah recalled, "not gruff, formidable George, uncomfortable in social settings that were not his natural environment, but what I thought must be the real George. I was struck by how humble he was, how at ease he made these families feel."

Jonah then flew him on a kind of dream tour to see other scientists working under the WCS umbrella. Landing in Kenya's Amboseli National Park to visit Cynthia Moss and Joyce Poole, George got closer to elephants than he had ever been. In Tanzania, he preserved in his journal a funny moment with former game warden Brian Nicholson, to whom he had carried the manuscript of *Sand Rivers* for Matthiessen. (The book recounted Peter's trip with the warden into Selous Game Reserve.) Nicholson had just read *The Snow Leopard*. "Though I don't know why anyone would want to. It has nothing about snow leopards." In Uganda, Schaller discussed contracts and budgets with monkey researcher Tom Struhsaker and heard from Ian Grimwood about the imminent reintroduction of white Arabian oryx to Oman. After watching impala, zebra, buffalo, and baboons, it "hurts a bit," he wrote, to hand the kob project to someone else. "I haven't seen wildlife like this since 1969."

But the pandas'* call was stronger, and reached back nearly as far. At the first glimmers in 1971 of a thaw in US-China relations, George had broached the idea with Conway, who had in turn reached out to Sir Peter Scott, chairman of World Wildlife Fund and creator of its panda logo. In 1979 a query had come from another quarter: Christopher Savage, the hydraulic engineer George knew in Pakistan, wrote to say that Beijing had invited him to make a panda proposal and "if I'm putting my name to it, I will move heaven and earth to have you." He never got that chance. Just months later, Sir Scott announced that he had the agreement with China, as well as a second one with Conway to borrow George. George immediately proposed to WWF that they also bring Chinese biologists to the US, since "different cultures have much to teach each other, particularly in their view of the natural world." That idea went ignored.

The announcement brought a flurry of curtain raisers, including a *People* magazine feature with a sultry picture of George peering through jungle Tarzan-style, and his anointment by *The New York Times* as "the envy of zoologists and writers of nursery rhymes alike." It also brought a flood of advice, some of it useful. William Sheldon, who in 1934 had shot a panda for the American Museum of Natural History—six years after Kermit Roosevelt shot one for Chicago's Field Museum—sent pages and pages on how to use honey as bait and dogs to tree the pandas. George took more seriously the guidance offered in an eleven-page letter from Nancy Nash, the Hong Kong–based journalist who served as WWF's chief intermediary. (She had addressed the letter to WWF conservation director Lee Talbot but copied George so he would know that "problems were big even before you arrived.") WWF leadership had created warm feelings, she wrote, by

* Scientists call the animal by its proper name, giant panda, to distinguish it from the lesser or red panda. But for brevity's sake the simple colloquial "panda" is used here.

sending Spring Festival greetings and displaying a genuine interest in Chinese art. But they had stepped badly wrong by announcing the project in London rather than Beijing, and by wrangling over a panda film before Schaller was even in country.

George soon had his own complaints. WWF-US was raising money for his panda study but directing it to other projects. "I will not have my name used that way." And the contract they sent not only gave WWF ownership of all data but stipulated that data remain confidential. George would not sign. "I would never ask a scientist to refrain from publishing. [And] I have as much or more responsibility to the Chinese who have a right to all information."

The China Schaller entered in 1980 was a traumatized place, only just emerging from the ten-year chaos of the Cultural Revolution. Mao's "violent struggle" against the old order had fallen hard on scientists, gutting a whole generation. Professor Pan Wenshi, with whom George would work for many years, had been sent to a labor camp for "reeducation." George's codirector on the study, Professor Hu Jinchu, had suffered unnamed humiliations. Few "elites" had been spared. Meeting Huang Zhou, renowned for his brush paintings of camels and chickens, George learned the origin of the artist's partial paralysis: Since he painted animals, Red Guards had made him work like one, pulling a plow. Though this campaign against "counter-revolutionary" elements had officially ended with Mao's death in 1976, four years later the atmosphere of suspicion remained. George was kept at arm's length by colleagues, many of whom informed on him and one another.

> *I had never been where relations were so cordial yet so inarticulate, where my freedom to do even the simplest task or veer even slightly from a rigidly prescribed course was so restricted, where all my actions were so unrelentingly scrutinized and reported, and where my presence was treated with such wariness.*

Remarkably, pandas had managed to float above the crazed political winds. Unlike the tiger and crane, the rarely seen creature had escaped association with China's now-scorned imperial dynasties. To the contrary: Since China was the only country to have pandas, Mao had elevated them to an emblem of the new nation and introduced controls on the export of live animals and pelts. In 1961 the publication by *The People's Daily* of Wu Zuoren's ink brush painting of a giant panda had marked it as acceptable content for the "new art." Field science, too—because of its physical rigors—had been spared some of the vilification aimed at the lab and its experts. The first-ever surveys of panda were permitted and published in the final years of the Cultural Revolution.

Still, this collaboration got off to a terrible start. Schaller's first vision of China was a meeting room, and then another, and another still, in each of which, accompanied by Nash, he sat through days and nights of lectures and grim negotiations. First came the botanists and mammologists, whose long presentations seemed intended to impress him with the depth of research to date, an impression that was dashed when he asked for the month's field records from his future study site and was handed a single sheet. Then came Wang Menghu from the Ministry of Forestry, wrangling line by line through the Protocol for Establishment of a Research Centre. Misunderstandings over that center, to which WWF had hastily and vaguely committed, would more than once nearly sink the project. Things only got worse when, on May 11, 1980, Schaller took a group of Chinese dignitaries to meet Peter Scott's plane, only to realize it was the wrong day. "Nancy and I are both sick at the mix-up, a tremendous loss of face." The deputy governor of Chengdu—the city closest to the reserve—had to postpone a reception for more than five hundred people, putting "everything into jeopardy."

Gathered at last, Schaller applied the acute attention that had plumbed the mysteries of gorillas and lions to understanding the two clotted bureaucracies he would spend the next years navigating. Where better to learn the art of politics than in this incomparably complicated school? He'd already won a sliver of Wang Menghu's trust by suggesting that his ministry require foreign scientists to sign a contract protecting China's interests. Now, in his journal, he drew the table, recording each person's name and taking notes on each performance. He saw that Wang Menghu's long soliloquies, "by turns voluble, humble, witty, forthrightly aggressive and bland" were played mostly for his colleagues. He saw that Scott's interruptions were registering as impolite; that WWF had erred in bringing a top-level person to negotiate with mid-level officials, forfeiting a face-saving way to back out of things; and that though Nash spoke Chinese, they needed their own interpreter, to convey and catch nuances.

When it was at last Schaller's turn, he revealed an aptitude for the gnomic way of speaking favored on such occasions, offering the first of what over the years would be uncountable toasts. "In the 1920s and 30s, Western expeditions came to shoot and capture pandas, to take something away from China." This expedition aimed not to take, but to give—"to help ensure pandas remain in their native home."* When Wang Menghu predicted that Schaller would spend half his remaining life in China, he replied, "I hope it will be a long life." His instincts extended to the deft political gesture. Learning that marine biologist Sylvia Earle was in Beijing and would dine with the powerful and revered Madame Sun Yat-sen, he sent along a gift: the landmark recordings, by his colleague Roger Payne, of the songs of humpback whales.

The meetings and banquets for now over, all piled into vehicles and headed for panda country. The first Westerners ever to be officially invited into these forests merited an entourage, especially given Scott's lineage: His father was doomed Antarctic explorer Robert F. Scott. More than four hundred people had helped prepare the visit, and twenty-six now followed them up the trail. Whenever Hu Jinchu stopped to point out the heaps of

* This story from his journal reappears slightly modified in *The Last Panda* (p. 10).

bamboo husks and torpedo-shaped scat that marked a panda's passing, or George paused to measure, the whole line halted too.

Their next stop was far less exhilarating. Taken to see the captive breeding facility at Yingxionggou, George was unsurprised to learn that it had never actually produced a baby panda. Five animals huddled miserably in barren, feces-encrusted cells. Never allowed outdoors, they could barely be persuaded to eat, let alone reproduce. Mortified at the thought WWF might play a part in building another such facility, he sent an SOS through Kay to Conway, whose transformations at the Bronx Zoo had set the standard for the world. A visa was expedited; Conway arrived and in short order made his recommendations. These solitary animals should be provided privacy, time outdoors, shelter from the wind, and nesting boxes made from wood to mimic the hollow trees they den in.

Most of those suggestions were ignored, George saw on his next visit, or realized in the cruelest possible way. To provide a windscreen and privacy, the staff blocked the cell windows with solid shutters, plunging the pandas "into perpetual night." Instead of cozy wood, they built the nesting boxes of bare sheet metal. This was not callousness but a failure to see the animals' capacity for suffering, it seemed to George. Just as Western thought included the Cartesian view of animals as automatons, so alongside the Buddhist reverence for all living beings was a view captured, he was told, in the written Chinese character for animal, which means "moving thing." When he expressed concern, he was asked, "What's so special about the panda?"—an understandable question given the recent torments inflicted on some hundred million human beings.

George was far less forgiving of the vet who—although his specialty was chickens—was conducting "nutrition experiments" on the cubs; one had already died. As a powerful Communist Party official, Sick Chicken, as George would call him (his name was Bing Ji), could do as he pleased. "And it had apparently pleased him to do feeding experiments on the baby panda, with tragic results."

The captive pandas' pitiable condition was on his mind when negotiations resumed back at the Beijing Hotel. Wang Menghu jumped to his top priority: the research facility WWF would pay to build. In a presentation lasting several hours, he detailed plans and itemized costs. There would be eight hundred square meters of labs, pens for twenty captives, a hydroelectric plant, five kilometers of fencing with 2,500 poles, and a perimeter tower and searchlight. "On and on. We were stunned. Total construction costs would amount to two million dollars." Though George thought the facility unnecessary, and urged a focus on study and conservation of pandas in the wild, he saw that arguing was pointless. No matter its scientific value, the facility was a grand political plum. Which, he noted, put WWF in the "curious position of paying for a research center so I can go into the field with a pair of binoculars . . . a token presence exchanged for an expensive lab."

Initial discussion of field equipment revealed a chasm just as great. Wang Menghu was adamant they start radiotelemetry right away, brushing off George's guidance that it was better to get data other ways first. Hu Jinchu wanted night vision goggles, which George knew would be useless in dense bamboo. They seemed, he wrote in his journal, to have "looked in magazines for fancy technology," putting one of everything on the list. Again, he wondered "whether WWF was to be only a source of equipment or a full partner in a conservation program."

That the Chinese negotiators were prepared to walk away was underscored by the late arrival of the man George dubbed Lumpy, a "political hack . . . with a face, figure, mind, and handshake like mashed potatoes." When Nash reacted impatiently to Lumpy's demand that they repeat all he'd missed, the party official turned to his colleagues, muttering in the Chinese he knew she understood: "We don't need foreign help." All knew how badly WWF wanted to showcase the project for its impending twentieth anniversary. And that the Smithsonian was at their heels, having won its own agreement for a panda study thanks to a word from Henry Kissinger to a vice-premier. In the end, Scott offered a million dollars and China accepted, with a nonchalance George couldn't help but admire.

Having agreed to begin work in November, George's heart sank when on his way to Wolong reserve he was again redirected to Beijing. For this round, WWF was represented by its director general, Charles de Haes. Another long list was put on the table, of equipment WWF must guarantee if work was to proceed. An adiabatic bomb calorimeter, atomic absorption spectrophotometer, refrigerated centrifuge, electrophoresis unit: Schaller himself didn't know what some of the instruments were for, only that most were irrelevant to pandas, and that such follies as hot-air balloons would be useless in these fogbound mountains. Again, he suggested they begin with basics and expand as they discovered needs. When De Haes deemed impossible the demand that the lab gear up immediately, Wang Menghu clarified. "I will tell you a joke. If the lab does not function, I have no job."

After nine excruciating days, Schaller felt like a pawn who might be flicked from the board at any moment but also half wished that it might soon "all be over" so he could go home. They at last finished the meetings, but still he sat for three weeks in a hotel, uncertain if they'd ever let him start. He didn't bother with planning, futile where agreements might be abrogated at any time.

George could make some sense of the Chinese officials' fixation on facilities and instruments. With the radically regressive turn of the Cultural Revolution now flipped to demands for the Four Modernizations, technology promised a quick catch-up. And after a long history of brutal treachery, China had good cause to be wary of foreigners.

WWF's bungling was harder to understand. First, from headquarters in Gland, Switzerland, came the demand for a lawyerly addendum to the original agreement, requiring various kinds of reporting and receipts. Communicating mistrust, it put the head of the Chinese delegation in a humiliating and untenable position. Next, WWF invited just one of three governing bureaucracies to an important event, creating allies and enemies without knowing which was which. Most destructive was WWF's relentless and obtuse publicity seeking, culminating with a mock-up of a commemorative panda stamp with characters that said "Taiwan." Even the award to Schaller in 1980 of their Gold Medal managed to slight the Chinese, with WWF billing him as "the man who would save the panda." David Attenborough would later marvel at how this most "intrepid, expert and experienced field zoologist" had turned artful diplomat to maintain this collaboration between "two grinding bureaucracies" with opposing priorities. He, too, laid the fault mostly on WWF. Suppose, Attenborough suggested, that "a Chinese delegation showed up in the US having taken the bald eagle as its emblem, and said that since the bird was, in their view, shockingly persecuted, they would send their experts to supervise its protection."

And yet, for all those disasters, the "propaganda"—as both sides took to calling publicity—may have saved the project, making it impossible for either to jump ship without unendurable embarrassment. When the Chinese again threatened to stop the project until WWF met their equipment demands, De Haes called their bluff. If they couldn't guarantee the work would continue through May "we might as well stop now." As George noted for future use, that brinksmanship worked. Assurances were given, and at last he headed into the reserve. The clink of iced bamboo leaves, a Ping-Pong game with Pan Wenshi, his first panda tracks in the snow: All gave him hope for the New Year.

•••

George had lived many places as high and cold and wet and impenetrable as the field station he settled into at Wuyipeng. He had rarely, however, faced all those miseries at the same time. And except when trekking, he had always had at least a hut to come home to, not a sodden little canvas tent like the one that became his home for two years in these panda mountains.

He did, for the first time in a decade, have Kay by his side much of the time; he'd made her inclusion a condition of his being there. Even in a tent so cramped with metal bed and rusty stove that one had to sit if the other wanted to move, she made them a warm home. The red blankets she hung on the walls blocked some drafts, and the panda feces she spent her days drying and sorting smelled like fresh-cut grass.

Hardest on George was the unrelenting intimacy of camp life, an obligatory camaraderie he had not endured since being shipped off to boarding school hideouts during the war. With their tent pitched right next to the others in a ramshackle huddle that reminded both Schallers of an Alaskan gold-rush town, and meals all taken communally, George had no privacy and never fully relaxed. Though he would complain at the disruptions of travel to New York for job and family, or to Europe for WWF, he clearly welcomed the breaks from the claustrophobia of a camp "crowded with misunderstandings and watchful eyes." A field biologist's greatest danger, he would warn himself, was not "fierce animals or treacherous terrain but in finding comfort and being seduced by it."

Outside the tent, the steep forests were reminiscent of Congo but rimed with ice and hiding an animal far more elusive than gorillas. As in Pakistan, George chose to be here in the toughest months of the year, when snow offered any hope of finding tracks. But while a barrel-shaped panda moves easily through bamboo bent under heavy drifts, a six-foot-tall man does not. Forty years later, Hu Jinchu would recall their long days together inching along in a low crouch. It could take a whole day to go just two kilometers, and "we couldn't straighten up, sometimes for hours." Dislodged snow fell down their necks, and the frozen stalks that caned their thighs and knees were sometimes so unmovable they had to back into thickets to tear through. Ice chutes and precipices obscured by snow or fog posed constant danger. If George stepped past a hidden drop-off he grabbed wildly; on good days he wound up suspended from a sapling. At nearly fifty, he had finally grown into some caution. Hiking with Hu Jinchu in the Jiuzhaigou reserve, he was stopped by a sheer earth and pebble slide dropping sharply to a lake far below. Hu Jinchu leapt the ten feet with a laugh, but George scrambled down to the water instead. For the first time in his journals, he wrote, "I was afraid."

Evenings offered little respite. In Congo, a day of crashing through jungle had ended with Kay's blackberry tarts or oatmeal molasses bread. Here "our rations were mainly steamed buns which froze unchewably hard," Hu Jinchu recalled. "We jokingly called them 'cold-hearted baozi.'" Since their electricity came from an in-stream hydrogenerator that quit when winter water levels got low, they had a dim five-watt bulb, or no light at all.

And week after week, nothing. "I've lost ten pounds in these two months of hard walking," George wrote in February 1981, "without seeing a single panda." Over the next four years, he would return home many nights so discouraged: soaked and chilled with throbbing knees but an empty notebook. Once he would go a full six months without a sighting. And his many near misses would seed more self-doubt. After coming up the trail just minutes too late to see the panda a colleague had been watching, and giving up on a spot where a day later a Japanese TV crew would find another, he would wonder why "I don't have the right feel or luck with animals these days."

A wild panda is in fact diabolically hard to see, thanks largely to the mismatched way it lives in the world. It doesn't announce its presence like gorillas with big, noisy families, nor does it roam like a tiger. Instead, it stays mostly alone and mostly still, inside a world that seems designed to hide it: of bamboo screens all around made still more opaque by near-constant mists and rains. There it sits, just quietly eating, day and night.

It must, because in one of the clumsier turns of evolution, it has become wholly dependent on a food it can barely digest. Though the purest of herbivores, eating *only* bamboo, a panda still has its carnivorous ancestors' gut. Lacking the internal fermentation vat and symbiotic microbes that enable cows, giraffes, and other grass and leaf eaters to access the nutrients in cellulose and lignin, a panda can assimilate just 17 percent of the bamboo it eats. It can't build enough fat to hibernate or even to sleep all night, but can survive only (like the orbiting humans in *WALL-E*) by combining gluttony with sloth. The rest of the sins, it must toss overboard: Lust and wrath in particular take more energy than a bamboo-eating carnivore can spare.

The panda's most endearing qualities in fact evolved to enable this endless eating. Their sweet, broad head provides a strong anchor for jaws powerful enough to snap, strip, crush, and grind woody stalks. Their roly-poly body serves as a big, bamboo-holding barrel: George calculated that his favorite panda ate on average eighty-five pounds a day, half her body weight. Their famous pseudothumb, an elongated wrist bone, allows them to grab and hold even the slenderest stem, and to eat with exceptional efficiency. As George counted, one big male bit into 3,481 stems, rhythmically feeding each into the side of his mouth like a pencil into a sharpener, levering it Bugs Bunny–style into pieces, and reaching for the next before the last was swallowed. Most passes right through: Schaller weighed a single scat pile at seventeen pounds.

Taking in such meager energy, pandas must spend just as little. Most barely budge in a day, traveling no farther than a few hundred meters. Like Roman emperors, they eat slouched or reclined; George watched one lie on his back and use his hindpaws to bend stems toward his mouth, saving both forepaws for shoveling in the leaves. They don't build beds, their plush bodies serving as both mattress and comforter. More than once, George saw a sated panda abruptly flop over onto its side or belly like a wound-down toy, fall promptly to sleep, then wake like Winnie-the-Pooh: raising arms overhead to yawn, rubbing their back end against a tree, even (when fed) licking a porridgy paw clean. Yet for all that adorableness, they were the most truly solitary animal George had ever known.

An exception must be made, of course, to create baby pandas, but even then, they husband their energy. Courtship lasts less than a day. Gestation is brief and birth weight startlingly low: rat-sized and rubbery, a newborn weighs five ounces, one nine-hundredth its mom's weight. That limits the maternal investment in case the baby is lost. However it also means that for four months, the mom can never put the hairless pink creature down but must carry it every moment in her paw or armpit or mouth. She can't do that for two, so she abandons any surviving twin. Only when the cub has grown into a "panda-colored beanbag with legs" will she leave it to forage, though a golden cat or fox might yet eat it.

Trying to find a motionless animal behind a bamboo curtain had nearly defeated George when, one March afternoon, the cook came running. Hu Jinchu had found a panda! Rushing to the scene—without binoculars; he had gotten out of the habit of carrying them—George followed Hu's gaze: first forty-five feet up a tree, where a little one lay on a branch making a distressed *huuu*, then lower, to where a bigger one was backing down the trunk. The pandas' meeting had clearly not been amicable, but for the American and Chinese scientists, it was a moment of tremendous shared joy.

Then, all at once, things began happening fast. George had finally asked Howard Quigley to bring his Tennessee bear experience to the pandas. In the weeks since his arrival, they had been setting out traps baited with burned bones: Like most animals, a panda will devour meat if given the chance. There had been the usual mishaps. Twice, a young colleague trapped himself while rebaiting. Then a weasel learned to run the trapline, stealing bait after bait.

George had shared with Howard his fears about handling this national treasure.

> He said, "Let's hope these animals don't die under anesthesia, because we'll spend the rest of our lives in jail." Here I was, a young grad student who idolized George. I watched every move he made, including how diplomatically he handled banquets and such things. He was who I wanted to be. So that rattled me. . . . But I'd done it many times. I knew the drug. And you get in a kind of trance when you have responsibility for an animal. If not for that, I'd have had a nervous breakdown.

Quigley's chance came on March 10, when Wang Lianke, whom everyone called Xiao ("little") Wang, ran in breathless. A panda was caught! They found the animal quiet, just looking at them with soft eyes. Moving quickly, Howard affixed a syringe to a pole and jabbed; the first needle bent, but a second try succeeded. The little panda was soon asleep, George massaging the paw that had been snared. They put on the radio collar and Howard took blood. ("What for?" George asked. "Don't know," said Howard. "Very scientific," said George.) They were releasing the animal when Kay reminded them to remove the magnet so the radio could transmit. George felt tremendous relief. "To catch is one thing. To have one uninjured, with no problem with the drug, no chomping, no lunges . . ." He shook hands all around and sent a message to WWF. "Cat belled, Schaller."

Two days later, they had not only picked up the first panda's signal but had a second one in a trap. This one was both bigger and less serene, honking in distress and warning them away with clacking teeth and a roar so explosive both turned to flee. It was too late in the day to handle her, so George and Howard slept by the trap, tense with worry, listening to the beep of one and the breathing of the other. They woke to find all nine campmates arrived, to watch. George asked the men to stay back and quiet as they put her under, explaining that the animal could get hurt if she flailed about. But the excitement proved too much and all crowded in. Seeing the panda grow frantic, George shouted. "Goddamned it, get back." He apologized that evening, explaining again the harm that can befall a drugged animal. But "get back" was translated to "get out," and reported to Beijing.

The next step was naming the pandas, an uncontroversial tradition in China. Hu Jinchu suggested they call the little male Long-Long. *Long* means dragon, and is the name of the serrated mountain from which Wolong got its name; the doubled name is an endearment. For the big female, Pan Wenshi suggested Zhen-Zhen (treasure)—aptly for George, who would forge with her his closest bond. Although she was apparently aging, with jutting hip bones and missing teeth, George knew from Zhen's dark, worn nipples that she'd had at least one baby, stirring his hopes for another. She wised up quickly to the traps, letting herself be caught four times so that she could eat the bait, then waiting patiently for George to release her. They soon collared three more, getting a glimpse each time of the panda's temperament. Ning-Ning (gentle) reached through the trap to let them hold her paw and pressed close to have her head scratched. Wei-Wei (grand) was more tragic: When trapped, he sat silent and cowed, bent over so far that his muzzle touched his hindfeet.

Unable to distinguish the pandas by sight, and still rarely seeing them, the team relied on the collars' distinct frequencies to track individuals. Once a day, they pinpointed each one's location; once a month, they monitored each at fifteen-minute intervals for five consecutive days and nights. As with the jaguars, George experienced the beep as contact, as if with aliens.

It was good old-fashioned noises, however, that led George and Xiao Wang to make "some of the first-ever observations of real, wild panda sex." On April 13, 1981, hearing a sudden outburst of whinnies, squeals, moos, and moans, George set off to find the male sounding his "amorous call." Though theirs is longer than a female's heat, which lasts just one to three days in an entire year, male pandas also have a limited period of sexual readiness, peaking in the first month of spring. When George later played recordings of this one's cries for an audience at the Barbizon-Plaza Hotel, *The New Yorker*'s Talk of the Town could not get over "the indescribably strange . . . medley of howling, screams and yelps [atop] tremolo chirps that seemed to come from an orchestra of species." George found the crooner up a tree. But having heard no answer, the big male slipped away.

The next day, another barnyard of noises, this time from two animals. Zhen had joined the male. Hurrying toward the din, George got close and pressed his body against a fir tree, piecing together what happened next from glimpses and the hard shaking of bamboo; a way of seeing not so different, he reflected, from how each of us perceives the other: in bits and gaps filled with supposition. A second, smaller male tried to join the party, but both times was driven off. At last, Zhen crouched in invitation. Her chosen one squatted behind her. But when he laid his forepaws across her rump, she changed her mind and slipped away.

To George's surprise, the male now turned and came his way. Relieved when the animal plopped down on soft moss for a rest, George did not know what to do when he stood back up and again approached, to within eight feet. Would he treat him as another male panda? The creature seemed to look right through him, and—half-hypnotized by the round head and masklike face—George's mind rewound to other close encounters: the gorilla joining him on the tree branch; his nose-to-nose meeting with the tigress. He scarcely noticed Xiao Wang's arrival until Zhen also came their way and he felt the young man step behind him. When Zhen was close enough almost to touch, George shooed her with the flick of a hand and soft *tch-tch* he'd used to chase off lions. But rather than flee she slumped softly to the ground, leaning against George's fir tree. All four held still for a time, until George ran out of film and crawled nearly into Zhen's lap to retrieve more from his rucksack. She stayed in her trancelike state.

Only when the male stepped from the bamboo did Zhen rouse herself. Going to him, she crouched with her muzzle so tucked it looked as if she might stand on her head. From just fifteen feet away, George watched as the male mounted again and again, lasting at most just a minute each round. Between times, she ambled about and he followed, getting scolded with a sharp bark when he tried to stop her by grabbing her leg. When she went up a hemlock tree to rest, the male bleated at her and paced, once passing so close that he had to step over George's tape recorder. Again, George froze, knowing the power of a panda's grip and crushing teeth. But both animals seemed to have tuned out everything but each other.

In seventy minutes, George counted forty-two mounts ("way more sex," as the BBC noted with the wink journalists seemed unable to resist, "than most humans get in a year"). Afterward, he watched the male gently bite her rump, the two "engrossed in each other" for half an hour. For the first time, he fully understood what the captive pandas were being denied. Male and female both needed the freedom to signal their readiness: he with a love song, she by responding. Given the marathon, she also needed a safe retreat. Forcing them into proximity was useless. George once watched a captive female solicit a male, backing up to him, writhing on her back, reaching with her forepaws, while he merely watched, utterly uninterested.

The following spring brought George more glimpses of the rare moments when pandas meet. He watched Pi-Pi ("brave"), weighing in at 235 pounds, fight another giant. Roaring as they stood and grappled in a fierce dance, Pi finally sent his rival head over heels off a twenty-foot precipice. (His exploits were becoming legend; a day earlier, George's colleagues had seen Pi climb a tree for an impressive bit of aerial mating.) George watched the poor incel Wei-Wei follow Zhen around like a lovesick teenager, finally giving up, "his social life . . . over for another year." Another male sought only chaste pleasures, happily sliding over and tenderly pawing a female until she lost patience and went off to have sex with several others, including an "exuberant" little guy.

Those encounters with wild pandas made all the more painful the ongoing abuses at Yingxionggou. Arriving one day in the spring of 1981, George heard an animal screaming. He found Bing Ji pushing what looked like an electrified corncob into the rectum of a goat that stood rigid with shock. The vet was practicing without anesthesia the electroejaculation used for artificially inseminating pandas. The goat later died. Despite their status as treasure, George saw pandas similarly manhandled: undersedated, dragged with ropes, still denied access to the outdoors because, he was told, they "don't like the sun." One female's teeth were broken off from biting the bars of her cage. One male ejaculated with Chinese equipment couldn't walk properly for a week. Meanwhile, Bing Ji had failed to monitor the captive Li-Li's estrus cycle so missed the one chance that year to give her natural access to a male. Just as two US colleagues were arriving to share their knowledge of artificial insemination, he had given her a shot of hormones at three times the proper dose, disrupting her entire cycle.

As chief veterinarian at the New York Zoological Society, Emil Dolensek was at the fore of captive animal reproduction. In 1985 he would open an animal medical center with an obstetrics ward, sperm bank, and capabilities in embryo transplants and endocrinology. Stephen Seager, a reproductive physiologist at the National Institutes of Health, had electro-ejaculated gorillas, and would later publish on the technology's application in humans. Seager objected to the facility's practice of ejaculating pandas in public view or, worse, for TV. (He had no doubt seen Wiseman's *Primate*.) But with George and four vets looking on, he demonstrated the procedure. Attaching the stimulator to a generator, he delivered thirty rectal jolts; at each shock, the panda's thighs jerked together. He then milked the penis into a cup. Small and burgundy, it looked to George like a deep-sea creature, a polyp or sea anemone. The first male was a bust—his sperm were all dead—but Shan-Shan was a champ. Looking through the microscope at the "crowd of little wiggles"—some three billion highly motile sperm—Seager sighed. "Beautiful, just beautiful." Although the female was past heat, they inseminated her for practice. The Chinese scientists later toasted Stephen for making the male pandas happy—a joke suggesting that polymorphous fantasies transcend cultural divides.

That comity was short-lived. Bing Ji threw up more obstructions, refusing to let an obviously ready pair "do their thing," as George put it in his journal. The Chengdu Zoo, too—once it had the $7,500 worth of equipment promised by WWF—abrogated its commitment to work with the American scientists. While Dolensek and Seager waited at the hotel, told that the work hadn't started, zoo staff inseminated two females. When the

Smithsonian's National Zoo research director, Devra Kleiman, arrived as agreed to observe reproductive behavior, her counterparts at the Chengdu Zoo said they'd never heard of her. Kleiman did manage, she wrote George, to get out twelve rolls of his film for *National Geographic*, and "all of Emil's pee."

Leveraging a visit by Lee Talbot, recently named director general of the IUCN, George requested a meeting with Wang Menghu. Aware that Bing Ji was Hu's superior at Nanchong Normal College, he trod carefully. But talking late into the night, he detailed the malpractice at Yingxionggou—from the missed opportunities for normal breeding to the maltreatment of animals. "For the first time," he wrote anxiously in his journal, "I have talked against a person for the pandas' sake." His political intuition did not fail him. The next day Bing Ji had a "sore back" and was replaced by young, attentive vets.

George called it "the downfall of the Nanchong gang." But Bing Ji soon returned, and the neglect continued, raising grave concerns for the animals' health. No one was using the medicines, vitamins, and disinfectants WWF had provided because, George finally learned, Beijing had ordered the staff not to use foreign supplies nor even to heed the advice of outside experts. So the pandas remained in their own filth and full of worms; George saw one pass a hundred. In a nine-page letter to De Haes, now in the Yale archives, he warned that China and WWF both risked scandal and reputational harm should a visitor report the deplorable conditions. "Is it justified for WWF to spend thousands of dollars and to ask recognized scientists to spend their valuable time if sharing their knowledge has no impact?"

George's heart ached especially for Ping-Ping, who during his capture had both ears ripped off by dogs. Teased by keepers, he had become aggressive. He was chronically malnourished, wanting not their rice gruel but bamboo. George pleaded for his release, arguing that they could learn how readily a captive readapted to the wild—and enhance China's image. Failing that, he contrived an experiment that required he spend two full days feeding Ping all the shoots he wanted. When the sated panda flopped over to nap, George went to gather more. His goal, he said, was to see how much a panda could eat. But as he left at the end of the second day, George could scarcely look back, knowing Ping-Ping sat waiting for his return.

Solving any of this was made vastly more difficult by the complexity of George's relationships with his Chinese colleagues.

Even in the remote fastness of their camp, the effects of the Cultural Revolution persisted. "Victims and possibly tormentors lived smilingly side by side . . . still keeping mental if not written dossiers, uncertain how soon history might repeat itself." Decades of radical swings in who or what was politically correct had taught everyone the dangers of committing to anything. "Fear of denunciation pervaded. . . . They had to become willows to survive—passive, afraid to take risks or excel."

Beyond that learned passivity, George knew that most of his campmates were there not by choice but by order of party secretary Lai Binghui. They simply lacked the political clout to be elsewhere. When negotiations with WWF faltered and Beijing warned that the project might collapse, he saw that many hoped for that deliverance. Especially since "I have been a tremendous intrusion in the slow-paced research community at Wolong. Suddenly a foreigner arrives, speeds up the program, establishes new lines of command, questions facts and procedures," and adds work with no extra pay. No wonder that when assignments were given to make the hard climb to check traps, the concept of *turns* proved "curious; there are five Chinese researchers but every other day it's my *turn*."

Harder still for George was the necessity that he stifle any effort at mentoring. Many evenings he gave talks, on such topics as how to interpret data, or territoriality. But the training in field skills his colleagues in India, Pakistan, and Brazil had so avidly sought felt unwelcome here. He couldn't correct or direct even a junior person's work without both paying a price: The other lost face; he was branded arrogant or proud, "a serious accusation."

He couldn't answer questions because his colleagues didn't ask any. When he finally asked his bright interpreter why not, Qiu Mingjiang explained that the young ones were afraid to "argue" and the old ones to show anything but mastery. George saw Hu Jinchu wrestling with these constraints. Not only did the party hierarchy block his efforts to effect change, but as a professor considered the consummate panda authority, he dared not reveal gaps in his knowledge. "That, at least, was my interpretation of the situation, not necessarily the correct one," wrote George.

What George called "the missing spirit of open inquiry" extended to data. He was not permitted to send bamboo samples out of the country for nutritional analysis. His requests to discuss local experiments in, for instance, seeding bamboo went ignored. And he struggled to make sense of turns like the sudden shutdown of trapping for "lack of bait." Bait was readily available in town. Had Beijing sent down orders and, if so, why?

On discovering that his Chinese colleagues were withholding relevant publications and lab findings, George began doing the same. When Hu asked for his April notes, he copied some but pretended to have sent his other notebook back with Dolensek. To ensure Beijing would have to let him back in, he purposely failed to teach the others how to change transmitter batteries or tranquilize an animal. "I feel cheap at such deceptions but here one learns to keep a bargaining position."

That he was a foreigner added mightily to the wariness, with China's Open Door policy still new and memories fresh of the punishments the Red Guards had meted out to anyone mixing with foreigners. "No one dared visit our tent alone," he wrote, "[and] we dared not express more than restrained friendliness to anyone for fear of exposing him or her to criticism, or worse."

We lived in one world, the Chinese in another. . . . Not that this was our choice. I have never made the ambiguous effort to . . . act like a

> *native, except to modify thought and attitude where courtesy demands it. I do not have the romantic notion that in another culture I can ever cease being a stranger at the periphery. . . . But I did hope . . . that acquaintance might lead to true companionship, a sharing of ideas, thoughts, and feelings, as it had in other countries. . . . I assumed trust would soon come. I was wrong.*

Never knowing "what anyone really thinks," he wished for even one colleague to take him into their confidence—though when anyone did share information or complain, he suspected it was a test to see if he'd join in. Having to please his hosts without understanding what they truly wanted was a stress without respite. A foreign scientist's "every human virtue and failing is scrutinized and magnified. He has no freedom to travel, his needs are little understood, his requests suspect."

One obstacle was of his own making. He never learned Chinese, at least not well—out of a lack of willpower, he said, though that is the last way anyone would ever describe George Schaller. Even when it came to book learning, he had no trouble applying himself to hundreds of volumes of natural and human history. He had not learned much Swahili or Urdu either, perhaps because humans were not yet central to his work. Or because, working in former colonies, he'd always been able to find speakers of English, or German, or French. Not to mention that in the remotest villages he frequented, the national language was often of little use.

But it may also have been a kind of choice—a way to maintain the distance at which he sees best, tapping his surest abilities to read and reach others. George knew that Hu Jinchu would be more forthcoming if they didn't have to talk through party-sanctioned interpreters, one of whom would not join discussions of panda sex but was happy to translate signs reading "Down with American imperialism"; another who corrected Nash when she praised an artist as a true Buddhist. ("He is a Marxist!" she said.)

But George did accurately read the power dynamics a surprising number of times—enough to win important victories for the pandas and his own freedom to wander the Tibetan Plateau for the next forty years. More than ever, he was served by the skills he gained as a boy—in how to navigate a dangerous political landscape of informants, and show trials, and abrupt changes in fortune; in how to be the object of suspicion and scrutiny and still carry on.

Most confusing were the swings in his collegial relationships, between hot and cold, showy deference and coarse dismissal. Meals were generally convivial, but then a night would come when he'd be put in a separate room, to eat alone. "I don't know why. It makes me feel so much an outsider." If he ever *chose* to sit alone, he was labeled secretive. So, he resisted his own panda nature and, whenever possible, joined in. The obligatory group photo shows up often in his journals; his always the only white face, bent low to get into the frame.

A few bonds did prove lasting. At age ninety, Hu Jinchu remembered both the attention George paid as "I taught him one by one, our plants and animals," and also how his "rigorous research and meticulous spirit" had changed them all. To underscore the point, he had dug back into his own archive to find George's data from his seven days and nights tracking Wei-Wei through a blizzard: The panda "ate 41 percent old shoots of *Bashania fangiana*," Hu read, "and excreted 537 scats."

Their mutual respect had ripened into tenderness. In 2023 Hu told *Giant Panda* magazine how—at their first Christmas, knowing that Hu had received a message from his ailing wife—George had spoken quietly to him: "Why don't you go home? I'm here to hold down the fort." Another night, awakened with the news that Hu had damaged his eyes tumbling into bamboo, George came to his bedside. "Hu grasped his hands," the article recounted in its breathless style. "'Doctor, I trust you. Do whatever I need.' And under Schaller's careful treatment, his eyes were healed."

George also grew close to Yong Yange, a young scientist who was one of the first to habituate a panda, at Foping reserve. George's excitement at his findings redoubled Yong's own motivation. Every evening, he came for lessons in how to find patterns and meaning in the data. Every day, the two walked for hours, Yong now the one following tracks while George relied on the receiver. "When he'd come and find me already with the panda, he'd give me a big thumbs up." Yong empathized with Schaller's frustrations. "Emerging from the Cultural Revolution, not many Chinese had any idea of modern science. They didn't really understand what he meant. And no matter how bad things got, George had no way to fix them."

That wasn't entirely true. George was finding a few leverage points, including those opened by visits from VIPs. He had been advocating for months for the two-year-old captive Jin-Jin, who spent his days at Yingxionggou weaving his head in his bare cage. But George had not won him even a toy when Lars-Eric Lindblad, who had recently launched the first cruises in China, and Russell Train, Nixon's first EPA administrator, arrived. With them in tow, George at last got Jin-Jin outdoors. He had never seen anything like this panda "exploding with joy," scampering up a tree and somersaulting, "an ecstatic black and white ball." He was also learning to defer to party members when it cost only him, not the pandas. Dragged out of bed one night to watch movies with county leaders, George neither yawned nor complained as the festivities stretched past one in the morning.

Along with Lindblad and Train had come Mark Halle, advisor to Prince Philip on conservation, as well as Lee Talbot and Charles de Haes. His local colleagues were touched, George saw, when he welcomed the illustrious visitors to his "second home." Wang Menghu suggested everyone hug and kicked it off by hugging Halle. George watched amused as Talbot hugged Train, knowing how they disliked each other. "A real love-in," he wrote in his journal. He then worked to steer the discussion. When Wang Menghu shut down talk of fines for killing pandas—warning that "the matter does not exist for WWF"—George tried with little success to get De Haes to leave it be. He had better luck pressing for other commitments. Kleiman would be included in efforts at natural breeding at Yingxionggou; she wanted females in heat set free to roam the facility to pick a mate. And Janet Stover from the Bronx Zoo would be allowed to hand-rear abandoned twins, who needed (for instance) something like their mother's tummy licks to stimulate defecation.

More than ever, in these difficult years, George relied on Kay—"to listen to my worries, to share insignificant thoughts, to reach out and touch." Kay's easy social graces helped warm relations with their Chinese campmates. Hu Jinchu and Yong Yange both recalled how George introduced her as our "lady of Giant Panda scats." When the young ones brought her soup, she often found bits of beak or bird foot in her bowl, a test she passed with such good humor that the men began to teach her Chinese in exchange for English lessons. She occasionally led the group in song: At an evening talent show, she sang "Home on the Range" with Quigley and "Yankee Doodle" with the camp leader.

At Christmas, she and George retrieved the little fir tree Zhen-Zhen had for reasons unknown leaned upright in her den when preparing to give birth to a (short-lived) cub. It was lopsided and tooth marked, but they decorated it with the paper-and-pipe-cleaner Santas the boys had made in Tanzania, Kay's chickadee, and George's tin chital and tigers. Their Chinese companions added glass golden monkeys. One sculpted a snow panda with coal eyes. The chef prepared a feast of spiced dofu, rabbit, sugared fried noodles, and steamed egg, and all exchanged small gifts. Kay told the Christmas story, and when she sang "Silent Night," everyone joined in. The warmth carried over: When George contracted laryngitis, a colleague gave him a box of three hundred pills. The ingredient list included musk (3 percent), bear gall (7 percent), and rhino horn (10 percent).

Together in the field without children for the first time since Congo, their life resumed some of the rhythms of those years. "Break the ice in the wash basin to take a quick sponge bath," an entry in Kay's diary began. "Breakfast of rice gruel, check the droppings in the drying oven, collect more firewood, write letters, heat water to do the sooty laundry." They foraged for mushrooms, which George drew, his first sketches of fungi since the little *Pilzie* he signed his notes with as a boy. Both found a lovely calm in measuring new bamboo shoots or gathering stems and leaves. Camped out to monitor a panda overnight, George slept while Kay listened to the beep, wishing it was always just the two of them with the animals. Then at 2:00 a.m. she climbed into the bag his body had warmed while he rose to the hot tea and chocolates she'd set out, the hours until dawn long but "not lonely, because Kay was there only a breath away."

Perhaps because they felt stranded together, the romance of George and Kay gained renewed intensity. "When at times we had to be apart for several months," he wrote, "my joy in the work diminished and I knew then how much she is the focus of my life." After evening strolls redolent of honeysuckle, lavender orchids, and wild strawberries, they returned to the tent carrying "the scent of moss and spruce with us, and crystal beads of frost sparkled in our hair." They noticed that silence had as many shades as the color black, the quiet of a snowy night utterly unlike the expectant silence of a tense animal.

Out one night alone near a panda, under snowflakes as "large as cherry blossoms," George was reluctant to leave. "But now I also want to return to the tent, to our home, [and] tell Kay of my night and hear of her adventures." (She had enjoyed visits from a brown weasel, and from the Qiang women who brought endless thermoses of hot water as an excuse to come look at her.) Kay wrote in her journal of another night camped together, with just peanuts and crackers to eat. Her sleeping bag was damp because George had climbed into it wet to dry his clothes, and her air mattress was collapsing. So she was freezing despite the down gloves, booties, pants, and jacket she'd worn to bed. "I finally burst into tears at midnight and woke George, and my misery made him miserable." They lay awake together, their breath forming ice crystals on the tent's low ceiling that twinkled in the lamplight like a night sky, and Kay began naming constellations while outside "snow slanted down, its silken curtain isolating us within the circle of light." Even George's metaphors got moony: He described the trill of a nuthatch in a birch grove, a long "downward crescendo like a girl descending a hillside with light and happy steps."

The rest of the family, meanwhile, seemed far away. When Mark wrote from Chapel Hill that he wanted to move off campus and buy a car, George replied that he'd have to pay for anything beyond standard room and board, that "a lot of distance can be covered on a bicycle," and that "your plans for running a blueberry farm sound promising but best not to spend money one hasn't yet got. Who knows if 1981 will be a monumental blueberry failure?" Kay advised her youngest that he would get home before she did and that for two weeks "you'll be on your own, my pet." To hamburger, "add 1/2 tsp salt per pound plus pepper, onion and Worcestershire sauce." Georg Sr. died that year, after a late-in-life rapprochement with Bettina. "Do you remember our first Christmas in Bad Orb?" she wrote in one of the gentle letters they exchanged. "How grateful we all were to still be alive."

Though he would follow her in a month for a summer visit home, seeing Kay off that first spring sunk George into gloom. He had tired of "fuming impotently" and saw that his single-mindedness only widened his distance from colleagues who had learned to ride lightly on life's unreliable currents. Since everyone now ate hurriedly and alone, he threw himself a solitary forty-eighth birthday party, playing cassettes of the Carpenters and Vladimir Horowitz while he ate the jam Kay had left for the occasion. He wondered again if he was equal to this project, felt his pleasure in the work dying.

It was Zhen, of course, who revived his spirits. One night, as George tracked her, she turned to him as she had once before, on that day of nearly four dozen copulations. This time, pushing through the bamboo, she seemed to actually see him, "her apprehension and mine a bond of shared feeling." Softly bleating like a timid lamb, she sat with her back leaned against tall bamboo, crossing her forepaws on her round white belly like a

Buddha. Watching her head sag and her body settle into a slow, rhythmic heaving, George was honored to see that she had fallen asleep.

Their next encounter, that fall, was not so placid. Monitoring her beep day and night, George saw that Zhen was staying in an exceptionally confined area, sleeping a lot but also restless: all signs that she'd had the baby he had hoped for. He stayed away until, after some three weeks, her pattern abruptly changed. Fearing she'd lost the cub, he and Hu Jinchu went to investigate her den tree. They were less than twenty feet away when Zhen suddenly charged, roaring. The two men split up—luckily. George scrambled up a maple tree. While Zhen contemplated following him and he resisted the temptation to snap a picture, lest it provoke her, Hu had time to run up the trail. Then, the sound they had hoped for: a small squawk from the hollow fir. Climbing from tree to tree, down one and then up the next, George made his way to Hu.

The year 1982 unfolded with more devastating acts of carelessness on both sides. Returning from a Christmas visit home, George found the costly equipment his Chinese colleagues had demanded idled or wrecked: telemetry gear in the dirt, blood samples spoiled. At Yingxionggou, he found Li-Li—aggressive after prolonged mistreatment—being held down for insemination by so many people that he could see only her head and Mei-Mei bleeding; someone had used glass tubing instead of plastic and broken off a shard in her uterus. Dolensek's instructions had gone ignored, for a ludicrous reason: The lovesick vet, spurned by the interpreter, refused to listen to her translations. When George asked why urine samples, essential for tracing the pandas' cycles, were rotting in a bucket, the vet told him that his job was to collect the urine; the person whose job it was to freeze it was on leave. At that, George lost it. "Did you even go to vet school? I don't believe it. No one can be that incompetent. If this country depends on people like you, I feel sorry for it. Look in a mirror to see something useless." Dumping the samples on the floor, George reported up the chain that the ruin of so much scientific work might doom the whole project. Getting in the local swing, he suggested to Hu that the vet should be stripped of his practice and sent to cut bamboo. At the same time, he made a point of publicly praising Bai Zhao, a Yunnan graduate in physiology, for his excellent notes on panda matings. And, year after year, he stayed.

WWF, meanwhile, was creating its annual fiasco. Though George had reported good progress on the construction of the research center, to be called Hetaoping, WWF gave official notice that "given the uncertainties" it could not guarantee its promised $600,000. The hard-won trust again shattered, China canceled George's work for the coming year. This time, their brinksmanship prevailed; WWF paid $400,000, and after nine lost months, George was allowed to return.

In camp, as the project's novelty wore thin, so did the camaraderie. George sometimes spent twelve hours in bed, to escape the cold but also for want of useful tasks and companionship. Even when Kay joined, both felt "tired and bored" and generally ate alone. George did his best to amuse Kay, offering such diversions as a cartoon panda rubbing his behind to scent mark bamboo. "To be in love / is hard at best / But none the less / Ours stood the test / of Time (at times) / oh Valentine."

Two years in, the Schallers also half welcomed a small "invasion" of Westerners. Botanist Julian Campbell was mostly a disaster, not least because of his proselytizing: first leaflets urging locals to use contraceptives, then an anti-Communist poster sent to his Chinese counterpart, which she had to report. He didn't last. More sympatico was a National Geographic television crew, into whose "new ears," George wrote, "my pent-up frustrations flow."

The crew had their own worries. ABC also had permission, direct from Deng Xiaoping, to make a panda film; former President Ford's son was involved. And, having failed to get wild panda footage, Nat Geo was at the mercy of the Chengdu Zoo, which demanded $6,000 to shoot their baby for twenty minutes. (They did offer to paint the ground green to simulate grass.) Schaller himself had gotten tangled up in such fakery the previous year, when German *Geo* published a cover story purporting to feature the first-ever pictures of wild pandas. That the photos were in fact of captives came to light, *The Washington Post* reported, when a "panda expert named George Schaller . . . questioned their authenticity." The first *genuine* pictures of pandas in the wild appeared in a *National Geographic* cover story in December 1981, taken, of course, by George Schaller.

The TV crew need not have worried. Their *Save the Panda*, which begins with Schaller measuring prints in the snow, became one of the all-time most-watched shows in PBS history. It preserves glimpses of daily life: Their porters, all women, hurry with heavy baskets; Kay sits cross-legged typing furiously; Hu Jinchu smiles and laughs. But its romantic language ("sequestered in this inaccessible universe . . . the panda remains mysterious") sits in uneasy counterpoint to images of the men handling the animals: scared, drugged, dragged from the trap by their legs.

George had gone six months without even a glimpse of a panda when in May 1982, as he prepared to leave Wuyipeng, he found Zhen-Zhen sitting on a boulder, her muzzle tucked into her folded arms. George sat with her for three hours, happy that his presence aroused no fear. Famous himself for his capacity to abruptly turn away from any human exchange that had exhausted its purpose, he admired the "startling self-assurance, a striking kind of freedom, in the way she ignored me."

These encounters remained of a different order, however, than the mutual recognition he'd shared with gorillas and pigs. Those beings' anxiety, curiosity, affection—all had been legible to him, giving him a portal to that transmutation he'd felt in his week camped with the snow leopard. Few if any biologists have come closer to achieving that metamorphosis. As Alan Rabinowitz once said: "Even as George records hard, clear, concise data . . . he goes far beyond . . . merges with what he's seeing. He knows as a scientist he must be apart and yet he dissolves into it at the same time." On one of his many return visits to India, Ullas Karanth took him to see a tigress who, after her regular afternoon swim, always walked the far side of the lake. Except on that day, when she instead crossed directly before George. Her tribute, Ullas believed, to their kinship.

That he could never plumb pandas in quite that way stirred George to his most searching reflections on the limits of our ability to ever truly know another. In Zhen, he met a consciousness he could not fathom. "Hopelessly separated by an immense space," he wrote, "her being eludes me. I am not even certain how to begin." He then summoned his own Zen koan, or perhaps Gertrude Stein. "The panda is the answer. But what is the question?"

His panda metaphysics accommodated the paradox that Zhen's bold coat served in fact to conceal. Seen through vertical bars of bamboo, her black shoulders and chest read as void, interrupting her outline; her head sometimes seemed to float free, "like a full moon on a frosty night." Her yin-and-yang coloring suited this world of hope turned to despair, and back again. Even another debate over taxonomy became an occasion for George to honor her irreducibility. Just as he hoped "that there is a yeti but that it will never be found," it pleased him that the latest molecular techniques had not resolved whether a panda is a raccoon or bear. Again, he echoed that other GS. "The panda is a panda."

One barrier between George and these animals was his greatest dependence ever on remote sensing. Looking at his notebooks full not of life stories but only symbols—the locating azimuths, the tempo of beeps—he felt as if he were "studying an abstraction." A second, again paradoxical, barrier was the panda's seeming approachability. With their plush, chubby bodies and clumsy way of sitting and tumbling, pandas have been imbued by humans who know them only from zoos with the sweet comedy of our own babies or the soft comforts of beloved toys.

But here in her universe, Zhen was anything but a clown. To George, she was a tragic figure, trapped in a time that no longer existed. Or she was a being beyond time, as monumental as an Olmec god, with a self-possession he struggled to name: "as durable as a fir tree . . . complete in herself . . . final and preordained."

The usual critique of anthropomorphism is that it grants animals too much—too much sentience, feeling, individuality, social complexity, culture. But to humanize another species is also to grant too little, dishonoring radically different perceptions and capabilities. Man is *not* the measure of all things. To judge the intelligence, the fullness of life, the very value of a nonhuman by how well they can mimic us ignores how clueless we are in, say, the language of vibrations. "Each creature has its own window on reality," George wrote, again distilling the concept of *Umwelt.* Casting back to the opacity of even the least private animals, like the lioness he watched eat her own cub, he quoted Wittgenstein: "If a lion could talk, we could not understand him."

Confronted with the impossibility of ever really seeing through a panda's eyes, George ventured something new: crafting a brief fable in which, like King Solomon, he could understand panda speech. The animal commended him for his dedicated measuring and counting. But remember, he cautioned

> *you cannot divide me into . . . fragments of existence. . . . I am, like any other being, infinite in complexity, indivisible. [Even] time is not the same for all living things. This fir lives more slowly than you, and I more quickly. . . .*
>
> *Some of you . . . hold that language is necessary before one can think, and that makes me and all others—except you—unthinking creatures. What frivolous nonsense! . . . I think mainly with smells. . . . Forget science now and then.*

Drawing Hu Jinchu into the fantasy, the panda continued. "Look at each other. Your ways of thinking are vastly different, yet you belong to the same species." With a nod to Heisenberg—"what we observe is not nature herself, but nature exposed to our method of questioning"—George conceded that Hu's different questions would conjure a different creature. He ended with a note of self-mockery. The magic panda vanished into a spark, leaving two droppings. "And these I collected."

•••

Meeting George on his return in March 1983, Wang Menghu pleaded with him to remain calm. Han-Han had been strangled by a wire snare. Her skin and meat had been taken, but they had found her collar hidden under a rock and her viscera under snow and leaves. Police with dogs had tracked the poacher to the closest settlement, a hamlet of six houses, and arrested him. Reviewing her activity records, George surmised she'd had a baby. Whom no one had looked for.

On April 8, going to see the forest where "beautiful Han-Han" had died, George stumbled upon the silver back of a sleeping panda. He approached quietly before realizing that she lay unnaturally still. Stepping around her, he saw her frozen grimace and severed tongue. She had bitten it off in the throes of death, her panicked circles only tightening the wire around her neck.

The next day, returning with colleagues and the police, he and Hu unwound the snare with pliers and turned over the rigid body. Hu was startled to see George nearly in tears on finding milk oozing from her swollen nipples. Somewhere in the forest was a second orphan, likely dead. This "motionless form . . . expressed such a suffering innocence that its power was almost religious," George later wrote. "Kneeling by this panda, its death so empty of purpose, I was filled with rage . . . at the officials who talked and talked while ever more pandas slipped silently into darkness. 'Damn. Damn. Damn them all.'"

Tying the panda's limbs to a pole, the men carried her out. When a paw slipped free as they passed through the poacher's village, she seemed to George to be waving at the families who had come into the street to watch. At the autopsy he was intensely interested to see her digestive tract and paltry fat stores; from her uterus, he could tell she'd had twins. No arrest was made. For two years, he and Hu had been reporting illegal snares and the forest department had done nothing. "Two of our study pandas dead. Acurizal repeated."

A few days later, searching for the dead mother panda's den, Schaller stumbled upon what looked like a fur coat stuffed into a hollow tree. After a quick backward step, he approached more carefully to find a young black bear asleep. At just thirty pounds, it was too small to be without its mother. She, too, must have been killed.

Fighting a newborn "savage pessimism," Schaller went this time to the political leadership. The party secretary insisted the problem lay with the Yi, Qiang, and other Indigenous inhabitants of Sichuan, three thousand of whom lived inside Wolong reserve. He then made an example of Leng Zhizhong, the poacher who had killed Han, moving his trial to the local community hall and inviting Schaller to attend. Before some four hundred villagers crowded into the aisles, the twenty-six-year-old farmer confessed that he had stolen wire from a bridge and set several dozen snares to trap a musk deer or wild pig. Scared when he found the dead panda, he had cut off her feet and gutted her, hid the viscera and collar, then carried the rest home. His wife had cooked the panda with turnips, but it tasted so bad they'd fed it to the pigs. "I am poor and did it to help my family," he told the court. "I caused a great loss to the people . . . Selfishness is the cause. As a result of being in prison, I will be helped to walk a new road." The schoolteacher representing him reminded the court that his client, with only a primary school education, had committed the crime accidentally, and that "there are several thousand snares in the reserve." Though prosecutors were satisfied with the two-year sentence, George spoke up to say the wrong person had gone to prison. They should have jailed the local officials whose negligence and "stout indifference" had made them accessories to both stranglings.

•••

After two and a half years, George was finally allowed the survey he had sought from the beginning, of five reserves scattered across the panda's range. Traveling for six weeks in spring, he was as fascinated as ever by animals and plants he had never seen: from a 110-pound salamander to the rare *Davidia* tree with a two-petaled flower like a white dove taking wing, found for him by botanist Qin Zisheng. He visited the "Catholic temple" near Baoxing from whence missionary Père Armand David, the tree's namesake, made his 1860s journeys collecting and describing hundreds of species unknown to the West. (The first European to see a giant panda, and instrumental in saving the deer that also bears his name, Père David would provide George his sorrowful epigraph for *The Last Panda.**) When Qiu Mingjiang caught a bamboo rat, George yearned to see its digestive tract but hadn't the heart anymore for killing.

Their walk through the Min and Qionglai Mountains followed a "definite progression: the Yi talking—a space—GBS communing with himself—a space—Hu and the other Han Chinese talking." Still George felt easy with this small group, "accepted as a person." When they reached a crest carpeted with *Helichrysum* flowers, all shed their wet socks to go barefoot in the sun-warmed grass. Charring bamboo shoots in the fire, his hosts showed him how to peel the sheath to get at the tender core, which tasted like asparagus. For his birthday, the maotai toasts went late into the night.

* "This Cosmos, so marvelous to those with eyes to see, is reduced to dullness by a blind . . . preoccupation with material things. . . . It is unbelievable that the Creator could have placed so many diverse organisms on the earth, each one so admirable in its sphere, so perfect in its role, only to permit man, his masterpiece, to destroy them forever."

Most of the villages they visited on this survey had not seen a foreigner since the 1930s, so George often drew a crowd. In Qingchuan, where he and Hu had been invited to speak to a local chapter of a new conservation NGO, he was told that a few students wished to meet him. Led onto a sports field, he found one thousand waiting. He urged them to be proud of their pandas—beloved by children around the world—then knelt so a little girl could tie the red triangle scarf of a Young Pioneer around his neck. One town prepared him a soft, fragrant bed in a storehouse of medicinal herbs. In another, an old man with no incisors sat timidly holding George's hand. When their driver, a party member, began amusing himself by running over families' chickens and dogs, George was the only one who dared reprove him.

As everywhere, George saw damage being done to the animals' world: deep wounds, for instance, in the old-growth firs pandas need for den trees, cut by shingle makers looking for good grain. But he also learned that no one had consulted with the local people when the reserve boundaries were drawn; that the promised electrification to ease their need for wood had not come; and that a badly designed resettlement plan had incentivized families to cut trees to build new homes so that the government would have to pay more to buy them out. Visiting an old hunter in his tidy home, where the two coffins he'd hewn for himself and his wife sat in the neatly swept yard, George was told that they would be among those removed from this reserve. Though he had urged the party secretary to move families outside of critical habitat, he wished this couple could die here amidst the walnut orchards and stone corrals they'd planted and built together.

The survey had revealed to George more cause for concern: In a few locales, the arrow bamboo was flowering. Bamboo blooms only rarely; it usually reproduces by sending up shoots. But when it does flower—at intervals of fifteen, sixty, even one hundred years—it then dies, sometimes across entire mountainsides. For three million years, pandas had taken those occasional die-offs in stride, wandering downslope or to the next valley to find bamboo that was still alive. Those movements were increasingly blocked, however, as humans fragmented the landscape, felling forests and replacing bamboo with villages and farms. More flexible animals often find silver linings in human changes to the land, like fat livestock to eat. But George found the panda to be as fragile as the similarly unenterprising gorilla, both trapped in their last refuge. When a decade earlier umbrella bamboo had died across two thousand square miles, so had 138 pandas, a tenth of the entire panda population.

George was not alarmed; the current die-off appeared highly localized. But he urged the team to map all areas where the pandas might be vulnerable, then tide them over by putting out food or bringing them into the temporary holding stations he had asked Conway to design. He also urged longer-term planning, for reintroductions and new reserves.

Instead, fanned by WWF, a media frenzy ("panda-monium") erupted, most of it mangling reality. Russell Train told UPI that a third of all pandas faced starvation. The *Los Angeles Times* quoted Lai Binghui saying it could be two-thirds. First Lady Nancy Reagan appealed to America's children to donate their "pennies for pandas." Tragic stories rolled in. *Time* reported that two pandas had died at Wolong and *China Daily* said several had starved in Tangjiahe.

George had recently been in both places and knew that the pandas and bamboo were both fine. But he was helpless to forestall the "rescue" madness that followed. Returning in November 1983 after a summer at home, he learned that China planned to build ten holding stations. And, no, they would not wait for him to assess the situation on the ground. Visiting the new research center, he was glad to see pandas relocated from the awful Yingxionggou but heartsick to see that a new male, Hua-Hua, had been added to the group. Why? The Hong Kong *Standard* said that Hua had been captured because he was "elderly and starving." But George could see that he was in full health and had been taken from an area with ample bamboo. Was Hu's prediction that pandas would be captured to "fill" the center coming true?

By 1984, though they had still not mapped the bloom, the government expanded its plans from ten to fifteen "panda refugee camps," one in each reserve whether needed or not. To fill those camps, the Ministry of Forestry promised one hundred dollars to any villager who brought in a "starving" animal. In Gansu, where a new million-dollar facility went up quickly, four cubs were captured to justify its existence. From Pingwu came reports of one panda starved to death and a youngster rescued. "I don't believe it," George wrote, having just been to the region. "No panda could starve with so much bamboo. The one died of other causes and the young was taken for the reward."

Even in Tangjiahe, George's new study site, the construction of a big new facility set the spiral in motion. When in June villagers trapped a panda in a cave, George sedated and collared it and was about to release it when a man he didn't know, a surgeon for humans, said that Tang-Tang (Sugar) would first have to be wormed and held for observation. George agreed to one night. When the caged animal refused to eat, the doctor said he was sick and could not be let go. Kay was nearly in tears, but George's voice went cold. "This panda is healthy now but will deteriorate if we hold him. This project is already responsible for several deaths, and any panda removed from the wild might as well be dead. If you keep this animal in captivity, my work here is finished." The panda went free.

George had been pushing back on the public story since at least January 1984, when he told *PBS News Hour*'s Judy Woodruff that few wild pandas were in peril outside of a single county, Baoxing. It took eight months for WWF to make a similar correction, and then only because they faced a "credibility problem," as their information director wrote Schaller, and risked being accused of sounding a false alarm. "You are the one person who can restore perspective," he wrote, asking George to come to Europe to deliver the message that just twelve pandas had died because "our knowledge helped prevent a more serious loss." George did go to Germany and say that the famine had been overstated but did not pretend that human intervention had improved things; to the contrary, most pandas would have been better off left alone. China, meanwhile, was revising its own story. "Pandas Saved, China Reports" was the headline in UPI. All glossed a tragic reality. Of the 108 animals captured for no good reason, 10 percent of the world's wild pandas, just a third ever saw freedom again. Thirty-three died in captivity. Forty were locked up in zoos or provincial holding stations forever.

All this time, George worried about Zhen. His colleagues sent occasional reports. In the fall of '82, they'd heard the squawk of a baby while tracking her, but she'd soon lost her collar, and it seemed the baby too. When she was seen in the spring with a yearling, George hoped she'd adopted the orphan of the lactating mother he had found strangled.

Before beginning work at Tangjiahe, George made a last trip to Wuyipeng with the two researchers who would replace him there: bear specialist Kenneth Johnson and botanist Alan Taylor. They found only disasters. Under the new camp leader, Wang Pengyan, and what George called his Gang of Three, traps were going unchecked for days. Twice, Long-Long had been left to nearly starve, growing so frantic he had torn his way free; though George needed to replace his collar, after those traumas he would never enter a trap again. A baby had been caught and locked up in the new research center, where it soon fell ill and died. That made four taken captive that year, always when George was gone. His worst fears were being realized, as this Panda Palace, as Chinese officials called it, became "nothing but a prison" and morgue. In four years, he had failed to win his colleagues over to the essential goal: to assure the panda's freedom and survival in the wild.

Wang Menghu had prepared him for the other big news: While George was away, a female panda, Bei-Bei, had moved into his tent. In one of his most famous pictures, she looks out at him through the canvas window, her round head close enough to touch. His campmates offered a gleeful tour of the havoc she had wreaked—pilfering ten kilos of pork, smashing the windows of the main hut, and sleeping atop his botanical specimens, ruining the lot. Worse than any of that, Wang Pengyan had turned her into a dangerous animal. Luring her in with food, then stabbing at her like a picador goading a bull until she charged, and finally pacifying her with gifts of sugarcane, he had effectively trained her to attack. "Finally, he's interested," George wrote bitterly, having discovered that "he could gain attention by these antics."

Work was out of the question. With Bei-Bei in camp, the researchers had to move furtively. In the event they had to buy her favor, they kept sugar in hand, though George hated seeing this "imposing and mysterious" creature reduced to a degraded panhandler. He also better appreciated the fear of Indian villagers with a man-eating tiger around. Twice, she charged him: Both times he fell, the first time smashing his hand on the ice, the second spearing his palm to the bone on a cut bamboo. She slashed Xiao Wang's pants; it was only a matter of time before she mauled someone.

George moved into the bunkhouse, soon followed by Ken and Alan, who burst in at midnight yelling, chased from their tents. Even after reclaiming a cot they had been using to keep plucked chickens out of Bei-Bei's reach, there were only ten beds for twelve people. And twice Bei joined them inside: once standing near, then in, the fire; once sitting by the root cellar, rump in the air, to pull out cane after cane as if devouring a bag of fries. She and he both, George thought, were in a place they didn't belong. "No matter what I do or say, it will not have lasting value."

When Wang Menghu proposed keeping Bei as a tourist attraction, George was briefly tempted. Like his wild pets, she could be a window into how pandas might adapt to their ever-less-wild world. And when she was feeding and calm, he found it comforting to have her nearby. But even if he hadn't been leaving for Tiangjahe, it was clear she had to be relocated.

The first attempt failed. Wang Pengyan took her only to the next valley, then "faced with public oblivion," laid a trail of sugarcane to lure her back. After months of slacking, he now monitored her signal, but only so he could feed her. He let her back into the shared hut, where she turned over a stove.

Ken tried protesting up the chain but was publicly denounced for "incorrect attitude." Though "the political exhortations and martial music that still blared at dawn from loudspeakers . . . in 1980 had by 1984 been replaced by 'Jingle Bells,'" George wrote, "the righteous irrationality and malice of Cultural Revolution tactics still lingered."

Schaller knew that he himself would have to make the appeal. It is "wholly wrong to turn a wild panda into a porridge-eating camp panda for no other reason than to amuse people," he wrote to the party secretary. WWF support was again at risk as staff abandoned real work, "happy to think their job is to sit by the fire and feed fat pandas." And if the world learned what was happening, it would severely tarnish China's image. The response broke his heart. Wang Pengyan was reassigned, but to the research center. And instead of releasing Bei far enough away that she wouldn't return and might again become wild and have another baby, they captured and imprisoned her there, too—to "eat bitterness," as George would write, using the phrase for those who suffered in the Cultural Revolution. Visiting, he found her fat-bellied and dulled by the porridge Wang Pengyan still spent his days feeding her with "barren zeal." Hua-Hua was there, too, still captive after nine months. Was their future to be drugged again and again to electroejaculate and artificially inseminate? The worst blow came in June, when Ken examined Bei's teeth and confirmed a suspicion that had nagged at George. This ruined panda was in fact his beloved Zhen.

To get Zhen and Hua released would demand of George the kind of patience he finds hardest, for months of discussion all the way up to Beijing. He discovered a quiet ally in Hu Jinchu, who in turn appreciated Schaller's "flare up," which he hoped might move scientific decisions out of political hands. Invited to yet another banquet, George was relieved that he had not been banished entirely.

He might have left of his own accord, had he not found at Tangjiahe all he had missed at Wolong: a rapport among colleagues, animals to watch, sweet camp life with Kay. China's growing openness seemed to have reached at least these mountains. Their lone political minder was relaxed. And a young staff eagerly pitched in, merrily sorting dung with Kay, patrolling so diligently that George found not one snare, volunteering for extra nights out, keeping equipment and notes in order, and attending closely as George taught skills like how to adjust for a radio signal's bounce.

Even those who came with him from Wolong soon shed the "protective coloration" they had assumed there—of doing as little as possible, or worse. Fresh out of college, Wang Xiaoming had "vanished into a spiritual black hole" but here sparked to life, especially when George suggested he set aside his ecology texts—which he read "over and over, repeating the principles like slogans"—and instead go outside to see what was actually there. Wang soon initiated several studies of his own, including of the foods favored by the local musk-ox relative called takin.

Some of this progress was the reward of time. After four years, Kay saw their colleagues "finally trust that George was totally sincere in trying to help the animals, not there to make a name for himself . . . or grab the data and run away." She was touched by their efforts to adopt American ways—"they began over-thanking us for everything"—and at their dedication to learning English. The Schallers often overheard their youngest assistant reciting vocabulary, or found him arrived at their door pointing silently to a word. George would say it out loud and the boy would repeat it until George said, "*Hao*, good." Then off the youngster would go until the next time. Wang Xiaoming came to Kay for lessons and began recording his sightings in English—for practice and for George's review.

Though George regularly scolded himself for lacking similar devotion to learning Chinese, he did work to close other distances. In the early months he had trapped small mammals, which Hu obligingly identified. Now he quit collecting, seeing that—after a century of plunder—his hosts rightly perceived the taking of specimens as an act of scientific imperialism. Out unexpectedly late one evening, his peers suggested camping and began preparing a separate place for George. He said that he preferred to sleep in the communal place, so all crowded together on a thin mattress of *Cyclobalanopsis* leaves, George in a borrowed sleeping bag that barely reached his chest.

These moments of connection began restoring George. When he and Kay found Tang-Tang, the panda for whom he had laid down an ultimatum, asleep in a tree, they spent the night monitoring him, cozy on a straw bed under a warm quilt eating *Vollkornbrot* and kippers from George's trip to Germany. They wandered unlogged forests full of pink peonies, white strawberry, and wild celery, savoring scent memories of Congo, and listened as hundreds of rare snub-nosed monkeys shook the treetops like a gathering storm. When a big one snatched an infant from its mother's arms, George was surprised to see that the blue-faced creature only wanted to snuggle and pet the baby. Collaring two Asiatic black bears brought excitement: When one charged with a roar, "furious at being detained," George saw that the trap held it only by the tip of its toes.

Most wonderful were the takin, who gave George his "first prolonged observation of anything in four years." The odd beast defied even his descriptive abilities: He proposed a grizzly's hump glued to a hyena's sloped hindquarters, but only really captured the face, which does indeed look "like a bee-stung moose." One male, recognizable by his broken horn, seemed to like being near them. When Kay washed clothes in the stream, he stopped to watch; when she went to the outhouse, he grazed so close she worried he might burst through the tar-paper wall. George named him Lao Pengyou, old friend. Gathering a bouquet from the meadows he and Lao both liked to visit, he asked botanist Qin Zisheng to identify not only the forage (nettle, mint, buttercup) but also the flowers, "whose beauty I admired and wanted to call by name." She showed him *Pleione* orchids, touch-me-nots, and a Jack-in-the-pulpit that "starts its existence as a male and then solves its midlife crisis by continuing as a female." One afternoon, he saw sixteen baby takin traveling like Madeline in one straight line, attended by a lone adult female. This "kindergarten," he guessed, gave the moms a chance to graze and socialize. Occasionally one of the cumbersome beasts bucked or bounded, and once George faced down a bluff charge. Kay always yielded, writing to Bettina: "I do not dispute with huge horned creatures that can outrun and out-scare me."

Kay's letters from Tangjiahe, written with "our half hour of electric light," ring with her own restored contentment. Sending "impatient" George out before breakfast to check traps, she banged the creosote out of the stovepipe and lit the fire to bake the rounds of *bao* dough George snuck out of the kitchen, preferring them crusty to steamed. When the water pipe froze, she walked to fill buckets. "But it's nothing like the quarter mile on a steep narrow trail I had to go in Wolong. This is modern living to have the creek so close." Their bamboo ceiling and cardboard walls made for draftier living than the tent Zhen had commandeered.

> *But it is serene and I prefer to be cold and with George than in the busy life in Roxbury. He looks well, Bettina, and seems happy about his upcoming work in Tibet, a place he has longed to work for as long as I've known him. Much as I hate to contemplate being apart so much, I am happy he gets the chance. Whenever I get depressed, I remind myself that at least he is not in a war.*

The interlude did not last, again smashed by bad news from Wolong. Arriving for his final months with the pandas, George had just checked in to his Chengdu hotel when Ken Johnson burst in, disheveled and rotten-smelling. Pi-Pi, the great fighter and lover, was dead. The official cause was an intestinal obstruction, but the "autopsy" and burial had been done in a hurry and out of sight. So under cover of night, Ken had dug up the panda and brought George the evidence: a thigh bone trailing tendrils of meat and several twisted neck vertebrae. Going to the scene of the crime, George coldly examined the spine, three broken ribs, and a severed paw. "One abandons pity when it is useless," he told himself. But having reached the end of his endurance, he feared he'd gone numb. Checking data sheets, he saw that Pi had been listed as active days after he died. "Such a failure. Years of effort and no one gives enough of a damn to do a simple reading. What am I doing here?"

A meeting, of course, had to be called, which George insisted must include Sichuan province as well as Wolong officials. He summarized the study so far. At the new research center, the delicate equipment bought at China's insistence was broken or unused, the labs and library empty. Hua, Zhen, and Zhen's adopted infant had been taken captive for no reason. At least three pandas had died in poachers' snares because officials had failed to establish patrols.

Nothing might have come of the meeting but for a *Xinhua* reporter who was allowed to sit in. The summary she wrote for Vice Minister of Forestry Dong Zhiyong—who just months earlier had been handed the keys to two jeeps by Nancy Reagan, along with the Pennies for Pandas—brought him to the scene. That changed everything: Zhen was released, camp leaders were demoted, and Party Secretary Lai apologized for his errors, with a cringing George found painful to watch. Dong vowed "reform through self-criticism" for everyone responsible. George said he was sure all would be solved, then lay awake thinking how he could leverage Dong's presence to get Hua freed too.

By December 1984, with their time in panda country coming to an end, Kay and George were getting on each other's nerves but also feeling the sadness of leaving a place "to which one will never return." They made a ritual of sorting the last scat, spent a final night in a tent near a panda, shared their last "family" Christmas with Hu Jinchu and the team. George appreciated the absence of forced gaiety and speeches. He meant it when he addressed his goodbye to "old friends."

Published in 1985, *The Giant Pandas of Wolong* gathered not only the first natural history of the species but also fundamentals for scientists who had been cut off during China's closed years—detailing, for instance, the workings of an ordinary herbivore's digestive tract. The collaboration on the text with Hu Jinchu and Pan Wenshi was not easy. While George's Indian and Pakistani coauthors had assimilated—another mixed legacy of colonialism—Western scientific protocols, after years of enforced isolation his Chinese colleagues had not. He gave in on disputes over the writing, but insisted on discarding bad data.

He then waited eight years to publish his own panda book—fearing that if he told the full truth China might shut down the work he had begun in Tibet. Pan Wenshi had lost his project in Foping after telling the State Council that pandas were dying not from bamboo die-offs but from poaching. George himself had been dressed down at a banquet by a forestry official who rose to denounce foreigners who came for research and then went home to criticize China; "such persons are not welcome anymore." Though he did publish a second cover story in *National Geographic*, he soft-pedaled the tragedies, leaving out the dubious rescues and breeding centers. The cover image of Bei-Bei (née Zhen-Zhen) looking out of the tent window gave no hint of that "infinitely sad" story.

The Last Panda, published in 1993, pulled no such punches, beginning with its despairing title. Though Schaller knew his "candor may not be appreciated . . . in an authoritarian state . . . where penalties for dissonant opinions can be swift," he could no longer evade his "duty." To fail to expose the dark influence of global politics and greed would be "a grave disservice to the panda."

Sparing no one, he named names.

On the Chinese side, he called out the vets who had harmed pandas or ruined work and the officials who sat idle as pandas died in snares, springing into action only when they could take an animal captive and get the fancy jeeps and facilities that came along. He also praised by name anyone he saw committed to their job, like Yong Tiazhong, a "quiet man with shoulders that a weight lifter might envy . . . a relentlessly hard and efficient worker" who had hand sawed and split wood for the team at Tangjiahe.

On the Western side, he detailed WWF's many blunders and breaches of trust and also laid into the latest menace, foreign zoos. The 1984 loan to the Los Angeles Olympics of two pandas had set off a frenzy, with zoos offering up to a million dollars a year to buy or "rent" a pair of the crowd-pleasers. Schaller at first supported such loans—if limited to animals already in captivity and past breeding age—to keep the door open to China, and to prod and finance improved management.

But the rent-a-panda business soon spun out of control. As deaths in captivity outpaced births, George saw zoos siphon more and more pandas from the wild, their rents going not to conservation in China but to expanded infrastructure for captivity. George devoted a chapter to what a reviewer called the "astonishingly shortsighted display of greed" by zoo directors around the world, "who outbid one another for pandas, heedless of the damage they might do to the species." With fewer than one thousand left on Earth, the panda was declared endangered in 1984—by both the US and CITES*—triggering binding import bans. But that just prompted higher-level meddling. Thanks to President George H. W. Bush, the Columbus Zoo got a post-ban pair. A congressman got the Toledo Zoo some, too, boosting its revenues by $60 million; as George wrote, the lucre to be realized from this "commodity" revealed "ignoble traits . . . in individuals and organizations." Especially painful was the arrival of a panda at George's home institution. Conway had explicitly barred the Bronx Zoo from seeking one after seeing how they were harmed by capture and how rarely they reproduced in captivity. But thanks to the "relentless" efforts of Mayor Koch, the zoo received a wild-caught adult female anyway. When she came into heat despite assurances that she was incapable of breeding, the only male within reach was too small and young to breed.

In the wholesale transformation of this ancient being into entertainment lay this question: Is a panda in a zoo the same as a panda living out its natural life in the bamboo forests? For George, the answer was of course no; a

* The Convention on International Trade in Endangered Species of Wild Fauna and Flora

caged animal has been robbed of its true nature. But the losses to humans are also tragic, he knew, when we reduce sacred beings to a "clownish face," encountered only across bars. As John Berger had reminded us, we once had (as many Indigenous communities still do) ancestors, oracles, deities whose gaze brought us into the world in a way we cannot do for ourselves. Debased captives can share no power.

Schaller's political calculus as he wrote *The Last Panda* is not hard to discern. He found a safe way to reference the Cultural Revolution—leaning on such sources as the *Beijing Review*—and praised Mao, "the century's greatest revolutionary leader," before criticizing him, for what Schaller dubbed in the house style his Four Errors. Error number two was the famine induced by the Great Leap Forward, which killed twenty million people and at least as many animals, hunted to relics by starving families. He used extensive appendices to hold everyone to account, reproducing key portions of the national conservation management plan published jointly by WWF and China's Ministry of Forestry, which called for expanded and linked reserves, as well as the position statement by the American Association of Zoological Parks and Aquariums outlining strict principles to govern panda loans.

George had hesitated to add another sad volume to his collected works, fearing that repetition might induce numbness. But forgetting would be worse, he wrote in language redolent of history. "Telling and retelling is a moral imperative."

So he summed up the failures that he felt in many ways to be his own. In the fifteen years that included his study period, the panda had lost half its habitat to logging and agriculture. The remaining few hundred free-living animals had been islanded in twenty-four separate patches of forest, sometimes with fewer than twenty others. "A blueprint for extinction." Spurious "rescues" had continued, condemning ninety pandas to captivity. And poaching had actually increased, spurred by the new official encouragement of *getihu*, private entrepreneurs. In 1988 alone, border officials intercepted 146 pelts—one in seven of all pandas still alive at the last census—bound for Hong Kong and Japan. Although China passed a wildlife protection law the next year, WWF undercover buyers continued to be offered panda cubs and pelts, as well as the bones and skins of tigers and golden snub-nosed monkeys.

In Wolong itself, panda numbers fell from 145 in 1974 to just 72 by 1986. Of the six animals whose lives George had followed there, only Long-Long survived. Han, Ning, and Pi had died in snares; Wei likely of roundworms; and Zhen, not long after her release, from meningitis contracted in captivity. Research had ground to a halt. With rare exceptions, the staff George and Hu had trained in the basics of field biology never used those skills again.

Worst of all, George had to face the possibility that the light he had shown on the panda's whereabouts and charms had done more harm than good. "Had the panda remained in the obscurity of its bamboo thickets . . . there might not now be so many captives . . . [or] breeding stations. Pandas might not have huge prices on their heads" alive or dead. He knew that absent his work, logging and agriculture would have soon doomed them. But "the realization that the panda has so suffered and declined in numbers while we chronicled its life burdens me painfully." A failure to act now would bring "an eternity of remorse."

As the most political of his works, *The Last Panda* also had the most far-reaching impact.

In the West it landed on many front pages, including that of *The New York Times Book Review*, where author and filmmaker Geoffrey Ward commended Schaller's exposure of "crimes against nature—that may already have insured the panda's disappearance from the wild." *Newsweek* also put the book on its cover, calling it a "shattering attack on the sham called panda conservation." "I am placing the blame on every institution, whether American zoos or the Chinese government or World Wildlife Fund," Schaller told the reporter, "anyone who did not think of the panda first." Writing in the British *Literary Review*, David Attenborough praised this "absorbing book" for making clear that, for conservation to succeed, "it is not only the habits and sensitivities of rare animals that must be understood, but those of the most abundant, the most variable, complex and difficult to manage of all large mammals, the human species." Deeming Schaller "truly a hero of our time," John Terborgh, who had just won a MacArthur for his own work on tropical conservation, found in the book "the most profound and articulate analysis I have seen of the dilemmas of international conservation." It leaves the reader "drained and unsettled, realizing . . . that the future holds great uncertainty, not only for the panda but for the welfare of all life on earth, including mankind."

Making the rounds of television, Schaller did not soften the message. On *Day One* with Diane Sawyer, he condemned the Toledo Zoo for its focus on profits at any cost and showed pictures of panda corpses. China has strict antipoaching laws, he told Charlie Rose. Three people have been executed and a dozen more imprisoned for life. "But that is after the pandas are dead."

The book was a national bestseller, selling nearly fifteen thousand copies in its first months, a debut for the University of Chicago Press second only to its 1976 *A River Runs Through It*. It cost at least one zoo its friends in high places. That fall, Secretary of the Interior Bruce Babbitt denied San Diego's waiver request. When the book was translated into Chinese (several times: Yong Yange read a Hong Kong version and it was also published in Taiwan), Conway prepared George for the worst.

> I said, "You'll never be able to go back to China." But I was wrong. It's one thing to be indignant and not know what you're talking about. It's something else when you know in a way no one in the world can argue with. He tells it like it is with such insight and integrity, such moral strength, that people accept his criticism.

In fact, the Chinese government responded to his lacerating portrait of their profiteering, misguided rescues, and cruel experiments with remarkably bold action. "George was the first to speak out against capture and captive breeding; the first to say that if we abandoned wild pandas, we had failed," recalled Yong Yange. "And the government finally agreed and banned further capture." In Beijing, the State Council approved a $77 million conservation plan, doubled the number of reserves along with protected corridors to link them, and increased patrols for illegal logging and hunting. And far from slamming the door, China invited Schaller to roam for the rest of his life over its remotest and often strictly restricted regions, the most extensive freedom ever granted a Western scientist.

Not everyone in China embraced Schaller's scalding attack. Bei-Bei's "trainer," unsurprisingly, declined to speak for this biography. As did Pan Wenshi, because (it seems) he thought George's book was too bleak and the man himself sometimes "Western arrogant," as a colleague passed on. Did Hu think George could have better managed team relationships? "It is a challenge to live and work in unfamiliar surroundings with people with different political backgrounds and customs. Just as a tree needs constant pruning as it grows, so must people grow in the course of study, work, and life."

Pan Wenshi and George did share a remarkable moment just months before publication. Having recently established the Department of Environmental Biology and Ecology at Peking University, Pan and his star student Lü Zhi took George back into the forest and presented a most extraordinary gift: a visit with a panda cub in the wild. Despite the baby's tiny roars and toothless chomping, Mother Jiao-Jiao stood by quietly as they approached to five feet; if she attacks, Lü Zhi had instructed George, just call her name. When the little one had exhausted herself and fallen asleep, Pan picked her up by the scruff to see if she had the kitten's "relax" reflex. She did. Lü Zhi had named the baby Xiwang. Hope.

Up the next day before dawn, George saw children walking to school, each carrying a little warming bowl of glowing charcoals. "Sparks in darkness," he wrote.

9.

Forty Years in Tibet

Tibetan Plateau, 1980–2020,
Tibetan Antelope, Tibetan Brown Bear, Snow Leopard

At last, George reached the land he had yearned for, ever since reading as a boy Sven Hedin's tales of adventure on the Tibetan Plateau—a million square miles with an average elevation of fifteen thousand feet, ringed by the "hard grandeur" of iced massifs including fifty peaks higher than twenty thousand feet. George had been mesmerized by the Swedish explorer's 1909 account of its "totemic loneliness." Is it possible, he wondered, "to be homesick for a world unknown?"

After five years in the overrun forests of Sichuan province watching giant pandas reduced to near wards of the state, George leapt at the Chinese invitation to extend his wanderings far to the north and west. Here was a last opportunity to study animals still carrying on something like their ancient lives, including the grand migrations that have elsewhere been blocked by roads, rails, croplands, settlements, drill rigs, border walls, and fences.

Until the end of his life, the steppe and mountain ranges that span the

Tibet Autonomous Region, Qinghai province, and the Xinjiang Uyghur Autonomous Region (from now on referred to simply as Tibet, Qinghai, and Xinjiang) remained the one landscape that never loosed its hold on him. The Tibetan Plateau is so vast—covering a quarter of China's territory—that it influences weather patterns across Asia. It is so high and cold that it is called the roof of the world, or the third pole. Its eastern edge, where the great rivers that water Asia originate, is temperate enough for forests, farms, and villages; in its southern reaches, feather grasses have sustained nomadic herdsmen for millennia. But its northern plains, or Chang Tang, are as inhospitable and unpeopled as any place on Earth.

Naturally they will become one of George's favorite places. And few will ever know them more deeply. In sixty-five trips he will cross thousands of miles, searching on foot, camel, yak, or donkey for Tibetan antelope (also called chiru), for Tibetan brown bear with five-inch claws, and—as ever—for his snow leopards. Beyond his few companions, he will sometimes go a month without sight of other humans.

As always, Schaller began his sojourn by reading old accounts—of expeditions, pilgrims, spies—who then "joined" him in his odyssey. He took particular pleasure in the imaginary fellowship of two British officers named Wellby and Malcolm, who in 1896 crossed northern Tibet to assess how well their empire was faring in its "Great Game" with Russia for dominance of central Asia. Noticing that in every picture the two men appeared jauntily dressed in caps, mustaches, kilts, and knee-high socks, he imagined them dirty and bearded and wondered if they ever called each other by their first names. He found equally fascinating two US intelligence agents who in 1950 crossed from Ürümqi to Lhasa with three White Russians fleeing the advancing "Red Chinese." Two of the Russians and Douglas Mackiernan were mistakenly killed by Tibetan guards. George made a FOIA request for the classified notes of the surviving American, Frank Bessac, and visited him twice in Montana. His ultimate lodestar remained Hedin's paeans to this "utterly God-forsaken region": to "march for weeks . . . and still find yourself the centre of a universe of mountains. I felt like an atom of dust . . . and fancied I could hear the swish of the planet as it rolled." These excavations accreted into a kind of palimpsest in George's mind—layer upon layer of geologic, zoologic, and human history, through which he made sense of all he saw.

George had made one brief trip into Tibet before ever working in China. In June 1980, he spent two weeks on a bus with an illustrious bunch led by Smithsonian head Dillon Ripley and his insectophile wife, Mary. A few in their party had been here before. In the 1930s Swiss geologist Augusto Gansser-Biaggi had snuck in disguised as a lama, smuggling out rock samples under his *chuba*. Ecologist Lawrence Swan had come with Sir Edmund Hillary in search of the Yeti; anthropologist Corneille Jest had written his PhD on Dolpo. Though nearly all were trailblazers, in fields like glaciology or permafrost studies, they come off in George's journals as eccentrics with nets, saying things like "Mary, I got you a weevil." (George himself gave her a dung beetle.) Or worse, as divas: When one found dirt in the water and said, "I can't drink this shit," George scribbled, "Some people should not be on field trips." Catching his first sight of the "shining galleon" of Lhasa's Potala Palace, he remembered Heinrich Harrer's tales, as they climbed in Alaska, of the Dalai Lama imprisoned there, watching life outside with a telescope.

George had made a second trip while still busy with pandas, to Qinghai's Anyemagen Shan (*shan* means "mountains"). Though he saw no wildlife, he found a culture familiar from northern Nepal that called to him like no other. Watching Tibetans in brilliant red blouses gather for a horse race, he wished he could linger. Though most of the temples seemed hollowed out, like stage sets "with a few monks thrown in for realism," he was charmed by the boy "monklets" at Gongsa, bowing to the lama in hopes he might pull from under his cloak the magic Polaroids George had taken, let develop before their eyes, and then presented as a gift. Ever more often, a sense memory stirred. From a patch of yellow *Primula* came a fragrance

George had last encountered as a child, walking through wet meadows picking bouquets for his mother.

Most fascinating were the nomads, who would soon be central to his life. "I never tire of looking at their low black tents, though I know the aura of freedom is my own romanticism." Often invited to stay the night, George marveled at the women's ceaseless work. With bent backs they gathered yak patties, tossing them over their shoulders into a basket, later squeezing them into bits to dry. Weaning the yak calves with a collar of four sharpened sticks that barred suckling, they milked their mothers into yak-horn vessels. Inside the tent—woven of dark yak hair, its pegs made of yak ribs—they tossed a yak-skin bag hand to hand to churn yak milk into butter, served yak yogurt and yak cheese, then sat twisting yak hair into rope, as they visited with their guests. George's hosts were just as intrigued by his technology, at least any of possible use to their lives, closely examining his air mattress and the clasps on his pack. Shown such hospitality again and again, George offered small gifts in thanks: a can of jam, warm socks for the children, once a down vest of Kay's to a shivering boy. Sitting one night by a yak dung fire with two herders, speaking their shared bit of Mandarin, George felt we were "all of us still antique people."

His Chinese colleagues on this 1984 journey were far less interested, including in wildlife. No matter his protests, they invariably stopped short of the crags where they might find snow leopard, so that they could sleep in a hut and have locals cook for them. George quickly tired of being forever on someone else's schedule. It was always rest time, teatime, mealtime. "Sit there." "Eat more." (Few things irritate him more than too much food put on his plate.) The one "field man" in the bunch had studied musk deer by shooting a hundred because "What other way is there?" He promised that villagers would show them snow leopard the "people's way": trapped for photos, or pelts.

Still, at trip's end, having smiled through three banquets held to "clean off dust" (meaning ease tensions), George reciprocated, hosting a feast at which he teased the husbands about the good time they had while their wives stayed home. Since his room had the TV, all gathered there, eating watermelon and watching weepy Japanese movies into the night.

•●•

It is in no small measure thanks to Schaller that wildlife biologists have focused for the past half century on "charismatic megafauna." Beginning with the gorillas, he had demonstrated repeatedly that big, gorgeous animals can capture the public imagination and win protections, carrying with them to safety the less-glamorous species that share their ecosystems. Pandas, seemingly designed by nature to be an icon in black and white, had marked the apotheosis of that approach but also revealed its unreliability. So, taking a page from Kenneth Grahame, George mounted a defense of a decidedly uncharismatic minifauna—a four-ounce relative of a bunny called a pika.

Near Gongsa monastery he had stumbled upon an expanse of burrows eerily absent any signs of life: no dirt freshly turned, no bright stores of edible flowers. His colleagues explained that officials were poisoning pika over tens of thousands of square miles—to stop them from eating grass and from causing erosion with their burrowing. Schaller knew this story; it was exactly the rationale given for the slaughter of prairie dogs in the American West, and just as wrongheaded. He explained to rangeland managers that pika in fact eat plants unpalatable or poisonous to livestock and keep soil healthy: porous and mineralized with their digging, fertilized with their feces. They do not degrade rangelands but move in after the damage is done, since on overgrazed land they can easily spot predators. They are also vital to other species' survival. Pollinators find refuge in their tunnels and Hume's ground jays nest in abandoned burrows. Raptors, foxes, wolves, and bears rely on them as prey, especially in winter when other fat morsels like marmots hibernate. The US had shown where this path led. In the wake of its success exterminating 95 percent of its prairie dog colonies, western rangelands, ungulates, and predators had all declined. Where poison was used, it moved up the food chain, ensuring prolonged and painful deaths for many creatures.

Not all pika, thankfully, were dead. And with animal intimacies ever rarer in his life, George fell for these hyperalert little characters who said hello by touching noses, sometimes boxed with their forepaws, and reciprocated his attentions. “Pikas and I have seen much of each other,” he wrote in one of several publications he devoted to their hectic lives. “We are both diurnal and acute observers of behavior.” One morning he watched a polecat chase several in and out of their holes, swallowing three in less than twelve minutes. Three hours later, when two at last peered out to see if the coast was clear, George was still there. He can wait with the best of them, including small prey paralyzed by fear. He matched his new friends in industry too. Having watched them carry home mouthfuls of aster, cinquefoil, and iris, and send up geysers of dirt enlarging tunnels, he succumbed to his own “fit of excess energy” and excavated a warren: “a male’s I think, as the interior was spartan.”

In place of awe, these writings tapped the identification most feel from childhood with small, helpless, furry creatures. Who couldn’t relate to the tattered youngster, bitten and chased by an adult into a burrow and promptly chased back out again by its irate owner, pressing itself to the ground in submission to avoid more punishment? George wrote a dozen Aesopian fables, translated into Tibetan and distributed in schools, animating the pika’s salutary role in the web of life. In one, Pika discovers that the flies he loathes in fact help him survive, by pollinating his flowers. In another, Wolf has to kill a family’s sheep when his pika prey are poisoned. The most delightful voice is Hare’s, recounting how a village that killed pikas to firm up their soil succeeded so well all their water ran off. When Pika asks where he’s bound, Hare replies: “I come from behind me and I’m going ahead." One Tibetan guide, who wore a cowboy hat atop his long hair, liked to read the stories out loud to George.

Summer 1985 brought Schaller for the first time to Xinjiang and its Taxkorgan Nature Reserve, the terrain he'd looked into a decade earlier from Pakistan. Returning to this land of troubled borders brought familiar frustrations. Officials withheld maps and barred him from areas close to Afghanistan or Kyrgyzstan—though he ignored the rule that foreigners must eat and sleep apart from nationals, shunning the tourist hotels from which the others were banned to join them wherever they slept. The team mix of Muslim, Han, and Tibetan added new challenges, as when the Han Chinese demanded pork or ordered the Tibetans about.

Still his team here was far better than the one the year before in Qinghai. It included his young panda colleague Qiu Mingjiang, returned from the University of Idaho, and a Uyghur named Talipu who became a treasured regular companion. Nearly all were eager to learn—how to age blue sheep horns or where to find cat droppings. It pleased George no end when he pointed out a spot that looked to be a snow leopard's ideal and then—sure enough—found a pile.

He found many occasions for gratitude. Rolex was by now a sponsor of the work, featuring George in ads. ("Inhospitable conditions pose no problem for George Schaller. Or his Rolex.") When he slipped one day and failed to realize until far downvalley that his watch had been torn from his wrist, he was embarrassed but relieved when Qiu and another young one climbed to retrieve it. Charged by a yak, George ran with her in galloping pursuit until Talipu heaved rocks to drive her off. One colleague invited them to his sister's wedding, where they watched the Kyrgyz version of polo, which chases a goat carcass in place of a ball. Other locals were just as solicitous. When a team member fell twenty feet and went into shock, herdsmen treated him with a paste ground from a pungent yellow root.

George, in turn, offered what he could. Seeing everyone's lips turning purple-brown from multiple sunburns, he gave each a ChapStick. When the camel driver came to his tent and pointed to his ten-day growth of beard, George gave him a disposable razor and soap. Five minutes later, the man returned. No progress. "So until dark, we sit by the river and I shave him."

This was Schaller's first trip with camels, which became something of an obsession. As with his pigs, he devoted pages to their charms: the prehensile lips and short baby-bird hair atop their heads, the skin behind their eyes that popped like bellows when they chewed with their spatula teeth. A yearling with humps as flaccid as empty bota bags high-stepped with an outward twist of foot that was "surprisingly graceful, a ballerina with rickets." Riding them, he enjoyed their rocking-horse gait. (He liked riding yaks, too, which picked their own footing so he could sit and think and look or sometimes be lulled to sleep by the rolling movement, "as if sitting on a bear.")

Even after thirty years and tens of thousands of pages of notes, papers, articles, and books, Schaller kept finding fresh language, less a writerly flex than a way of maintaining his acute attention. On the long drive from Ürümqi across a desert so windy it was said that a man's beard never hung straight, the white hills with brown rippled tops became hot-fudge sundaes. A man in a hat trimmed with curly black wool called to mind a Marx brother (he thought Chico but perhaps meant Harpo) and inspired a marvelous sketch. Small observations often revealed larger meaning. Seeing domestic yaks standing on aufeis, he realized that—having grazed the last sagebrush to stumps—they were waiting for patches to melt so they could grab whatever green blade might be exposed.

Returning to Qinghai that fall, George was confronted still more starkly with the near impossibility of survival at these furthest edges of existence, even for animals adapted over millennia to forces far older than human beings.

His team was again far from ideal, as careless as his colleagues at Wuyipeng. The driver was wound tight: When a car nearly sideswiped them, he spit in his hand, downshifted, and took off after the offender. Their guide's sole qualification was having caught twelve snow leopards in steel traps that, he assured George, he had checked once a week; when an owl hooted, he swore it was a lynx. Their logistics manager brought no tent stakes and crouched with his back turned to eat an entire jar of jam, one of four they'd brought for the whole trip. (He was discovered by George, who never quite shed the habits of a starved boarding school boy.) Waiting and more waiting frittered away the days. The truck had not come. The others were still asleep. The gear was not organized. The meal would take hours to prepare and more hours to eat. Qiu was again along, though in these circumstances, George found him ill-served by his experience abroad; he now overreacted to the "laziness" and inefficiencies. He was also heedless in ways reminiscent of George in his youth—who nonetheless now paced all night when Qiu and another young biologist lost their way and were out until dawn without jacket, matches, or food, in temperatures below the lowest reading (minus forty degrees Fahrenheit) on George's thermometer.

All was redeemed when finally—after mandatory EKGs—they crossed a 15,639-foot pass in the Qilian Shan and there, to the farthest horizon, was "the Plateau I've waited years to see." George was elated by the crystalline space and still more by the sight of his first white-lipped deer and hundreds of wild asses glowing golden in the lowering sun. "After years in bamboo thickets, after the past three weeks climbing to see small herds here or there, we casually counted more than 1,000 animals. For 20 kilometers we were always in sight of animals, like a snowbound Serengeti."

Most important was his first encounter with chiru, the animal that would preoccupy his years to come. The elegance of the males' tall, slender, lyrate horns had for centuries stirred men to near hallucinations. Hedin saw armies of thousands bearing glittering bayonets. Through shimmering heatwaves, horns sky-lined above the steppe, George saw a "great spiny creature, submerged just below the surface of the sea." The vast herds moving in single file, nostrils steaming like trains, ice crystals glinting in the air, carried him back to Alaska's caribou migration, though here there was no

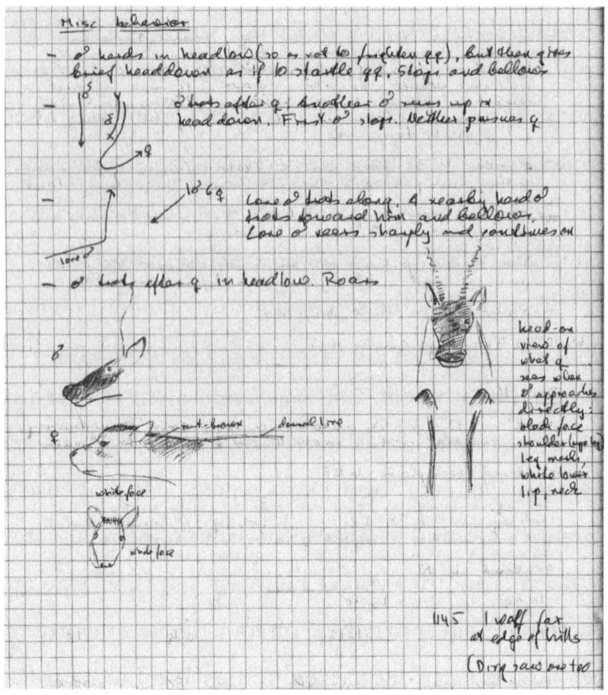

A single journal page might encompass even a chiru's point of view, in this case "what a female sees when a male approaches directly."

clicking of hooves on gravel but only silence. For three hours he gave up counting or classifying and just watched; "my best wildlife experience in years; for a brief time I was free again." The chirus' movements became a driving mystery. Where were they going? Why?

George's writing in these Tibet years is his most uneven. Going here to there and back again, his life had lost the classical unities of time, place, and action, and featured an everchanging cast of human and animal characters.

And yet, like the artists of the Renaissance who strove to create beauty so transporting it would compel people to live virtuously, he pressed toward and sometimes reached new heights. His images can be as vivid and concrete as those of William Carlos Williams: We go "into the white void in an old red tractor." And as otherworldly as William Blake's: Plowing alone through deep snow toward a hill covered with chiru that "floats like an iceberg on a low layer of fog. . . . I am in a dream landscape of unicorns." To approach them was to enter "the center of this consecrated space."

It was on the final leg of this journey that George came face to face with life's extreme fragility at these edges of the world. An unusual windless blizzard had sealed away all the grass under a blanket of snow, starving livestock and wildlife. The governor, come to Tuotuohe to assess the emergency, asked Schaller to join a rescue team. He agreed, and set off with four Tibetans and four People's Liberation Army soldiers in two vehicles: an ancient army truck and a red tractor pulling a wagonful of firewood for nomad families. George opted to ride in the open wagon to get the best view, though it cost him several frostbitten toes when the felt liners of his Sorels froze to the rubber boot.

They progressed steadily until late one afternoon when, crossing the ice on a shallow pool, the truck broke through. They always kept the tractor idling all night lest it refuse to start, so in the morning they yanked the truck free. Its wheels were frozen. So out came the blowtorch, which George had discovered was nearly as multipurpose as a yak, able to thaw boots, warm engines, or heat food—though when its roar became too much to bear, they boiled water over tin cans half filled with gasoline. Once the tractor itself broke through ice, but it just backed up and roared forward until, like an Arctic icebreaker, it had rammed its way through. A surreal moment came mid-journey when officials from Beijing showed up in a new Toyota for a photo op—and just as quickly disappeared.

The nomadic families, as George had anticipated, were fine: warm in their *chubas* and yak-hair tents clustered by twos or threes every few miles, eating their starved livestock. They always invited the motley visitors in for salted buttered tea, the guests seated as was custom to the right of the fire. After cold nights in a tent, his sleeping bag iced where his breath froze, George was grateful when invited to stay, warm on blue-sheep-hide rugs, the babies cozy nearby in baskets of kiang skin.

The wild animals were another story. Though the Hume's jays could sneak in to pick fat off the livestock carcasses, for the grazers, the energy required to paw away the snow often exceeded that in the plants they uncovered. The delicate Tibetan gazelles suffered most. George took a picture of a young female reclined, unable to rise as snow slowly drifted against her cold body; his autopsy found gelatinous marrow, a sign of malnutrition. He was touched by a kiang guarding his dead companion, trotting around her as they passed. A raven hacked at a dead female chiru, next to whom lay her soon-to-die orphan, snow on its eyelashes.

Some animals met more violent deaths. Weakened chiru were easy to kill, by dogs or men, often for sheer pleasure. The snow told the story: desperate leaps, splattered blood, empty shell casings. When George asked the governor to stop the illegal roadside hunting, his response again recalled Alaska: To "protect" the game, he would send in helicopters to kill wolves. The team could do as they wished, however, with the feral mastiffs. When they came upon two killing a chiru, and the driver shot them, "somehow," wrote George, "we are all pleased."

George stayed mostly quiet when his Tibetan companions joined the killing: shooting choughs (a crow relative, which George shared in eating) or unleashing a "bored fusillade" at the tiny birds that gathered near the herders' tents where soil was exposed. When one raised his gun to shoot the first manul cat George had ever seen, he shouted and waved the tractor on, but later apologized for interfering.

All were moved, on occasion, to acts of mercy. Seeing a man perched on a kilometer stone, his bicycle beside him and rifle in hand, they shooed away the oblivious gazelles drifting near. It took them longer to pity a thin hare, conspicuous in its dark coat in the deep snow. George and Qiu chased it for pictures until, too fatigued to flee, it simply sat. When a trucker stopped and grabbed it by the ears, George asked Qiu to retrieve it. They hid it behind a mound of earth. "I'm not proud that I took part in the hazing. But at least it will die on its own."

By trip's end, George had fallen ill. Shivering from fever, he agreed to ride in the truck's cab while the tractor went ahead. When they again broke through the ice, his Tibetan companions walked six hours to fetch the tractor back. George and the soldiers waited all night in the bed of the truck, everyone's feet toward the center, rescue coats piled atop them, eating frozen bread. But by the time the tractor came, the truck was again iced in. So after a dinner of kebabs—hacked-off hunks of a sheep jammed onto a wire and grilled over a dung fire—the men spent a second night together, this time stretched side by side. In the end, they had to abandon the truck; its gearbox was broken. George hitched a ride in another truck collecting kiang bodies for glue. He sat atop the frozen corpses.

Finally back at the military base that was housing them—a new oil pipeline from Tibet to Qinghai had one guarding every pump station—George was thawing gazelle and chiru heads to cut the jaws free, eating the chocolate he'd saved from a stopover in Germany (to receive another award) and reading his last book (Robert Ludlum) when Qiu came to say that the driver did not want to go on. And that since he was on the same political level as the lead scientist, he got to decide.

Fortunately, the driver was at that moment in George's room watching *Garden of Allah* starring Marlene Dietrich. George persuaded him to go one last stretch but immediately regretted it, filled with the sudden desire to go home. As he prepared to leave, in what seemed a throwback to 1980, Qinghai forestry officials demanded his notes and film. He refused, but again doubted his purpose. "Qiu is the only one I feel we've given some direction and sharpened integrity and interest in a valuable goal." He was happily surprised when, in response to his blizzard report, the Chinese authorities banned hunting in the region for three years.

The pattern was set. For the next thirty-five years, George would visit China every year, often more than once, traveling for up to four months at a stretch in Qinghai, Xinjiang, or—soon—Tibet.

The year 1986 brought his deepest commitment yet: seven months in

China with, for the first time on the plateau, Kay along. Arrived in May in Xinjiang, they found a willing team. Up most of the first night crimping dumplings to bring an auspicious journey, the entire group, even the cook and drivers, cheered at finding their first snow leopard scrape, and jumped up at dawn to see argali. Only Talipu peered out late, "like a tousle-haired marmot." As languid as Aman, he was similarly forgiven, George unable to resist his eagerness to *pa, pa* (climb, climb) with him, often carrying the tripod. When Talipu invited them all home for dinner, George guessed that he'd spent two weeks' pay on the feast of kebab, fish, cakes, and fruit—though he would still not toast "China's panda," reminding all that they were in the Uyghur Autonomous Region. Their Mongol guide was equally helpful: excellent at spotting and, always valued by George, a baker of delicious bread. The young ones were more trouble. George caught two poking his tripod into a den, taunting wolf pups. One rode the yak across every stream, leaving the rest to wade through; when he got it wedged into rocks, he stayed on till the yak fell over and George had to rush in to right it. They drank far too much: That same youth ended one night face down in melon rinds, a second night in a scuffle with a herdsman that left him "crawling and bawling," a third with the angry smashing of noodle bowls.

Returning to the farthest edge of Taxkorgan Nature Reserve, George was sure he would find the Marco Polo sheep that had gone missing from Pakistan. He did in fact meet his first in the wild: A herd of two. Just two. In valley after valley, no polii; only horns scattered about herders' huts, taken with the guns locals got from officials to kill wolves. Plus a dozen lambs who had starved the previous winter, their bones cracked open for the marrow by a bear. George's only other sighting was of four baby marmots. It had been a long while since he had witnessed a kill, so he took careful notes as a dog crunched each little head, cut the small bodies in two with his carnassials, then swallowed first the front and then the back half. When nearly fledged magpies startled and jumped from their nest, he lifted them back before the dog got them too.

Kay stayed mostly in camp, burdened by back pain and low blood sugar. She was glad when May's sleet turned to hot June sun, though as rivulets of meltwater began braiding every slope, she and George had to build stone dams to keep the water out of their tent. When warm nights brought still-more-treacherous glacial melt, locals warned that in the Raskam Valley ahead they would have to cross the dangerously rising river at least two dozen times. Better, they said, for Kay not to go on. Reluctantly, she agreed to be driven to Kashi,* from where she would fly to Beijing and on to Alaska. She gave George a parting haircut, watched as the team made their first crossing, and waved goodbye. "The tent is lonesome and empty without Kay," George wrote that night.

Kay tried to lift both their spirits with an ebullient letter from the road. At the hotel in Kashi, she'd met an American couple who reported that Galen and Barb Rowell had just been through on their way to Kashmir as the guests of Pakistan's President Zia-ul-Haq. While Galen climbed, Barb would stay at base camp. "Bless her for not being a strong hearty climber type." Kay herself would rise at 5:00 a.m. to curl her hair, "to look OK at the last check post." Tired of her own books, she had tried to persuade the American couple to take them in trade for their copy of Peter Fleming's *News from Tartary* but got only a Dick Francis thriller. At breakfast, a polite man who was being ignored by the waitress asked Kay if she'd mind asking for sugar. Kay obliged and learned that he was from Lahore and they knew people in common. "This talking to strangers is getting to be enjoyable; so many know of you. One American, hearing my name, knew us from your gorilla work. How nice to be related to you! I won't do it in New York, but in selected areas I think it is worthwhile to chatter. So there!"

George, meantime, had been visiting with Xinjiang families. Though he was not generally drawn to domesticated landscapes, he found an Arcadian peace in the white Mongol yurts "sprung like mushrooms" from the lush

* Kashi is the city's Chinese name; its Uyghur name is Kashgar.

grass, ringed by grazing sheep and roses in bloom. The hospitality was just as lyrical. An eighty-six-year-old Tajik who had served with the Kuomintang told him that snow leopards "are like the wind, sometimes here, sometimes not." Seating them in his yurt, he brought a black goat to the door and asked, "Is it satisfactory?" When they nodded yes, he said, "We thank god for this goat and it will go to paradise." The others said amen and moved cupped hands toward their faces, a gesture of gratitude. Everything was impeccably clean: from the slice of the throat to the use of the warm hide as a surface on which to cut up the carcass. A girl whose beauty all had noticed carried off the viscera and meat, returning after a time with bowls of broth; a boy came with a kettle to pour water three times over their hands. At 10:30 p.m., their hostess served a platter of boiled goat; in the morning, she kneaded them a mush of naan, butter, and milk.*

In counterpoint to these traditional rhythms was the faster beat of changing lives. The wife of a road builder showed George their new summer home, sewing machine, and cassette player; Deng's new economic policy had made them rich. In Ürümqi, an exhibition of new market opportunities included among "medicinal and forest products" the skeletons of several snow leopards.

The trip ended, as they now always did, with a meeting with officials—high-ranking ones, in deference to Schaller's stature. All of these went pretty much the same way. Schaller would describe the importance and dire state of (in this case) Marco Polo sheep and the snow leopard. He would then offer suggestions and investigative leads: One shop in Kashi had eleven snow leopard skins including a small cub; another in Gansu had six tiger hides and 154 snow leopard pelts, "as many as were alive in the

* These rituals of hospitality, and sacrifice, seemed to endure as others fell away. On another occasion, they watched their elderly host pour water from a small blue teapot into a sheep's mouth. If the animal swallowed, it was ready to give its life to feed them. This one swallowed, at which point the man tied its muzzle shut until it suffocated. The method ensured that not a drop of blood was wasted so all could be carefully scooped out and consumed.

entire province." If officials broached trophy hunting, he would urge that they limit it to abundant species and give all fees to locals—along with aid like better shears so the families could waste less wool. The county leader and party secretary would explain that snow leopards kill livestock, people need to kill wild sheep for meat, and the deer eat all the grass. "Give us money, and we'll hire guards and pay villagers for their loss of livestock and food." It wasn't always fruitless: The head of forestry did post three guards in Taxkorgan to protect the Marco Polo sheep.

Back to Qinghai and yet another indifferent team. While George worked on his field notes or greased boots, the rest played checkers. If he took a walk, no one asked what he'd seen. There wasn't much, anyway, mostly carcasses left by the 1985 blizzard and females with no young at heel; they had been too starved to reproduce. He began leaving camp early to escape "the rising noises of spitting, snotting, and urinating. I am tired of the company of men."

He had discovered one reason the Qinghai teams lacked the gusto of those in Xinjiang. There, they were paid eight yuan a day plus meals; here, just three yuan, and they had to buy their own food. Still, it was maddening to have his every suggestion—on what route to take or how to resuscitate a broken-down vehicle—ignored. The rides themselves were long and tortuous: pounding his sore back and noisy with loud arguments, pranks, or off-key singing. If he made a companionable comment or pointed to something of interest, he met silence or a condescending *ya, ya, ya* from the young ones. Eating his supper in an ice-cold room at a base, he knew the others had made friends whose rooms had heat, "but I am not asked to come."

While enduring these worldly woes, seeing flickers of life in scattered monasteries revived thoughts first seeded at Shey Gompa, where the lama protected blue sheep to honor a core Buddhist principle: that "kindness to all living things is the true religion." With the help (it seems) of the Dalai Lama, George had cards printed to offer as gifts. On one side was a message in Tibetan from Milarepa, the tenth-century monk he had learned about from Matthiessen: "To all life is dear . . . One should neither kill nor cause to kill." On the other, a hunter lay down his gun and animals gathered around.

His hosts were pleased to receive it. One placed the card on his shrine, a transistor radio flanked by pictures of the Buddha. An old scribe, seated at an outdoor table writing letters for a line of waiting people, read it aloud, setting off a "mini-riot": Thirty Tibetans begged George for one. Monks held them as they circled mani stones or blessed followers with a tap of the card to the brow. When a six-foot-tall Tibetan in Chinese blue cotton took one, however, George grew cautious. Had the DL, as he calls him in his journal, put a secret message into the drawing or script? "Will the tall chap inform on me? Best I not give them in big groups."

Everywhere he found this mix of tradition and change. The wait for permits or truck repairs or a July blizzard to end often trapped them in administrative towns, charmless clusters of cement buildings thrown up only since 1959. The young "urbanites" who drifted through wore green plastic cowboy hats or long hair, bell-bottoms, and high-heeled shoes; George once watched them scatter when gunslingers in *chubas* galloped through. The abundant trash—fish guts, rubber soles, decomposing dogs, a "glitter of broken glass"—was informative if it included wild yak heads or gazelle hooves. Most worrying was a fast-growing trade in animals and their parts, capitalizing on expanding roads and the opening of the economy. A towering German named Oswald told Schaller that he was in exports, shipping ten tons a year of blue sheep meat to Europeans craving wild game; in one county alone, the export quota was one hundred thousand pounds: several thousand sheep. Why survey, George wondered in his notes, "if they're all going to be shot? I'm just pinpointing good hunting." A young colleague worked the market closer to home, paying three months' wages for a pound of *dongchongziazhow*, a two-inch fungus that grows from the head of a caterpillar and is said to have medicinal properties. Schaller drew the oddity in his journal, not yet knowing the critical threat it would pose to snow leopard habitat.

George was back in his Xining hotel preparing to leave the country when the phone rang. Qiu was in the lobby, having traveled in secret from Chengdu, 750 miles away. Sitting up with him most of the night, George saw the twenty-four-year-old's agitation, his sentences drifting off as he ran his hands through his hair. Qiu was in despair: at the continued mistreatment of the pandas, and at his own future. His parents insisted he marry, and his forestry bosses had rejected more study overseas. George advised him to put aside what he could do nothing about, set a goal and timeline, and burn no bridges. He would take him along on surveys, to get data for a thesis. The next day, George flew to Austria to spend the winter there with Kay. He gave her the blue poppy seed heads he had collected, a flower he would later describe as beautiful but barbed.

•••

At winter's end, in April 1987, twenty-four-year-old Mark joined them, returning with George to China while Kay headed home. The two flew first to Hami in northeast Xinjiang to explore the Mongolian border. They then spent six weeks crossing the Tian Shan (Celestial Mountains) and the southern edge of the desert called Taklimakan, a Turkic word meaning "enter and you will not come out."

For the first time since Brazil, Mark got a sustained taste of his father's life. That included the quotidian miseries: from blistered lips and ears to naan so frozen they had to smash it with stones. Keeping his own journals, Mark cursed the "malicious" wind, fierce enough to blow gravel out of river bottoms, geese out of formation, and the chocolate-and-peanut-butter treat he'd assembled for his dad into the sand. He saw truckers drive in reverse so windblown rocks would break their rear rather than front windshield, and was awed at his father's nonchalance as their small plane bucked approaching an abandoned airstrip, the windows of its tiny control tower opaque with crusted dirt, its rollaway steps nearly vanished in high weeds. He was awed again when a mail call brought George a long, handwritten letter from actress Glenn Close, describing her "raucous azaleas," recent separation, and move back into a tiny Village walk-up. (Though George has met many celebrities, when asked about them he looks blank, less discreet than oblivious.)

Mark felt the unease his father lived with all the time, of being an oddity in places that had rarely seen a foreigner. Gestures intended as gracious were sometimes isolating: The proprietor of one roadside eatery led them into a private room to dine sitting on the bed, indicating they should drop their mutton bones on the floor. Always the object of fascination, George decided to make use of the attention, carrying a book of color plates to show villagers the local mammals and birds that were theirs to treasure. One afternoon, twenty men sat nibbling sunflower seeds, watching them eat noodles. "You have an audience," said George. "Here's your chance to do something." "Apparently," Mark answered, "I'm doing enough."

When his dad bristled at the big obligatory teams—"born a century too late," as he said, "to be free of the minders and hangers-on"—Mark saw why. George had next to no agency. Others decided where they would go, when they'd set out, who would join. And again he was saddled with useless companions. One forestry staffer would finish breakfast, light a cigarette, and with a "supercilious" hand in the air flicking ashes, say, "It's 10:30, we may as well wait here until lunch." Once, for no apparent reason, the team took on as guide one of the long-haired city boys who knew nothing, only babbled on about getting drunk and thrashed by a relative.

Schaller continued to learn from those who managed to subvert the hierarchy, at least for a time. When the Han berated the Uyghurs for their choice of camping spot, George saw Mark's distress and reassured him. "Let's just watch. They're not about to listen to what a Han Chinese says."

Sure enough, the Uyghurs ignored them the way "the Chinese often ignore me, as if I'm baggage."

Still, power plays and politics kept intruding. Due to "student unrest," they were rerouted, traveling five days instead of one. They wasted more time on a redundant survey, because the local leader had arranged horses and would lose face. In several closed towns, neither Schaller was allowed out of the military compound; at one airport they were locked into the waiting room. Worst was the harassment by the Chinese Environmental Protection Office. Because they were traveling with the Ministry of Forestry, the rivalrous EPO sent a telegram, then a truckful of men 650 miles, to tell them they could not enter Arjin Shan, an EPO reserve.

They had not violated that prohibition a week later when yet another delegation arrived, bursting into breakfast with great bonhomie. George returned their smiles. "It always pays to be friendly with your enemies," he told Mark. "You might learn something." Mark then watched his father apply lessons learned with the pandas. When the forestry officer insisted George back him up in his "counter-fight" against the EPO, George glowered with "undisguised contempt." He was equally indignant in a meeting with high officials in Ürümqi, as Mark transcribed:

> Anyone who claims we went into the reserve is a liar . . . My team is careful and honest. I am a guest and internal political matters don't normally concern me, but you have impeached the honor of every member of my expedition. You have wasted time, vehicles, and money, and jeopardized our future work. If you had sought truth through facts [a favorite precept of Mao's] rather than truth through rumor, this would have been a minor matter. . . . I would like to know the real reason for this great effort to obstruct and discredit us.

Then, a tactical softening. "Good can come from the bad," George continued. "The Arjin and Kunlun Shan are the most important wildlife areas in Xinjiang. I'd like to continue to work here in collaboration with forestry and EPO. Our research could support a master plan."

As officials scrambled to undo the damage, Mark continued taking down the dialogue.

> **EPO:** "We have full confidence in every team member and sincerely hope this has not been misinterpreted and that Dr. Schaller will come back."
>
> **GBS:** "It's the past. Let's think of the future."

Beyond all the hardships, Mark also saw the pleasures. Talipu was again along, still charming with his zeal. Mark and George both wrote of the day they bought a sheep from herdsmen who struggled to lasso it. Suddenly, Talipu, wearing a suit, dashed after the startled animal, hurling his body to pin it to the ground. They tied it alive to the top of the green truck, its big head lolling, a sight Mark found unsettling—though he did that night help eat "it down to its stomach lining, feet, brains, eyes, nose." Talipu also caught a red-spotted bluethroat, a small bird of the far north, the first George had seen since canoeing Alaska's Colville River thirty-five years earlier. Qiu Mingjiang (Xiao Qiu, Mark called him, using the endearment) was also along, entertaining with tales of his childhood. His grandmother had cured his stutter by taking him into the fields in a thunderstorm and beating him over the head with an eggplant. On the rare days when the wind quieted, their companions played the three-stringed *rawap* and sang Uyghur songs. Li Hung—"another loner," in George's words, who "sincerely wants to get things done"—answered with Chinese ballads in a sad, sweet voice. Evenings, they watched Li flick noodles into the soup, fast and accurate from six feet away; a skill he learned, he told Mark, when sent for two years to a reeducation camp.

Local hospitality was as generous as ever. For breakfast they were served milk so fresh it was still warm, and flaky naan spiced with a black seed

Mark remembered from Lahore (likely nigella). Passing through Tula at the end of Ramadan, they were invited to every home to sit on a sleeping pallet and eat *sangza*, a deep-fried twist of noodles. Mark noticed that a herdsman who invited them for lunch wore the same high-top blue Keds as his dad. When the man's beautiful, "soft-faced" daughter gazed on Mark in fascination, his dad teased him. "Maybe you ought to stay."

Animals were few, but on May 26 they sighted chiru, wild ass, wild yaks, fox, and a wolf, "a nice present for dad's 54th birthday. That and the two marzipan bars I'd been carrying as a present from mom."

Most memorable for both was the June trip into the desert on Bactrian camels. George was glad to be released from the vehicle. "This is the way to travel, not encased." Even if, in this land unmarked by a single bird or shred of green, they rode hours yet seemed to stay in one place.

George again indulged his camel obsession. He noted how the wind stripped their molting wool in long strands, as if unraveling them, and how with their inner thighs bared first, they seemed to wear a short, frilly frock. He recorded their expressive vocalizations: When annoyed, they clapped their lips like castanets or squawked like his baby blue heron, Siegfried. Mark, too, watching a shearing, was riveted by the necks whipping round, "spewing and braying in a nightmarish, almost-human shout as their purplish skin emerged." In a rare warm sun, he found riding the soft beasts with spongy hooves soothing, the gentle rocking like a boat riding a tide. George, half hypnotized by his mount's nude neck bobbing like some "grotesque bird," was often last in line, getting spritzed with blowing urine. Never one for physical vanity, he took to wearing snow goggles.

Though he is also a PhD scientist, Mark fell somewhat farther from the tree than his older brother Eric. Mark would sanctify his second marriage in shorts with a Celtic handfasting, and name his kids Jasper Quincy Experience Schaller and Maddox R. Traveler Schaller. That countercultural streak was already evident here: In his journals he often experiments with raw accounts of his bodily urges and indignities à la William Burroughs, and he is thrilled to find, at a mail stop, that he has sold a piece to the *Journal of Polymorphous Perversity.*

Yet for all their differences, father and son had few conflicts. George records just one moment of annoyance: When he celebrated their arrival at last on the Tibetan Plateau, Mark said that it seemed no different from the place they'd just been, five hundred meters below. "I replied sharply, 'It's not that different 2000 meters below, so why bother coming?'" Mark recalled his dad's disappointment at the books he had brought to share. George had packed James Michener and Larry McMurtry. Mark had brought *Don Quixote*, which despite his best efforts ("men have . . . learnt many important things from the beasts," he read out loud, "for example from storks, the enema, from the dog, vomiting and gratitude") George deemed boring. He got downright upset when he tried the other novel his son had carried in, Julio Cortázar's *Hopscotch.* (Telling this story, Mark pulled out a picture of George on a later visit to see his grandchildren, sprawled in a chair, happily reading *The Adventures of Tintin.*)

Mostly, they enjoyed each other. Waiting for the cars one day, "Dad took a stick and cleared a trough for water to flow into the garden. 'If we had a boat,' he said, 'we could play.'" Climbing a desolate glacier in blowing snow and finding no wildlife, George said there was little point to going on "except for the hell of it." So on they went. Eight years later, when his first marriage was breaking up, Mark would go backpacking alone for sixteen days, wanting again to be "governed by the elemental challenges of existence: to get from here to there, come up with a meal, find a dry place to sleep." After the first week, he abandoned the trails, confident he could find his way.

Unearthing in George's pack a treat of peanut butter "lovingly packed by your mother a year ago," they shared a moment of homesickness, dreaming of Kay's pot roasts. On what Mark thought might be Father's Day, he tended George—shivering from flu or "low-altitude decompression" sickness—with one last marzipan bar held back for emergencies. They reminisced about the Pantanal, Pakistan, the Serengeti, and India, Mark summoning memories and nudging his father to fill them in.

They even dabbled in mild debauchery, joining the team in a game involving eggs dyed red. Each player tapped the narrow end of his egg against his rival's; whoever's shell broke first had to drink a shot of *baijiu.* (The grain liquor can be 60 percent alcohol.) George tried to keep track, but since Qiu's face burned red after one drink and Talipu was always disheveled, it was difficult to say how much anyone had downed. Li Hung, unaccustomed to letting decorum slip, kept saying, "My English is very elementary at present." George was glad to see Mark being razzed, and responding in kind. He dissolved everyone in laughter climbing on a donkey and having to walk as he rode, his feet touching the ground. When he got back to the tent, he did thirty push-ups—"Maybe a macho thing to show Dad. Maybe I was just drunk."

Comedy was their most consistent bond. Seeing a road sign at a sharp curve that was simply a giant exclamation mark, Mark wondered out loud whether there were others with, say, a semicolon or apostrophe. "'At least it's not a question mark,' said Daddy." At a party in their honor, a tambourine brought everyone to their feet for a dance in which you chose your partner by passing them a rose. George gamely accepted and entertained all with what Mark described as a "stork-like salute."

Like anyone who spent time with George, Mark was often on the receiving end of his dad's "juvenile" pranks.* One morning, having agreed to walk out early together, Mark was annoyed to find that his dad had left while he was still brushing his teeth. He jogged after, imagining saying, "Thanks for waiting, you bastard," then saw him on the far side of a stream. Crossing the tippy branches that counted as bridge, Mark smiled and

* I was too. Out together on a tiger walk and picnic in India, he brought me a paper muffin cup holding what looked to be small truffles; in fact, they were chital droppings.

thanked George for letting him catch up. "'Sure,' Dad said. 'I wanted to see you fall in.'"

In their two months together, Mark knew this feeling of closeness would sometimes slip away. Getting stuck in town wasn't always the worst thing. After an aimless day with his dad eating kebabs skewered on old bicycle spokes and talking about books, Southern food, James Earl Ray, and Hitler, "I feel strong and not lonely," he wrote. Even if "tomorrow we head for the silent hardships of the field again."

Inevitably, once back in the wild, George would cross over into his more-than-human world. Returned to slow tasks like "collecting an animal's artifacts . . . the mind is more receptive." He saw that a sand fox had marked with feces a frayed, half-buried skull, reincarnating it as "landmark and signpost, a place of identity in this immense space." An unafraid yak, his mantle of hair almost to the ground, had the "aura of a Stonehenge dolmen in a land of power and magic."

One night, the wind grew so fierce that it tumbled Mark's tent—with him in it. After much wandering in the blasting sand to find the stakes, George helped Mark secure it again. Then, even though it was 4:30 a.m., he went off for a walk alone, and Mark

> *again felt that brooding melancholy drift over me. Alone in my tent I felt so lonesome, as if I hadn't talked to Dad in days. I suddenly felt the need to find him. I took off across the plain, climbed a hill and looked all around but didn't see him; blinked back odd tears. Where is he? Finally I saw a figure descending the valley. I began to run down the hill and he seemed to know and be happy that I'd come looking for him, happy that his own lonesomeness was broken. I said little and let him talk of the difficulties of the project. He seemed to need to talk. I didn't; just felt good being with him, swinging a bear skull he'd found.*

George had suggested Mark return to town to recover from intestinal woes, but he said no.

> *To spend the next week away from bàba would fill me with such loneliness. "When chewing maize, camels kneel in a circle facing each other like lions at a kill." He's always saying stuff like that. . . .*
>
> *On a river broken by sandbars into narrow channels, I winged a good flat stone. It hit the near channel. Skip. Over the sand into the next channel. Skip! Again floated cleanly over a sandbar into a big channel, where it did a few soft tadpole skips and disappeared. "Hey," said Dad behind me, "that was pretty good." I felt a completeness. "I tell you," I said, "I am an accomplished stone skipper." "Yeah," he said, "but that middle channel was luck."*

On what they didn't realize would be their last night out, a flood drove Mark into his dad's tent. "After two months in separate tents, we slept the last two hours side by side." Two days before they were to head home together, George decided to stay on in Gansu, news Mark met with a "quiet feeling of acceptance." In a brief passport hassle at the airport, "I forgot to shoot him one last thankful complicitous look. Ah well, he's all too experienced with such omissions." He did find one bright spot in going home separately. "Mom can enjoy two homecomings and my presence without having to deal with the work and mental readjustment of Dad returning."

•••

In the summer of 1988, after three years on its fringes, Schaller became the first Westerner allowed to work deep within the Chang Tang, the highest and wildest piece of the Tibetan Plateau. Larger than Germany and Poland combined, it is almost Martian in its harshness. Streams that flow down from the mountains can't escape but get trapped in shriveling brackish lakes; George once found fifteen pika mummified in the alkaline waters. Few plant species survive, none of them easy to eat: the *Carex* too sharp and rigid, the flat cushion plants "impervious to herbivory." Most mammals adapted to this extremity can live nowhere else. Chiru, wild yak, and

Tibetan argali are all endemic, as are Tibetan gazelle, Tibetan wild ass, and Tibetan brown bear. George sketched a chart comparing the species here with those in the other great grasslands he'd explored. Alaska had a bear and wolf; the Serengeti a leopard and zebra (genus *Equus*)—only the Chang Tang had all four.

The trip began with the usual wranglings. Pushed to pay for "security," Schaller refused. "If they want to watch us, they can pay." He also refused the $50,000 "fee" demanded to work in the still-off-limits Arjin Shan: "We are collaborating, not buying our way in." With Kay along, plus Shell executive Tim Brennand and his wife and teen son, they faced familiar trials. The team leader infuriated them all: racing ahead scattering animals; driving past dark, both dangerous and useless for surveying; cutting off George's inquiries of officials and locals. But don't criticize him, George warned the Brennands, lest he turn on us later. The roads, or rather tracks, were the worst yet; hours of lurching and shaking left all exhausted. After an entire day driving up a streambed through raging waters churned the color of chocolate, one by one they sank their vehicles in the mud. First, the truck carrying the gear foundered, separating them from their food and sleeping bags. Then their own jeeps were mired, though when forty Tibetans materialized as if from nowhere offering to push them through a kilometer of muck for forty yuan a vehicle, George suspected they kept the road impassable for just this purpose. With night and a storm approaching, he and Kay walked to a village where a herder let them stay—bringing them a hot thermos and water to wash. They lay alongside a cow who, after some nervous huffing, settled into rhythmic belching as she chewed her cud.

All was worth it to reach the Aru Basin, the fulfillment for George of another dream. He had read the accounts of the British travelers who "with imperial restlessness" had passed this way a century earlier. Since the last of those, in 1903, no outsider had beheld the marvels now opened out before them. In every direction around the strange turquoise of Aru Co (*co* means "lake"), herds of chiru, kiang, argali, and gazelle grazed. Best of all, "wild yak excitement!" Hundreds milled about, the sparring bulls bellowing and stabbing the ground with their thirty-inch horns, their tails waving like pennants. All were black but for one golden calf.

Kay had fallen ill on the trip, and even after ten days rest in the Holiday Inn Lhasa remained fluish. In mid-September, agreeing that she could not spend six more weeks in altitude and cold, "we made the awful decision for us both." Taking her to the Lhasa airport, George saw newly posted signs. "Foreigners are not allowed to crowd around watching and photographing the disturbances manipulated by a few splitists. And should not do any distorted propaganda."

Still jet-lagged a week later, Kay wrote him at 2:00 a.m.

> *It seems only yesterday that I saw you so unexpectedly and so happily from the bus window as we were driven to the airplane. [The doctor] said we were absolutely right to have me come home; it is too easy to have things go very wrong at such altitudes. [And I] came home only 116 pounds . . . Did I tell you that it cost only 9 yuan to order the beef platter, (8 slices of fillet around a mound of potato salad and two hard rolls) delivered to the room at the Lido? To make you feel better . . . So now I shall toddle back to the mail (actually, I WALK.) . . . Goodnight my love; just a week ago I was driving out to the Lhasa airport with you, and morose to say the least. But I'm better now that the break has occurred . . . And at least I'm not lying and freezing in a sleeping bag all day. I tell myself.*

Four days later, she wrote again. "Dearest, I'm lonely [and] still so tired from the virus. I got in the habit of being with you and miss you so much more than usual." She reported on the family: "Eric and Paulette sound more and more committed to each other, which is sweet to hear." And the garden: She was planting dill and licorice basil, and cosmos for bouquets.

These letters began a habit that would persist for two decades, as Kay's increasingly unreliable health and George's race against time (his own and the animals') make for ever-longer separations, now without the boys to ease her solitude. In every letter, she will tally chores done—mowing, fertilizing, crawling under the house to drain pipes, prepping for power outages, overseeing repairs, investing their money, managing family strife, supporting sick or heartbroken friends—as if to justify her days. Her tone will often be antic and flirtatiously pouty, but over time a weariness and even resentment will seep in—though of course letters document only their times apart, and thus provide only a partial view.

George stayed on another month, making his first reconnaissance of the plateau's eastern edge where grasslands give way to deep gorges and forests. He witnessed a few marvelous sights. In the sky, one hundred black-necked and demoiselle cranes. On the ground, a dozen young Tibetans going three hundred miles to Lhasa the traditional way: lying face down with arms forward and forehead touching the ground, standing back up and moving a few steps to where their hands had touched the ground, then repeating it all, like an inchworm. But as always, those thrilling moments were bought with days and days of nothing.

•••

The year 1988 ended with a brand-new experience for George: serving as expert witness in a trafficking case that exposed the complicity of Smithsonian scientists in the killing of endangered animals.

At the center of the story was Richard Mitchell, a zoologist in the US Fish and Wildlife Service's Office of Scientific Authority, whose duties included enforcing the Endangered Species Act. Mitchell was then temporarily posted to the Smithsonian, to develop projects in partnership with Chinese researchers.

George had been hearing through the grapevine of Mitchell's activities. In 1984, despite his official roles, Mitchell had founded an NGO, rewarding its funders with trophy hunts in Asia. In 1987 he had brought his boss, Robert Hoffmann, then director of the National Museum of Natural History, to China with hunter Donald Cox, soon to be enshrined in Safari Club International's Hall of Fame for killing 340 species, including forty kinds of wild sheep. On that trip, Cox had shot one of just 350 surviving Przewalski's gazelle, which by 1996 would be listed as critically endangered.

Mitchell's luck seemed to have run out when in April 1988, US Fish and Wildlife enforcement agents met him at the San Francisco airport on his return with four hunters including oilman Clayton Williams and his wife, Modesta. The agents confiscated the heads and hides of four *Ovis ammon hodgsoni*, Tibetan argali listed since 1976 as endangered. The Williamses had paid $25,000 each for the privilege of killing one.

But just three months later, Mitchell was back in business, as George learned passing through Ruoqiang County, where he bumped into a team from the Chinese Academy of Sciences. After some prodding, the academicians confided that they were there to meet Mitchell; in addition to Cox, his party this time included James Conklin, a plastic surgeon who called Asia his "playground" and had killed fifty species of goat and sheep. On this trip, for which the men had paid $172,000, both shot a chiru, though it had been listed as endangered since 1979 and had the highest degree of protection by international treaty.

Upon his return to the US in October, George was himself visited by Fish and Wildlife agents seeking his help verifying the taxonomy of the sheep they had confiscated in April. Provided with photos and hides, Schaller confirmed that they were indeed *hodgsoni* in nuptial pelage, "one of the easiest subspecies to recognize." Two other experts concurred. George also had in his possession (it is not clear why) telexes and letters proving that all involved knew they were *hodgsoni*, and that Mitchell was also seeking a permit to hunt Marco Polo sheep. "The bastard. It's difficult to make headway on conservation when foreign 'scientists' indulge their propensity to kill." Crossing paths with Mitchell's staff, he gave them reprints of his article on the rarity of Marco Polo sheep and the need for a total hunting ban to pass on to their boss.

Mitchell tried to muddy the science, claiming both that the endangered subspecies did not occur where Mr. and Mrs. Williams had hunted—a place Schaller knew well from his 1985 travels—and that the trophies were in fact the subspecies *dalai lamae*, which was not endangered. The only problem being that few credited the existence of that subspecies, as George reminded officials. Valerius Geist had shown that *dalai lamae* was simply a taxonomic misnomer for a juvenile *hodgsoni*. And in an unpublished book chapter, Mitchell himself affirmed that these were simply two names for the same animal. In November, George wrote Hoffmann: "That at this critical period in the world conservation movement, the Smithsonian seems to promote killing of animals known or thought to be rarc should be embarrassing."

The Washington Post, beginning with a page one story in December 1990, turned a bright light on the case, with the result that Mitchell was investigated by the Department of Justice, Fish and Wildlife, and the inspectors general at the Department of the Interior and the Smithsonian. The evidence was damning: It included a January 1988 letter to Cox listing animals that would be "available," including the endangered chiru and argali; solicitations from Mitchell on Fish and Wildlife letterhead offering the opportunity to hunt protected urial; and testimony from a Safari Club member that Mitchell had assured him that most inspectors couldn't tell subspecies apart, and anyway, he was the guy who cleared in animals.

Thanks to Williams's political connections, including several US senators, the government dropped the case and even returned the horns and hides to the hunters. But that was not the end of it; in 1992, Mitchell was indicted for smuggling endangered species. That the Smithsonian then spent $350,000 for his legal fees spurred angry congressional hearings.

For Schaller, his involvement brought several chastisements from the officials in China tasked with issuing his permits, upset that he'd "told US

Fish and Wildlife too much," leading them to ban all argali imports from China. Each time, George explained that if asked by his government to identify specimens he had to give an honest answer, and that if a scientist couldn't weigh in on such matters without political interference, then science in China was in deep trouble.

George had never objected to hunting, for subsistence or for science, so long as an animal was not overhunted. He himself had "taken" plenty of birds and other specimens, traipsed off in college on many a fruitless moose hunt, and on one occasion even killed a large mammal. Out hunting with Cousin Ed, he had come upon two caribou bulls in a clearing "sleeping, dreaming, their heads lowered under the weight of ebony antlers and white necks gleaming." They were just a hundred feet away, an easy shot, and failed to notice him even as he walked into the open. "One bull collapsed and lay there, legs tucked under, puffs of breath visible in the cold air. Instead of fleeing, the other bull stood over his dying companion." This was the animal whose tenderloins he'd wanted to share with his mother, if only she'd have visited him in Fairbanks.

Though he would come to feel nothing but regret for a killing that seemed more execution than hunt, Schaller was open even to trophy hunting. He had championed it where it could provide income and incentive for villagers to keep animals alive. But he had also seen its complexities. Hunting can wreck other tourism: The animals learn to run. It can alienate locals if they are exiled from parks, barred from hunting wildlife reserved for rich foreigners, or shut out of the proceeds. And it sometimes killed the golden goose. Between 1967 and 1989, Mongolia sold more than 1,600 licenses to hunt Gobi or Altai argali at $25,000–30,000 apiece, earning tens of millions of dollars. But that decimated the animals so thoroughly that in 1994 they could sell just fifteen.

Though George had lost any pleasure he'd found with a rifle when young, he had watched enough lions and wild dogs enjoy killing to still half understand hunters' lust for blood and the jolt of power it seems to bestow.

He scorned, however, any claim to it being a "sport." The Tibetan boy he met hunting with a muzzle loader had to get within a hundred feet of his quarry. But the advanced guns carried by trophy seekers, along with brilliant local trackers, erased any need for courage or skill.

George had also seen far too often what these great adventurers were made of. In Tajikistan, he watched a Marco Polo ram shot through the thigh and chest drag itself for half a mile, pursued by guides who finished it off and fetched it back to camp so their client could straddle it for the conqueror's photo. On a flight from Mongolia to Beijing, hearing an American loudly boast of shooting an argali, George asked how long it had taken. He shot it on the first day and left at once, the man replied, happy to get back to Chicago. "Where precisely has he been? He has no idea."

Though he abhorred such characters, George never missed a chance to gather intel. Meeting a hunter from Laguna Beach, California, he asked conspiratorially how to smuggle illegal horns into the US. On a flight with WCS supporter Edith McBean, he sent her to chat up yet another boaster, to find out where he had killed his trophy and what he paid. And though he didn't realize it at the time, the summer of the Mitchell case brought George his first clue in a mass-murder mystery it would take him years to unravel. Passing through the market town of Gertze in western Tibet, he saw traders plucking the fine underwool from chiru hides. When he asked why, they told him it was much sought-after in India, but because the animals were protected, they would have to smuggle the wool through Nepal. It would be two years before he understood that conversation.

For George, 1989 was a lost year in China. The May declaration of martial law and June 4 crushing of the student uprising in Tiananmen Square foreclosed all possibility of continuing his work. Though the Schallers passed through Beijing in August on their way to Mongolia, "we heard no more toasts to eternal friendship." Returned in late September, George was advised for political reasons to attend the fortieth anniversary of the Revolution at the Great Hall of the People. With his ticket came a notice to stay in his seat if bombs went off or shots were fired. The featured performers were all People's Liberation Army: soldier-acrobats and a baritone from the PLA Song and Dance Ensemble. Things eased only slightly in the two and a half months he stayed after Kay went home. Foreign newspapers and magazines returned to the kiosks, but flipping through a rack full of *US News and World Report* he saw that the China section had been ripped out of every one. Permitted into Lhasa for two weeks, he met the head of the Tibet Autonomous Region People's Congress, who had fought for the PLA against Tibet and at twenty-eight been appointed mayor. He told Schaller he must abide by the Five Principles, including noninterference. "A clear warning."

Often meeting Pan Wenshi and Lü Zhi for dinner, George shared his fear that his time in China might soon end. He would publish *The Last Panda*, even if it barred the door. He could not go on wasting time and money, yanked about by bureaucrats' internecine wars. And he continued to worry over the true impact of the work. Told that a tiger breeding station run jointly with the Ministry of Forestry had applied to CITES to export eighty kilos of tiger bones, he wondered if his own organization had been an unwitting accomplice to this trade. To stop China's capture of wild tigers—done in the name of diversifying the captive gene pool—WCS had arranged a gift of eight tigers from US zoos. His fears were later confirmed: The eight had been bred to produce valuable penises and bone.

George worried for Kay, knowing that to continue in China would mean more long stretches away. He questioned his own motives: Was he in thrall to a romantic idea, following Hedin?

And then, without consulting anyone, he committed to three more years. As he explained in a Christmas letter to Bettina, China now wanted his help to "set up a 100,000-square-mile reserve I suggested and have been pushing for many years." He simply could not turn his back on this chance to protect the Chang Tang.

Stopped on his way home by airport security alarmed by his jab stick,

George pulled out the copy he always carried of his *National Geographic* panda story. They let him through.

On July 15, 1990, Kay wrote Mark and Eric from Lhasa. Finding no one to greet them at the government guest house, George had gone off to find their hosts, leaving her in the lobby with their packs, tents, and boots. For the next several hours, she had done her best to return the smiles of the hundreds of Tibetans streaming past; it seems she had crashed the annual meeting of the Tibet People's Congress. "Like dumping me and a pile of scruffy bags on the Senate steps in D.C.!"

George was looking forward to three months in Zhongba County, directly across the border from Dolpo, where he and Matthiessen had been seventeen years earlier. But a year after Tiananmen, China felt to George as it had when he first arrived in 1980, in the wake of the Cultural Revolution. The cynicism, lassitude, and self-dealing had returned. One colleague's sole focus was securing boxes of his favorite food: flattened pig heads. The driver traded their gas coupons for rifles, and said *bu xing* (not possible) so often they began calling him that when he wasn't around. Since he also drove for an important party leader, he could not be overruled.

Most difficult was the team's taste for sleeping away much of the day. Dinner rarely started before 11:00 p.m. Though the Schallers then went to bed, the rest sat up drinking for hours. Waking early, George and Kay would wait: he burying the tins and cigarette butts from the night's revelry; she (feeling the altitude and sleeping poorly) clipping vegetation plots or reading under the shade of her red polka-dot visor. Around noon, someone might stir to rustle up "the usual rice, pea and gelatinous pork-fat mixture." All then lingered over the meal, followed by elaborate grooming and desultory packing. Finally, in midafternoon when the frozen ground had become impassably soft, they began the day. After weeks of this, Kay told the most conscientious of the bunch that sleeping until noon was unprofessional and useless for animal research. The next morning, the Schallers were fed early, on separate plates rather than from the communal pot, while the rest sat tensely by. The dictionary vanished from the car. "We are being isolated."

Animals interested the team only as meat. Finding a still-warm young chiru, George had to press the others to leave half for the wolves who had killed her—though the steaks were, he conceded, far tenderer than mutton. At a stream black with fish, all waded in to catch them by hand. Though George found these, too, to be delicious, especially in spicy hot pots, he stood behind the overzealous driver, picking up flopping bodies and throwing them back. He and Kay watched ravens feast on fish trapped in eddies and a spawning female churn sand to cover her eggs. But no one else cared to look. More often they scared animals off, talking loudly and wandering about.

Tensions were highest with a China Central Television photographer who had tagged along. He regularly killed birds "to show the foreigner he could do as he pleased." When George found a snow leopard skull, the man demanded it to add to the hundred chiru horns he had gathered to "give to friends." George buried it. The last straw for Kay came when the journalist returned from the market with four snow leopard skins he'd bought to resell in Lhasa. She refused dinner and walked off into the dusk, spurring the team into a frenzied search.

The real gem was their trip leader. In white shirt and tie, Gu rarely stirred from what the young ones joked was his "reflective mind." When he did, he assigned one of the Tibetans to take his notes for him, while insisting Schaller credit him as coauthor of any papers. His juniors didn't bother working because, they said, he took all the data if they did; when one took the initiative to collect scat, Gu shut him down, saying it was of no value. The LEADER—as George and Kay began calling him, imagining the *White Waters*–style spoof they would write—did rouse himself to shoot seven tasty sandgrouse and two fat hares, frustrating George by failing to get "even basic data on the animals before or after he killed them." He "collected" two eggs, but then threw them away, embryos and all. When a truck sent by Shuanghu officials shot seven protected wild yak for meat, Gu told the Schallers not to tell the Forest Bureau, because the county bosses "help us." (When on a later trip a party secretary boasted of ending poaching while a just-shot gazelle bled in the back of his truck, Gu told him, "You better hide it; a foreigner is here.")

Worst was Gu's performance as general of their little convoy. Of the forty-nine pages in Schaller's journal chronicling this trip, more than half record travel disasters. They had begun ill-prepared: carrying hundreds of wasted pounds including broken stoves and a huge earthen crock of something pickled that no one ate—but too little gasoline, too few containers to carry it, and none of the binoculars George had supplied. They were off course and arguing: about how to pick their way through the rock fields, swollen streams, and devouring mud, and where they were even headed in the first place. They were stuck: digging out the truck, jacking it up to put rocks or cushion plants under the wheels, winching it out with the ever-heroic Toyota.

George invariably suggested they unload it first, but was just as invariably ignored. Instead, a hard jerk with the cable ripped off bolts or the bumper. One day, tents packed and ready to go at 3:30 p.m., they drove a hundred yards, got stuck, and after hours of digging put the tents back up where they started. Another day, the Toyota extricated the Nissan, then got stuck itself. Instead of repaying the favor, the Nissan inexplicably roared back into the mire. At that point, everyone lost it. "I call the driver stupid. Gu, who never lifts a shovel or carries rocks, stands on the roof and yells at me. Kay tells him he doesn't understand cars." Redirecting their fury, George dug and Kay carried rocks all afternoon until the Toyota was ready to be pulled free. Instead, the driver backed into even deeper mud.

> *But I say nothing. Gu had taken me aside and said the drivers were angry with me. Gansu, Qinghai, Xinjiang, and Forestry do not want to work with me. Only he supports me (hand under elbow to illustrate). In other words, I better not dispute things, or no project.*

These trips were never without rewards, especially when in August they again reached the Aru Basin. After years of searching, George met his first Tibetan brown bear. Prized for their gallbladders, which are used in traditional Chinese medicine, they had been shot almost out of existence. Watching the solitary three-year-old, George wondered how it felt to be the last of your kind, searching for a mate but finding not even a scent of hope. Kay had her own apt run-in, with a mother bear guarding two cubs, who with head low and hair bristling charged her car three times. "I have never, ever, seen such an angry animal."

Those close encounters, like the vision spread before them, carried them back to their bliss in the Serengeti. Chiru raced like Thomson's gazelles alongside their vehicles. Yaks moved like wildebeest in dark lines across the yellowing grass. Against glaciers blue-gold in the evening sun, wild ass posed like zebras, white bellies flashing. The memories were so strong George made highly uncharacteristic mistakes in his journals, writing "zebra," for instance, when he meant "kiang."

Bathed in that glow, George softened toward Gu. He knew he was posted to Tibet by no choice of his own and saw his family in Sichuan just once a year. Feeling him grow distant after the trip, George had a pang of guilt. "It was my fault too. He had no friend on the team, and they scarcely took his leadership seriously." But that empathy withered when at their ritual end-of-trip meeting with county and party leaders Gu pronounced the Chang Tang poor yak habitat, adding that chiru should be shot because they eat too much grass. He appeared to be providing cover for the officials' continued hunting. George nonetheless committed his time and $22,000 from WCI to the next survey.* "I wanted to give enough so they work to get me back into Tibet."

They did, though he half regretted it as 1991 began with the familiar

* Wildlife Conservation International (WCI) was George's division of what in 1990 was still the New York Zoological Society and would later become the Wildlife Conservation Society.

torturous delays. On the fortieth anniversary of what China called its "liberation" of Tibet, political tensions ran high. Adding to those tensions, George's own papers were being slow-walked in continued reprisal for his role in the Mitchell affair. "Never have I felt less like returning to China. On my knees I must beg for permits. I wait weeks in hotels; go months getting nothing done; feel the years slipping away." Kay had stayed home, "so I feel without spark, not unusual heading for China but this time worst of all. All I want is to get this year's fieldwork done and my stuff out of Lhasa; I'd leave tomorrow if our cars and equipment weren't there. After that I don't care. I'm depressed fighting each year to do a little fieldwork. It's not worth the time away from Kay."

And then, breakthroughs. A forestry official, persuaded by Schaller's work in the Mitchell case, asked the Caprinae Specialist Group at the IUCN for help reining in overhunting, beginning with rectifying the inflated estimates of sheep and goat populations used to set foreign trophy quotas.

And thrillingly, incredibly, the Tibet Autonomous Region approved the letter of intent signed with Schaller a year earlier, committing to protect nearly a fifth of Tibet in a new Chang Tang reserve.

George had been pressing for such a reserve since his first glimpse of this "wilderness unlike any I've seen outside of Alaska," as he had written in 1987 to the Minister of Forestry. He'd kept up the campaign ever since, appealing to national pride. China could protect a treasure that would dwarf the Arctic National Wildlife Refuge. Or they could lose it to the same pressures that had erased tens of millions of pronghorn and bison from the American frontier.

To avoid the mistakes made in Yellowstone and the Serengeti, where boundaries were drawn by people who didn't know the extent of the herds' migrations—harming both nearby communities and the wandering wildlife—he urged them to continue surveying the animals. And just as importantly, to extend protections to the people who had lived there since before recorded time, including of their traditional nomadic grazing and hunting. The celebratory banquet hosted by the party secretary, a "jockey-sized Tibetan wearing a Mao button," was as dissonant as ever. Genially calling George "the Imperialist American," he laid a feast of frogs' legs and a whole softshell turtle.

Though the reserve existed as yet only on paper, things were looking up in the field too. It proved useful to have befriended a Shell executive: When the airline said no flight was available for weeks, Brennand picked up the phone and got George on the next day's plane. With the post-'89 malaise easing, teams were improving. This year's bunch—which included Qiu at George's insistence—stirred early and kept camp clean. The drivers chose mostly solid routes. Instead of "loud competitive males" they had two women scientists along. And no one brought a gun. Though—blocked by meltwaters—they saw few animals, George spent the days teaching Qiu how to quantify plant biomass and protein, necessary lessons but also his way of trying to overcome some estrangement in their near-filial relationship. George didn't understand why the young man had never brought his family to meet him, nor why, having once leapt at the chance to go with George into the field, he now rarely even included him in conversations. Perhaps it was the attentions Qiu was being paid by one of the young women researchers (Miss Motormouth, George called her), which also led her to ignore everything else. Qiu finally came to George upset, saying he'd been too critical. George replied that Qiu had been acting like a tourist. And "it all blew over."

In George's journals, these early trips to this or that corner of the Tibetan Plateau seem at best haphazard and ill-starred. Repeatedly turned back, rerouted, stalled: It's hard to see how he could extract any coherent picture. And yet, year after year, he distilled a record as enduring in value as his gorilla and tiger studies. His detailed descriptions of every ridge and ravine mapped uncharted topographies a decade before the launch of Google Earth. From Neolithic flint sites he surmised where thousands of years earlier, animals had been abundant enough to hunt; once back in Lhasa, he bicycled to Norbulingka Palace to have its director of archaeology date the artifacts. His bird and plant lists were comprehensive in a way no narrower scientist could have assembled, recording a moment in natural history that will be mined for decades. Approaching sixty, he still evinced the quality that most moved Matthiessen, a never-jaded astonishment at the marvelous beasts.

But fatigue was also setting in—with the months of political games, close living with unkindred spirits, punishing physical ordeals. After a careless move that left his knee badly smashed, he allowed his pack to be carried, then skipped a climb because he "didn't need to see a few more scats." On the eleven-hour nights he spent in his sleeping bag snuggling his boot liners, camcorder batteries, and shaving cream to prevent them from freezing, he struggled to quiet his thoughts and sleep.

Letters from Kay brought diverting dispatches from the larger world, plus family sorrows and joys.

> *Miscellaneous news: ten-million-year-old fossil jawbone found in Namibia, perhaps common ancestor of chimp and man; Lee Remick died; Demi Moore posed nude. And locally: Kay was out planting grass seed by 6am; saw a huge woodchuck, who has been eating the parsley and raspberries and breaking the frame. Miserable wildlife! Like you I have dug and carried rocks to stuff in holes. Would rather be with you digging out vehicles. I am nearly voiceless with laryngitis. Poor timing on your part to miss me at my best!*

To celebrate a reprieve from home renovations—"Peace! Privacy!"—she had eaten a package of chocolate chip cookies for dinner. She had called Migs (Cousin Ed's second ex) "to assure her we care for all of them," and Eric, who was off to visit Paulette. "She'll have to give him a haircut as soon as he arrives because he won't pay money for something she can do; shades of you know who." She had called at exactly 5:40 p.m., "the moment 30 years ago he was being born. But I was lonely for you then too." Updating him on a tragedy that had befallen Mark and his first wife, whom George would soon visit, she offered gentle advice. "Do give her an extra hug and tell Mark again how sorry you are, even if he seems not to want to discuss it."

Though Kay hoped that *some* New Year's Eve they might be together, George went to Tibet alone again that winter, to see the chiru rut. A thousand males greeted him, croaking like toads as they herded females. A joyful pack of wolves romped past, tails wagging. And George got his first look at the insides of a Tibetan bear. A herder had killed it when, starved out of hibernation, it had entered his family's tent; they had kept the gallbladder and fur, but let George examine the carcass and, with a hatchet, take the head. Team solidarity remained strong: Even Gu rose before dawn to light the fire. When others spit bones and George swept them up, he saw that one team member wanted to relieve him of the chore, but by Tibetan custom could not take the broom from his hand.

Now in his seventh year in Tibet, George began bumping into old friends. A herdsman showed him his shrine to the Dalai Lama, adorned with a small box George had given him years earlier. Another brought out a lighter, now empty of fluid; George gave him a new one. In Lhasa, he had become a regular at the Holiday Inn, landing him in Alec Le Sueur's best-selling *Running a Hotel on the Roof of the World*. Their usual spiel for guests who heard rats scrabbling (you are "very fortunate to have witnessed a rare Himalayan Hamster") did not work, Le Sueur groused, when Schaller was around. George's bathroom was popular with the team: One man brought his whole family to bathe there.

As always, he had politics to navigate. Knowing how much he depended on Gu, George toasted him at a banquet full of VIPs. Next summer "he will become the first biologist to study chiru births." Another WWF misfire required his attention: Planning a meeting on Tibet in DC, they had invited both "Dalai Lama people" and young Chinese scientists studying in the US. That foreclosed any candor and put everyone at risk, Schaller told them. And they needed to remove his name.

The most important event for Schaller that winter was happenstance. Chancing upon a patched gray tent, he was welcomed by the three men camping there. They proudly showed him their take: two dozen skinned chiru carcasses, a pile of skull tops with the elegant horns still attached, and the ingenious leghold traps they had fashioned from those horns—using one dead animal to kill the next. When the men said they were from Gertze, a hundred miles away, it clicked. Gertze was where George had seen men plucking wool two years earlier, and where last year his Tibetan colleagues had taken him to the bazaar where that wool was sold. "A Class I protected species!" he'd written in his journal. "Which Gu had seen and never said a word." Now one of the poachers smiled for a picture, which George took along with a trap as "proof for what people in the US don't believe." When he gave the hunter his Milarepa card about laying down arms, his own crew laughed.

The last piece of the puzzle fell into place on his return home in January 1992. A letter was waiting, from California Cashmere president Michael Sautman. An Italian fashion house had asked Sautman if he could supply *shahtoosh*, a wool said to be at once so warm that a pigeon egg wrapped in it would hatch, and so fine that a shawl woven from it could be pulled through a wedding ring. Could Dr. Schaller, he asked, confirm the fiber's origin story? Purveyors told their elite clientele that it came from Tibetan ibex, which shed tufts on thorn bushes that nomadic women then gathered. When Schaller told Sautman that there were few if any ibex in Tibet, let alone bushes, it didn't take long for them to sort out the rest. This finest of all wools came from the lithe animal who had evolved to survive the world's most extreme climate and altitudes. Because it was the chiru's undercoat, it could not be gathered or sheared; its bearer had to be killed.

When Schaller passed this all on to officials, their response surprised him. "The Chinese often try to hide a problem. But they said, 'You have to let this be known internationally, because we can't handle it just ourselves.'" Knowing that the wool was smuggled to Kashmir—the only place with weavers skilled enough to handle the short, delicate fibers—Schaller reached out to his old friend Belinda Wright, who had founded the Wildlife Protection Society of India (WPSI). With her colleague Ashok Kumar, plus forensics labs in France, Italy, and the US, they confirmed the bloody provenance.

Returned to Tibet with Kay that summer, George saw the impact of the new *shahtoosh* craze. (Though it had begun in the sixties, when the jet set discovered India, demand had lately taken off—ironically—as antifur activism became radical chic.) In the Aru Basin he found rutted tire tracks where two years earlier there had been no vehicles at all. He met five families who had given up nomadism and settled permanently to hunt; they got sixty dollars a kilo for the chiru wool, they told him, versus nine dollars for goat cashmere and just a dollar for sheep wool. With those earnings, the families had purchased a truck, cutting their trip to market from weeks (by yak caravan) to hours or days. To keep it in gas, they needed, of course, to keep selling hides.

Far worse was the carnage overseen by officials or their agents. Police loaned high-powered rifles to nomads in exchange for hides. One county leader organized the killing of one thousand chiru a year, in winter when frozen ground made driving easy and the meat did not spoil. When those officials stopped hunting, local families reasonably told George, they would too. Back on the main road, he saw the emerging market infrastructure. A truck driver in a motel parking lot explained that he'd had to shoot the bear whose skin he was hauling when it came into his tent. But early the next morning, George saw the same driver unload three freshly killed chiru. The motel, it appeared, served as middleman for various trades. Team discussions of how to stop poaching invariably settled on this: You had to go after officials and middlemen, beginning with confiscating any truck carrying skins.

Most distressing was George's realization that poachers had located at least one of the chiru's four prime calving grounds. Nothing was easier than killing females giving birth or nursing their newborns—skinning them on the spot to leave their flesh to rot and the orphans to starve. The gangs often hunted at night, their headlights blinding the animals, or slaughtered entire herds by driving through in trucks with logs sticking out the sides. The team had to find the calving grounds, before more poachers did.

It was with that aim that George and Kay, again led by Gu, set out that summer of '92 from Aru Basin to what one of the "British intruders," Captain Henry Deasy, had in 1896 named Antelope Plain. For all the miseries of previous trips, this month outdid the rest, with blizzards and hailstorms repeatedly marooning them as time ran out; the Tibetans in the group celebrated the June solstice by building a snowman. Moving north from steppe into desert, they reached the limit of existence for almost every species. In three thousand square miles they saw just seventy-three wild yak, one lynx, a few pika, and a lone saker falcon. Without a single tree or bush to supply twigs, the birds built their nests of chiru horns.

As he and his human companions struggled, George grew ever more baffled by the chiru. First, at their uncanny ability to elude him as few animals ever had. "Where are they? They must have gone over the ridge. If not, I don't know where to look." And beyond that, a larger mystery. Why would females leave the fine greening steppe in the south to come hundreds of miles to these naked lands—offering little but a dry, sharp-tipped sedge—when they needed nutrition most? Were they after some mineral? Escaping insects, or predators who could not themselves survive here? His strongest hunch was that they were retracing trails set in the last glacial period, ten thousand years earlier, when ice and lakes still covered the basins that now seemed to offer a better maternity ward. Reenacting patterns, that is, that had persisted not as biology, but as culture.

When the cook spotted hundreds, Schaller rejoiced that they were not too late! The northbound females were only now arriving, and close to birthing: At a wolf kill he found a full-term fetus.

But just as quickly, they vanished again. He took some comfort in finding other creatures. From a snow leopard scat he pried five little blue sheep hooves: two lambs. Seeing a gazelle spronking, he suspected she had young and found it, crouched with its neck stretched along the ground.

And then, "a Gu Day." Since they were low on gas, which driving in snow and mud burned quickly, George proposed they scout on foot. Instead, determined for no clear reason to reach a big volcano, and oblivious to the boulder field and sheer escarpment that guarded it, the LEADER took them again and again into the mire. Once, they got stuck so badly that after hours of fruitless pulling, digging, and jacking—the Tibetans working all the while with their feet in ice water—they finally dug a bypass for the river, built a dam, and bailed the resulting pond with dishpans. By that evening they had traveled twenty-seven brutal miles to arrive less than two miles from where they'd camped the night before—in the wrong direction. "I now know how troops feel when led into battle by an incompetent officer." Nearly out of gas, they returned in defeat to the Aru Basin.

Kay now stayed mostly in camp, enjoying their neighbors. A horned-lark nest holding two eggs was often covered by snow but for a small hole; "peering down, I would meet the bright gaze of the lark looking up from her igloo." A jay nest was attended by three adults, an intriguing ménage.

Only when the team spotted the return migration—thousands of females, many with young at heel—did George hurry to get Kay so she could witness the stunning spectacle. Under a brilliant sky, with ravens swooping overhead and wildflowers abloom, sitting close enough to hear the animals grunting, he called out numbers and she recorded. They saw how the journey exhausted the babies, who grabbed a quick suckle or flop in the grass whenever their mother paused for a bite. It all felt glorious but also tragic: Just decades earlier there would have been ten times more.

There was much to celebrate at the post-trip banquet, a feast of camel footpads and yak tendon. That month, at the first Earth Summit in Rio de Janeiro, China had presented the Chang Tang reserve to the world. Expanded as George had advised to include the Aru Basin, it became the second-largest protected area on any continent. Only Northeast Greenland National Park is bigger and, as George liked to point out, most of that is ice.

Still, with the demand for *shahtoosh* unabated, it was not enough. When, the following summer, *National Geographic* published twenty-five pages of Schaller's text and photos, he used the opportunity to toss the first of several bombs into the midst of the fashionistas still driving the slaughter. A dozen gorgeous pages swept readers through the romance of this "high and sacred realm," smack into the pictures that followed: of mountains of mangled chiru bodies.

•••

To the end of his life, George would never quit seeking the "desolation . . . that suits my inner landscape." But here on the Tibetan Plateau, his flights into the "boundless emptiness" were becoming ever rarer, as he completed the essential turn in his life. The expanded George, as Jonah called him, no longer sought above all else solitary communion with animals. Instead, his deepest commitments increasingly centered on humans. On humans, because animals live or die at our will. On humans, because as animals ourselves we are utterly entangled and interdependent with the other beings that live inside and around us. On humans, because decade after decade, continent beyond continent, he had experienced the transformation effected by what Lewis Hyde calls "the Gift." Since childhood, George had known how readily our species will take up the pitchfork to drive out or annihilate the other. But far more often, in what he calls his "vagrant" life, he had been carried along by the currents of generosity and hospitality that still flow in these worlds where no chance meeting is complete without offerings all around—of moose meat or mustard greens from the garden, a can of Red Bull or spare socks, a place at the table or inside the tent to sleep.

A first step in supporting nomads of every species was to understand more fully the fast-changing rangeland ecology. George got early help on that task from two new colleagues: a former *Xinhua* reporter named Ding Li, ace at busting through bureaucratic logjams, and Kathmandu-based rangeland expert Dan Miller, who not only spoke Tibetan but had the good sense to bring chocolate, pancake mix, and maple syrup to satisfy the Schaller sweet tooth. (Their Tibetan drivers refused even to try the strange fare.) Schaller and Miller shared many affinities. Finding a chiru kill, Dan milked the lactating corpse for a sample. He shared local fables: When three cow sisters needed salt, one went to Tibet wearing her sister's hair for warmth and became a yak. The bare one went to the warm plains and became a water buffalo.

For millennia, humans and wildlife had shared the Chang Tang without threat to the other's existence. The wild herds had easily absorbed subsistence hunting—whether with Neolithic spears or muzzle loaders. The nomads' animals, kept moving, had helped the wild grazers sustain the health of the grasslands, which need both disturbance and rest.

That harmony was now crumbling as China rushed to modernize. Every year George found new roads and settlements, built to administer this "autonomous" region, tap its oil, and open markets. To increase livestock production, officials had begun pushing herders into the barren north, until then the province only of wild animals. More radically, they began shutting down their traditional ways. "Governments don't like people that move," Schaller explained. "They can't control them, don't know what they're up to." So this Communist government privatized the steppe: cutting it up into household plots, subsidizing fences, and settling families into mudbrick huts. Transforming nomads, that is, into "small-time ranchers." George saw those changes unfold in the life of his friend Pujung Narla. Living in a tent and moving with the seasons when George met him in 1991, he had on every visit something new to show: a three-room hut, a motorcycle, a truck. His modern guns, steel traps, and a bale of chiru hides made clear how he had financed it all.

George was himself a beneficiary of the comforts many families now acquired. One hospitable herder, whose new wealth included a generator, invited George to sleep next to his fine new stove, piling the family into other rooms. But others were faring less well. With livestock now confined to the same small pastures month after month, year after year, herders struggled not to overgraze. Those whose land played out were relocated into towns, to become, as one told George, "empty sacks that can't stand up by themselves." Penned livestock became easy pickings for predators, as did permanent homes. Having rarely seen bears when they carried their lodgings with them, families now often returned from summer pastures to find their huts ransacked.

If modernization was a mixed blessing for humans, the new roads, ranchettes, guns, and markets were for wildlife an unmitigated disaster. Since the 1949 revolution, animals of all varieties had faced periodic assaults. First came Mao's campaign to exterminate any creature, from sparrows to wolves, seen to menace crops, livestock, or people. Then came the famine precipitated by his Great Leap Forward, which turned everything that breathed into a desperate meal. By the 1990s, hunting from cars with high-powered rifles had become so commonplace that chiru fled from a vehicle a mile away. Visiting the Tibetan Medical College, George saw newly affluent herders bringing animal parts by motorcycle or truck to trade for treatment: yak heart for blood diseases, snow leopard bones for depression. Poverty, that is, brought one kind of pressure—for money and meat to survive; wealth another—for skins, exotic medicines, and trophies. Either way, the wildlife lost.

Beyond overhunting were the threats posed by rising livestock numbers. Seeing a herd of wild yak without young, Schaller suspected they had contracted abortion-inducing brucellosis from domestic animals. Proximity also enabled hybridization, costly all around: Half-wild calves could be dangerous for families to handle, and interbreeding threatened the wild gene pool. Wild grazers faced ever more competition: Where the grass was best, Schaller and Miller found twenty-three domestic animals to every wild one. Without rest, the land deteriorated; grasses gave way to mats of *Kobresia* (bog sedge), then slid off hillsides in chunks. Tolerance vanished;

herders who had lived forever alongside wild grass eaters and predators now wanted both shot. Fences delivered still more fatal blows, obstructing migrations and entangling individuals to suffer slow, agonizing deaths. As Jonah Western had witnessed in East Africa, land fragmentation—"capitalism foisted on traditional communities who had owned the land collectively"—hits wildlife harder than almost anything else.

With all these changes it hardly seemed to matter that these animals, and the Chang Tang, were protected. In 1990 Schaller found 681 yaks in the Aru Basin; a decade later biologists found just 70. He captured this larger picture in the final pages of *The Last Panda*, tallying (as he could, he would later say, for almost any country on Earth) the Chinese animals driven near or to extinction by overhunting and habitat destruction: rhinos, elephants, tigers, civet cat, cranes, gibbons, wild horses, saltwater crocodiles, freshwater dolphins, pangolins, golden monkeys.

Whether the Buddhist reverence for life could counter these forces, even with the religion officially suppressed, was again on George's mind. He had recently met Terris Temple and Leslie Nguyen, a couple at work on a 108-foot-tall silk *thangka* to replace sacred textiles destroyed during the Cultural Revolution. This one would hang at Tsurphu monastery, the twelfth-century seat of one of the four major schools of Tibetan Buddhism. George urged them to include imagery of Tibetan animals; they in turn invited him to meet Tsurphu's spiritual leader, recently reincarnated as a nine-year-old boy. Enveloped by chanting, cymbals, and drums, George stood on line with other pilgrims for the seventeenth Karmapa's blessing. The handsome boy, sitting cross-legged on a raised dais in a golden jacket—his mother in herder's clothes prostrated nearby—touched George's bowed head. Later granted a private audience, George knelt to touch his forehead to the ground, then presented his *National Geographic* article on the Chang Tang. The young Karmapa pored over it, fascinated, while the Rinpoche described for George the protections they offered blue sheep and musk deer.

The appearance of a Schaller totem underlined the complexity of this evolving interplay between people and wild beings. Driving back to Lhasa, George watched a raven surf the updrafts created by the new elevated road, effortlessly matching their pace. But stopping in one of the new cement towns, he found a second raven being stoned by Tibetan boys. He lifted it onto a wall, but it flopped off and the boys went after it.

Though he was now in his sixties, George was traveling more than ever. With his commitments in Asia expanded to include the countries that border China, plus frequent returns to India, Tanzania, Rwanda, Kenya, Uganda, Congo,* Chile, Brazil, and Argentina, he would be away six to eight months of almost every one of the next twenty years.

Kay occasionally still went along. Far more often, her wobbly health kept her home, as did her near-full-time job as George's stand-in. She fielded calls from colleagues—speaking with Howard Quigley about Qiu Mingjiang's future, with Craig Packer about "our lions," with Rodney Jackson about Tibetan red deer, and with Lü Zhi (at the National Institutes of Health on a postdoc) about life in general. She accepted accolades on George's behalf. PETA cofounder Alex Pacheco called to voice his fervent admiration—a surprise, given the usual divide between animal rights advocates and conservationists.† Novelist Michael Thomas wrote to say that his dinner with Peter Matthiessen had "consisted almost entirely of encomia directed at the head of GS."‡ Eric called excited to find his dad in the latest *Jurassic Park*. A certificate arrived from Webster Groves High announcing George's enshrinement on the alumni Wall of Fame.

Kay fact-checked articles by or about George, ran down photos and cap-

* The country's name changed several times: from the Republic of the Congo to Zaire to the Democratic Republic of the Congo.

† Animal rights champions focus on individual and domestic animals, conservationists on the survival of wild species.

‡ Also present was Farrar, Straus and Giroux editor Elisabeth Sifton, who had read all of George's books and had "always wanted to publish you. . . . I told her you were off to Laos and Tibet for the nonce, but that on your return I'd fix up lunch."

tions ("You have no idea how time-consuming it is for me to research things you'd have at your fingertips"), edited manuscripts. "I've not enjoyed the Tibet typing; too many symbols that take 7 keys to bring up! And I just had to read 15 pages in the manual on how to do tables. I am annoyed at how badly you've scribbled. Your sons and I agree it's time you learn to use the computer!" When her spirits sagged, she added a doomy coda. "The report will be on the Quicken disk in case I don't survive the Anchorage trip." Or "I'm telling you what we owe the contractor in case I totally lose my mind." Once, a journalist asked to interview her for a book on women's careers, but she demurred, insisting that she was only an assistant to her husband's career.

His family also became hers to manage. Bettina often called late at night—to say she'd received George's birthday greetings ("You are just too good," Kay teased), to complain, forty years on, about Uncle Talcott and Aunt Paula, or to fret over Chris. Kay suggested George call his brother and "throw in a comment about the difficulties of getting back to home routines and independent wives." In advance of an Ed and Bill visit, she spent six hours making beds, ironing linens, vacuuming (*behind* dressers), scrubbing tubs, picking bouquets, doing her hair, and simmering chicken for soup. "Ed is recommending silicone implants for my derriere."

Winter brought storms for her to dig out of, and spring her own animal adventures. "Our nestling robin has disappeared; still pin-feathered, so maybe a predator. Damnation. Must be 2 weeks so maybe old enough to leave? I will try to think so." A huge gray squirrel moved into the stovepipe, growling at Kay every time she tried to "encourage" him out. She finally put a pinch of sleeping powder in some peanut butter and left both the stove and outside doors open. She'd brought in a live trap but couldn't figure out how to make the stove door small enough to funnel him in. "I need more hands! And have missed the whole basketball game today because of squirrel arrangements."

Mourning Michael Jordan's retirement ("you had better be kind") she had thrown herself a solitary birthday party: a margarita and tandoori chicken. She was "flaccid" with flu, though not too flaccid for sarcasm. "How I miss your nurturing and sympathy at such times." She had given her sister the skirt she was wearing "the night we met Robert Redford, who was so gracious to you, and you shared a urinal with Dick Cavett. I told her to suit her recollection of its history to the occasion."

Kay did get back to Tibet in the summer of 1994 for their least troubled trip yet. They met concerned officials: One vice-governor volunteered livestock numbers; another had shut down fence building. They felt uncommonly appreciated: A colleague told George that his ability to get leaders to listen was "worth forty vehicles" and that his "reputation for being difficult" was undeserved. They met locals who would not show them animals if they were there to hunt. One old lady came to their tent to search out any guns before she would let her son guide them.

Antipoaching patrols were no longer just ignored or bought off. Coming upon a youth with a freshly skinned chiru, the team spotted his two brothers across the gully, hurriedly packing a tent. The three had killed a wild yak—it took forty shots—which carried a twenty-thousand-yuan fine. They also had seven chiru hides, two heads tied together by the nostrils, and three fetuses laid in a line. The forest officer photographed the young men and their contraband, confiscated rifles and cartridge belts, and sent them to Shuanghu to confess their crime and be fined (not jailed, said Ding Li, which might attract the scrutiny of human rights organizations). "Our people are excited," George wrote. "Men like to hunt men."

Live animals brought still more excitement, especially the discovery of several hundred female and yearling chiru traveling a previously undiscovered route. Kay again stayed mostly in camp, with the cook or policeman, eager at day's end to show George her own finds. The saker falcon that had flown in with a broken leg, she guessed, had likely missed a stoop at prey and hit the ground.

Still George was grateful when after two tremendous days out—spotting argali, kiang, bear, and a wall of mani stones where Buddhist herders had hidden during the Cultural Revolution—the driver volunteered to go fetch Kay; George walked out to meet the returning car. Together they watched a wolf kill (George saw only a burst of dust, but Kay saw the jaws lock on the chiru throat) and a bear digging holes to bob for pika, swallowing them whole. A white wolf followed a few feet behind the bear, to pounce on any that escaped. Behind the wolf came a courting male bear, a strange procession.

George's journals sometimes read like a serialized adventure. This one ends with a month-old snow leopard cub in peril. An American had chanced upon Tibetans in a medicine shop trying to sell the cub for its bones. The American bought him, then passed him on to a couple who promised to smuggle him to Darjeeling for rewilding. The couple in turn tracked down George for advice. Though impressed by the excellent care they had provided—even rubbing the baby's full belly to simulate his mother's tongue—he told them it could never return to the wild, and that no reputable zoo would take an endangered animal from a smuggler. It must be turned over to the Forest Bureau.

George and Kay offered to deliver the cub to the Beijing Zoo, shielding the couple with a cover story. Taking turns on the flight holding the makeshift carrier in their laps, giving the baby a finger to suck and bite, they lulled him to sleep. But they got a distressing surprise on arrival, when officials insisted it go to a facility that had no experience with cats. Haunted by panda memories, George faxed the zoo for guidance on inoculations and diet. But all went ignored. Paying a visit, they found the cub locked into a plastic crate and soaked in his own urine. Kay begged the keepers to "clean and feed and love him like a human baby"—giving him toys, something to chew on, space to move—though she spoke gently, lest they lose access. Taken for the usual fancy lunch, they wondered if they should have insisted on feeding and cleaning him, even if it meant others lost face. George alerted an ABC reporter in hopes his knowing about the cub might help protect it. Though the little snow leopard disappeared from George's journals, he later recalled that it wound up at the Xining zoo.

Returned to the US, George found more of the old sort of frustration waiting, beginning with the latest Forest Bureau proposal. "Do they think foundations are stupid?" he wrote Ding Li. George had proposed $150,000 for a survey of Eastern Himalayan endangered species; the Chinese wanted $518,000. "The budget must be moderately close to actual costs [and make] clear that the work, not the money, is the principal goal." Ding Li should convey that the funds would go elsewhere if forestry wasn't reasonable, but "informally, as a friend. Please do not put it in official letters." To the funder (MacArthur) George offered a reality check—two Toyotas cost $70,000, not twice that—and advised they begin with a start-up grant.

The other frustration originated closer to home. William Bleisch, part of George's global conservation team at WCS, had written to explain why he'd been detained in southeast Tibet. Bleisch had been working in China since 1987, on snub-nose monkeys. But when the vice-director of agriculture said she would need to do "significant research" before granting a new permit, he had blown off what he took to be a demand for a bribe and gone into the forest anyway. Escorted back out by armed Forest Bureau Police, he had pushed further, angrily calling their boss a *hutuchong* (a confused insect—sometimes knowing the language is not an asset). "In hindsight, we walked into a bitter dispute over who controls the lucrative rights to research permits, but it is my responsibility if there is fall-out." George thanked him for his detailed report, then gave him a dressing-down all the more blistering for its self-control.

> *By breaking several regulations in a politically sensitive region, you caused problems for [our Forestry liaison], which I regret to say will in all probability seriously hinder WCS's future involvement there. Certainly I wish you had followed my express instruction not to enter Tibet without official permission. . . . I can only ask that under no*

> *circumstances should you send a letter of complaint as you threaten to do. Such a formal act of confrontation would make matters worse, reaching various political channels. Please follow this order. WCS will salvage what it can.**

With George gone again by December, Kay contented herself with the latest Rolex ad: "It's nice to have such a large photo of you to improve my day." At his request, she sent the menu for the Christmas he would miss: ham, nut bread, spiced peaches, and green beans. She loved his gift, a brooch made by Paulette, now Eric's wife, from a Paleolithic scraper; joked that she was "glad you're not here to need, but refuse, my nurturing," and panned Disney's *Lion King*: "It makes the hyenas the bad guys . . . the lion cub grows up into a wimp, and the warthog is an idiot."

But as the months of loneliness multiplied, her letters grew more melancholy.

She had heartbreaking news: Mark's marriage was ending. "So terribly sad for them both. He will be feeling guilty for hurting her, and unsuccessful as a husband." She was glad he had his adored brother to lean on, "though Eric says they stay away from personal stuff, which keeps their relationship better, he thinks." Kay's own sense of helplessness was wrecking her sleep. "I keep dreaming of Mark as a child, when it was easier to comfort him. I need to feel I can."

There was more pain to relay. Photojournalist Nick Nichols had called to report that the Rwandan genocide had not spared the gorillas: Three had been shot through the heart, and so many Hutu were camped "where we used to start our Kabara climb" that the forest was gone. Closer to home, their friend Jack Selsor was shattered. "We lost Ken by his own hand. They'd been together 40 years and made all those wonderful trips. They

* The damage was repaired; a few years later, Bleisch served as resident foreign expert in China's Forest Bureau.

loved your saying you'd like to have gone somewhere they'd not been first. Illness and old age are just not worth it."

She had weed whipped again—"SUCH A GOOD WOMAN"—but had no heart for gardening. "I'm not in a sunshine and flower mood." In the thick July heat, she longed to be with him in the cool, thin air of Tibet. Her head, neck, and wrists ached, her back pills made her groggy, and she was "over-napping. . . . You will have to accept the more and more noticeable fact that I AM DISINTEGRATING IN ALL MUSCLES AND JOINTS, which is no doubt one of the problems with my BRAIN. I could really use you here at such times."

George, meanwhile, was being allowed into ever-more-restricted parts of China, but at a price.

Reunited after a decade with panda protégé Wang Xiaoming, who after four years in France peppered every sentence with a *merde* or an *ooh la la*, George made his first extended trip into the forested gorges in Tibet's southeast corner, just north of the Indian border. This was forbidden territory for foreigners. He could take no pictures, had to wait twenty kilometers from the border while his colleagues went on without him, was accompanied at every turn by police—whose uniforms scared off informants—and had to pay $30,000 for the "costs" of the monthlong trip.

A year later, in 1996, he and Wang were allowed into the equally tense Inner Mongolia. Beyond a few camels and kulan (the local wild ass) they saw no animals; hunting had been unrestricted here since 1960. Unused to foreigners, his local companions masked their discomfort with a familiar heartiness and fussing over him he found exhausting. People on bicycles swiveled to look, laughing and shouting. Police again shadowed him, even sleeping in the next room. He longed to walk in the forest but it was "always teatime or not permission or something." Covering three thousand miles on bad roads jarred his damaged back to the point of nausea; he often snapped at drivers for going too fast.

Again barred from border zones, he spent many days waiting while the others went on—carried as ever through such doldrums by the wonder of small things. He noticed that the cicadas here sang like reedy birds. In these quiet realms, they could be heard without their rainforest cousins' screeching. Finding stone chips, he imagined the people who had camped here thousands of years earlier, when there was water and no dunes. A mob of European cuckoos on the banks of Ulansuhai Lake carried him into the nearer past, as did the black-speckled eggs he found in the little houseboat-nests the terns had built atop floating plants. He chilled watermelon in an icy rivulet and got sticky eating it. "A flower's scent," he scribbled, "the mind reaches for the past like a hunger, but grasps only fragments that slip away."

A darker echo of his boyhood came that night, another first for him in China. Awakened by a late knock at his door, he sat mute while for two hours three secret police grilled Wang about their trip: where they slept each night, who they spoke to. George had managed to quickly stash most of his exposed film, but soon after the first trio left, two more policemen came, demanding it all. George blustered—"I've worked in China sixteen years. You can't confiscate the film of an invited foreign guest"—until his colleague pleaded with him "to help me as a friend" by handing it over. The police returned the next day to say they had printed the film and had "problems. . . . One, of mountains, could have a military installation." Grilled again until midnight, Schaller asked each "inquisitor" to write down his name. Still they weren't finished: At the airport, nine security people surrounded him, searched his body and luggage, went page by page through his field notes, and confiscated the last four rolls he'd tucked away.

The whole thing felt surreal, especially when a ninety-minute flight carried him back to the future in Beijing. He was overjoyed to see Pan Wenshi, Lü Zhi, Ding Li, and several sharp young PhD students. "It feels good that one can openly make friends here now. A new generation."

Schaller has won all of the highest honors in conservation, including the Order of the Golden Ark, the Explorers Club Medal, National Geographic's Hubbard Medal, WWF's Gold Medal, and the Indianapolis Prize. The accompanying awards—hundreds of thousands of dollars—he has mostly passed on as "field grants" to young scientists working in their home countries, often pressing cash into a hand, Uncle Talcott–style.

He rarely mentioned any of this in his letters or published works. But when in October 1996 he was awarded Japan's International Cosmos Prize, the audience he and Kay had with the crown prince and princess affected them both. Kay was stirred by the young princess's (much-publicized) depression. George's connection ripened a few months later, when on returning alone to Tokyo he was asked again to visit with the young couple—for ninety minutes, just the three of them—and then with the emperor himself. The two had met briefly in 1983 when, still a prince, Akihito had come to his lecture on pandas. Now ushered into the royal quarters, George found the slight, graying monarch alone. For an hour they discussed the emperor's own scientific work and the lost mammals George had recently turned up in Laos.

Kay had carried the medal home with her, writing George that "I've just gone to look at it again, and am newly impressed." She had also heard from everyone she knew, and many she didn't, when *Parade* magazine—then with a circulation of thirty-three million—ran a three-page story on "the man who has seen more of nature than anyone else alive." Never one to miss an opportunity, George had used the story to skewer American sanctimony. "We go all over the world like missionaries saying 'save this, save that.' Then people say, 'what about your rainforests in the Northwest? You're not only cutting those down, but doing so with taxpayer subsidies."

As yet another year turned, Kay resumed her chronicles for faraway George. She was doing footnotes for his article on muntjac, answering letters to him from schoolchildren, talking with Qiu—then in New York's Chinatown investigating the diaspora's medicinal use of endangered species—and managing the final stages of George's first book on the region, and the first to include people in the title, the stunning *Tibet's Hidden Wilderness: Wildlife and Nomads of the Chang Tang Reserve.* Regarding the last, he asked her to please make sure the Dalai Lama quote was not too prominent. She was doing their taxes, getting the car serviced and the tub re-porcelained, dealing with renters at the Vermont farm, and, after twenty-five years, patching it up to use themselves. "Maybe you should come home? We could use your love of weeding and mopping and waxing, and, and . . ."

Her animal adventures continued, some glorious ("the spruce is absolutely decorated with gold finches"), most not. She had greased the bird-feeder pole but had to livetrap the persistent squirrels, then drive them across town to release, "though only when the weather is nice since they won't know where to find shelter." She was wrangling the mice overtaking his writing shed, or "hovel" as she called it. But should she lure in the baby woodchuck with cabbage or peas? Another beast growled in the chimney. "If I'm bitten, I will deem it your fault for not being here to rescue me. But you would open the draft and then we'd have it in the house." It escaped as it was, thumping like a telltale heart along the upstairs hall. She shut it into Eric's room, opened the window, propped up a pine branch for it to climb up and out, and put out water "in case it was dehydrated." After four "nerve-racking [days] all but locked in with the frantic, fierce creature," it was finally gone.

The boys were fine. Mark was moving in with his girlfriend, Quincy, which "we talked about from our differing generational point of views." Eric sometimes added a cartoon; in one, a birdlike creature as skinny and beaky as his dad looks glum. "All he got for Christmas was a magazine he'd already read and a candy cane someone had sucked all the stripes off of." Kay still lobbed the occasional tease: "I like having even these one-sided conversations with you not listening; at least there's the excuse you're not here." She was reading a biography of two brothers; one who talked to entertain; the other to inform. "And the second was far more censorious than the first. Ahem."

But the shadows were lengthening. "I wonder if you will even get this far in this letter? 17 pages so far, since I fear I'll have an accident and you won't know what needs doing." Her body continued betraying her. After a storm brought down the phone and power lines, she had gone out at midnight to shovel before all turned to ice, slipping and banging up her spine. Now her back was in spasm and her arm numb, "which I assure you is more of a nuisance for me than it is for you to hear about." She had tennis elbow from "rototilling, banging stakes, attacking poison ivy, etc. etc. <u>It is all your fault for not being here to do those things for me</u>." Squatting to reach a low file, she had lost her balance and sprained her wrist. "It's irritating to be so easily damaged. But will you please not leave again? All hell breaks out and that is not nice during the Merry Holiday." Though she had always been gamine, she was now dieting, and complaining of not losing weight. Kay was happy, George had told *Parade*, "to lead this footloose life." But it seemed she had landed, after all, in the ordinary life she'd once been exhilarated to escape.

Tucked into the materials Kay collated for the Yale archive is an essay dated April 5, 1997.

> *Dreams, thoughts on a depressing day*
>
> *My always-dream was to marry a man who loved the woods, and to travel and work with him . . . to know one person so intimately you become almost one . . . the times you share in work and family adding to that intimacy. . . . For much of my life, to allow my husband to follow his dreams meant following my own as well. But as the need for him to go without me became a more frequent pattern, I now live with regret. . . . All my dreams seem to have been unimportant, ephemeral, fleeting wishes that meant little to anyone but myself. I have given my whole adult life up to others' dreams, and now there's nothing left of mine, unless I again become a greater partner for my husband in the field, which seems less important to him than his being there wherever, whenever, with whomever. All of this has become more important than family, than spouse, than*

the basic needs we have for him. How ironic to marry someone whose dreams seem to match so well with your own and then see them carry him away. I rue it and feel it was not that worthwhile if it ended by separating the family. [Even] my boys learned to do without much family time, because they were so often without a part of theirs.

A week later, George came home. Midway through his six-week stay, they traveled to the University of Southern California, where alongside Jane Goodall and orangutan researcher Birutė Galdikas he was awarded the Tyler Prize. "I guess I'm a stand-in for Fossey" he'd written Kay, referring to the three women Louis Leakey called his trimates. "Please come with me for moral support, etc."

That "etc." seems to have contained a great deal. On his return to Tibet on May 24, Kay wrote nearly every day. "It's amazing how terribly I miss you. This being back as in old times is harder somehow but so wonderful." Two days later: "I feel so lighthearted to have had the fax from you." And a day after that: "You just called from Lhasa. I could hardly focus on the Bulls after so much euphoria and comfort again. You have been wonderful about staying in contact and I've been serene and well, far better than either of us expected, due to your compassion and kindness before leaving and since. It is so reassuring to be connected in spite of the separation. A lovely sense of warmth and love." On May 30, having done all his laundry except a jacket she wore for the scent of him, she was "thinking of you in our favorite odiferous Nagqu hotel. I showered and thought how nice to have hot water rather than a glacial stream, the only positive part I can think of being here instead of in the field." On June 3, "this new (the old) me finds your absence difficult . . . but better to feel it so much. No letter but it is rather soon."

She would in fact not hear from George until mid-July, when he finally made it back to Lhasa after the first real peril he'd faced in some years. With Liu Wulin of the Tibet Forestry Bureau and Lü Zhi, by then leading the species program at WWF-China, he would again try to follow chiru, this time to their Kokoxili calving ground. The trip began hopefully enough. To bring good weather, the driver Lo Te used George's saw to cut and then burn a slice of argali horn, choosing a reddish piece signifying the sun. The salt rim around Xijir Ulan Lake brought to mind Kay's favorite cocktail. And on June 12, they spotted four hundred pregnant chiru headed north through a sea of purple iris blooming in snow.

Soon, however, the weather turned—worse even than in the summer of '92—condemning them to hard labor, pummeled by hail as they dug vehicles out of wet sand or mud, once eight times in a mile. Like Gu, Liu was indecisive or cowed by well-connected drivers, who in turn feared meeting illegal gold miners, with whom there had been deadly conflicts. They made it to Hoh Xil Lake at sixteen thousand feet but it snowed and snowed and they could not go on, though George knew the time of birthing was near:

One female who circled rather than flee led him to her hidden three-day-old. Like spirit creatures luring them on, the chiru appeared by the hundreds, then disappeared.

It becomes hard not to see, in some of these visitations, animals restored to their ancient role as avatars. A house swallow, all alone and obviously tired, landed right next to George. Then on she flew. A solitary kiang approached to inspect a fellow isolate. This "feral biologist," as he called himself,* was tracked by a feral dog. "When I last saw her, she sat at the edge of town and howled."

At sixty-four, George's susceptibility to mortal afflictions—so rare when his blood-filled sneaker had heartened Matthiessen twenty years earlier—was also showing up more often. In one of the odder instances, George drank hot Tang and fainted. "When you lose the joy of discomfort," he had told *Parade*, "you lose the joy of work."

The travails only worsened. On June 28, they beat their record, getting stuck eleven times and moving less than a kilometer. After three weeks of this, George wondered how many more days the drivers could work this hard. Certainly not for the hundreds of kilometers still ahead, and with food stores so low they had begun rationing. They would have to leave the trucks and take the Toyotas to get help. Liu said George would have to pay $20,000—for rental of the trucks and guards to watch them. George said no, he'd just buy one of the trucks, though he would pay the policeman, an exceptionally hard worker, to stay until help came. For the man's little daughter, he added the gift of a watch.

Setting out to find rescue, George told Lü Zhi to save her crackers for the lean days ahead. They considered hunting but, hungry as they were, would not kill protected species. That left only marmots, pika, and hamsters, which the others feared carried plague. To rally everyone's spirits, George offered a wager on whether the river would rise. Winning the bet when it fell, their Tibetan colleagues were emboldened to drive through;

* George uses "feral" precisely, to mean a domesticated animal that has reverted to its wild state.

the rest waded, carrying the gear. From the headwaters of the Yellow River they crossed into the Yangtze basin, at which point the drivers abruptly announced that they must walk the remaining eighty-seven miles to the road. None were at full strength. Liu and Lü both had heart issues; George could no longer carry heavy loads. And all had only Chinese army biscuits left to eat. When in ninety minutes they had gone less than two miles and were stumbling, Liu began to cry, calling it all his fault.

Fortunately one of the locals on the team had traded wool in the area and knew that a day's walk to the southeast would bring them to a herders' camp. Sure enough, reaching a crest at 7:30 p.m., they saw far across the plain two white dots. Nomads' tents. In the morning, while the others went to bargain for yaks, George recorded the flowers in bloom—pale yellow *Thermopsis*, tiny *Arenaria*, deep purple *Oxytropis pauciflora*. Investigating ruddy shelduck nests he found in one only an egg lining and bit of down. In the other were two cold eggs: one destroyed; the other pipped but with the chick dead inside. Why had it been abandoned?

Their hosts fed the famished travelers generously, piling on tsampa and flatbread, fresh butter, yogurt, and roasted meat. They then loaded the yaks, who—unaccustomed to carrying—promptly bucked everything off, including Liu, who after landing on his head complained of double vision and roaring ears. After a long hike to the road and a twenty-hour bus ride, they arrived in Lhasa to find the truckers' wives mourning husbands they were sure were dead. Schaller did pay $20,000 for the military to retrieve the vehicles and men, getting a truck out of the deal. He later learned that Liu had cheated him, billing him for rent and guards long after the trucks were out. That particularly stung since George had given Liu all of his chiru data to publish under his own name.

Pasted into this journal are fragments from Kay's letters. "Remember me when you see Orion. . . . I will look at the stars each night and tell them I love you." On July 14, he could finally write back. "Dearest Kay, I'm so sorry for the long silence. An awful trip. You're lucky not to have been along."

A month later, George was able to bring Kay on a (far more coddled) donor trip to see WCS work in the Russian Far East. Traveling via Anchorage to Petropavlovsk and from there by helicopter, they celebrated their fortieth wedding anniversary in the company of Kamchatka brown bears, cousins to the grizzlies that had chaperoned their first meetings. One stood, nine feet tall, to bat a gull feather like a balloon. Another, upstream from where Kay fished, grabbed each passing salmon behind the head and neatly stripped out a choice filet. A little one, distracted by cloudberries, looked up to find George just fifteen feet away. Displaying fine cross-species etiquette, each stepped sideways in opposite directions.

On their way back to Connecticut, George and Kay visited Mark and Quincy, Chris and Diane, and Bettina.* George was home for Thanksgiving, for Christmas, and for New Year's Eve.

As soon as he left again, however—in February 1998, for Sichuan and Tibet—Kay's spirits plummeted. She planned to visit a friend in Australia, "sure you'd be glad to have me do something to make myself happy." But she felt "more than usually bereft and also frightened. I dreamed this morning you'd returned and awakened me reaching down to hold me. It was wonderful to feel your arms even if imaginary." Sunk still further by the news that an Alaska friend had dementia, "I just went to bed and cried till I fell asleep. I am so bitterly lonely and frightened." And always, the dark coda: "If I crash my car or something . . ."

George was nearly as miserable. Near Batang, he saw locals dynamiting fish. Near Yushu, he visited a monastery said to have three snow leopard cubs. Two were dead and the third was gaunt and catatonic. Mastiffs lunging at their chains and "dull-eyed" monks lent the whole place an "evil aura."

His grim mood did serve him in negotiations. When he told the Forest

* After the death of her second husband in 1971, Bettina had moved to Eureka Springs, Arkansas, to care for her father, then a decade later moved again to northern California to be near Chris. When she died in 2006, at the age of ninety-six, the family requested that memorial gifts go to WCS.

Bureau that if they kept upending plans and demanding exorbitant funds, he would return the MacArthur Foundation's money and quit China for good, "I meant it all. My life's work is winding down. I feel adrift in so many ways." An epigraph in his 1998 monograph, *Wildlife of the Tibetan Steppe*, captured his malaise. "Hard is the Journey / Hard is the Journey / So many turnings / And now where am I?"

George was never so defeated as to turn the world gray. Approaching a spruce that seemed aglow with Christmas lights, he delighted to find a dozen bright white pheasants roosting. Chasing butterflies, but clumsy with the net, he enlisted the help of an eight-year-old boy; turning over a rock together, they found an earwig with tiny pearl eggs. Few things ever revived George more reliably than children, who haven't yet forgotten that animals can talk and be friends.

And in this ever-contradictory country, he still saw progress. He met a general who had ordered his troops to stop shooting wildlife and villagers who wanted to hear about the communities he'd visited in Nepal who were restoring their forests outside the national parks to bring tigers and rhinos back—so *they* could reap the tourism dollars. Qiu's growing expertise especially pleased him: Warming frozen musk deer droppings with his breath, Qiu knew from the odor that they were males.

But it was in his lost state that George wrote a letter the contents of which can only be guessed at from Kay's response. Her hands had trembled as she opened it. "I get frightened hearing from you, and frightened when I don't." But she had then reread it many times.

> *It is a good thing to learn more about yourself, dear heart, even if it is to acknowledge a weakness. But it really is your only character flaw. And just as I have been disillusioned in you (and myself), you must forgive yourself as I now have, and weigh it against all the positives. You have a unique strength and determination. There must*

be a time you relax that strength. But loving you through that flaw has in a way strengthened our bond. I do not want you or me to dwell on it. Learn from it. Acknowledgement is the first step. Move to the promising future in which your whole self counts. Not only your professional excellence.

You like to say that nothing matters in your work but what comes next. Do not fear the future—you will adjust and contribute as always. You have long been more than a field researcher. I take such pride in what you have done for our earth.

It is strange to hear Kay without her usual vinegar—stranger still to hear a hint of the self-help shelf. But neither lasted. It was unclear, she next wrote, whether Cousin Ed had made it to his daughter's high school graduation.

Being a Barnes he's no doubt protecting himself from anything emotional, going elsewhere to please himself, putting family needs at the bottom of his list. Seems to be a family failing, with Alice, Bettina, etc. . . . I am trying to handle everything here but finding it more and more difficult to say the least. Too many peoples' personal problems as well as all the necessary business and two homes [and] WCS junk and photos and such. . . . It would be such a relief if you didn't have Mongolia on your list. 5+ months seems unending and not reasonable from any viewpoint, much less peace for you at the expense of everything here. Try to at least feel guilty each day?

To the typed letter, she added a scrawled addendum: "It really is urgent. I'm not managing at all."

George was by then back in southeast Tibet for a 250-mile walk along the Yarlung Tsangpo, the river called Brahmaputra when it flows into India and

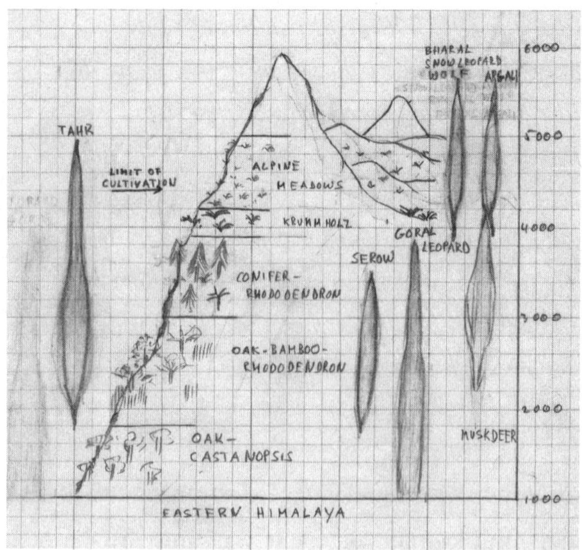

Schaller's homage to the revolutionary Tableau Physique drawn by Alexander von Humboldt in the Andes c. 1805.

Bangladesh. Flanked by 25,446-foot Namcha Barwa and 23,461-foot Gyala Peri, this is the planet's longest and deepest gorge; in a day George could walk from low rainforest to alpine glaciers, as if from the equator to the Arctic. Lü Zhi was again with him, against Liu Wulin's objections: He blamed her for alerting George to his truck-rental scam the previous year and saw travel to this sacred region as "man's work."

Beyond a bit of birdwatching—without binoculars for George, who saw how tense he was making local officials—he and Lü Zhi spent no time looking for animals. Instead they visited several of the hundred tiny villages scattered through the gorge, inhabited by the Indigenous Lopas* or more recently arrived Monpas and Tibetans. From Lü Zhi's interviews, they learned that life was growing harder for everyone. With fifteen thousand people now farming the steep terrain, soil erosion and landslides had become commonplace. Little game was left to eat; musk deer, takin, red goral, Assamese macaque, and black bear had all been hunted out. Tibet's last twenty tigers still lived here, but with wild prey gone they killed livestock and were reviled; by 2009, the last of them would be dead. The head-

* Lopas (sometimes Lhobas) and Monpas are the Tibetan names for what are in fact many different tribes inhabiting the southern mountains.

man of Gyala village spoke for many: He wanted his five children to leave, for an education and better future.

Lü Zhi would have learned more were it not for the "forestry official"—wearing camo with a pistol on his hip and automatic weapon on his shoulder—who intimidated villagers and jumped in to answer questions before they could: "Life has been very good since 1949." She finally closed her notebook: "If I wanted the party line, I'd read the newspaper." But George had already begun compiling her questions into a template others might use: How long has your village been here? How many hectares do you cultivate, and how far do your animals graze? Are you growing your herd? What other kinds of development have you seen? And the wild animals: What do they mean to you? What damage do they do? As they prepared to leave, a monk's wife brought George a strangely dark barking-deer hide. He snipped off a bit for DNA analysis, a move he would soon regret.

Though he would spend nine months of 1998 away from home, Kay was ever on George's mind. His tribute to her in *Tibet's Hidden Wilderness* had nodded at their changed reality. "Whenever one works in an alien place, one needs human intimacy, someone [to] . . . relieve strangeness. . . . Whether she is beside me in a tent, deep in her sleeping bag with only her blond hair visible, or awaiting my return in the United States, she is my emotional center." He again taped a note from her into his journal. "You may not hear from me, because I love you too much and want to cry too easily. So no words are loving words." Noticing newly sprouted willow and birch trees greening a slope stripped bare by an avalanche, he thought about how cataclysms can clear the way for revitalization.

But Kay grew frustrated at their uncertain communication—"let me know when you're leaving, where and who working with, etc."—made worse by George's choice to forgo a smartphone. And she had more sad news: His climbing buddy Tom Lane had died. "Take good care of yourself until I can do it for you. And please let's do everything we can to spend these years together?"

In October, George again managed to bring Kay along on a brief trip. This time it was a WCS meeting in Sarawak, where nearly four decades earlier she had discovered she was pregnant with Eric. The colonies of flying foxes they had watched in 1960 were all gone: eaten or killed to protect the durian crop. They'd hoped to see wild bearded pigs and did spot a few, as well as some too-tame monkeys and an intriguing worm harvest—women pounded the mudflats like a drum to drive them out. That was all. And yet for weeks after, with George in Mongolia and Kay returned home, she was elated from this "regression to youth and love. Thank you for giving this back to me." In November, she expressed her "great sense of relief at setting a deadline to put an end of some sort to all this sadness. And this is all the discussion of that. I just need you to know how much and continually I love and will love you." Whatever deadline she was referencing, it seems not to have come.

•••

In February 1999, George agreed to return to Yarlung Tsangpo on a National Geographic trip led by mountaineer David Breashears. Also joining would be Ian Baker and Hamid Sardar, whom Schaller had met there the previous year. Though generally averse to travel with other outsiders, George was keen to learn from Sardar, then finishing a doctorate in Tibetan studies. But all went immediately wrong. On short notice, Breashears dropped out and two others joined, landing George in a "mob" of five foreigners. Then their permits were canceled, a rebuke to National Geographic for its grandstanding about Baker's "discovery" of a waterfall that Chinese Academy of Sciences researchers had in fact reached first. To cap it off, George was charged with smuggling out a new species. It took a moment for him to realize that his accusers meant the snip of barking deer, and longer for him to reassure them that, while unrecorded in Tibet, it was found in several Chinese provinces.

With time to kill, he did get his wish to learn more about the sacred significance of this landscape. In the eighth century, Sardar told him, the Indian sage Padmasambhava had brought Buddhism to Tibet and created eight hidden sanctuaries called *beyul*: earthly paradises so filled with mysterious power that no one ever wanted to leave. This gorge was one of the eight: Dechen Pemako, Hidden Land of the Lotus. To follow the river was to move through the reclining body of the deity Vajrayogini: the peaks her head and breasts, the pass her hips, Rinchenpung gompa her navel, the center of bliss.

Since they could not enter this *beyul*, Hamid suggested they go to another in Nepal, where a decade earlier he had spent three weeks meditating in a cave near Serang monastery. Stopping first at several temples in Lhasa, George found a new, or renewed, fervor. A blessing ceremony at Sera devolved into a riot, with the faithful breaking down doors and monks hitting them with sticks. At Tsurphu, George was half hypnotized by the dancers, spinning in their masks and crowns of skulls. Approaching the Karmapa—now fourteen, wearing leopard skin and seated on a throne—George bent his forehead to the ground three times. Each time, the Karmapa tapped him with a wand, which George hoped would help his "depression, personal and professional." At Jokhang for the new year, he felt his arm clutched and looked down to find an old woman seeking protection from the crush. "I nearly carried her up the steep stairs, until everyone gushed like released water into the courtyard." Another woman then handed him her two-year-old grandson. The proprietor of a little shop, though closed, gave George a bowl of thick barley soup as a holiday gift. Later, at Xigazê, he and Sardar stopped at a restaurant that turned out to be a brothel. "We smiled and went next door for noodles."

George was happy in Nepal to again travel with Sherpas, though for the first time he grumbled at their early starts; by 6:00 a.m. they had his tent down. "No leisure." Meeting a woman fluent in various local dialects, they asked her to come along. She began introducing them as "secret lamas," her pilgrimage friends.

Recounting that journey for the collection *Extreme Landscape*, George struck rapturous notes not unlike those of fellow contributors Terry Tempest Williams and Barry Lopez. On a high pass

> *at this convergence of snow and sky, I lift my face and feel like a passing cloud. Spirits soar in such infinite space and the silence speaks to the soul. The mind has a heightened awareness and clarity. Mountains are said to be the abode of distant gods. But at moments such as this they become intimate. . . . I have since read that during transcendent experiences . . . some neural traffic . . . is shut down . . . the brain ceases to make a distinction between self and surroundings.*

However heightened his awareness, his powers of observation remained superficial, he continued, compared to those of local people, who could see and hear the "hidden and intangible forces." Even leeches, for those so attuned, were agents of paradise, drawing out karmic impurities. Entering the Valley of Bliss, he and Sardar met a glorious sight. Flanked by torrents and sheer cliffs streaked with snow, the three-tiered temple seemed a place where "human and divine can merge." While monks danced, a flock of silver snow pigeons banked "in winged unity."

And then, he exploded the story. Just beyond those dancing monks, Schaller revealed, was a massive construction site: barracks going up to "expand the dharma"; monks felling two-hundred-year-old hemlock trees for shingles, discarding most of the wood and burning kerosene for fuel; nuns hoeing up meadows to plant corn to ferment into drink. Grass fires burned, set by locals to bring new shoots; thick smoke and charred stumps of old-growth fir testified to how often those fires ran amok. When they asked the lama why the Buddhist prohibition against harming other beings did not apply to trees, he said it was because they lacked sentience—"trees don't meditate"—leaving George to wonder if new findings on tree communication might move him. As in the West, where God's command to Noah to save all creatures was mostly forgotten, the gap here "between Buddhist precepts and practice about the sanctity of and compassion for life is considerable." In his journal he was more sardonic: "A typical day of local people living in harmony with nature."

This trip George hoped would "release various stresses" had only left him more depressed, as did renewed dealings with the Tibet Forest Bureau. In the Hongla Golden Monkey Reserve he found no monkeys, only oak forests being felled to make fertilizer. "A waste of bear food." Why not, as in Sichuan, generate local income by using the oaks to cultivate a valuable tree fungus? Returned to the US, he awaited a long-planned visit to the Bronx Zoo from Liu Wulin and colleagues. "We looked forward to hosting you. Apparently, you hadn't time because you did not come." He still had no accounting for the $30,000 he had given the bureau in May. His calls and letters about permits went unanswered. Please tell them, he told Lü Zhi, "that the time for doing nothing is over."

Liu finally addressed the truck scam. Though corruption for personal gain was punished, he explained, "when staff cheats foreigners on behalf of our department, we get praised by the leaders." George replied that WCS could not continue with no planning, no reports, and no replies. He took his equipment out of forestry storage, growling for all to hear that "twenty years in Tibet is enough."

No matter his own future here, George would not abandon the chiru. Turning again to those buying their lives, he wrote to a hundred top designers and stores laying bare the source of their treasures: You are selling not shawls, as he put it, but shrouds. That caught the attention of *International Herald Tribune* fashion editor Suzy Menkes, whose April 1995 piece, "New Age Luxury Isn't So 'Correct,'" also ran in *The New York Times*. Quoting George, she exposed the tale of benignly shed tufts as "fantasy." Some houses responded. Hermès and Yves Saint Laurent discreetly removed *shahtoosh* from their stores. Bergdorf Goodman coöperated with US Fish and Wildlife, handing over their inventory and confirming that their supplier was a New York socialite.

Though chiru had been on the CITES Appendix I (most endangered) list since 1979, making illegal any commercial trade in them, there remained two important holdouts. Of 142 signatories to the treaty, only Switzerland refused to recognize chiru's status. And with its *shahtoosh* earnings topping $150 million a year, the Indian state of Jammu and Kashmir refused to outlaw production—until forced to by a lawsuit brought in 2002 by Belinda Wright's WPSI.

Still, almost no one anywhere was enforcing their laws. So with $1,000 and a foreword supplied by George, WPSI published an exposé that also served as a manual for interdiction. They identified the key smuggling routes through Nepal, the dozen Kashmiri families in control, the five-star hotels and boutiques around the world selling the finished products. They detailed identification techniques, in the lab and with a trained touch. (The diameter of a *shahtoosh* fiber is ten microns; ibex, the next finest, is fourteen. Human skin can sense a one-micron difference.) They exposed corrupt officials—one Chinese trade officer in Lhasa had offered their undercover investigator three thousand kilos, worth three million dollars—and explained how buyers harmed two endangered species: Indian poachers often killed tigers to trade their bones and penises to Tibetans for the fine wool. WPSI ran workshops for border police and assisted a raid of an exclusive shop in Mayfair, London. The seized goods from that store alone had cost the lives of six hundred chiru, plus several hundred more orphans left to die alongside their flayed mothers.

In October 1998, with shawls still in boutique windows and the pages of *Harper's Bazaar*, George took the fight directly to Park Avenue. An old-money couple who had long supported WCS invited friends to the Asia Society for a luncheon and lecture on "Saving Inner Asia's Wildlife." "It was hysterical," hostess Catherine Cahill later recalled, because "George's talk started out with wonderful pictures of scenery and wildlife, and everyone was oohing and aahing. And then there was this nomad with a leghold trap, and there was an audible gasp and all over the room you could see shahtooshes sliding surreptitiously under the tables." Several attendees excoriated George for "slandering" Tibetans, who they knew lived in harmony with all living beings. They called him a shill for the Han Chinese.

People, *Time*, *The Wall Street Journal*, *Newsweek*, *Vogue*—all did stories about George's efforts to get the "Fashion Herd," as the *Los Angeles Times* put it, "to Save Tibetan Beauties." The biggest splash landed in November 1999, with Bob Colacello's feature in the Society section of *Vanity Fair*, cleverly headlined "O.K., Lady, Drop the Shawl."

> *It is probably safe to say that the last thing the dowagers, heiresses, and trophy wives of New York's high society were expecting as they lunched on lobster salad at the Bathing Corporation in Southampton this past summer, or worked out with their personal trainers by their pools in East Hampton, or packed their Vuittons for the Paris couture shows, was a flurry of subpoenas from the United States District Court in Newark, New Jersey, of all places. But that's exactly what more than one hundred socialites and celebrities, including fashion icon Nan Kempner, supermodel Christie Brinkley, and arts patroness Beth Rudin De Woody, had hand-delivered to them at their country houses or Manhattan apartments by U.S. marshals.*

Beyond testifying before a grand jury, the women had to surrender their contraband, and endure *VF*'s cartoons—of the perfectly coiffed creatures behind bars, in a shawl tug-of-war with police, or digging spike heels into a bearskin rug with mounted heads all around.

Colacello was merciless in capturing socialites in high umbrage. "'What do you mean, people have to turn in their shahtooshes?' huffed Pat Buckley. . . . 'I haven't heard of anything so ridiculous in a long time. Some of our friends will have to call a moving van.'" Publicist Peggy Siegal dared the law to come after her set. "Tell me they're going to bring in the closet police. Or have undercover agents stationed at Le Cirque." Kempner found the thought of not wearing hers "perfectly horrifying. . . . [But] there's no way you could make me buy one now. Not after what I've been through." Others were not so traumatized. Colacello learned of at least eight other people who had sold or were still selling *shahtoosh*, including a British magazine editor, the wife of a prominent Park Avenue doctor, and an American art dealer. "Darling, I have ladies calling me from the Aegean Sea," a private seller told a friend. "They're afraid this is their last chance. What I really should do is double my prices."

The piece included a photo of George and several by him—of chiru alive and dead. Calling it emblematic "of how some fashion item coveted by the rich just kills off a species in a remote area they've never heard of," he estimated that of a population that once exceeded a million, fewer than 75,000 remained.

•••

The turn of the millennium found Kay, now in her seventies, still struggling. Having contracted Lyme several times ("your leaving seems to send a message to the bacteria and ticks"), she was dizzy and unsteady, often lying awake at night and sleeping away the day. "So good that you're not here to be driven crazy . . . and leave me longing for sympathy." She grieved for friends fighting with spouses or children or mortal illness. "Marge was moved to tears that you'd spun all the wheels by the Dalai Lama's home." And she continued caroming through every possible feeling, including about what was driving George.

> *The doctor says I am too stressed and urged me to ask you to come back for even a few weeks before Tibet. . . . I love you so totally and want to help you and am full of guilt that I am losing such control of*

myself. I usually get pleasure from being part of your work but have become fearful of the fax reminders that arrive. . . . Some days I don't eat; other days too much. No matter . . .

You blame your unhappiness on others rather than what may be a continued unhappiness within yourself. This may explain your need to control (picnics, movies, "type something now" when not needed for a week). And the threat, if any hesitation: 'Then I won't ask you again' . . .

I saw a film about John Denver, how his focus on his work ended with his wife leaving. I couldn't help but compare how much he needed from his parents that he did not get. . . . I have always felt it exceedingly important to reassure you that you were absolutely the most important person to me; you need to know at all times that you are totally important to someone. . . .

But perhaps one person over a long time is not enough. . . . You may need new admirers finding you important, wonderful, exciting.

I am here as long as you want me to be, but I need you to help me hang on. . . . You called and said so many good things and we will go on to the future and contentment we both wish. So good to hear your voice and know you found your antelope and to hear your voice and to hear your voice and to hear your voice.

George's trip that spring of 2001 was one of his saddest ever, his third failed attempt to find the western chiru calving ground. He had again broken his own taboo about travel with outsiders, in this case climber Jon Miceler, who the year before had helped him survey wildlife in Arunachal Pradesh; surgeon Jeff Boyd; and photographers Pat and Baiba Morrow. Having been turned back twice trying to migrate *with* the chiru from south to north, this time he would start to the north in Xinjiang and travel south in hopes of intercepting them.

Breaking a second principle, he had allowed the others to do the organizing, which already on day one had him concerned. Miceler had hired two camels and twenty donkeys, but George thought they needed at least ten more. The local handlers ignored him when he offered his scale to weigh the loads, and again when he asked them not to also ride the overburdened beasts. He did manage to jettison some useless baggage, including crate after crate of old army tins of eggplant. But he only later realized just how lax the planning had been: The expedition had brought a whole donkey load of cabbages but left sugar and rice behind, and for a month of travel had just three cylinders of gas.

Having rarely consorted with donkeys, George found them "not easy to like." The males bared their teeth, slashed with their forefeet, and mounted each other. They formed no friendships, never moved in unison, and ignored any quiet command. He could see why the men kicked or whipped them—though he thought they hit far too hard. And, with little to graze on and the supply of oats running out, he saw them weakening: the camels shivering in the vicious wind, the donkeys' legs bleeding from the rocks. "I would not have agreed to this route if I'd realized how rough it would be on the animals."

At sixty-eight, George was himself struggling, with more cracks emerging in his legendary stoicism. He complained to his journal of the misery of walking in hot sand, of how thirsty he was. "I hate this feeling of physical lessening." He flared at his younger companions, "strong impatient walkers" who (recalling George's own vanishings) often scouted ahead and returned only past dark, causing "needless worry." Trudging with head down over a 17,300-foot pass through hours of blowing snow, he was grateful when Pat backtracked to take his pack and feed him a PowerBar—though his tally of every such occasion on which he needed help seemed a kind of self-punishment. Finally catching up, he told Jon it had been stupid to push on in late-day storms instead of camping. Jon apologized, saying they had wanted to reach a site used in 1908 by archaeologist Aurel Stein. The desire to relive bygone glories can be a potent driver.

Alongside George's faltering body came the first signs of a faltering will.

When Baiba spotted five hundred pregnant chiru, they knew they had found the calving area, but ten days too early. George wanted to wait, to finally see the animals give birth. But Jon and Jeff wanted to complete the circuit, and with little argument, George yielded. "My chance lost. . . . I'm just going along now."

His worst fears for the pack animals were soon realized. First, two donkeys lay down and, refusing to rise, were left behind. That finally forced the team to toss the now-frozen cabbage, though at day's end George discovered that someone had stuffed two heads in his pack; he fed the leaves to the donkeys and ate the raw core himself. Two weeks later another donkey died. When a fourth lagged, George, lagging too, removed its load. It moved a few feet, but then stopped again.

Ever since Balaam's ass turned to reproach his Biblical master, the sight of a mistreated *Equus* has acted powerfully on men. Dostoevsky's childhood witness to the beating of an old horse found its way into *Crime and Punishment*: In his dream, Raskolnikov tries to protect a little sorrel mare, finally throwing his arms around her bloodied muzzle to kiss her as she dies (then going off to commit murder). The author returned to that scene in *The Brothers Karamazov.* Nietzsche's collapse in Turin has been linked to a similar encounter; Kundera thought the philosopher "was trying to apologize to the [brutalized] horse for Descartes." In *The Secret Agent*, Conrad alludes to the continuum of violence and pain explored by writers from Rousseau to Frederick Douglass. On seeing an animal abused, Stevie is "newly opened to . . . sympathies of human and equine misery in close association."

George stayed with the exhausted donkey, speaking gently, stroking its neck to brush away the sleet icing its hair. "Be a good donkey, you can do it." The animal walked a bit, George's hand on his rump. Then it turned, walked back down the trail, and stood, ears back and head down. With the others far ahead and their tracks drifting over, George had to leave the donkey, and then a second who gave out. Looking back, he saw the tiny silhouette in an expanse of white and hoped it would not die slowly, ravens taking its eyes—but that the locals, or even a wolf, might find it.

The camels held out longer, luckily, since each carried three hundred pounds. But now one refused, kicking and biting till it was abandoned, the other moaning for its lost friend. They later rescued it and, finding easier travel on a road built in 1949 by Kuomintang prisoners, reached grass, with five animals lost and little to show for it. Calling Kay, George learned that a Lyme flare-up had caused her to cancel a trip, which "saddens me because I'd thought she was in Italy having a good time."

That George kept trying again, and failing again, inspired climbing legend Rick Ridgeway, who had followed the chiru saga since reading Schaller's 1993 *National Geographic* piece. Moved that, though "nearing seventy, he considered many of his unrealized goals . . . to be open projects," Ridgeway gathered three other famed mountaineers—Conrad Anker, Jimmy Chin, and George's friend Galen Rowell—to complete the expedition. First, he called George, who invited him to the Bronx Zoo for what turned into hours of planning. George was skeptical that even this superhuman foursome could walk the nearly three hundred uninhabited miles unsupported, with no place to resupply and fifty-mile stretches without water. When they told him their plan—each would drag an aluminum rickshaw loaded with more than 250 pounds—George declined their invitation to join.

As chronicled in Ridgeway's 2004 book, *The Big Open*, the trip was brutal even by the standards of men who had summitted K2 and located George Mallory's body on Everest. Toughest of all was what they nicknamed the "gorge of despair," where they had to muscle their leaden rickshaws over sofa-sized boulders as icy waters rose and rocks rained down from overhead; still, Ridgeway refused to toss overboard George's 383-page *Wildlife of the Tibetan Steppe*. Nearly defeated after sixteen arduous days, they came upon barely perceptible tracks and realized whose they were.

"This is classic," Galen said. "One of the wildest places I've ever been, and still I'm following the tracks of George Schaller." (They later found a donkey skeleton.) Ridgeway finally used his satellite phone to call George, not realizing he was in Tajikistan. Kay answered.

"Where are the chiru?" Ridgeway asked.

"Keep looking," she answered. "They are there."

The men did ultimately find the calving ground. Though they'd run out of food so could stay just two days, Rowell and Chin managed to document the birthing and newborns in still photos and video, giving George what he needed to persuade China to add another five hundred square miles to the reserve. That protection came just in the nick of time. George had been running into processions of gold prospectors on their "handle bar" tractors (*shou fu tuolaji*), and hearing of mad foxes, poisoned by the mercury they used. But Ridgeway and company had found something far worse: a vast placer operation, with hundreds of miners, nearly as many guns, and bulldozers building a road that would put the chiru in easy reach by anyone with a four-wheel drive. George had also gotten wind of another alarming development. Though he had warned that animals adapted to extreme cold and low oxygen would not survive captivity, officials were hatching plans to "farm" chiru.

Two weeks after the foursome's homecoming, the Schallers and Ridgeway visited the Rowells at their home in Bishop, California. Two weeks after that, Galen and Barbara died in a plane crash. In his tribute to the friend he had first met on the flanks of K2, George wrote that "we followed each other's footsteps—to Annapurna, Denali, Mustagh Ata, Anye Machin—and envied each other." George never made an eight-thousand-meter summit. Galen once wrote that "I'd rather have seen a snow leopard than have climbed Nun Kun." All four climbers considered the chiru trip among the most fulfilling of their lives.

George did finally get to see chiru calve, retracing the Ridgeway route in the summer of 2005, joined by a young scientist named Aili Kang. The two found scenes of remarkable peace. Females moved in waving lines "like golden katas in the wind," the dark red placenta they left behind "glistening like stranded sea creatures." A wet, black baby slid out headfirst and soon unfolded itself to suckle; another slept so deeply that when Aili touched it, wearing surgical gloves rubbed with aromatic *Ceratoides* to hide her scent, it raised its nose to nuzzle what it took to be Mom.

They also saw how often it went wrong. Tiny hind legs only half emerged attested to the breech birth that had killed one mom. A second had labored so hard she had dislocated her own leg. That her rumen was empty, its contents digested, meant it had taken a day for her and the baby to die. When Aili brought him all that remained of a newborn—just head, neck, and tattered legs—George showed her the punctures on its throat: a Tibetan fox. Several dead babies flummoxed him: While they were outwardly fine, he found their insides crushed. Again he reflected on the tenuousness of life here at the planet's extremes; how little these creatures could bear all the roads and guns pressing down on them. A female licked her rumpled little newcomer, "alone on the wide-open flats, the baby so visible and vulnerable."

Many of Schaller's final trips in China were to Sanjiangyuan, a nature reserve (later park) created in 2000 to protect the headwaters of Three Rivers (as it is also called), the Yellow, Yangtze, and Mekong. This focus on Qinghai province was partly imposed: Politics had put both Tibet and the Uyghur regions off-limits. But with fifty thousand inhabitants, the reserve also provided a laboratory for the problem now central to his work: How to reconcile—in a radically changing world—human and animal lives?

He and Miller had been exploring that question in the Chinese context for a decade. Beginning with their rangeland studies, they had expanded to the full array of relationships that collectively made life possible in the Chang Tang—"how the deep-rooted *Carex*, white-flowered cushion of *Androsace*, snow finches, pikas, Tibetan brown bears, and broad-mouthed yaks function together to create fertility and diversity in this seemingly barren land." They now dug deeper into human-wildlife relationships, including conflicts. Wild asses, they found, did in some places take grass that would have otherwise fed livestock; herdsmen wanted those kiang driven out or dead. But *in some places* was key. Rather than sweeping policies, like poisoning every pika, what was needed were solutions tailored to the most local realities. How many kiang grazed right here? How fecund were they? What was the state of these particular rangelands and the true extent of competition? The picture had to be high resolution. It also had to be continually refreshed as the world changed. While Schaller pressed the Chinese to restore nomadic practices that had sustained the grasslands, he had no illusions that anyone could simply rewind the clock. Conservation had to contend with the world as it is, including locals' evolving aspirations. Whatever outsiders might want them to be, he wrote, Tibetans "are not part of a pastoral theme park."

He continued developing that work with Lü Zhi, fast emerging as China's foremost voice for the approach that increasingly defines modern conservation: locally led, melding traditional knowledge and practices with Western science, and concerned in equal measure with the needs and cultures of wildlife *and* humans. In a 2005 paper, Schaller and Lü Zhi tackled the thorny reality that conservation success sometimes exacerbates conflict. Since the Chang Tang became a protected reserve, wild yak numbers had increased tenfold. A triumph, except that wild bulls now absconded with domestic cows and menaced women going to milk or collect dung. The paper laid down a principle. "The nomads' . . . livelihoods are not expendable." Their knowledge, traditions, and interests must be part of any solution. But "the rights of wild species [deserve] equal weight."

Finding that "dynamic stability" required, again, specificity. In his 2008 foreword to Richard Harris's *Wildlife Conservation in China*, Schaller critiqued sweeping hunting bans that included still-abundant species used for subsistence. Such bans both harmed and alienated local people. He urged consideration even of a commercial chiru harvest, counting it a far more valuable use of the plateau's sparse grass than raising domestic sheep. He regularly spotlighted models of trophy hunting, both those that failed: In Xinjiang, 65 percent of the $26,000 fee to kill an argali went to "locals," but that meant local officials, who spent it on banquets and Land Rovers. And those that worked: In Pakistan, 80 percent of ibex trophy fees actually went to families, although some now wanted to kill snow leopards lest they kill potential trophies!

It was hard not to see in China's stunning economic growth—its construction cranes and cement plants, new truck routes and overnight cities—nothing but a human steamroller. As Schaller and team limped along with their ragged tents and empty gas canisters, they kept running into crews from the Department of Mines in fancy camps with towers of fuel barrels. And far from being damped by modernization, the appetite for animals and their parts had been supercharged. In just a few years' time, the price of caterpillar fungus spiked one thousandfold, to $5,000 a kilogram, bringing whole families piled high on motorcycles to harvest it. Coming upon a dead kiang stallion with a mutilated groin, Schaller was told that its "sex parts" had been taken for medicine.

As always in China, the destruction ran in parallel with hopeful changes. Many officials, up to the ministry level, now had biology degrees; the scientific pipeline smashed in the Cultural Revolution was finally restored. Post-trip debriefs went from empty ritual to substantive: discussing better positions for guard posts or the need to develop materials in local languages. Most exhilarating, a brilliant new generation of wildlife scientists now joined George in the field.

Two of the brightest stars, Aili Kang and Yanlin Liu, both came to George through *Wildlife of the Tibetan Steppe*: Aili, because at the urging of Endi Zhang—her thesis advisor and WCS's first China program director—she had been the one to translate it into Mandarin; Yanlin because he'd pirated it, photocopying it in his last year of college when the National Library would not let him borrow it.

Their first trip all together, in the summer of 2005, was a shock for Aili, who had never been at altitude nor even outside the "seminature" of zoos and breeding centers, where she studied saiga and crocodiles. Still George found in the twenty-seven-year-old a kindred spirit: somewhat stiff ("I'm not a person willing to be close with many people; to joke with them is close enough") but precise and tireless in her work. His respect for her grew daily, whether from watching her brace her small frame against freezing winds to hold up the antenna for locating collared chiru, or catching a glimpse—behind a bearing as stoic as his own—of tender feeling. Stumbling upon a two-day-old chiru, she rushed to him in pure delight: "I saw a baby!" Working together one day to catch and weigh calves, they found a still-wet newborn struggling to stand. "'Shall we weigh it or leave it in peace?' I ask Aili. 'Leave it in peace,' she replies." On a trip that fall to western Xinjiang, her persistence in searching out Marco Polo sheep left George feeling "downright negligent."

By Aili's account, she was only emulating his habits. When at day's end George left it to others to unpack the kitchen and cook dinner while he fussed with his tent and notebooks, she teased him that he needed to do more teamwork. But she soon understood that he was fulfilling his duty as lead researcher, keeping, amidst the messiness of camp life, their data ordered and secure. She found his reputation as a committed mentor well-earned. "We wanted to work with him not only because of his stature and experience, but because he was so good to young people. He treated everyone fairly; the most honorable of field biologists."

Aili could tell when he was frustrated, sometimes seeming—like her mentor—to observe humans as if she wasn't one. "He never yelled, but because I study animal behavior, I would catch his changing face." On six more trips together over the next five years, the twentysomething and seventysomething found many affinities. When the others crowded into a room to smoke and drink and play mahjong, George and Aili chose quieter pastimes. One night they watched a DVD of *Charlotte's Web.*

Yanlin, who on that trip was just twenty-four and finishing a giant panda thesis under Lü Zhi, was a kindred spirit of another sort. A skilled mountaineer, he climbed ridges "for the hell of it and ate huge amounts," George noticed. "I envy him and think of my younger days." Madly curious, Yanlin spent hours digging out a lizard hole and returned elated from meetings with fox kits or bears. He found in George a model of endurance—"at seventy-two he insisted on carrying a forty-pound pack like the rest of us"—and of a nonideological way of seeing the world. "George does not manipulate outcomes; he was open to whatever results we found." Watching him jot all day in his pocket notebook, then "at night, no matter how late, copy the data into one book and his feelings and stories into another, I began keeping notes that way too." Like the Sheenjek team fifty years earlier, all three loved roaming alone, coming together in the evening to share what they'd seen.

Self-selection was of course at work. The young ones that attached themselves to George—even as remote sensing and genetics conquered wildlife science—were the few still persuaded that life can be understood only through direct experience. That required the field skills in which George still had no equal. After peering through scopes one day in the Chang Tang to count a galloping herd of kiang, which George liked for their habit of "wheeling like well-trained cavalry," he asked Miller how many he'd seen. Fourteen adults and six foals, Dan replied. But how many females, George persisted. "How the heck can you tell when they're running 200 meters away?" "Easy," George answered. "The males have only the small black circle of the anus ringed by white; the females' black area, with the vagina, is larger."

George did not reject his young colleagues' machine skills. To the contrary: Instead of waiting months to analyze data, he looked over their shoulders as they opened laptops to do it on the spot. With Google Maps recording their route, he gave up his hand-drawn atlases. And he was often fascinated by the stories caught on digital camera traps, like the snow leopard who followed one fox while being trailed by another. But with machines came the distance that above all else he had sought to close. Data collecting had been secondary to his true search—for communion beyond the bounds of his small self and species. Now, every year that experience became more attenuated.

That may have been what impelled George on a last great expedition in 2006: a two-month traverse across the full one-thousand-mile breadth of the northern Chang Tang. No one had made this journey since Wellby and Malcolm in 1896, and they had crossed in summer. This would be the first-ever winter survey of this true no-man's-land. For the first twenty-three days, the team saw not one other human being, only chiru, wild yak, and wolves who—having never seen a human—showed no fear. Nineteenth-century travelers had been terrified by the emptiness. To George, it was the essence of its beauty.

The trip almost hadn't happened, thanks to lingering bad blood with Liu Wulin. The "lying and cheating" forestry officer had proposed another "nonsense" budget that included a $20,000 "fee" and a requirement that George deposit all funds in advance to bureau accounts. Liu had also tried to jam his people onto the team. It fell to Aili, now on staff for WCS, to navigate the obstacles. George suggested she go to Liu's more reasonable boss and say they would pay $12,000—a quarter when their permits came, the rest when the trip was complete—but directly to the reserves. Aili should make clear that a European Union project George was overseeing was also at stake. "No cooperation on the traverse; no EU for Tibet. We'll move to Qinghai where people are not so obstructionist." Hearing that Liu was badmouthing him at the foreign office, he offered a last bit of guidance. "No sense confronting him. Better to show sweetness, then ignore him." Tied up in Alaska at a fiftieth anniversary of the Sheenjek expedition, he apologized. "I hate to leave this mess to you."

In the end, the team was as strong as any he'd ever traveled with. Of the fourteen, ten were Tibetans. Expertly managing logistics was Karma

Damchen, who later recalled his surprise when their truck broke through lake ice and sank to its hood—and seventy-three-year-old George pitched in to help. "Most scientists can't do much of that kind of work." On rest days, the two men shared long walks, and over their years of working together Karma often invited George—"as great a man as I've ever met"—to his home. His son adored the elderly scientist, read all his books, and ultimately went to mainland China to study zoology.

Moving east into more populated regions, the team began hearing from locals of new kinds of challenges. Vanishing glaciers, more violent sandstorms, springs and wetlands disappearing as the permafrost melted: Like the Arctic, this third pole was warming at twice the rate of the rest of the planet. As alpine meadows withered, families moved their livestock into new areas.

Many of these, already critical for wildlife, were only becoming more so as those animals also responded to the impacts of climate change. Yet it was in these peopled regions that George also found his greatest cause yet for hope. Among his teammates was Zhaxi Duojie, called Zhaduo, an original member of the Wild Yak Brigade. For a decade, the brigade had patrolled its community's lands to protect chiru, apprehending hundreds of poachers and thousands of hides. Two leaders had been assassinated. Yet Zhaduo had gone on to found the Snowland Great Rivers Environmental Protection Association. George described its work in Zhaduo's home village in his report to the National Geographic Society, an expedition sponsor.

> A very positive development is the increasing effort by nomad communities to protect their own wildlife. For example, Cuochi village set aside a mountain range of about 2,000 square kilometers to protect wild yak. A dozen families resettled themselves to reduce disturbance in the area. Cuochi also cooperated with another community to end all livestock grazing on a huge plain, 1,400 square miles, to provide chiru a safe haven. These wholly local Tibetan initiatives are the

> best means of establishing long-lasting conservation efforts, and should be encouraged in every possible way.

As the team wound their way across the eastern plateau, Zhaduo interviewed families, learning of other initiatives. Near Qinghai Lake, the last home on Earth for the endangered Przewalski's gazelle, one herder had become "locally famous" for cutting his sheep numbers to leave grass for the wild creatures. His village had removed the top strand of its barbed wire fences to prevent often-fatal entanglements; another had organized themselves into anti-poaching patrols and rented one hundred hectares of grassland just for the gazelles. The Global Environment Fund had subsidized them for a time, but now seven families collectively spent $5,000 of their own money each year.

Kay's spirits continued to rise and fall with George's attentions. "I sent off your CV to Xie for your visa. Which visa for which trip?? I hope you will obey my request that you be home from now on Thanksgiving to January." His foot-dragging toward retirement buried her in new paperwork. "Do not leave me in these messes or I swear I will pack up and move to the farm while you sit here and try to figure out what you've done or indicated or agreed to. . . . It is all such fun being your PA; I think we need to inaugurate a program of DO IT YOURSELF." That he kept going, seeming sometimes like the chiru to be repeating journeys past their point of making sense, only made it harder to bear. In June 2010 (she is eighty): "I will check on the phoebe to see if she's sitting on the eggs you saw a week ago. I feel you are desperately trying to find reasons to remain in the field."

Her account of her own life remained as acidic as ever: A book she was reading about Lyme's "foggy brain truly has me in tears (surprise!)." She had faced the "heartbreaking" truth that he would never notice that she'd tossed out piles of catalogs, only that the closets and files were still full. The *Cleome* that had not been growing "is now popping up! My life is so full of drama!" Sometimes her needling was indirect, and arced like a boomerang.

A friend's husband was off skiing and shooting "when he should be home with her, though I suspect he does so to get away from her being more noticeably unhappy."

At the same time, she saved every birthday, Christmas, and anniversary card George drew for her, each more marvelous than the last. In one, titled "5,000 years of stone tool evolution," topless cave Kay in leopard-skin skirt scrapes the bloody belly of an ibex; next to her modern Kay wears that scraper as earrings, brooch, and belly button ring. Several are self-aware: From a maple tree hung with a birthday-gift bucket marked "for Kay," she walks bent under the weight of two brimful with sap. A darker one has her under a black cloud, kicking away a Valentine heart. She wrote her own wry endearments ("Happy Fourth to you, Former Enemy Alien") and assured him that his more frequent calls, even on a bad line, left her "exhilarated and in tears." When *South Pacific* came on TV, bringing back her own star turn from their college days, "I think I shall forgo that happy/sorrowful remembrance. These emotions are so overwhelming and confusing."

•●•

Of all the insights gained in three decades in China, most important to George in these final years was how important Buddhism, or its atrophy, remained in determining how communities treated nonhuman beings. He met one man who had given away his yaks rather than kill or even sell them. Then another, laden with traps, who said he didn't care if snow leopards or wolves existed at all.

Even at monasteries, which he now visited regularly, George saw that spectrum from reverence to alienation. At one, abandoned by its top lamas for the US and Taiwan, poaching gangs roamed unhindered. Shown the corpses of four snow leopards, he saw signs of poisoning. Another was like a taxidermist's fever dream: Embalmed snow leopards, argali sheep, and a small brown bear hung from the temple ceiling, seeming to tread the air. Several monasteries had more mastiffs than monks, the half-feral dogs menacing both people and wildlife.

Yet where Buddhist beliefs remained strong, so did attunement to what Barry Lopez called the "numinous animal." When one village killed two snow leopards, and soon both men and yaks fell ill, the Rinpoche said that the gods were angry. Leopards attacked livestock, another taught, only if their owner had sinned. When in 2006 the Dalai Lama spoke of his shame at followers who still trimmed their *chubas* with otter or snow leopard fur, many gave up their skins. Some monasteries mingled darkness with light: Among Serang's massive construction and ruined forests, herds of Himalayan tahr grazed without fear.

At Xia Ri monastery, George found "an ideal vision of nature and humankind as one." Though just a month had passed since an earthquake had killed three thousand people and reduced the monastery to rubble, the sixty monks made room for the team in their blue refugee tents, the first time George had slept in a displaced person's camp since 1947. That context made especially poignant the lama's dedication to the threatened Himalayan griffon, a key scavenger at traditional Tibetan sky burials, where human corpses are set out to be eaten. The monks fed the giant birds by hand, offering bits of yak fascia and meat, sometimes smashing bones with sledgehammers to help them reach the marrow. When the old Rinpoche emerged from his tent with a hunk of butter and a knife, George watched two dozen of the bald birds run to him or plummet from the sky. The carrion eaters clustered like hens around the (also bald) monks, a scene of remarkable trust and intimacy. George also caught a glimpse here of the complex political terrain monasteries navigated. For centuries reachable only by raft across the Yangtze, Xia Ri now had a bridge and orders from the Chinese government to develop ecotourism to pay for its own rebuilding.

At Thartung monastery, George met the remarkable Tashi Sangpo. Sent to the monastery at thirteen, Tashi had, like young George, found solace in birds and mountains, drawing the ruddy shelducks he missed from home. After years at prayer wheels, he had realized that devotion to birds could itself be his Buddhist way. Filling notebooks with detailed renderings of hundreds of species, he created the region's first field guide. When he learned that one of those species—Kozlov's bunting—was rare, he committed his life to its protection. With training and tools from Shan Shui, the conservation NGO Lü Zhi had founded, he mapped their distribution, then enlisted a fellow monk to watch their ground nests for forty-six days, recording the insects they ate and the nestlings' development. With that data, Tashi persuaded herders not to graze in nesting areas during the critical months. He also talked to the birds themselves, urging them to nest among thorns where predators couldn't reach them. Other lamas might be reincarnated Rinpoches, the senior monks told him, but he was a reincarnated sparrow. Out walking together, Tashi and George both noticed the glittering dirt at the mouth of a burrow: the crystallized breath of hibernating marmots.

None of these monasteries were removed from the world—quite the contrary. Devices claimed as much attention here as everywhere: George found monks lost in their cell phones, or glued to televisions watching war movies or cartoons. Indolent young men, usually arriving on motorcycles, often loitered about. The "roadhouse," George nicknamed one popular gompa. Among the offerings at the photo of a great Rinpoche, he spotted a can of Red Bull.

Yet these temples were essential to snow leopards, as George's youngest, greenest protégé would soon prove. Out with George for her first-ever field experience, twenty-five-year-old Juan Li had two transformative encounters: first with a big-bellied lama who hosted them in his hermitage high on a holy mountain; then with a snow leopard, who bolted out of a cave she'd wandered into, nearly tumbling her as he leapt past. That pairing sparked the question she would spend years unraveling, with George's guidance: How much do snow leopards rely on the sacred lands that ring Buddhist monasteries? After many surveys together, she had her answer: Of the more than three hundred monasteries in the Sanjiangyuan region, 90 percent were in places the cat roamed. If all were to uphold their taboos against killing and grazing, they could offer more shelter to snow leopards than the entirety of Three Rivers National Park. If extended to other Tibetan Buddhist regions, such monastery-based conservation could encompass 80 percent of the snow leopard's global range.

The culmination of this work came soon after George's eightieth birthday, when Tashi and Lü Zhi gathered 250 lamas from thirty-one monasteries to camp at the foot of the sacred Nyanpo Yutze mountain. George began his talk with a bow to the many ways that Buddhism had prefigured the principles of ecology, beginning with the understanding that all beings are one. But why, he challenged them, do you not protect pika, or wolves? Did Buddha sanctify all beings except these?

The following year, George gathered his thoughts on Buddhism in the only novel he ever wrote. (He wrote it for young adults, an audience that has also seen two nonfiction books about his life.) A mastiff pup named Deki must fend for herself: her father gone, her mother too busy for the forlorn creature. Left to guard the tent belonging to her human family, she watches a wolf amble by; though she is well fed and lustrous and he scruffy and hungry, she feels the chain round her neck and envies his freedom. Soon forced to choose between comfort and companionship, or hardship and loneliness, she opts to roam wild "without plan or purpose." For a time, she finds both freedom and love, mating with a wolf and nurturing their two male pups. But that joy ends in pain and terror thanks to guns. The story mixes high adventure—magic light wands vanquishing night demons, corrupt monks, a fight to the death with an evil treasure hunter—with lessons in natural history. Its Buddhist strands include Siddhartha

Gautama's journey to enlightenment, the inner warming fire some monks can summon through meditation alone, and the unanswerable riddles posed by a hermit who lives in solitary penance for felling an ancient tree.

•••

It seems fitting that the last animal George would study in China was the only one he had always feared. If his empathy for villagers' dread of tigers had been abstract, he felt the danger of bears in every nerve. When Ridgeway had asked what to do if they met a hostile one, he'd had no good answer; only a warning that the biggest tree was just twelve inches high. Sending pepper spray to Lü Zhi, he wrote long instructions, rehearsing it all for himself. "Read the directions on the can. Talk to warn a sleeping bear. Don't go into tight places or wander about at night. If close, raise arms to look big and tell it in a calm firm voice to go away. Do not stare or run."

The rarest mammal on the plateau, bears had only recently begun frightening locals. Traditionally, they were seen as protectors, intermediaries with the gods; some herders still believed that any attack was punishment for man damaging the land. And conflict was mostly recent, thanks to the new opportunity offered by settlement: houses often left empty, ripe with meat and trash.

Adding to the pressure were climate change—delayed winters, earlier springs—and extermination of the pika on which bears had always fattened themselves, both of which shortened hibernations. The two bears George eventually collared emerged with depleted fat stores in early March, when the ground was still too frozen to dig for prey and the grass too dry to eat. The starved animals scavenged winter-killed livestock but also raided huts to burgle butter and flour. In another instance of the complexity of success, a hunting ban had made them less afraid of humans.

The bears were anything but stealthy, bashing in doors, windows, even walls. Wading through torn furniture, toothpaste tubes, a broken basin, smashed dolls, and gnawed cardboard, George wondered if some was sheer vandalism: One bear bit off a headboard, tracked all over the furniture, and scattered flour. Looting half the houses in Sanjiangyuan, often several times, the bears exhausted people's tolerance. One village cut a gas storage tank in half, buried it to its rim, and tossed in bait. Any bear that jumped in couldn't get back out. In a year they dispatched three troublemakers.

Others sought nonlethal solutions. Zhaduo's Snowland group tried digging dry moats around huts. Some families took all their food with them and left their doors open. George offered solutions from grizzly country, describing the caches Native Alaskans set up on stilts. He fought the "modern" solutions insisted upon by their funder: The EU's costly six-foot fences lent the aura of a prison to a hundred homes, and left the rest unprotected entirely.

All of these solutions were shots in the dark, uninformed by any systematic knowledge of the bear's fast-changing behaviors. So with a team that included Yanlin Liu and Dajun Wang, another essential collaborator from Lü Zhi's Center for Nature and Society, George set out to collar a few bears to answer baseline questions like: Do all bears maraud, or just a few hooligans?

The project began in late 2010 with a fascinating round-robin, a window on the advancing technology and its challenges. Tom McCarthy, who had been relocating bears in Alaska, advised on sources for traps (either one highly skilled Idaho welder, or snareshop.com), the best tranquilizer (straight Telazol), and whether a blowgun could deliver the heavy dart required. Ben Jimenez from the Montana Department of Fish, Wildlife and Parks, who would join them in the field, recommended traps that pinged when sprung so they could reach the animal quickly or—if the snare got dragged off—find it and "the unhappy bear it was attached to." He also estimated the bears' neck size and how much weight gain the collars would have to accommodate; a three-hundred-pounder preparing to hibernate would plump up to four hundred pounds. As for cables, "it would be interesting to calculate the max E a charging bear can generate and compare it to the load certification."

George in turn answered Ben's questions about camp life. "I never worry about food. It happens. And Chinese go in flocks so numbers are always uncertain. If I sound indefinite I am. Being definite in China just causes you tension and I've found that things usually work out. As for how long: depends on you. I went to China 30 years ago to help out and I'm still here."

Trickiest were the new collars themselves, which cost $5,000 apiece and sent their signal not to field antennae but to a satellite, which beamed it to the German manufacturer, which relayed it to New York, where a colleague generated maps of the bears' wanderings and emailed those back to the team. George relied on his younger colleagues to lead him through the GPS user manual, which "might as well be written in Chinese characters," again bridling at his "complete lack of control." Finding frequencies China would permit was a challenge. As was figuring out how often to take a fix to capture diurnal, crepuscular, and nocturnal activity without burning through the batteries, and how to be sure that a lost signal meant the bear had denned. "No snow yet," Lü Zhi would write to the departed team, when both signals stopped. "But the marmots have hibernated."

Just finding the bears stumped them. They came upon grass mattresses in caves—but not their owners. They chased reports of break-ins, spending hours to set traps that only got tripped by dogs. Having not snared anything since the early nineties, George was tense, practicing with the syringe as he waited and worried.

When at last they got a bear, it was a furry bundle of danger: a cub, with frantic tracks all around but no mom in sight. "That was really scary," Dajun recalled. "But George calmly got us through. While he watched for the mother, I threw a blanket over the baby. It bawled, but I held it down while Ben released its hindfoot." George feared the sow had bolted and the baby would never find her, so he was relieved when an hour later they saw a mother and cub. A camera trap later showed her with two cubs. Both fine, the babies tumbled about, then sacked a house.

Nearing eighty, George tolerated arrogance less than ever. Tristan Dickerson had trapped cats at home in South Africa, so had come to help with the bear team's parallel effort to collar snow leopards. George now often deferred to the younger scientists—letting Ben, for instance, guide Yanlin on darting and proper dosage. But when Dickerson, who was not a scientist, darted a ten-month-old with three times the needed dose of tranquilizer, George got furious. He got angry again when the brash young man began showing off his slides of an African leopard cut by snares, giving "locals . . . the wrong impression of our aims." When Dickerson began supervising George's handling of traps and scat, and spouting misinformation, Schaller finally corrected him before the team. "Tristan went off in a huff. Instant world authority on snow leopards."

The bear trapping, meanwhile, was not going well. One snared animal pulled free, leaving two toes behind. With that wounded bear about, George dressed down a young colleague for staying out alone past dusk, endangering herself and the project. Then one of their collared bears got trapped again, pulled the snare loose, and went off carrying the spring and wire in his mouth to keep them from catching on roots. How could he possibly hunt? George felt the helplessness of no longer having the speed or stamina to track, dart, and free the bear.

Next came the saddest trapping story since his attempts with jaguars. Finding a mother and cub both snared, George and Yanlin left them while they went back for help—in hindsight, a terrible choice. Ben went and sedated both, but returned with reports of a bloody mess: In her terror, the mother had mangled the cub, biting through its foot, splitting its lip to the nose, and ripping out its lower teeth. George was aghast; how would it eat? Released, it limped off on the back of its torn paw, using its nose to balance, a tragic echo of heron Siegfried. "We can do nothing, haven't even a penicillin injection," George wrote. "I've had enough of mauling bears. I said, 'Let's close them all.'"

With each passing year, Schaller felt more peripheral, including to the institution he had so defined. Though in WCS official histories he is the heir to Hornaday's and Beebe's world-changing legacies, his journal memorialized a most painful exchange. A colleague suggested "in a friendly way . . . 'I'm sure you won't mind'" that he retire to save the organization money. George reminded the younger man that he was largely self-funded, through his lucrative prizes—even after he gave most of that money away—and the chair he held,* endowed by the Ella for whom the dog-eating jaguar was named. But "the decades of effort are discounted."

Still, no one accused him of irrelevancy more harshly than he did himself.

> *Here I am in my seventies, and still I lie in cold tents as I did in the Arctic in the '50s, the '70s in Nepal, the '80s in Sichuan. . . . Because I selfishly stuck to what I like and know how to do—given the freedom by WCS, and Kay's support—I've not advanced. I'm hanging on to something, without the new scientific or conservation achievements to justify the absences from Kay.*

The country to which he'd given half his life could carry on without him. "I'm not doing anything Chinese biologists aren't doing or could do, and that's as it should be." Reading Peter Hopkirk's *Foreign Devils on the Silk Road* about treasure hunters, "I am reminded that I am one too, seeking new natural history." He copied a line: "The Chinese, they were told, did not need any help from foreigners to explore their own country."

George did still venture the occasional act of bravado: jumping between tussocks as on Alaska's boggy North Slope; carrying 101-pound filmmaker Kathy Yin across a river; leaving Peter Matthiessen's son Alex, thirty years his junior, gasping to keep up with the eighty-three-year-old as they retraced portions of the 1973 journey, again crossing seventeen-thousand-foot passes. He never stopped carrying in his hip bag everything he would need to survive, as if he might still take off alone.

But he often railed at his decline. "I hate this aging body." The muscles

* The Ella Milbank Foshay Chair in Wildlife Conservation

that had reliably grown ropier in the field now refused to harden. He lagged far behind, skipped ridge climbs and overnights, returned early to camp, having lost "the willpower to suffer." The mishaps he had always shaken off now demanded long recovery: A fifteen-foot fall left his ribs aching for weeks; an insect bite in Guyana required multiple IVs of antibiotics and a surgeon's incision to remove "an egg-cupful of yellowish rotten meat." The storms he'd once taken satisfaction in surviving became joyless; "it's superfluous to mention another heavily-clouded day." He grew ever more curmudgeonly. Staying at a monastery, he barked at any youngster who left the door open to the cold wind.

His beautifully legible handwriting grew spidery, though its periodic return to clarity suggested less a wobbly hand than ennui. He lost all interest in photography, feeling it pointless as all took digital photos, and finding "a sameness to the landscape and wildlife and people." Though he still had no peer at reading animal sign—and never will, in this world emptied of animals and full of technologies—he no longer spotted them first.

He bristled most at the babysitting: "Dr. Schaller should not walk so far. Dr. Schaller must not sleep outside." Descending alone from a peak, he was hailed by a monk and three teens, who bounded down to share their picnic. He soon realized that they had been dispatched to escort him. Watching the others neatly lick tsampa mixed with butter and milk from little porcelain bowls, he was embarrassed by the mess he made of his own face. River crossings brought the worst indignities. At one, a colossal monk insisted on carrying him across. At another, thigh-deep in white water wearing oversized hip waders, he held fast to a pole anchored by Yanlin. But when his feet slipped—as on that long-ago trip with his father—the water that filled his rubber pants made him too heavy to get up. Yanlin never forgot that moment. "All I could think is, 'I can't drown the greatest naturalist of the century.'" Strangely, though all else was soaked, the notebook in his breast pocket remained dry.

The echoes of times past now often sounded in this melancholy key.

Hunkering down in a gully to escape the wind, he remembered hiding in trenches. His later journals recalled his first—he noted that breakfast was included in the room charge and that his plane flew at "9500m, 940 km/hr"—but with an altered sense of the world below. Crossing the Arctic, he saw dark fissures compromising the ice. How do polar bears manage even now?

And always, at the back of his mind, were Kay's struggles. Though George continued well into his eighties traveling five months or more most years, he began calling home regularly, even daily—because he could borrow a cell or satellite phone, but even more because he needed contact as much as she did. Three weeks with no word left him near frantic with worry; a missed trip into town left him terribly sad. "I so want to call Kay." When he did reach her, he often found her down with a flu or vertigo or the heartache of watching their beloved poodle Delilah lose the use of her legs and slowly die. She said she wanted to move into the senior living condo they'd bought near Eric in New Hampshire but in three years had made no move to pack. "I don't. I will be cut off from zoo contacts. And it's the last step to decline."

And yet, for all his old-man ailments, Schaller's status as an elder statesman carried enormous weight, especially in China.

On nearly every trip now, to his dismay, he was followed about by film crews; at one conservation festival, a nineteen-car convoy of journalists and officials trailed him, all wanting pictures hugging him or holding his hand. His appearances drew standing room crowds; after a ninety-minute talk at the Peking University Library he was kept for ninety more minutes of questions—then donated back his honorarium so that they could buy books on natural history. An exhibition at the Beijing Zoo on the history of panda studies featured a Schaller diorama, complete with his tent and battered boots. Magazine writers sustained the hagiography. An article on the travel website Sichuan Fun (by Tan Kai, winner of the "Coal Mine Literature Award") recounted a visit to Wolong all those years ago by a photographer who handled a bamboo rat until it fainted in shock. "Dr. Schaller took off his overcoat and held the rat with it. He did not speak for half a day."

Every one of Schaller's books was translated: In a redacted mainland version—he never knew what they left out—plus editions for Hong Kong, and in Tibetan and other regional languages. To young wildlife biologists he was a national hero; even middle school students knew him as *Yeye Xia*, Grandpa Scha. Artists continued to find inspiration in his work. Asked who he admired most in the world, BBC Wildlife Photographer of the Year Xi Zhinong named George. "The difficulties I've seen aren't worth mentioning compared to his."

George sometimes wondered if he ought to have cultivated this celebrity. "As a different kind of person, more outgoing, keen to become a public personality, I could have accomplished more." But he had also seen how publicity seeking, especially by outsiders, could damage projects; and how, in a place like China, he was often more effective under the radar. Nor did he have any illusions about his gifts as a speaker. "I get most nervous at such events, which I dread and avoid, feeling insecure and incompetent."

He did get dragged into things. Filmmaker Joel Bennett opened his 1995 documentary on Mongolia with Schaller stepping Indiana Jones–style from a Soviet prop plane into a shoving crowd—though with his fanny pack, lumbering gait, and wooden delivery he proved a resistant subject for such glamorizing. Lü Zhi once insisted he delay going home so he could be honored at a snow leopard conference. When Kay said no, he made the eighteen-hour flight home to stay for thirteen days, then flew eighteen hours twice more for the conference week.

But Lü Zhi also seemed to be growing "exasperated with me. She says I'm not needed, should not use her staff for my trips, is annoyed that attention is on me rather than the younger field staff." At least for the last,

George was too. He turned down countless requests from reporters, pushed them to write about Yanlin or Juan Li, even began turning his back to the relentless cameras.

At the same time, ambivalent or not, he made use of his iconic status. Awarded China's top environmental award in 2008, the only foreigner so honored, he tapped Aili's help to deliver his speech in Mandarin, announcing that he would give all the prize money to local conservation groups and hoped that future awards would go to them too. He used his access to high-ranking officials to advocate: for protection of a snow leopard landscape spanning Tibet, Sichuan, and Qinghai; for a requirement that any land use change—whether a new mine or a new protected area—be approved and governed by the community; and for doing scientific studies *ahead of* making radical changes like settling nomads, which even some officials now conceded had been a loss for both culture and grassland ecology. He pressed the vice-governor of Qinghai to restore herders' freedom to move, arguing that they would be far more resilient to climate change, and that—once unfenced and released from overgrazing—the enclosed lands would recover quickly.

To any substantive request for help, he said yes. He advised the head of Sanjiangyuan National Park on plans for a research station; contributed to a book in Tibetan on local wildlife; spoke to college students on the need for diverse, place-based strategies, from small gorilla reserves to a transnational park for Marco Polo sheep. He even judged a high school science fair, again handing out grants on the spot from his own pocket. Asked to present awards to guards who had shut down poachers, he honored them with the deep bow and prayer hands he always offered those on the front lines of research and protection, voicing a deep regard both genuine and meant to inspire.

When asked in 2016 to write the foreword for *Snow Leopards*—the first compilation of knowledge gathered on the "cryptic felid" across its twelve home countries—he celebrated an evolution in wildlife biology that had followed his own: expanding beyond single species to encompass whole landscapes and ecologies, humans included. Citing new studies of signposts, he suggested that other species were far ahead of us in recognizing our intertwined fates; at these crossroads, the whole "carnivore guild" (snow leopard, wolf, fox, bear, marten) and their various prey read one another's scents, exchanging information. On the human side, he urged a redoubled focus on securing the safety and livelihoods of families whose "obvious option" was to kill predators that took livestock they could not afford to lose, and quoted anthropologist Shafqat Hussain on the need to align with a region's traditional cosmologies and moral values. In a second foreword to a collection of snow leopard folklore and oral history, he lauded the gathering of local voices, including the memories of old men. This expansion continues: Ethnography is becoming "multispecies" and so, slowly, are conceptions of justice.

How does an explorer grow old? Alexander von Humboldt made his last grand journey at sixty; Sven Hedin at seventy, to Mongolia and Turkestan. In *Tibet Wild*, George quoted Dryden: "What has been has been, and I have had my hour." But he lived T. S. Eliot:

Old men ought to be explorers
Here or there does not matter
We must be still and still moving
Into another intensity
For a further union, a deeper communion
Through the dark cold and the empty desolation

On one of his last trips in Tibet, George found a bearded vulture lying on its side, wings drawn in but head alert, "watching us with that beautiful red-rimmed eye." It did not struggle when he picked it up, and seemed to be neither hurt nor ill. Later, he saw a strange black creature crouched on a slope. Though he knew every animal on Earth, Schaller could not identify the slouching beast.

EPILOGUE

North, South, and West of China, 1989–2020,
Gobi Bears, Mongolian Gazelle, Wild Bactrian Camels,
Indochinese Warty Pig, Saola, Asiatic Cheetah

Schaller took many journeys not chronicled here: returning often to India, East Africa, Alaska, and Brazil, and to advise on work in Argentina, Chile, and a half dozen other Latin American countries. To the very end, however, it was Central Asia that claimed the largest measure of his devotion. Following its animals, he repeatedly crossed borders, interleaving his five dozen trips to China with months or years traveling north into Mongolia, south to Vietnam, Laos, and Myanmar, west to Afghanistan, Tajikistan, and Iran. His priority remained overlooked places. "The maestro," as colleague Joel Berger put it, went "where the ratio of biologists to species is small."

• • •

MONGOLIA

To Mongolia, a land of few people and many wild animals, he returned seventeen times.

Arriving as was his habit in a time of traumatic change, he witnessed the multiple shocks endured by its satellites as the Soviet empire crumbled. First came chaos: the gleeful destruction of anything Russian-built; the empty shops—in one he found only tins of North Korean cucumbers; a sudden lawlessness. To thwart thieves, his driver had to remove their car-door handles and drive into the steppe to sleep. Several on the team got mugged by the packs of "half-drunk young men with necks like telephone poles"—though when one snatched the notebook from his breast pocket, George snatched it back. "I'll knock your teeth out if you try that again."

Then came the shock therapists, bearing capitalism as the cure. George agreed to be a science expert for the UN Development Programme but immediately clashed with the "martinet" in charge, who demanded a national land-use plan in three days with no field study or data, and had allocated none of his three-million-dollar budget for Mongolian biologists. Minister of Nature and the Environment Zamba Batjargal was stunned when George told UNDP that he would stay on only if his local colleagues were paid, preferably out of his and the other consultants' exorbitant fees. "I'd never seen anyone say 'Give me less money.' He changed forever the terms on which we worked with international institutions." Batjargal later awarded Schaller a medal never bestowed on a foreigner: Outstanding Employee of the Year.

George was mortified to be part of "the mob" importing yet another "alien value system," and still more at the eager converts who leapt to "privatize" wildlife. In one year, these entrepreneurs killed two million Mongolian lynx, fox, wolf, bear, leopard, eagle, and saiga for their hides, feathers, fat, meat, gallbladders, bones, penises, paws, and horns. A day after George found a nest of saker falcons, prized by falconers, the babies were gone. Their confused parents lingered. "Damn the poachers. They could have left them one."

Never one to be derailed by geopolitical chaos, George managed despite all to collar and track three of the last forty Gobi bears on Earth, and to publish with Mongolian colleagues the first scientific paper on the shaggy creatures. For the first time, he held a (sleeping) snow leopard in his arms, stroking its ears and plush pipe-cleaner tail. To the great migrations he'd witnessed—of caribou, wildebeest, white-eared kob, and chiru—he added another: watching tens of thousands of rose-colored Mongolian gazelles flow across the steppe like a sea of flowers on the move. He implored the global funders of billion-dollar infrastructure projects: First, understand these migrations.

Most mesmerizing to this camel lover were the wild Bactrians, with their neat, conical humps and small dinosaur heads. The critically endangered beasts faced impossible pressures. Forced to cluster as climate change erased their oases, they became easy pickings for wolves and men—though George couldn't help but marvel, watching them mingle at the water with wild ass and sheep, at this live Nativity found nowhere else on Earth. Breeding efforts made matters worse: both the crossbreeding by camel racers hoping to speed up their steeds, jeopardizing the wild ones' genetic integrity, and the state's payments to ranchers to capture and raise newborns on foster mothers. Few of those captives survived. Those that did often grew dangerous: One pubescent male draped his neck over George's head in a dominance move made all the more effective by the camel's wild-onion breath. Having seen the same animal sink his canines into the thigh of a bigger male, and not knowing "what sets a camel off," George ran, throwing his hat when the teen gave chase.

When some proposed—as George had for pandas—moving the problem animals deep into the reserve, he surprised even himself by advocating not only for continued captive breeding but also for the culling of baby-camel-eating wolves. He'd not forgotten the green tongues of Alaska's poisoned packs nor the lessons of Aldo Leopold. But "we can't wait for more data. The population is crashing, and too few young survive."

Though admirably undogmatic, Schaller's embrace of wolf "control" was short-lived. Even here, far beyond the reach of the Brothers Grimm, he found "men inflamed with the irrational hatred" that has made the wolf the most persecuted of all wild creatures. In short order, the only ones he could find were eight corpses tied to the roof of a border guard's jeep, facing forward like an obscene figurehead, and six not yet dead waiting in a filthy enclosure to be killed for their tongues as a cure for cancer—another mythic holdover.

Kay, who could tolerate these altitudes, made some of her last field trips here, still sharp-eyed enough to spot sixty-million-year-old bits of dinosaur armor. Whether he had Mongolia or married love in mind, George began one summer's journal with Wordsworth. "Though nothing can bring back the hour / Of splendour in the grass, of glory in the flower; / We will grieve not, rather find / Strength in what remains behind."

SOUTHEAST ASIA

While Mongolia kept calling George to China's northern flank, for more than a decade he alternated those expeditions with nearly as many trips to its south.

Fifteen years after the last American bombings, he found ruin beyond anything he'd seen since childhood. Agent Orange continued to do its deadly work, on farmlands, forests, and families. Vast areas of the mountains remained too littered with unexploded ordnance to even explore.

Yet two last firsts remained for him here: the exhilarating discovery of species new to science, and the pain of seeing one wink out forever before his eyes.

In Vietnam, he and three local colleagues found evidence that Javan rhinos, thought to be gone from the Asian mainland, still survived. As was often the case, it was local hunters who enlightened them. One showed George the horn from a recent kill, a small, hairy bump just four centimeters long; he'd eaten the tasty meat, the hunter said, and sold the hide for a pound of gold. A second, recalling the herds of his childhood, brought out a strongbox to show another nubbin, a worn tooth, a toenail—the relics of saints. When George then found tracks, the government set up a rhinoceros conservation group. But it was too late. After millions of years, the last one was soon gone. George's sketches of their tracks had recorded their final wanderings.

The story in Laos had a happier ending, and starred a favorite animal. George had been stumped when he asked about wild pigs and villagers asked *which* pig he had in mind. His Lao colleagues knew only a common black one; they thought the red one the locals described must be a juvenile. George wasn't so sure. If the villagers shot one, he asked, could they please save him the parts?

When next he came through, they had it waiting. Pulling down a tray hung to smoke over the fire, they showed him a broken jaw, half a head, two legs, and a shoulder blade, all of which George bought for three dollars. Saving enough flesh and bone for DNA and morphological analysis, they then sat together and ate the rest.

William Robichaud—who had come to Laos for the International Crane Foundation and stayed for WCS—wheedled news of their find out of George's guide, biologist Khamkhoun Khounboline. When Bill then found his own red pigskin in the private menagerie of General Sayavong Cheng, he tracked the two down, at a noodle shop. The game was on.

> I had to pretend, because George didn't know I knew. I said, "I've been talking to villagers who say there are two species of pig." George calmly choked on his noodles. "Yes, there appear to be," he said, as if it were common knowledge. Still playing dumb, I said, "You might be interested to know there's a skin at the menagerie." "Oh," he said nonchalantly, "maybe we'll go look later." But then two minutes later: "OK, Khamkhoun, let's go."

With WCS geneticist George Amato and taxonomist Colin Groves—who had found a similar skull at Beijing's Institute of Zoology, collected by

Jesuits in 1892—George and Khamkhoun confirmed the pig to be a species Western science had believed extinct.

The Indochinese warty pig was just the first, as the four wrote in *Nature*, of a "steadily enlarging clique of primitive large mammals" that scientists now realized had survived the last glacial period, but only in these mountains, the Annamites, which form the border between Laos and Vietnam. To be part of the first big wave of "discoveries" since the early part of the century felt to George like finding the Lost World—the corner of the Amazon where, in Arthur Conan Doyle's 1912 fantasy novel, prehistoric animals still lived.

His juices stirred, George turned to the next race: to determine if the unfamiliar antlers he had seen in villages—from both an unusually big barking deer and a tiny, black one—were also from "new" species. Again, General Cheng's menagerie proved key: The team found both kinds alive there. The little one proved to be *Muntiacus rooseveltorum*, a species last collected in 1929 by Harold Coolidge, George's gorilla patron. As for the giant one (*Megamuntiacus vuquangensis*), Robichaud later grumpily recalled that George's skepticism as to whether it was new caused them to be scooped by a team in Vietnam.

No one roused Schaller's rivalrous streak more than Alan Rabinowitz: "It's like this," Robichaud recalled Kent Redford saying. "Alan wants to be George. But George isn't through being George." The two nonetheless joined forces to make the most thrilling find—of the largest land animal identified since 1937.

Unknown to science until 1992, the serene, sweet-faced creature had shown up first in Vietnam, where two local biologists visiting a hunter's home had spotted long, straight horns unlike any they had ever seen. Noting their resemblance to the African oryx, they had named it *Pseudoryx*. Locals called it saola, their word for the upright posts on a spinning wheel.

Hoping to be the ones to find saola on the Laotian side, George and Alan walked hundreds of miles, chasing any lead. Hearing of a market where someone was selling saola meat, they arrived too late but were directed to a Hmong village, where they were shown a fresh set of horns. A man asked what they would pay for a saola: His dogs could drive one into a river, to capture or shoot as they wished. Another smoothed ashes and used a hoof to make a track. The animal was so distinctive and rare that all remembered just where they had seen or killed them, even years earlier.

The paper George published with Alan tackled the puzzle of what this dreamlike animal might be. An in-between beast like blue sheep, saola was neither wild cattle nor antelope. It was in fact not just a new species but the sole member of a new genus. How marvelous, George wrote in his journal, that a 220-pound animal with rapier horns could have for so long eluded scientific detection.

In January 1996, two people on the WCS Lao PDR team told Robichaud that General Cheng had a live saola in his zoo. Bill immediately flew down and took up vigil with the calm creature: "The most important captive animal in the world," as George told CNN. Named Martha, she lived just eighteen days. Opening her up, they found a beautiful little male fetus. The general then ate her.

Though it filled him with shame to see how locals feared him, George also found in Laos the kind of human connection he manages best: half-wordless, at least for him; centered on life's recurrent rhythms—"5am chickens get fed (rice); pigs get fed (greens); we get fed (rice and greens)"; and communal in a way that shelters his shyness. When he asked to buy a chicken in a Hmong village, the headman said he must buy eight, one from each household. He then threw the visitors a party. George envied the closeness of village life "where everyone is needed," felt again "nostalgic" for a life he'd never known—of extended family, an enduring sense of place.

Admiring how the local people made tidy use of everything, he was most moved by their repurposing of the detritus of war. US fuel canisters were now canoes; mortar shells (their markings still often legible: "Loading Date 2-69") made sturdy fence posts or watering troughs. On remnant bits of chicken wire, women hung bright yellow cocoons, the silk from which they spun, wove, and embroidered into beautiful skirts, "full production, from worm to clothing."

Such skills never ceased to dazzle George. Bill Leacock, who was in Laos on a cooperative forestry project and volunteered to translate for George, recalled a day out with two Hmong brothers. Spotting a rufous-throated partridge, George watched as the brothers quickly knotted vines into a snare, then stood on either side, singing the birds' courting duet. In moments, they had lured one in. When George bought it for 800 kip ($1), pausing to feel its breath swell its delicate bones before releasing it, Leacock laughed to hear the older brother say, "Is he crazy?" Dinner that night was fish instead of fowl: a tadpole and frog stew. The frogs had spent their last moments on the riverbank in what to George sounded like an existential debate, one croaking "What?" and the other one "Why?"

Red pig and saola aside, however, these mountains presented to George the same disjunction he had seen in the Amazon: an intact forest with almost no animals in it. Vegetative life was thick enough on the steep paths that he could swing from sapling to sapling, proceeding "almost by brachiation." He found marvelous invertebrates: six-inch orange centipedes and two-foot-long earthworms that looked like sea serpents; river snails that made a tasty soup.

But vertebrates were few, for reasons that soon became clear. The water at the bottom of their canoe often sloshed red around a yearling sambar or marten taken by his guides. "Two squirrels courting. Naturally the men shoot. The female is hit in the foreleg; still the male stays near. Took 15 rounds to get both to drop. Lovely glossy black, large as cats." Finding dhole on a deer, his companions shot the pregnant doe out from under the pack. George took the head, and for dinner they roasted the meat and boiled the six-ounce fetus and stomach lining. "Giant Barking Deer! We are eating a CITES Class I protected animal. At least it was the wild dogs who killed her."

The men hunted birds and fish just as intensively, often trading them for more bullets. George barely had time to admire a wreathed hornbill's blue gular sac before a shot rang out and down it fell. "I don't like the shooting but to stop it would alienate the locals." He found her stomach full of a purple fruit the Hmong called *singdong.* "So I at least get a little data. Boiled for dinner, her neck very tough." He slipped feathers left by hunters into his journal: pink trogon, yellow minivet, peacock pheasant with black eyespots on their wings. In the rivers they sometimes had to cross by swimming, he watched women flush bottom-feeders for men to spear with sticks. Once he found a gibbon skull, and once a muntjac dying in a snare. But on night drives he did not see the eyeshine of a single mammal. The brilliant green paddy seedlings drew no herons or other waders. Ripe figs fell to the ground, with no monkeys left to eat them.

George again had to rely on markets for his sightings, though even there, mammals were few: a lone leopard cat, one pygmy slow loris, a brush-tailed porcupine smoked and ready to eat. Birds were a bit less scarce. Among heaps of green-eared barbets and puff-throated bulbuls George found a defiant accipiter on its back, still flexing its talons, and scops owls tied into a bouquet by their feet. He wanted to buy and release the owls whose yellow eyes were still open but saw that their wings were damaged. He did buy one some children were leading about like a toy, and cut its string.

As in China, a new openness to foreign markets drove much of this commerce. The cynically named US Carnivore Conservation Trust had made a deal with General Cheng to capture sun bears and tigers to sell to zoos. On the road that crossed Laos from Thailand to Vietnam, Leacock recalled, women took orders for mouse deer, "tiny things, hooves the size of your little finger," which they then drove into nets, and boys sold sacks of cicadas they'd gathered on bamboo poles tipped with sap. While no crueler than the treatment everywhere of most animals destined for eating, these markets were and are complicated by the endangered status of many of these animals, their crucial roles in ecosystems, and the risk of disease spillover.

Those emptied forests were front of mind when George waded into a debate that has roiled conservation ever since, centered on two questions. Had these jungles brimmed with life until European colonization put an end to age-old harmonies? And—tropical forests survive best when protected by the people who have long lived there; do forest animals too?

On the first question: George himself often wished to have lived in the before times, when tigers roamed by the thousands from Turkey to Indonesia and jaguars were commonplace in Texas. But he also knew the growing body of work by ecological historians documenting just how long *Homo sapiens* had been disrupting other species' lives—with fire and the plow and ever-more-ingenious killing technologies. Colonization was not new. For millennia humans had arrived on islands and continents bearing novel plants, animals, microbes, and ambitions, often wiping out the prior inhabitants—people included. While their impacts were no match for what was to come, as George told an audience at San Francisco's Commonwealth Club, "various island birds, such as the moas of New Zealand, were eaten to extinction."

As for the present: George had been chronicling overhunting by locals since Congo. Now, colleagues were finding "empty forests" throughout the tropics. In an essay that sparked decades of rancorous dispute, Kent Redford—who had begun his career by finding no armadillos in Paraguay—warned against letting a "forest full of trees fool us into thinking that all is well." Many large animals were already "ecologically extinct"* in vast reaches of the neotropics, far more of them taken by local hunters than by colonists.

In one of many forewords George was asked to write from his singularly global perspective, in this case to a volume on *Rainforests of the World*, he tapped a memory of sitting atop a 140-foot *Anisoptera* tree in Malaysia. Describing how the flying *Draco* lizards snagged insects and the fairy blue-

* "Ecologically extinct" means a species is too few in number to interact meaningfully with other species; that is, it no longer fulfills its ecological role.

birds harvested the high-hanging fruit, he reminded readers of the consequences when such creatures vanish, the inevitable unraveling when you pull at anything and find it, as John Muir said, "hitched to everything else." Lose the fruit eaters who fly about dispersing seeds and you lose those fruit-bearing plants, then the insects that rely on their nectar, then the lizards that eat those insects, and so on. And those are just the obvious interdependencies. Most remain hidden to us.

George never missed an opportunity to excoriate the "overdeveloped nations" for their starring role in this destruction. "The average American uses a thousand gallons of gas a year, the average Laotian two," he told the San Franciscans. "With 5 percent of the world's population, we use 40 percent of its resources." As for "sustainable development," its "principal aim is to promote access to resources [extracted by] multinationals beyond the reach of law . . . so developed nations can continue their profligate lifestyles." Clear-eyed on the power of money, he sometimes joined those tallying the "services" other species provide man. But he loathed the idea that those beings' existence had to be justified that way. "An ancient Greek philosopher said that if horses had gods, those gods would resemble horses. So in a society that worships billionaires, Nature's virtue is measured in dollars."

Yet even as he railed against the expropriators, Schaller challenged the widely embraced credo that all Indigenous or traditional peoples—the nomenclature itself is complicated—are natural stewards who would keep wild animals alive if only they were restored to their charge. Though he was frequently attacked for suggesting that local people could be anything but simple paragons of harmony, his views were based on a rare degree of direct experience across four continents. Having lived most of his life in places where nomadic herders or subsistence hunters were his only neighbors—often in intimate proximity, sharing shelter, food, and any danger—he was among the first Western scientists to honor their depth of knowledge and the protective power of traditions that see at least some animals as sacred, or kin. But he also witnessed how rapidly cultural structures were giving way, especially where people were purposely severed from the land and animals. "Yes, if there are few people with limited technology and no commercial markets, you can have a certain harmony. But traditional lives in most places have long ago unraveled under the pressure of market forces."

Beyond obscuring reality, he thought it patronizing to resurrect what Redford had dubbed the myth of the "ecologically noble savage"—a myth not far, as he'd argued in a second cage-rattling essay, from European Romantic visions of New World "Indians," nor from the pieties regarding "*Naturvölker*" George had encountered in the Belgian Congo.

The people George had come to know over half a century were not icons, flat and perfect as Byzantine paintings. They were not sylvan creatures living outside of time, but individuals with history, agency, politics, aspirations—as varied as the rest of us in how they see animals and respond to change. Some Alaskan Natives oppose drilling for oil or natural gas terminals, worried at the harm they will do to the caribou; others are eager for them.

Rather than broad-brush characterizations, George argued as always for getting on the ground for as long as it takes to understand *this* moment in *this* place, including everyone who lives there. In an introduction to *Hunting for Sustainability in Tropical Forests*, he lauded the several dozen studies collected from around the world for adhering to such rigorous empiricism, their data gathered not via remote sensing—which can read forests as intact no matter the state of the fauna—but through prolonged fieldwork. "By studying the demographics of duikers and measuring the grams of meat eaten per day by a Wana villager, the way we think about and treat the world is affected."

Schaller got his own last lesson in the complexity of these communities when in 2016 he returned to Ringmo, Nepal, home to the porters who had ditched him and Matthiessen forty-three years earlier. His guide, Tshiring Lhamu Lama, had grown up here herding her family's yaks, before becoming the first Ringmo girl to pursue a master's degree—on human–snow leopard conflict. She had written George a fan letter and been amazed by his immediate reply, inviting her to join the return trip to Dolpo he would make with Peter's son, Alex.

That trip confirmed Tshiring's calling. "By the end I knew I wanted to work on snow leopards in my home region. I watched George go to every family and listen intently to their views. My heart said: 'If a man of his stature comes all this way to talk to my community, why not do that myself?'"

The trip also gave Tshiring the chance to teach George about her community's ancient religion. Adherents of the Bon cosmology revere mountains and trees. But some view snow leopards as demons who ought to be killed, as Tshiring later captured in a most disturbing film. "My neighbors would ask, 'Why are you studying our enemy?' Some saw me as the enemy."

Still, one needn't homogenize nor beatify local people, George argued, to honor their "rightful claims to land and . . . cultural continuity." Contrary to a persistent critique, it has been a long while since the dominant paradigm in conservation has been a "fortress" or "pristine" Eden, purged of people. George's own influence was crucial, including through his leadership at WCS. From Alaska to Pakistan to the Chang Tang, he worked not to set animals apart as if in oversized zoos but to rejoin our sundered worlds, recognizing both the sovereignty, and the interdependence, of every being, human and otherwise.

•••

On two occasions, in support of Rabinowitz, George worked briefly in the most complicated country of all. Myanmar stands among the world's most oppressive nations, ruled then and since by abusive juntas. But it also holds Southeast Asia's largest remaining forest, a forest that sustains a million people, rivers critical to farmers feeding many more, and three endangered species—tigers, Asian elephants, and rhinos.

George went first in early 1994, to help the country fulfill its Rio commitment to expand protected areas and shut down trade in rare mammals. Alan had secured permission to go up the Chindwin River into places not even the Forest Department had been in decades, kept out by the long-running Kachin insurgency. That permission came from the notorious SLORC, the State Law and Order Restoration Council, which had seized power in 1988. Soldiers flanked them front and behind as they walked, slept nearly on top of them, and used a hand-cranked radio to report their location hourly. When "Rambo," as the lieutenant colonel called Alan, jogged on the soccer field, men stood with automatic weapons at each corner; if George stepped out at night, a soldier held him in his flashlight beam. He had made no preparatory study: All the bookshops were empty.

George and Alan rebelled against the prerogatives of power exacted on their behalf, but in vain. They were forbidden to pay the men conscripted as their porters, nor for the thirty-five boats commandeered for their "armada." The officers jeered when George failed to get seats on a flight. "This is Myanmar," one said; he'd simply bump others. As their retinue grew—to eighteen porters and twenty-one soldiers—George found himself trapped in the kind of expedition he'd loathed since Congo, and with the same ludicrous luxuries: After washing their shirts, a "batman" wheeled out a VCR to play a movie: *Tarzan the Ape Man*, starring Bo Derek. So encased, they caught only glimpses of ordinary life. Women panned for gold using (toxic) mercury to pick up flecks; a boy fished with a bicycle spoke and rubber band.

Despite the heavy military hand, the new wildlife law had not ended the hunting of endangered species. George and Alan found snares armed with sharp, poisoned sticks on which a tiger would impale itself. A hunter told them quietly that nine rhinos had been killed here since 1980, one of them by him. "I seem to do more and more of this," Schaller wrote, "recording the last survivors." But the survey also confirmed the exceptional biodiversity that remained. George saw far more birds than in Laos, smelled fresh tiger urine, and found tracks of leopard, wild dog, and Asiatic black bear. Where elephants had squeezed "for how many centuries?" through a cleft in the sandstone, he saw how their bellies had worn the sides smooth.

Schaller returned to Myanmar just once more, in January 2002. Alan had just been diagnosed with leukemia, so instead of returning together to the Hukaung Valley, "my mentor and friend . . . agrees to go alone." Traveling with seven local WCS scientists, George was glad to find all of them there of their own accord.

As the only foreign conservation group in Myanmar in the SLORC years, WCS was both criticized and praised. Some charged them with abetting "green militarization": giving the regime new excuses for land grabs and a still-heavier police presence. Others credited them—for opening dialogue with the Kachin Independence Army, building a team of nationals that would survive many political turns, and helping to create the world's largest tiger reserve.

•••

THE 'STANS

Turning finally to the arid countries to its west, Schaller completed his circumnavigation of China. He had, by the end, mapped his own life atop his avatar's: Of the snow leopard's twelve-country range, only Uzbekistan remained outside George's own.

Afghanistan, into which he had peered longingly for three decades, brought the greatest danger and most profound hope. It was still a war zone when he arrived in 2004, with Apache helicopters throbbing overhead, car bombs occasionally exploding, and assurances from the US embassy that they'd flown his dinner in from Dubai, to be sure it wasn't poisoned. When he promised a report at trip's end, the Minister of Irrigation and Environment smiled. "I'll be here. Or, I'll be dead."

Joining him on this trip were photographer Beth Wald, who on assignment the prior year had forged crucial contacts with local warlords, and writer Scott Wallace—both sent by *National Geographic* to do a Lifetime Achievement story about George. Over six weeks they surveyed the two-hundred-mile-long Wakhan Corridor, the slender claw of land in northeast Afghanistan that reaches to touch China. By now Schaller walked Central Asia like others might visit the old neighborhood. From the northern border, he looked into a familiar ravine in Tajikistan; heading south, he reached Baroghil, the Hindu Kush pass he had climbed to from Pakistan in 1973.

The dicey environs plus team tensions made for an exceptionally tough trip. At seventy-one, George was as impatient as ever with anyone less stoic than he. He bristled particularly at Scott's attachment to the satellite phone, which "instead of immersion here" kept his mind back in the US and lent a false sense of security. "He demands certainty where only uncertainty is possible, seems scared of wilderness though we are with local people. And no one would rescue us anyway." Too little food worsened tempers; one night, to still the gnaw, George drank a cup of spaghetti water with Tabasco.

In this region where half the land was given over to poppies, supplying thirty heroin factories, Schaller also had a close call more frightening than any with apes or tigers. Frustrated that they weren't finding wild sheep, he went out alone and was miles from camp when he saw three men hurrying his way. Their satchels were empty, strange for mountain travelers. In a flash, they surrounded him, snatching at his pack and binoculars. Pushing past them with a bluff, "I must catch my team," he scrambled off over steep talus. That evening, when the trio turned up at their fire and his guide affirmed that they were dealers of *afim*, opium, Scott asked if maybe the whole idea of being here wasn't crazy. "If you waited for the world to be quiet," George answered, "you'd end up staying home."

He did return twice more, though as conditions deteriorated, he was given mandatory training in the color codes marking land mines, and

Prince Mostapha Zaher, head of the National Environmental Protection Agency, insisted on driving him about in his six-ton bulletproof Mercedes. In 2007 George made the rare decision to cut short a survey in Herat when his translator overheard the local commander tell their fifteen-soldier motorcycle escort to peel off if they met trouble. Kay was relieved. "People ask me how I'm doing," she emailed George, "and I of course tell them fine but even more of course I am not." The dangers, Mark thought, were harder on her. "I don't think I ever saw my dad afraid. He's busy navigating the moment as sensibly as he can. Imagined fears are far more intense." Kay had also had a car accident, her fault, due to "too much bad past making the present nearly impossible."

As with so many of his seeming dead ends, these trips bore remarkable fruit. Accompanying George to Herat was twenty-six-year-old Zalmai Moheb, a research assistant in WCS's new Kabul office. Zalmai had traveled with other foreign biologists but been ignored when he'd ask, "Why are we watching, not eating, these animals?" George was the first to answer. Soon Zalmai, who had grown up in a refugee camp in Pakistan, was urging his countrymen to take pride in their wild national heritage, and dreaming of his own study. Afghanistan had no graduate programs in wildlife science, so George helped him go first to India and then to the University of Massachusetts Amherst, where he became his country's first zoology PhD. "If I'd been with anyone else, I'd still be unsure about my path."

Kabul, meanwhile, agreed to protect the entire Wakhan Corridor as a national park, only the second in Afghan history, and that the communities long resident within its boundaries would not only stay but be comanagers of the park. "George had taught us," said Zalmai, "that to make people illegal in a place is neither moral nor realistic. If you impose imported models, you fail. We built according to our traditions, and with everyone involved."

That again and again George had to navigate war, corruption, and repressive regimes reflects a tragic reality: Since 1980, most major conflicts have erupted in the world's most biologically diverse and threatened places.

As he had seen in Congo and many times since, the same borderland forests and mountains that shelter animals also offer haven to armed factions and their refugees. Those dangers can be protective, as in Myanmar, where insurgents helped secure the Hukaung Valley and its tigers for thirty years; or devastating, as in the Virungas and Annamites. But they obligate conservationists to work beyond stable countries. Animals should not be condemned by the accident of living where human darkness descends. Regimes change; extinction is forever.

And rather than await peace as the condition for conservation, dooming many creatures, Schaller's wildest dream reversed the arrow. His success in the 1970s at getting Pakistan and China to create contiguous reserves had set him questing after a grander vision: a four-country Peace Park offering safety to people and animals. That dozens of villages had begun collaborating across borders to protect their transient snow leopards and Marco Polo sheep fanned his hopes for a far more durable path than top-down action. Although they lived at great distances, and on the edge of survival, these herders were sustaining a "flourishing . . . international diplomacy."

With Afghanistan now in, George needed only Tajikistan to complete the quartet. In 1989 he had used a chance meeting in Beijing to lobby a Soviet delegation to protect the Pamirs. On five trips to Tajikistan since, beginning in 2002, he had resumed the campaign, inspired anew by his first-ever sighting of a truly vast number of Marco Polo sheep—more than a thousand—at a hunting camp, and his discovery that its proprietors were about the only ones protecting wildlife in the chaotic wake of the collapsed USSR.

The Tajik scientific establishment was completely broken; when his host—a "small and smiling" bat scientist with flashing gold teeth—absconded with George's car and equipment, he couldn't help but empathize. Visiting the E. N. Pavlovsky Institute of Zoology and Parasitology, he'd seen that its three dozen biologists had just one computer among them, and no working phone.

Tajik herders were in still worse straits, able to procure essentials like matches and salt only by trading away their livestock, leaving them without meat—except wildlife. When one invited George to share his family's Marco Polo mutton, he felt guilty eating "even a little of something so precious to them."

Such subsistence hunting was the least of the animals' troubles. Several Ministry of Nature Protection officials held direct stakes in hunting or wild-meat enterprises, to which military men rented their Kalashnikovs. Their profiteering roused George to a rare political blunder: a letter sent over the ministry's head to the president. When the president in turn delivered a "big kick" to the director of Tajik National Park, who ran a hunting concession inside park boundaries, the director turned on George: spreading rumors that he was "washing big money" (meaning cashing in), then dispatching "bland-faced KGB chaps" to reject his permits and threaten arrest.

With a session of "self-criticism and apology," George seemed to set things right. In 2006 all four countries agreed to cooperate across nineteen thousand square miles to protect both grazing lands and wildlife. The agreement made clear that each nation would design its own land-use plan. But that bit never reached Tajikistan, leaving a vacuum that filled with rumors—that the park would be a supranational authority and ban both trophy hunting and grazing. Again, George scrambled to correct the misinformation and seemed to right the ship. The Tajiks sent out invitations to a second conference, filled with inspiring visions of how conservation could promote peace. But just a month out, they canceled, and in the twenty years since, no meetings have been held.

•●•

IRAN

George had alighted briefly in Iran twice in the 1970s, first on his drive from London to Lahore, then on a visit from Pakistan to biologist Raul

Valdez and his two housemates—an Iranian wolf who liked to play-hunt his best friend, a urial sheep. All four had gone together to the Great Salt Desert but found only ghosts: Iran's last Caspian tiger shot in 1953; its last Asiatic lion in 1942.

Thirty years passed before George was invited to return, to help save the world's most endangered cat. Skinnier than its African cousin, with a punk's spiky mane, the Asiatic cheetah had survived for centuries from the Red Sea to India, coveted by royals as hunting partner rather than trophy. Revolution and war had put an end to all that; by 2000 fewer than forty remained. George's only sighting on his first visit was of the captive Marita, who as a cub had been the sole survivor of an attack by villagers on her mother and siblings.

That he was there at all reflected, as colleague Luke Hunter put it, the authorities' recognition "that he was credible no matter the politics." Chief among those facilitating his work was Masoumeh Ebtekar, head of the Department of Environment, who had become known to the world in 1979 as the nineteen-year-old spokeswoman for the students holding fifty-two American hostages.

The perils seemed insurmountable. The cheetahs' reserves, tainted by association with the shah, had been opened to livestock, and their favorite prey to hunters, who used motorcycles to run the gazelles to exhaustion. Though cheetah rarely depredate domestic stock, the herders' half-feral dogs tore apart the slender creatures. Still more cheetahs died on roads: In one murderous fifteen-kilometer stretch, two females and two cubs were struck in a matter of days. All told, between 2001 and 2012, the biologists had eighty-two sightings of live animals, and found forty-two dead.

Just finding these remnant animals was nearly impossible. With camera traps among the technologies put off-limits by economic sanctions, George's tracking skills again became essential. As did the bridge he provided to the global scientific community for his gifted but isolated young Iranian collaborators. "His timing has always been right," said Hunter, "arriving when a country needed him most."

Which was also, often, when all hell was breaking loose. On September 11, 2001, six days after his arrival in Tehran, Kay called to "tell me of the disaster." UNDP said he must leave at once, but he rebuffed them. "Our loyalty is to Iran's Department of Environment and the cheetahs. Not even the US embassy could force us out." Three days later, still ignoring their frantic calls, he left for Kavir National Park with Mohammad Farhadinia, a twenty-year-old who had just quit medical school to devote his life to the cats. They ended their survey on October 6 at the border where, twelve hours later, US retaliatory strikes would begin.

Twice more, in 2003 and 2006, Schaller returned for surveys and trainings, greeted as a "superstar . . . a hero to our young activists." He formed his closest bond with Houman Jowkar, the "capable, committed, knowledgeable" young scientist who ran the national cheetah program, "the best cat biologist and conservationist the country had." They published together, including a chapter in the first comprehensive volume on cheetah, which opened with George and Kay's foundational 1968 work on the Serengeti subspecies. In 2008 Jowkar helped launch the Persian Wildlife Heritage Foundation (PWHF); its surveys shaped the national conservation plan. Public support grew: In 2014 a cheetah adorned Iran's World Cup team jersey.

Then all came to a horrible end.

The work had never been entirely free of political danger. On June 29, 2009, a week after state security forces killed and imprisoned citizens protesting disputed elections, Kay had written George in Tibet to say that a friend from Tehran had sent word of the birth of his son. "I can't imagine anyone sending an email on such a day of confusion and suspense, and wonder if he was testing whether it would get out. I'm not about to send in a message no matter how innocuous."

Conservationists were spared the worst until the fall of 2017, when billionaire Thomas Kaplan—chair of Panthera, a funder of the cheetah work—went public championing a group hostile to Iran. A week later, Kavous Seyed-Emami, a managing director of the PWHF who had traveled with George, wrote his board of directors urging that "although we avoid mixing politics and environmental activities," because "an individual named Kaplan . . . is involved in anti-Iran activities . . . we must completely discard Panthera."

A letter to Panthera followed, but it was not enough. In early 2018, security forces arrested nine of the cat biologists, including Jowkar and his wife, Sepideh Kashani, on charges of spying. The army chief of staff described their acts of "animal espionage," including using desert reptiles whose "skin attracts atomic waves" to infiltrate the nuclear program. He entered their camera traps as evidence, ignoring George's desperate explanations that, given the cameras' low resolution and three-meter range, Google Earth offered far more actionable intelligence.

Sent to the notorious Evin Prison, the biologists endured "enhanced interrogations" that included threats of sexual assault and long periods of solitary confinement. Seyed-Emami died after two weeks in what authorities called a suicide. Despite appeals from around the world, in November 2019, a Tehran court found all guilty, including of "contacts with the US enemy state," and sentenced them for up to ten years.

George publicly denounced Kaplan, resigned from his position as vice president at Panthera, and in an open letter appealed to Tehran on behalf of the

> *dedicated conservationists, the kind of persons Iran greatly needs. All prisoners should be promptly released. Holding them is wholly against the Islamic principles espoused by Imam Khomeini. Please return the prisoners to their families in the Name of God, the Merciful, the Compassionate.*

As the prisoners' colleagues fled or hid, the cheetah returned to the brink. A lone reported sighting came in March 2023, of a female cheetah pregnant with three cubs, dead on the road. WCS had been criticized for working with Myanmar's repressive regime. But the purist posture is at least as problematic: Kaplan's stance cost conservationists and animals dearly. By the time Houman and three others were released in April 2024, after six years and three months in prison, less than thirty cheetah remained alive.

George had made his own last trip to Iran in February 2016, traveling with Houman to Miandasht reserve, where they had an encounter as rare in our age of extinction as George's communion forty-five years earlier with the snow leopard. They were in the car when a reserve guard traveling with them got an excited call from his father, who was out herding goats and had just seen gazelles fleeing a predator. By chance, the guard's father was close by, so they detoured down a dirt track to where he sat on his donkey, pointing to a spot three hundred meters away. There stood a cheetah with two large cubs, their color so like the brush they seemed imaginary. George had woken that morning, his last in the field, to find that a small, striped gecko had snuggled in to sleep with him in his warm blanket. Now, these "pale animals in a plain habitat of short shrub, almost a vision" tolerated his presence for half an hour. "No meeting with a wild creature, these fragile survivors and the last of their kind, has touched me more deeply."

•••

The disaster Schaller witnessed across decades and continents grows ever more dire. Of the migratory species he followed, a quarter were in 2024 deemed endangered; worldwide, a million species spiral toward extinction. The perils remain the same: overhunting, habitat destruction, and, increasingly, climate change—sometimes all at the same time. In the Pantanal, a mass reengineering of the Paraguay River to ship more sugar and soy is desiccating the jaguar's wetlands, already burning in unprecedented fires. A third of panda habitat is too hot for the bears. Heat-fueled microbes periodically fell Central Asian ungulates by the hundreds of thousands. In Pakistan's Hunza Valley, glaciers collapse as temperatures reach 120 degrees Fahrenheit, a threshold temperature for survival of all mammals.

There will never be another George Schaller. He is himself an "endling." The last of his kind. Most immediately because societies in thrall to disembodied, artificial ways of knowing no longer support his kind of fieldcraft. "George could say, 'I'll be gone for three months and tell you what I did when I get back,'" said Kirk Olson, who after apprenticing with George in Mongolia wound up making his life there. "Now to go for six days you have to present a detailed plan." More irrevocably, said Bill Conway, "these worlds, these animals, are simply no longer there."

And yet—nearly every one of Schaller's species is at least a sliver more abundant and secure than when he started. The maps he generated with his singular vision—both an astronaut's, encompassing the planet as a whole, and a miniaturist's, learning it inch by inch, individual by individual—have guided conservation ever since, by meters and by continents. Even after years with jaguars, Rabinowitz turned to George. "After so many thousands of hours . . . understanding so many animals, he could look at a landscape and tell us where the cat will move: Will it skirt that meadow or take that forest path?" When asked to lead science and conservation efforts in sixteen countries from Afghanistan to Fiji, Aili Kang also turned to George. "He showed us where to focus, the last and most crucial wild havens." The entire field "works where you have gone before," longtime WCS conservation and science programs director John Robinson once told him. "Our commitment to feet on the ground, understanding the local, all is from your example."

The generations of biologists who gained their first field experience at his side, or were trained by those he trained—"his scientific grandchildren" as Juan Li calls herself and her peers—continue to lead wildlife science and conservation globally. Felix Ndagijimana hadn't envisioned himself a zoologist until, as a young assistant at Karisoke Research Center, he met George. "His appreciation of the work I was doing gave me confidence. And he told me it would be people like me, who were born here, who would ensure the gorillas survived." In 2012, Ndagijimana became the center's first Rwandan-born director. Z. B. Mirza grew from his early hapless apprenticeship to chair Pakistan's Snow Leopard Foundation and supervise the fieldwork of master's students at the Kinnaird College for Women. For China, "Schaller established the foundation of wildlife ecology," said Yong Yange. "The decades since have all been an extension of his thought." Some of his students there have even stepped into his transnational role. Yanlin Liu has helped Tibetan Plateau communities access wildlife-monitoring technologies; he has also advocated in global fora to secure their authority to evict poachers and illegal miners. Schaller protégés support one another: After George brought Jonah Western to Tibet to advise on rangeland erosion, Jonah brought Yanlin and Lü Zhi to Kenya to learn from Maasai herders how to avoid overgrazing.

Beyond all else, Schaller's work will ever remain a reminder that we live in a world of kin, in which—as he was first taught by the Gwich'in—every being has an equal claim on life and freedom. We call people animals when we wish to degrade them. To treat a human "like an animal"—caged, bought and sold, torn from family, beaten, or killed—is the ultimate barbarity. But a whole noble literature, from Schopenhauer to Isaac Bashevis Singer to Toni Morrison to Che Gossett, recognizes that all are lifted or debased together. Perhaps with the deepening understanding Schaller set in motion—of animal perception, language, desires—we can aspire to Percy Bysshe Shelley's dreamed-of world, in which "man has lost / His terrible prerogative."

Including over other humans. "The most telling expression of Schaller's reinvention of the field," wrote a WCS history, is his "revulsion against claims of human superiority . . . his fascination with and treatment of . . .

reviled creatures." It does not seem chance that George gave his life to shielding—from degradation, imprisonment, mass extermination—the most hunted and vulnerable beings on Earth.

Kay died on March 7, 2023, at the age of ninety-three. She had chosen her urn in Nepal and an epitaph from Pushkin. "Where I lie / May heedless Nature still be shining / With beauty that shall never die." Several monasteries in Tibet lit fires for her. George's grief weathered him terribly. He struggled ever to feel again as he had on first seeing Ringmo, that he might emulate the Buddha: "When you realize how perfect everything is, you will tilt your head back and laugh at the sky." Perhaps, as living Buddha Jamyang Tongyon told him on his eightieth birthday, he is a reincarnated Rinpoche and in his next life will be yet more powerful.

ACKNOWLEDGMENTS

In the nine years that have elapsed since my first visit to George and Kay Schaller in 2017 to propose this biography, I have received unstinting help of every variety from many, many people.

This book would have never happened without my earliest and steadfast supporters: Jane Alexander, Edith McBean, Bradley Goldberg, Darlene Anderson, David Yarnold, Michael Massing, Gail Ross, and Beth Wald.

I thank the Sloan Foundation, the Leon Levy Center for Biography, and Yaddo for support that came at crucial moments and connected me with fellow writers and artists who have sustained me ever since, especially Kai Bird, Thaddeus Ziolkowski, Susan Morrison, Nicholas Boggs, Lance Richardson, Francesca Wade, Tirtza Even, and Molly Hart-Zuckerberg.

I thank the incredible archivists, researchers, and documentarians who preserve our history in all its complexity. From the very beginning, Madeleine Thompson and Natalie Cash have shared their acute historical insights and bottomless knowledge of the New York Zoological Society/Wildlife Conservation Society archives. Nathan Utrup at the Yale Peabody Museum of Natural History saved me when COVID-19 closed down their archives, year after year digitizing tens of thousands of pages of Schaller's field journals and correspondence so that I could keep working as we all quarantined. Researcher Kate Kelley helped me organize those mountains of material, translated German documents, and offered an astute first read. Tom Veltre generously shared the transcripts from all of the interviews he completed while making his *National Geographic* documentary about

Schaller, giving me access to voices of people I never got to meet, including Peter Matthiessen and Alan Rabinowitz. Alex Matthiessen allowed me to tap his father's extraordinary chronicles of his time with George, James McGrath Morris offered essential practical guidance, and Kathy Moran brought her amazing eye to selecting images.

I thank everyone I interviewed (more than four dozen in sixteen countries), all named in the text itself. Special thanks to Ullas Karanth, who arranged my first trip into the field with George, and to our hosts in India, including Belinda and Anne Wright, Harshawardhan and Poonam Dhanwatey, Priya Davidar and Jean-Phillipe Puyravaud, and Nidhi and Tushar Kothari. Thanks to Stephen Emlen for sharing his father's self-published book, to the late Peter Crawshaw for generously sharing his memories, expertise, and unpublished manuscript, and to Dajun Wang, who not only tracked down Schaller's elderly and long-retired panda colleagues but facilitated our interviews, including by interpreting in real time. Thanks to Aili Kang, Yanlin Liu, and Stephane Ostrowski for connecting me to people I would otherwise have never found. Thanks to Pamela Turner for her wonderful book and helpful early conversations, and to Megan "Turtle" Southern, Valer Clark, Valerie Gordon, Aletris Neils, and Chris Bugbee for inviting me to join them as they introduced George to their inspiring jaguar conservation work in the Borderlands.

My deepest gratitude to my editors at Penguin, Ann Godoff and Will Heyward, who gave me the freedom to vanish for years and then, when I resurfaced, offered exactly the guidance I needed to pare the book to its essentials. Thanks also to Victoria Laboz and Casey Denis, who provided constant help navigating the maze.

Thanks to my expert readers: Jonah Western, Joel Berger, Amy Vedder, Bill Weber, Ullas Karanth, Anthony Sinclair, Andrew Laurie, Dajun Wang, Dan Miller, and Yanlin Liu.

Thanks to my friends, family, and neighbors, who hosted me, fed me, held me together body and soul, and listened and listened and listened: Rod and

Kay Griffin, Denise Ribera, Lucrezia Reichlin, Guillaume Malle, Ilan Averbuch, Alka Mansukhani, Ross Cheit, Marie Christine and Paul Katz, Corrie Yackulic, Sindri and Danae Anderson, Linda Ribera, Ruth Rominger, John Hoffman, Patricia Cohen, Catherine Cusset, Vlad Jenkins, Esther Perel, Jack Saul, Maryam Sharif, Sierra Campbell, Adah Frank, Sunny Bates, Philip Chadwick, Kimberly Sharky, Charles Sabel, Claire Sabel, Christina Malle, Glenn and Toni Murray, Jeanine Tesori, Katie Mark, Ruth Friedman, Gil and Holly DeShazo, Paul and Donna Schnitman, Shannon and Todger Anderson, John Barabino, Rachel Katz and Jim McFadden, Dany Celermajer, Eva Ansley, David and Karen Horn, Ric and Marsha Thom, Frank and Holly Bergon, Chris and Harriet Burrow, Houston and Michelle Hall, Frank and Barby Harvey, Katie and Constantine Triantafyllides, Michelle Frank, Leah George and Jeff Stampfer, Gilda and Richard Sherwin, Shari Levine, and, especially and always, Francesca Sabel.

Above all, thanks to the Schallers, from mother Bettina, who preserved everything, to cousin Bill Barnes, brother Chris, sister Renate, sons Eric and Mark, and especially George and Kay, who against all odds carried through their wild lives a treasure trove of letters and photos and memories—and fearlessly shared it all.

NOTES

FREQUENTLY CITED MATERIALS

Unless otherwise noted, the recollections of George and Kay Schaller here are from his field journals, field notes, and family correspondence held in the George Schaller Papers at the Yale Peabody Museum of Natural History archives, or from interviews and personal conversations with the author.

Additional material comes from the private collections of George and Kay Schaller, John (Chris) Schaller, Mark Schaller, Eric Schaller, and many of those interviewed across sixteen countries.

Author interviews with George and Kay Schaller were conducted on the following dates: April 12–13, 2017; April 30–May 2, 2018; October 10–12, 2018; March 11–14, 2019; and September 17–19, 2019. The author traveled with Schaller in India January 11–22, 2020, and in Mexico's borderlands May 5–9, 2024.

ABBREVIATIONS IN CITATIONS

YOG	George B. Schaller, *The Year of the Gorilla* (University of Chicago Press, 1964).
D&T	George B. Schaller, *The Deer and the Tiger: A Study of Wildlife in India* (University of Chicago Press, 1967).

GSFH	George B. Schaller, *Golden Shadows, Flying Hooves* (Knopf, 1973).
SOS	George B. Schaller, *Stones of Silence: Journeys in the Himalaya* (Viking, 1980).
TLP	George B. Schaller, *The Last Panda* (University of Chicago Press, 1993).
THW	George B. Schaller, *Tibet's Hidden Wilderness: Wildlife and Nomads of the Chang Tang Reserve* (Harry N. Abrams, 1997).
WTS	George B. Schaller, *Wildlife of the Tibetan Steppe* (University of Chicago Press, 1998).
TW	George B. Schaller, *Tibet Wild: A Naturalist's Journeys on the Roof of the World* (Island Press, 2012).

All interviews with Tom Veltre cited below were conducted for *Nature's Greatest Defender*, Veltre's 2009 documentary on Schaller for the National Geographic Channel (The Really Interesting Picture Company, directed by Cathe Neukum; transcripts shared with the author by Veltre).

Unless otherwise noted, references to Peter Matthiessen's journal entries and letters are from the Peter Matthiessen Papers, Harry Ransom Center, University of Texas at Austin. These citations are abbreviated below as "Peter Matthiessen Papers."

Letters and reports from the Wildlife Conservation Society archives in the Bronx, New York (abbreviated below as WCS) come from the following collections, unless otherwise noted:

Fairfield Osborn records, 1933–1973, collection 1029
Center for Field Biology and Conservation Records, circa 1962–1982, collection 4062
Unprocessed George Schaller records, accession 2014–145

Prologue

1 **Fearing that not even:** Paul B. du Chaillu, *Explorations and Adventures in Equatorial Africa; with Accounts of the Manners and Customs of the People, and of the Chace of the Gorilla, Crocodile, Leopard, Elephant, Hippopotamus, and Other Animals* (John Murray, 1861), 101.

1 **"a plain darn fool":** Carl E. Akeley, *In Brightest Africa* (Garden City Publishing, 1920), 196.

2 **C. R. Carpenter had observed:** Donna Haraway, "A Semiotics of the Naturalistic Field, From C. R. Carpenter to S. A. Altmann, 1930–55," in *Primate Visions: Gender, Race, and Nature in the World of Modern Science* (Routledge, 1989), 84–111.

4 **"I thought he must":** David Attenborough, interview by Tom Veltre, October 22, 2008.

5 **reached for archaic phrases:** "Pillar" is Siberian tiger scientist Dale Miquelle in a book assembled by the Wildlife Conservation Society (hereafter abbreviated as WCS) for Schaller's ninetieth birthday in May 2023. "Sterling character" is Eric W. Sanderson, in an interview with the author at the Bronx Zoo, December 5, 2018. Entrust with one's life is Peter Matthiessen in *The Snow Leopard*, p. 43.

5 **the "selfie emotions":** Mark Schaller, interview with the author in Mark's Vancouver, B.C., home, February 2020.

5 **"the most important animal researcher":** Michael Crichton, *The Lost World* (Knopf, 1995), 232.

6 **"'out-Schaller Schaller'":** Farley Mowat, *Virunga: The Passion of Dian Fossey* (Doubleday Canada, 1988), 32.

6 **zoologist Desmond Morris:** John Bulwer, "Desmond Morris: My Six Best Books," *Express*, October 23, 2009, express.co.uk/entertainment/books/135636/Desmond-Morris-My-six-best-books.

6 **"micro-documentation [of the] frequency":** Ted Gott and Kathryn Weir, *Gorilla* (Reaktion Books, 2013), 71.

7 **"George saved more":** Eric W. Sanderson, interview with the author, December 5, 2018.

8 "awful lot of 'I'": Kathleen Jamie, "A Lone Enraptured Male," review of *The Wild Places*, by Robert Macfarlane, *London Review of Books* 30, no. 5 (March 6, 2008), lrb.co.uk/the-paper/v30/n05/kathleen-jamie/a-lone-enraptured-male.

9 "Not easy to know": Peter Matthiessen, *The Tree Where Man Was Born* (Dutton, 1972; repr., Penguin Classics, 2010), 126. Citations refer to the Penguin Classics edition.

10 "tend to be shy": Vrushal Pendharkar, "The Life and Times of an Intrepid Biologist: An Interview with George Schaller," *Current Conservation*, December 21, 2018, currentconservation.org/an-interview-with-george-shaller/.

10 It was through this writing: David (Jonah) Western, interview with the author, New York City, October 23, 2019.

10 "that real poetry": Vladimir Nabokov, *The Gift*, trans. Michael Scammell (Weidenfeld and Nicholson, 1963), 136.

Chapter 1

22 "I'm glad my parents": Translations from the German are by Kate Kelley.

28 only bit of autobiography: George B. Schaller, *Tibet Wild: A Naturalist's Journey's on the Roof of the World* (Island Press, 2012), 167–201 (hereafter abbreviated as *TW*).

30 by a "blockbuster": The Allies' unprecedentedly destructive bombs were the origin of that word.

30 estimated thirty-five thousand dead: Jason Dawsey, "Apocalypse in Dresden, February 1945," The National WWII Museum, February 13, 2020, nationalww2museum.org/war/articles/apocalypse-dresden-february-1945.

37 Paula had tapped: Frank L. Beals to Paula Barnes, July 31, 1947, George and Kay Schaller private archive.

Chapter 2

43 grandfather Edward Hegeler: Hegeler had immigrated and set up shop with Frederick Matthiessen, a classmate from the Freiberg Mining Academy, whose descendant Peter would figure importantly in George's life.

52 Taking that many lives: Cade, who on their trip shot four peregrines, including a female with two downy young, would go on to play a critical role in rescuing that iconic species from near extinction. In addition to helping win a ban on DDT, he founded the Peregrine Fund, which has worked for half a century to preserve birds of prey worldwide.

56 "He'd just take naps": Dave Klein, interview by Tom Veltre, August 11, 2008.

62 manage a factory: *Kind v. Clark*, 161 F.2d 36 (2d Cir. 1947), law.justia.com/cases/federal/appellate-courts/F2/161/36/1566799.

68 physical anthropologist Harry Shapiro: For a fascinating discussion of how Shapiro not only reversed his own views but also provided the definitive challenge to then-prevalent pseudoscience on the dangers of "race-mixing," see Sonia Shah, *The Next Great Migration: The Beauty and Terror of Life on the Move* (Bloomsbury Publishing, 2020), 122–25.

69 "tall warm-blooded one": Margaret E. Murie, *Two in the Far North* (Knopf, 1962), 259.

72 "What wonderful neighbors": In Homer's *Odyssey*, the offer—or betrayal—of hospitality is what separates the civilized from the barbarian.

72 in *Outdoor Life*: Titled "New Area for Hunters" (an important constituency to claim), the March 1958 essay was part of the multipronged campaign by all members of the Murie team to win protections.

72 George's trip report: George Schaller, "Arctic Valley: A Report on the 1956 Murie Brooks Range, Alaska Expedition" (master's thesis, University of Wisconsin, April 1957).

73 "river of life": George B. Schaller, "Saving America's Last Great Wilderness," *Defenders Magazine*, Fall 2010, defenders.org/magazine/fall-2010/saving-americas-last-great-wilderness.

73 Frank Fraser Darling's: Schaller cites Darling's ecological study of red deer in the Scottish Highlands as essential inspiration. Frank Fraser Darling, *A Herd of Red Deer: A Study in Animal Behaviour* (Oxford University Press, 1937).

73 and Adolph Murie's: Adolph's influence on Schaller went further still: He not only made the first serious field study of wolves, he also gave them names and attended to their personalities and relationships. Adolph Murie, *The Wolves of Mount McKinley* (University of Washington Press, 1985), 28.

75 "Good-by, grizzly bear": Margaret E. Murie, *Two in the Far North*, 352, 337.

75 "loneliness that is joyous": Douglas Brinkley, *The Quiet World* (Harper, 2011), 385.

76 named him Siegfried: George Schaller, "A Heronry in the House," *Audubon* 69, no. 4 (July/August 1967): 66–67.

78 "he was still an undergraduate": John Emlen, *Adventure Is Where You Find It: Recollections of a Twentieth Century American Naturalist* (self-published, 1996, shared with the author by Dr. Stephen Emlen), 234.

78 thwarted her own ambitions: After Aldo Leopold died, the university told Kessel they would not accept a woman into their wildlife management program; transferring to Cornell, she completed her PhD in two years.

78 He's still very young: Murie to Coolidge, January 31, 1957, Fairfield Osborn records, 1933–1973, Collection 1029, WCS Archives, New York.

Chapter 3

83 "eerie shadow-world": George B. Schaller, *The Year of the Gorilla* (University of Chicago, 2010), 33 (hereafter abbreviated as *YOG*).

84 "recognized the kinship": *YOG*, 35.

84 As US directors: Coolidge to Osborn, February 27, 1957, WCS.

84 "old Cambridge friend": Coolidge to Osborn, February 14, 1957, WCS.

85 Perhaps, suggested Leakey: Leakey to Coolidge, March 16, 1957, WCS.

85 study of homicidal baboons: *From Apes to Warlords*, Zuckerman titled his autobiography, though his findings were later discredited due to the baboons' wholly unnatural conditions of captivity. See Haraway, *Primate Visions*, 162–63.

85 Leakey was "not beloved": Zuckerman to Coolidge, November 1957, WCS.

85 resolve the disputes: Emlen, *Adventure Is Where You Find It*, 235.

85 "to remain here": Schaller to Coolidge, December 19, 1957, WCS.

85 Emlen help George: Coolidge to Osborn, March 8, 1957, WCS.

85 illuminate man's nature: "Simian Orientalism," historian Donna Haraway calls this: Just as, per Edward Said, the West had defined itself by how it had differentiated from its source culture (the Near East, birthplace of written language and civilization), we would understand ourselves as a species by how we had differentiated from our evolutionary ancestor. See Haraway, *Primate Visions*, 10–13.

86 "normal, authoritative practice": Haraway, *Primate Visions*, 199.

86 "contemporary infra-human primates": "The Importance of Primate Studies in Anthropology," *Human Biology* 26, no. 3 (September 1954), 185.

86 by Franz Boas: Stanley M. Garn and Eugene Giles, "Earnest Albert Hooton, 1887–1954," in *Biographical Memoirs*, National Academy of Sciences, vol. 68 (National Academies Press, 1995), 167–79.

86 "termination of permission": Coolidge to Schaller, February 26, 1958, WCS.

86 "tagging a doped gorilla": Emlen to Coolidge, June 26, 1958, WCS.

87 "will be unthinkable": Emlen to Coolidge, June 26, 1958, WCS.

87 "A wild idea": Emlen to Leakey, June 26, 1958, WCS.

87 "make a choice": Emlen to Washburn, July 1958, WCS.

87 "brilliant, original guy": Washburn to Emlen, July 1958, WCS.

87 "George's draft board": Emlen to Osborn, December 16, 1958, WCS.

88 "welfare of the United States": Douglas A. Dixon, University of Wisconsin Committee on Occupational Deferment to Selective Service Board No. 100, December 16, 1958, WCS.

88 "fired the imagination of man": *YOG*, 1.

88 "biology and myth combine": George Schaller, foreword to *Ulendo: Travels of a Naturalist in and out of Africa*, by Archie Carr (University Press of Florida, 1993), xi.

88 the racial overtones: Gott and Weir, *Gorilla*.

88 That included Jews: Jay Geller, *Bestiarium Judaicum: Unnatural Histories of the Jews* (Fordham University Press, 2017). See especially the introduction and chapter 4. The musical *Cabaret* recalled this trope: "If you could see her through my eyes," the Emcee sings as he dances with a gorilla-suited partner, "she wouldn't look Jewish at all."

88 "fearfully like hairy men": Du Chaillu, *Explorations and Adventures in Equatorial Africa*, 86.

88 "look of piteous appeal": Carl Ethan Akeley and Mary Lenore Jobe Akeley, *Lions, Gorillas and Their Neighbors* (Dodd, Mead, 1932), 167, quoted in Gott and Weir, *Gorilla*, 59.

89 "of a hostile": "A Linguist in a Cage," *Our Church Paper* (New Market, VA), March 30, 1892, virginiachronicle.com/?a=d&d=OCP18920330.1.4.

89 "as might be expected": *YOG*, 6.

89 "correcting" her vegetarianism: Gott and Weir, *Gorilla*, 71, 103–4.

89 hid when she saw gorillas: Rosalie Osborn's reports, "Advice and Relevant Facts for an Expedition" and "Observations on the Mountain Gorilla, Mt. Muhuvura, S.W.

Uganda," are in the WCS archive. Both are undated but were enclosed in an October 14, 1958, letter from Coolidge to John Tee-Van, who shared them with Fair Osborn.

90 "having the proper support": *YOG*, 24.

91 watched the porters: *YOG*, 25. In 1929, Yale psychologist Harold C. Bingham had "deemed it essential to track the animals with guides, gunbearers and porters." George B. Schaller, *The Mountain Gorilla: Ecology and Behavior* (University of Chicago Press, 1963), 1.

91 first to spot gorillas: Emlen, *Adventure Is Where You Find It*, 244.

92 in Kisoro, Uganda: Baumgartel's hospitality was also legendary; one visitor recalled arriving at midnight to be greeted by the "stocky, muscled" Walter rushing outdoors in a billowing nightdress. He was just as flamboyant in his writing, describing a "bachelor" gorilla "making love." Fritz Dieterlen, "Memories of Walter Baumgärtel," *Gorilla Journal*, no. 16 (June 1998): 14–15.

92 a gorilla tracker: *YOG*, 42–43.

92 train several generations: Schaller's eighteen-month study bore out everything Reuben taught him, with one exception: Reuben went too far in assuming gorilla family life was like his own, with males bringing home food and females secluding for birth. Schaller, *The Mountain Gorilla*, 19.

92 "like pensive birds": *YOG*, 43.

93 "glancing at Reuben": Emlen, *Adventure Is Where You Find It*, 241.

93 summited Mount Muhavura: *YOG*, 45.

93 "I tried to hide": Kay Schaller, interview with the author.

94 Africa's first national park: Raf De Bont, "'Primitives' and Protected Areas: International Conservation and the 'Naturalization' of Indigenous People, ca. 1910–1975," *Journal of the History of Ideas* 76, no. 2 (April 2015): 215–36.

94 two endangered groups: Raf De Bont, "A World Laboratory: Framing the Albert National Park," *Environmental History* 22, no. 3 (2017): 404–32.

94 treated the Batwa properly: De Bont, "'Primitives' and Protected Areas," 225.

95 "If you call the animal": *YOG*, 51.

95 "gurgling of water": *YOG*, 52.

96 "they believe them": *YOG*, 53.

96 lowland neighbors out: Though he doesn't always avoid this kind of *National Geographic* exoticism, Schaller never seems easy with it. Stumbling upon a ritual—men drumming while a woman in a loincloth "gyrated" on the earthen floor—he calls her "a wild ebony creature caught in the love of rhythm" and falls into a nostalgic reverie: "I did not find it difficult to imagine the explorer Speke" here a century earlier with Burton, "the gentle Livingstone and the cocky Stanley sitting beneath the mango trees. . . ." But he immediately mocks himself, quoting Arctic explorer Vilhjalmur Stefansson on how a country is said to be discovered "when for the first time a *white* man . . . sets foot upon it" (emphasis mine) *YOG*, 84.

97 condemned Africa's colonizers: Fairfield Osborn, *Our Plundered Planet* (Little, Brown, 1948), 116.

97 did not take "more": Osborn, *Our Plundered Planet*, 115.

97 Aldous Huxley wrote: Aldous Huxley, praise for *Our Plundered Planet*, by Fairfield Osborn, first edition (Little Brown, 1948), book jacket. Brother to Julian Huxley,

Aldous wrote that Osborn lays out lucidly how man's "ignorant or wanton disregard" brings nature's "terrible revenges."

98 **Waking in the morning:** *YOG*, 98–99.

98 **"arms over their heads":** *YOG*, 101.

100 **"We are tired":** *YOG*, 74–76. Kay's transcription of the Swahili was slightly off; she wrote: *Tunachoka kuta futa ile ngagi.*

100 **"Elephants, hello," he said:** *YOG*, 78–79.

101 **"like a worm":** *YOG*, 98.

102 **"affairs outside the forest":** *YOG*, 96.

102 **"Africans would think badly":** Kay Schaller, interview.

102 **his AMNH diorama:** *YOG*, 28.

103 **"ceased to be anonymous":** George Schaller and Michael Nichols, *Gorilla: Struggle for Survival in the Virungas* (Aperture, 1989), 12.

103 **"Junior sits and peers":** Schaller, *The Mountain Gorilla*, 155.

104 **"I just had to wait":** Kay Schaller, interview.

104 **"demure little fellow":** *YOG*, 186.

109 **"when I held her":** *YOG*, 161–62.

109 **serenaded by four gorillas:** Emlen, *Adventure Is Where You Find It*, 245.

109 **inadvertently surprised one:** At times, Schaller seemed to envy in other primates the social graces he often lacked. Writing of baboons, he marveled at "the amazing number of relevant pieces of social information the animal can store in its brain and quickly retrieve so that it can respond with finesse to any of a thousand potential social interactions in a large group." Foreword to Shirley Strum, *Almost Human: A Journey into the World of Baboons* (Norton, 1987), xi.

109 **included their roar:** *YOG*, 117.

109 **"loneliness of our species":** John Berger, *Why Look at Animals?* (Penguin, 2009), 15.

110 **"The name 'guard'":** *YOG*, 161.

112 **arrangement in Rwanda:** Philip Gourevitch, *We Wish to Inform You That Tomorrow We Will Be Killed with Our Families: Stories from Rwanda* (Picador, 1998).

112 **"the tall Watutsi":** *YOG*, 163.

112 **"Surround the Watutsi!":** *YOG*, 164.

112 **"about 350 mountain gorillas":** John Hillaby, "Gorillas Periled by Tribal Moves," *The New York Times*, February 24, 1960, 5.

113 **his "killer ape":** Dart theorized that humans were launched on their distinctive evolutionary path by their appetite for murder, especially of their own kind. Raymond A. Dart, "The Predatory Transition from Ape to Man," *International Anthropological and Linguistic Review* 1, no. 4 (1953): 201–17.

113 **"How indebted science":** Dart to Merck, November 21, 1959, WCS.

113 **"some kind of ethological record":** Robert Ardrey, *The Social Contract: A Personal Inquiry into the Evolutionary Sources of Order and Disorder* (Atheneum, 1970), 141.

114 **dissected twelve female baboons:** Amanda Rees, *The Infanticide Controversy: Primatology and the Art of Field Science* (University of Chicago Press, 2014), 61.

114 "the well-known Oedipus fashion": Carleton Coon, *The Origin of Races* (Knopf, 1962), 84.

114 ideas about gorillas: *YOG*, 124.

115 mounting the male: *YOG*, 119–21.

115 their various postures: Schaller, *The Mountain Gorilla*, 285.

115 "big male baboons": George Schaller, interview with the author. To his credit, Washburn seemed more open than most to having his mind changed. Two months after Schaller's first bombshell report, he wrote Merck reporting a similar picture emerging among their baboons: no harems and dominant males, after all, but peaceful sociability. Washburn to Merck, November 5, 1959, WCS.

115 "father as the master": *YOG*, 132.

116 "stoicism and patience": David Attenborough, interview by Tom Veltre, October 22, 2008.

116 they spoke to him: *YOG*, 112, 162; Amanda Rees, "Reflections on the Field: Primatology, Popular Science and the Politics of Personhood," *Social Studies of Science* 37, no. 6 (2007), 888.

116 other, said Kay: Kay Schaller, interview.

117 important moral judges: See Bathsheba Demuth, *Floating Coast: An Environmental History of the Bering Strait* (Norton, 2019).

117 "For the first time in history": John O'Reilly, "The Amiable Gorilla," *Sports Illustrated*, June 20, 1960, vault.si.com/vault/1960/06/20/the-amiable-gorilla.

117 "recent success of *Born Free*": Korda to Osborn, August 3, 1960, WCS.

117 "we'll get renewed offers": George to Kay, September 5, 1960.

117 the "Primate project": Irven Devore, ed., *Primate Behavior: Field Studies of Monkeys and Apes* (Holt, Rinehart and Winston, 1965).

118 The full fourteen pages: Schaller, *The Mountain Gorilla*, 221–35.

118 In *Sociobiology*, E. O.: E. O. Wilson, *Sociobiology: The New Synthesis* (Harvard University Press, 1975), 220.

118 eight pages of: *Current Anthropology* 6, no. 3 (June 1965): 295–302.

118 Zuckerman in *Oryx*: S. Zuckerman, review of *The Mountain Gorilla: Ecology and Behavior*, by George B. Schaller, *Oryx* 7, no. 5 (1964): 253–54.

119 "models of excellence": Sir David Attenborough, acceptance speech, "Sir David Attenborough Honored at Annual Gala | WCS," posted June 9, 2016, Wildlife Conservation Society, YouTube, https://www.youtube.com/watch?v=ox9ZQZY2kx4.

119 "This wonderful creature": David Attenborough, interview by Tom Veltre, October 22, 2008.

119 "what we were seeing": Ian Redmond, interview by Tom Veltre, October 22, 2008.

119 "George was willing": Amy Vedder, interview with the author, November 24, 2020.

120 sometimes mind-bogglingly quantitative: Amy Vedder, interview.

120 a new genre: Mary Sanders Pollock, *Storytelling Apes: Primatology Narratives Past and Future* (Penn State University Press, 2015), 10–11.

120 "English nineteenth-century novel": Ian McEwan, "Literature, Science, and Human Nature," in *The Literary Animal: Evolution and the Nature of Narrative*, ed.

David Sloan Wilson and Jonathan Gottschall (Northwestern University Press, 2005), 11.

121 "stripped the soft meat": *YOG*, 66, 96, 116, 210.

121 transgressed the twentieth-century norm: Amitav Ghosh and Richard Powers have both described this kind of encounter with the "uncanny"—natural forces beyond our control, the mystery of nonhuman beings—as the essence of the stories humans told for most of our history. See, for instance, Amitav Ghosh, *The Great Derangement: Climate Change and the Unthinkable* (University of Chicago Press, 2016); and Richard Powers, "Kinship, Community, and Consciousness: An Interview with Richard Powers," interview by Emmanuel Vaughan-Lee, *Emergence Magazine* Podcast, January 16, 2020, emergencemagazine.org/conversation/kinship-community-and-consciousness.

121 rim of a volcano: *YOG*, 145, 244–46.

121 his car's approach: George to Bettina, July 16, 1959, Yale Peabody Museum.

122 sensitive portrait of gorillas: Naomi Bliven, "Good Neighbors," review of *The Year of the Gorilla*, by George Schaller, *The New Yorker*, July 18, 1964: 96–99.

122 "produce a book": Desmond John Morris and Harrison Matthews, "Bestial," review of *The Year of the Gorilla*, by George Schaller, *The Times Literary Supplement*, no. 3312, August 19, 1965, TLS Historical Archive.

123 "an eerie feeling": Dan Richter, *Moonwatcher's Memoir: A Diary of* 2001: A Space Odyssey (Carroll & Graf, 2002), 50–55.

123 "They don't carry off": "*To Tell the Truth*: Amnesia victim for 20 years; Sculptor-with-rifle" originally aired on CBS, December 6, 1965, posted August 3, 2017, *To Tell the Truth* (CBS), YouTube, youtu.be/OESiPX6qYBU, beginning at 16:40.

124 sustain water supplies: Raymond Dart, "The Urgency of International Intervention for the Preservation of the Mountain Gorilla," *South African Journal of Science* 56, no. 4 (April 1960), 85–87.

124 "I think they're scared": Kay to Dick and Thelma Smith, George and Kay Schaller private collection.

125 Captain Frank Poppleton: *YOG*, 236–37.

125 "isolated for days": Schaller to Doc Emlen, July 13, 1960, WCS.

125 the soldiers puffed: *YOG*, 238.

126 "13 drunken soldiers": George to Kay, July 22, 1960, Yale archive.

126 all clear from Verschuren: Telegram, Verschuren to Schaller, July 30, 1960, Yale archive.

127 "date the letters": Kay Schaller, interview.

129 "the white race": Harry Gilroy, "U.N. Guard Asked for Congo Wilds," *The New York Times*, August 3, 1960, 11.

129 "Osborn, Merck, Emlen": Telegram, Emlen to Schaller, August 3, 1960, WCS.

129 "tradition was too strong": *YOG*, 245.

130 "a large extent": Schaller to Osborn, August 9, 1960, WCS.

131 "Only by shooting": This is how he tells it in his field journal. To Kay, he wrote: "I sent a note to Verschuren to hurry and come with rifles and we will go on a cow-shooting spree."

131 "It was peaceful": *YOG*, 234.

132 "plunged his long knife": *YOG*, 248.

132 "Things have been popping": Schaller to Osborn, August 14, 1960, WCS.

133 "try to finish the filming": With a movie camera supplied by NYZS, Schaller had been capturing some of the earliest footage of mountain gorillas, shipping the film canisters to New York for processing. "All movements of your friends can be seen very clearly," Merck assured him in a June 1960 letter. "The nest building, sliding down tree trunks, feeding and throwing branches around are excellent and represent something that has never been done before."

133 "They would probably": George Schaller, interview.

134 Emlen wrote in French: Emlen to Miruho, August 17, 1960, WCS.

134 "for many years to come": Translations from the French are by Guillaume Malle.

134 "He is in a unique position": Emlen to Osborn, August 20, 1960, WCS.

134 and written Osborn: Schaller to Osborn, August 21, 1960, WCS.

135 Again he wrote: Osborn to Schaller, September 1, 1960, WCS.

136 "I'll judge when": George to Kay, September 5, 1960, Yale archive.

137 slamming the Belgians: Schaller to Osborn, September 5, 1960, WCS.

137 "frustratingly slow procedures": Emlen to Osborn, September 9, 1960, WCS.

137 "We found no cattle": Schaller to Van Straelen, September 20, 1960, WCS.

137 "Project finished leaving": Cable, Schaller to Osborn, September 23, 1960, WCS.

137 "tact, common sense and bravery": John Hillaby, "Wildlife Region in Congo Is Saved," *The New York Times*, Oct 26, 1960, 3.

138 met gruesome ends: Jacques Verschuren, "Schwarze Wildhüter geben ihr Leben für Verteidigung der Tiere im Kongo" ("Black gamekeepers give their lives in defense of animals in the Congo"), *Das Tier: Die internationale Illustrierte für Tier, Mensch und Natur*, February 1, 1961. *Das Tier* was founded by Bernhard Grzimek and Konrad Lorenz.

138 "not to a single clan": Jacques Verschuren and S. Mankoto Ma Mbaelele, "La renaissance du premier parc national d'Afrique (1925–1960)," in *Virunga. Survie du premier parc d'Afrique*, ed. M. Languy and E. de Merode (Lannoo, 2006), 85–95.

138 "an affectionate letter": Alex Shoumatoff, "#2: A Report on the Wildlife of Eastern Congo," *Dispatches from the Vanishing World* (blog), October 10, 2001, dispatches fromthevanishingworld.com/home.html (site discontinued).

138 "congenial nature of gorilla": *YOG*, 256, 269.

139 "was just extremely kind": Dale Peterson, *Jane Goodall: The Woman Who Redefined Man* (Houghton Mifflin, 2006), 202. Peterson quotes Kay saying "biggely-browed," but she said she'd never used that word.

139 they could not swim: Islam had been introduced by traders to the African Great Lakes region in the tenth century and its spread accelerated under German colonial rule.

139 "George Schaller showed": Peterson, *Jane Goodall*, 200.

140 "whole year's work": Jane Goodall, *Africa in My Blood* (Houghton Mifflin, 2000), 155.

140 commending her work: Schaller to Osborn, October 18, 1960, and November 7, 1960, WCS.

141 "not too horrified": Goodall, *Africa in My Blood*, 163.

141 writing home in May 1961: Goodall, *Africa in My Blood*, 180, 185.

141 "mother/child relationship": Though George says he did not glean parenting tips from the gorillas, Jane has said she learned to mother from her chimps. Goodall, *Africa in My Blood*, 185, 180.

141 "Better than George's 'Junior'": Goodall, *Africa in My Blood*, 180.

141 "You make mistakes": George Schaller, interview.

141 knowledgeable local guide: Schaller found significantly less to admire in his host Tom Harrisson, a zoologist and British intelligence officer who was also an erratic and bullying drunk. His 1998 biography is titled *The Most Offending Soul Alive* (Judith Heimann, University of Hawai'i Press.) Harrisson is known for parachuting live cats into Borneo to kill an infestation of plague-carrying rats, which he rightly traced to the spraying of DDT; his testimony helped push the US Congress to a ban. He also grew close to Charles Lindbergh, who emerged from seclusion to help Harrisson persuade Ferdinand Marcos to save a species of wild buffalo and a monkey-eating eagle in the Philippines.

142 Sarawak's orangutans, the least known: Gaun anak Sureng, "Orang-utan on Mt. Kinabalu," *Sarawak Museum Journal* 10, nos. 17–18 (1961): 262–63.

142 hit their target: George Schaller, "The Orang-utan in Sarawak," *Zoologica* 46, no. 6 (1961): 73–82.

142 predecessor William Hornaday: Schaller, "The Orang-utan in Sarawak," 81.

142 In one visit: Schaller, "The Orang-utan in Sarawak," 75.

143 "I liked having": Kay Schaller, interview.

Chapter 4

145 "greeted with this tableau": M. K. Ranjitsinh, interview with the author in Ranjitsinh's New Delhi home, January 23, 2020.

147 "animal is there": John Stein, "George Schaller on Saving the Species," *Omni Magazine*, October 12, 2017, omnimagazine.com/interview-george-schaller-saving-species.

152 "no postcoital display": George B. Schaller, *The Deer and the Tiger: A Study of Wildlife in India* (University of Chicago Press, 1984), 84 (hereafter abbreviated as *D&T*).

154 2 percent of the terrestrial planet: Arjun Srivathsa, Mahi Puri, Krithi K. Karanth, Imran Patel, and N. Samba Kumar, "Examining Human–Carnivore Interactions Using a Socio-Ecological Framework: Sympatric Wild Canids in India as a Case Study," *Royal Society Open Science* 6, no. 5, May 29, 2019, royalsocietypublishing.org/doi/10.1098/rsos.182008.

154 last four thousand tigers: Gayathri Vaidyanathan, "India's Tigers Seem to Be a Massive Success Story—Many Scientists Aren't Sure," *Nature* 574, October 30, 2019, nature.com/articles/d41586-019-03267-z.

155 "on other animals": *D&T*, 64–65.

155 their "bar code": Ullas Karanth, interview with the author, June 8, 2019.

158 "make him happy": Kay to Bettina, February 4, 1964.

160 Annie Dillard's famous frog: Annie Dillard, *Pilgrim at Tinker Creek* (HarperCollins, 1974), 7–8.

162 "irresolutely back and forth": *D&T*, 265.

166 "swarm[ed] from their": *D&T*, 33.

170 from 168 to 16: *D&T*, 16, 102.

171 relationships George developed: Schaller did briefly have a colleague, R. De, helping him in the field; "Though it will not be easy," Southwick had advised, "we would expect you to obtain an Indian zoologist to work with you." De published twice with Schaller but did not continue with wildlife study.

171 *Wild Life of India*: The Kaziranga sanctuary later became the center of a controversy over "green militarization." See Justin Rowlatt, "Kaziranga: The Park That Shoots People to Protect Rhinos," *BBC News*, February 10, 2017, bbc.com/news/world-south-asia-38909512.

171 Gee watched George: Michael L. Lewis, *Inventing Global Ecology: Tracking the Biodiversity Ideal in India, 1947–1997* (Ohio University Press, 2004), 62.

171 go out alone: Bittu Sahgal, "Meet Anne Wright," *Sanctuary Asia* XXXII, no. 4 (August 2012); and Anne Wright, interview with the author, January 6, 2020.

172 Most were "anglicized Brahmins": Karanth, interview.

172 twice in *Cheetal*: George Schaller and R. De, "The Shedding of Antlers by Cheetal Deer," *Cheetal* 8, no. 1 (1964): 15–17; Schaller and J. Spillett, "The Status of the Big Game Species in the Keoladeo Ghana Sanctuary, Rajasthan," *Cheetal* 8, no. 2 (1966): 12–16.

172 cover story in *Life*: George P. Hunt, "Tense Assignment: Stalking the Tiger" and George B. Schaller, "My Year with the Tigers," *Life* 58, no. 25 (June 25, 1965): 3; 60–66.

173 Kipling called it: Rudyard Kipling, *The Jungle Book* (Macmillan, 1894), 4.

173 A children's book: George B. Schaller and Millicent R. Salsam, *The Tiger: Its Life in the Wild* (Harper & Row, 1969).

174 India's Wild Life Act: M. K. Ranjitsinh, "'A Memento of a Lifetime': An Excerpt from Wildlife Conservationist MK Ranjitsinh's Autobiographical Account of Wildlife Conservation in India," *The Caravan*, May 28, 2017, caravanmagazine.in/vantage/life-with-wildlife-excerpt-mk-ranjitsinh.

174 were already extinct: George Schaller, "The Vanishing Wildlife of India," *Audubon Magazine* 70, no. 3 (May–June 1968): 80–89. As happens far too rarely, the hispid hare was rediscovered in 1971 in Assam.

175 to poison predators: Schaller, "The Vanishing Wildlife of India," 80–89.

175 ten thousand export licenses: Wright, interview.

175 "wealth of factual data": Quoted in Lewis, *Inventing Global Ecology*, 62.

175 parasites to apex predators: In a letter to Charles Southwick at Johns Hopkins, Schaller had cited Frank Fraser Darling's *A Herd of Red Deer* (1937) as a model of the "intensive behavioral-ecological study" he had in mind. Fraser Darling included a section on the role of insects and vegetation in determining herd movements. He also recognized human politics and game policy as ecological factors in the lives of animals.

176 a tigress cuts the skin: *D&T*, 266–67.

176 "it is the bible": Valmik Thapar, interview by Tom Veltre, October 22, 2008.

176 "a wildlife revolution": Ullas Karanth, interview by Tom Veltre, January 18, 2009.

177 series of interviews: Lewis interviewed researchers at the Indian Institute of Science, Wildlife Institute of India, and BNHS. See *Inventing Global Ecology*, 59–63. The influence of *The Deer and the Tiger* went beyond India: It inspired George Archibald to establish the International Crane Foundation "to do for big birds what [Schaller] was doing for large mammals," and Luke Hunter, given the book for his tenth birthday, to become a cat biologist. (Archibald quote from an April 24, 2018, letter to Jane Alexander shared with the author with his permission; Hunter from an April 8, 2020, interview with the author.)

177 as "absolute rubbish": Wright, interview.

Chapter 5

181 "something spiritually satisfying": George B. Schaller, *Golden Shadows, Flying Hooves* (Knopf, 1973), 32 (hereafter abbreviated as *GSFH*).

183 "In Ngorongoro Crater": Eric Schaller, interview and email correspondence with the author, September 2019–August 2022.

185 such forlorn babies: *GSFH*, 52.

186 sort the rows: Shot by Alan Root for the National Geographic Channel, the skull sorting appears in selects assembled by Tom Veltre for his 2009 film, shared with the author.

187 "such piercing wails": *GSFH*, 176.

188 tipped-up snout and tusks: *GSFH*, 173.

188 "joy and prolonged affection": *GSFH*, 176.

191 In the 1930s Schäfer: Vaibhav Purandare, "When Nazis Tried to Trace Aryan Race Myth in Tibet," BBC, September 14, 2021, bbc.com/news/world-asia-india-58466528.

191 a Yeti Schäfer: Janaki Lenin, "Searching for the Abominable: Is It a Man? Is It an Ape? It's a . . . Bear?," *The Wire*, February 20, 2018, thewire.in/science/searching-abominable-man-ape-bear.

191 documentary on "Africa": Katherine E. Flach, "Eliot Elisofon: Bringing African Art to *LIFE*" (PhD diss., Case Western Reserve University, 2015), rave.ohiolink.edu/etdc/view?acc_num=case1427999641.

192 *National Geographic* magazine: See *National Geographic* 128, no. 6 (December 1965); *National Geographic Specials*, episode 2, "Miss Goodall and the Wild Chimpanzees," directed by Marshall Flaum, featuring Jane Goodall, narrated by Orson Welles, aired December 22, 1965.

192 "'pride of lions there'": Attenborough, interview by Tom Veltre, October 22, 2008.

192 "The Man Who Was Eaten Alive": George Plimpton, "The Man Who Was Eaten Alive," *The New Yorker*, August 15, 1999, newyorker.com/magazine/1999/08/23/the-man-who-was-eaten-alive.

193 "curt and loath": *GSFH*, 29.

193 "tumbl[ing] end over": Ernest Hemingway, *Green Hills of Africa* (Vintage, 2004), 26.

193 "George took me home": Tony Sinclair, interview with the author, February 28, 2020.

195 Toni Harthoorn and Sue Hart: The Harthoorns became yet another famous wildlife couple with release in 1965 of a film based on their work. *Clarence, the Cross-Eyed Lion* (directed by Andrew Marton) was followed by a TV series called *Daktari* (1966–69, aired on CBS).

195 "stabilizing, gentle and very beautiful": Susanne Hart, *Life with Daktari: Two Vets in East Africa* (Bles/Collins, 1969), 202.

197 six radio collars: The beta versions had until then been tried mostly on rabbits and skunks, though John and Frank Craighead had put a few on grizzly bears. See Etienne Benson, *Wired Wilderness: Technologies of Tracking and the Making of Modern Wildlife* (Johns Hopkins University Press, 2010).

197 dead set against it: *GSFH*, 122.

197 "Scientists are in charge": Matthiessen, *The Tree Where Man Was Born*, 121.

199 "pressure to amass": George B. Schaller, "The Pleasure of Observing," in *Field Notes on Science and Nature*, ed. Michael Canfield (Harvard University Press, 2011), 23.

199 "behaviors become apparent": E. O. Wilson, "The Natural History of Lions," review of *The Serengeti Lion*, by George Schaller, *Science* 179, no. 4072 (February 2, 1973): 466–67.

199 a kind of "natural experiment": Niko Tinbergen, *Curious Naturalists* (Basic Books, 1958; repr. Doubleday, 1968), 288.

200 "To grab a zebra": George B. Schaller, *The Serengeti Lion: A Study of Predator-Prey Relations* (University of Chicago Press, 1972), 265.

201 "longer than it usually": Schaller, *The Serengeti Lion*, 343.

201 a "degree of cooperation": Wilson, "The Natural History of Lions," 466.

201 "group hunting is": Wilson, *Sociobiology*, 504. Subsequent long-term studies by Craig Packer and John Fryxell found that the maximum group size is four; beyond that, each huntress winds up with too small a portion of the kill to make it worth her while. Tony Sinclair, email to the author, February 9, 2025.

201 "nothing so much": Ardrey, *The Social Contract*, 325.

203 Since Alaska, Schaller: As in India, Schaller underscored the interdependencies from top to bottom of the trophic chain. A certain tapeworm larva lives inside the sacrum of wildebeest and topi, which only the hyena (with the most powerful jaws of any predator in East Africa, and blunt conical teeth) is strong enough to crush. Hyena intestines are the only place that worm can then complete its life cycle.

203 "such proficient killers": George Schaller, "Deadly Dogs of the Savanna," *Life*, May 10, 1968, 100–105.

204 down a whole pack: *GSFH*, 220.

205 letting anger sabotage: *GSFH*, 226.

205 "shoveling in on": Hans Kruuk, interview with the author, May 9, 2019.

205 books and a film: Jane Goodall and Hugo Van Lawick, *Innocent Killers* (Houghton Mifflin, 1971); Hugo Van Lawick, *Solo: The Story of an African Wild Dog Puppy and Her Pack* (Collins, 1973); Hugo Van Lawick, dir., *Jane Goodall and the Wild Dogs of Africa* (Metromedia Producers Corp., 1973). George did include *Innocent Killers* in his picture book bibliography.

206 "I am unable": Schaller, *The Serengeti Lion*, 298.

206 "What struck me": Kruuk, interview; Sinclair, interview.

206 William Conway, the NYZS director: William Conway, interview with the author, June 25, 2018.

207 "merely stupid," wrote: Matthiessen, *The Tree Where Man Was Born*, 157.

207 "He says they are playing": Martha Gellhorn, "Animals Running Free: Two Weeks in the Serengeti," *The Atlantic Monthly* 217, no. 2 (February 1966): 70–76.

207 "As a scientist I am": George B. Schaller, *Serengeti: A Kingdom of Predators* (Knopf, 1972), book jacket.

207 "I saw a starving cub": *GSFH*, 58.

207 meet the other scientist: As PhD supervisor for Sinclair, Kruuk, and Douglas-Hamilton, Tinbergen visited East Africa several times.

208 the "happy event": Tinbergen, *Curious Naturalists*, 264.

208 "long periods of relaxed": Tinbergen, *Curious Naturalists*, 153, 287.

208 "the entire behaviour": Richard Burkhardt, *Patterns of Behavior: Konrad Lorenz, Niko Tinbergen, and the Founding of Ethology* (University of Chicago, 2005), 407.

209 "Killer ecologists," Gellhorn called them: Gellhorn to Raymond F. Dusmann, 1968 (exact date not known), in *Selected Letters of Martha Gellhorn*, ed. Caroline Moorehead (Henry Holt, 2006), 337.

209 "a place would explain itself": Lily Huang, interview with the author, February 8, 2019.

209 "the calm rhythm": *GSFH*, 31.

209 "lacks a discrete category": Schaller, "The Pleasure of Observing," 22.

209 "When a lion sees": Schaller, *The Serengeti Lion*, 241.

210 Tinbergen concluded that: Tinbergen, *Curious Naturalists*, 15.

210 secure not dominance: This is the subject of philosopher Martha Nussbaum's 2022 book *Justice for Animals: Our Collective Responsibility* (Simon & Schuster).

210 "recorded who licked": *GSFH*, 90.

210 "plains at night": *GSFH*, 130.

211 says Jonah Western: Western, interview.

211 "existence revolved around": *GSFH*, 90, xxi.

211 "a bold and somewhat derisive": *GSFH*, 121.

211 "in a world older": *GSFH*, 62. Quote from Henry Beston, *The Outermost House: A Year of Life on the Great Beach of Cape Cod* (Henry Holt, 1992), 25.

211 "mystery of a lion's mind": *GSFH*, 110.

211 buffalo standing belly-deep: *GSFH*, 131.

211 "a Victorian who has successfully": Wilson, "The Natural History of Lions," 466–67.

211 this fertile hybridizing: Gellhorn, "Animals Running Free," 72, 74.

212 1971 *Science* article: Thomas R. Blackburn, "Sensuous-Intellectual Complementarity in Science: Countercultural Epistemology Has Something of Value to Contribute to the Science of Complex Systems," *Science* 172, no. 3987 (June 1971): 1003–7, science.org/doi/10.1126/science.172.3987.1003.

212 at some length: *GSFH*, 82.

212 "patiently, inexorably, heads": George Schaller, "Life with the King of Beasts," *National Geographic* 135, no. 4 (April 1969): 510.

212 "participate emotionally, knowing": The first quotation is from Schaller, *The Serengeti Lion*, 233. The second is from Schaller's field journals.

212 disrupting the lives of the Maasai: For a comprehensive history, including the role played by Grzimek and the then-prevalent fantasy of a prehuman paradise, see Thomas M. Lekan, *Our Gigantic Zoo: A German Quest to Save the Serengeti* (Oxford University Press, 2020). See also Anthony R. E. Sinclair, *Serengeti Story: Life and Science in the World's Greatest Wildlife Region* (Oxford University Press, 2012).

213 "mania for killing": *GSFH*, xvii.

213 detailed observations of lions: Stephen Makacha and George Schaller, "Observations on Lions in the Lake Manyara National Park, Tanzania," *African Journal of Ecology* 7, no. 1 (August 1969): 99–103.

215 men were legendary trackers: Matthiessen, *The Tree Where Man Was Born*, 101.

215 Turner's reports detailed: *GSFH*, 244.

216 decades would explode: David Western, afterword to the paperback edition, *In the Dust of Kilimanjaro* (Island Press, 1997), 283.

216 "To cause unspeakable pain": *GSFH*, 119–20.

216 "may or may not": Pamela Turner, *A Life in the Wild: George Schaller's Struggle to Save the Last Great Beasts* (Farrar, Straus and Giroux, 2008), 51.

216 "I cupped the blood-encrusted": Schaller, *Serengeti*, 67.

217 "Big Business" types: Gellhorn, "Animals Running Free," 73–74.

217 as many cheetahs: George B. Schaller, "This Gentle & Elegant Cat," *Natural History* LXXIX, no. 6 (June–July 1970): 30–39; and reprinted in George B. Schaller, *A Naturalist and Other Beasts: Tales from a Life in the Field* (Sierra Club Books, 2007).

218 "but also a reincarnation": *GSFH*, 183.

219 Lem Billings, who: David Walter, "Best Friend," *Princeton Alumni Weekly*, April 12, 2017, paw.princeton.edu/article/best-friend.

219 covered at length by *Life*: "Bobby Jr. in Africa," *Life*, February 14, 1969, 42–49.

220 overcoming "cross-cultural barrier[s]": George Schaller, foreword to *Born Free: A Lioness of Two Worlds*, by Joy Adamson (Mcfadden 1960; repr. Pantheon Books, 1987).

221 "large wild animal": *GSFH*, 185.

221 "hounds of hell": Matthiessen, *The Tree Where Man Was Born*, 118.

222 "You don't want them": Matthiessen, *The Tree Where Man Was Born*, 185–86, 206–7.

222 been "social carnivores": Schaller, *The Serengeti Lion*, 378, 10.

223 noted writer John Vaillant: John Vaillant, *The Tiger, A True Story of Vengeance and Survival* (Knopf, 2010), 112–13.

223 Schaller's second Serengeti book: Schaller, *Serengeti*.

223 "a little masterpiece": All correspondence between Schaller and Knopf can be found in "Schaller, George—Lion Book," Alfred A. Knopf Addition to Editorial Files, Box 22, folders 4–5, Harry Ransom Center, University of Texas at Austin.

223 "But with exquisite timing": Schaller, *Serengeti*, 62.

224 "in moist clefts sometimes": Schaller, *Serengeti*, 3.

224 "lean viscous creatures": *GSFH*, 171, 35, 50.

225 "I cannot see": The Owen letter is in the Yale archive. More than fifty years later, *The New York Times* was still referencing the Schaller study. See Anthony Ham, "Why Don't All Lions Climb Trees?," *The New York Times*, March 4, 2022, nytimes .com/2022/03/04/science/lions-trees.html.

225 "Just go, and hurry": "How a Couple of Zoologists from Georgia Have Taken on an Army of Wildlife Poachers in Zambia and Are Winning—Not with Weapons, but with Jobs," *The Chicago Tribune*, published on January 21, 1993, updated on August 10, 2021, chicagotribune.com/news/ct-xpm-1993-01-21-9303170936-story.html. The Owens' role, if any, in the unsolved killing of an alleged poacher remains unresolved. See Jeffrey Goldberg, "The Hunted," *The New Yorker*, March 29, 2010, https://www.newyorker.com/magazine/2010/04/05/the-hunted.

226 "a world of swift": Loren Eisley, "Tiger, Tiger, Burning Bright," review of *GSFH*, *The Washington Post*, December 30, 1973.

226 a "cherished notion": Wilson, *Sociobiology*, 247.

226 "we are among": Wilson, "The Natural History of Lions," 466.

226 Wilson was then: Wilson's *Sociobiology* built on the insights of Charles Darwin, the first to posit that evolution had shaped emotions and social structures as well as physiology. Wilson's work has been controversial from the day of its publication, particularly its chapters on humans: His Harvard colleagues Stephen Jay Gould and Richard Lewontin criticized it for resurrecting the kind of "biological determinism" that underlay eugenics and "more recent claims for a genetic basis of racial differences in intelligence." See Elizabeth Allen, Barbara Beckwith, Jon Beckwith, Steven Chorover, et al., "Against 'Sociobiology,'" *The New York Review of Books* 22, no. 18 (November 13, 1975), and see Richard Rhodes's *Scientist: E. O. Wilson: A Life in Nature* (Doubleday, 2021) for a persuasive challenge to this critique.

226 the rare step: Boyce Rensberger, "Sociobiology: Updating Darwin on Behavior," *The New York Times*, May 28, 1975, 1.

227 "distemper had hit": Ardrey, *The Social Contract*, 147.

227 "A lioness came": Ardrey, *The Social Contract*, 141.

227 "the antidote to Elsa": Cyril Connolly, "Carnage in Eden," review of *The Serengeti Lion*, by George B. Schaller, *The New York Review of Books*, January 25, 1973.

227 "jurisdiction" of their "overlords": *GSFH*, 98; Schaller, "Life with the King of Beasts."

227 "remind one of American": George Stade, review of *The Serengeti Lion*, by George B. Schaller, *The New York Times Book Review*, December 3, 1972, 6, 88.

228 first "not surprisingly": S. A. Barnett, "Pride and Prey: The Delicate Balance," review of *The Serengeti Lion*, by George B. Schaller, *The Washington Post*, December 10, 1972. He added, "The sex life of flesh-eaters is energetic but not (in the moral sense) edifying."

228 rivalry and respect: Kay Schaller to Maria Eckhart, April 8, 2014. Schaller private collection.

228 book on the Kurelu: Peter Matthiessen, *Under the Mountain Wall: A Chronicle of Two Seasons in the Stone Age* (Viking, 1962). In 1965 Matthiessen had published one of his finest novels, *At Play in the Fields of the Lord* (Random House).

228 "the world-renowned George Schaller": E. R. C. Davidar, *Whispers from the Wild* (Penguin, 2012).

229 "lanky and solitary": Priya Davidar, interview with the author, January 9 and 10, 2020, on site at the land her father protected as an elephant corridor, the Sigur Nature Trust in Tamil Nadu, India.

Chapter 6

234 not eager at first: Zahid Beg Mirza, personal correspondence and interview with the author, July 18 and 19, 2020.

234 "behind me played restlessly": George B. Schaller, *Stones of Silence: Journeys in the Himalaya* (Viking, 1980), 139 (hereafter abbreviated as *SOS*).

234 the ostensibly fierce: *SOS*, 146.

234 "handsomest . . . their pelage": George Schaller, "Stalking the Wild Sheep of Kalabagh," *International Wildlife* 5, no. 4 (1975): 46.

235 regularly birthed twins: Schaller, "Stalking the Wild Sheep," 44–46.

237 "his fur-lined trousers": Mirza, personal correspondence and interview.

238 *National Geographic* published them: George B. Schaller, "Imperiled Phantom of Asian Peaks: First Photographs of Snow Leopards in the Wild," *National Geographic* magazine 140, no. 5 (November 1971): 702–7.

238 "beyond the subjective": *SOS*, 9, 24.

239 "let the leopard interfere": Peter Matthiessen, *The Snow Leopard* (Viking, 1978; repr., Penguin, 1987), 226. Citations refer to the 1987 Penguin edition.

239 the "colonial types": Hemanta R. Mishra, quoted in Benson, *Wired Wilderness*, 99.

242 Tibetan Buddhist world: In coming years, Schaller will spend significant time reading the growing literature on how Buddhism and Western science do, and do not, mesh, returning again to the work of D. T. Suzuki and Aunt Paula's father, the publisher Paul Carus, who had collaborated with Suzuki in introducing Buddhism to the West.

247 was called Colt: Masood Hasan, "Remembrance: True Blue," The News on Sunday, *Daily Jang*, December 14, 2008, jang.com.pk/thenews/dec2008-weekly/nos-14-12-2008/foo.htm.

247 George was "Teutonically punctual": *SOS*, 133

247 publishing their findings: George B. Schaller, S. Amanullah Khan, "Distribution and Status of Markhor (*Capra falconeri*)," *Biological Conservation* 7, no. 3 (April 1975): 185–98.

248 regent for twelve years: Christopher Buyers, "Chitral," The Royal Ark, May 2022, royalark.net/Pakistan/chitral9.htm.

250 confirming a decline: Markhor were listed as endangered by the IUCN in 1994. Mary Bates, "Pakistan's Comeback Kid: The Wild Goat," July 17, 2012, American Association for the Advancement of Science, https://www.aaas.org/pakistans-comeback-kid-wild-goat.

250 founder C. J. McElroy: In *Stones*, Schaller misspells his name "Mikelroy," ostensibly transcribing Hassan's mispronunciation, but perhaps also forestalling retaliation by McElroy (safariclub.org).

251 endemic wild goat: Also called the Sindh ibex.

251 "can't break away": Andrew Laurie, interview with the author, London, May 10, 2022.

252 "acquaintances are difficult": *SOS*, 124.

253 "like desert rodents": *SOS*, 119.

254 a Pashto song: George seems to have been reading Peter Mayne's 1955 book, *The Narrow Smile: A Journey Back to the North-West Frontier*, for which these lyrics served as the epigraph.

254 "Something to expiate": *SOS*, 123.

256 famously dangerous road: "Lowari Pass Is the Grand Daddy of the Passes," Dangerous Roads, dangerousroads.org/asia/pakistan/4183-lowari-pass.html.

256 "five jeeps fell": *SOS*, 36–37.

259 he would quote Virgil: *SOS*, 33.

260 "National Geographic book": *Lion Cubs: Growing Up in the Wild* (National Geographic, 1972) included George's pictures of Kay bottle-feeding Ramses and the boys crawling and wrestling with him.

264 grandest and rarest sheep: Laurence Bergreen, *Marco Polo: From Venice to Xanadu* (Knopf, 2007), 74.

264 "head of heads": Rudyard Kipling, "The Feet of the Young Men" (originally published 1897), The Kipling Society, kiplingsociety.co.uk/poem/poems_feetofyoung.htm.

268 Raja of Gupis: All such titles, Schaller notes, had been abolished the previous year.

268 who "with precision outlined": *SOS*, 71.

268 circling "like jackals": *SOS*, 64

268 "A desecration of": *SOS*, 65.

268 "survey had been a failure": *SOS*, 69.

269 "as local people do": *SOS*, 51.

269 "pitiless walls of ice": *SOS*, 59–60.

270 "no animals left": Peter Matthiessen Papers, box 48.

271 "in fine shape": Peter Matthiessen Papers, box 48.

271 about visiting Peking: Nixon's historic visit had occurred just a year earlier.

271 "no experience of climbing": Peter Matthiessen Papers, box 48.

272 story in legend: Peter would tell the story in *The Snow Leopard*, though without naming House.

273 "pagoda-shaped stupas": *SOS*, 225–26.

273 "someone, probably Gyaltsen": *SOS*, 254.

274 "I see his earth-colored figure": *SOS*, 217.

274 "oldest and slowest": Matthiessen, *The Snow Leopard*, 97.

274 "constraints of Kathmandu": *SOS*, 207, 222.

275 "dwindling lamp fuel": Matthiessen, *The Snow Leopard*, 147.

275 "finance and frugality": Peter Matthiessen Papers, box 45.

275 did not trust them: *SOS*, 216.

275 The "dirty Kamis": Matthiessen, *The Snow Leopard*, 81. "'Kami' people of the blacksmith caste, soot-faced familiars of the smelt-fires and the iron stolen from the rock," he added, are "feared and despised as black magicians by primitive people throughout Eurasia and Africa."

275 "we are stranded": Matthiessen, *The Snow Leopard*, 87, 88.

276 "scene of utter desolation": *SOS*, 220, 221.

276 described him as "disgusted": Matthiessen, *The Snow Leopard*, 100.

276 "pulled up high": Matthiessen, *The Snow Leopard*, 118.

276 "'short like a pelt'": *SOS*, 213.

277 "A *Bauhinia*, I think": *SOS*, 208.

277 The "human chamois": Peter comes back to this theme of George's "marvelously sure" step and balance many times: "One of my first impressions of this man, in East Africa, in 1969, was the sight of him standing casually on the very edge of a towering granite outcrop, in hard wind, scanning the Serengeti Plain through these same binoculars." He noted that on these narrow trails, the scope strapped across George's rucksack could—if caught on a rock—nudge him off the edge, and that GS walked upright across log bridges that Peter had to "hitch ignominiously across" on his backside. Matthiessen, *The Snow Leopard*, 99, 54.

277 "he is mortal": Matthiessen, *The Snow Leopard*, 33, 98.

279 "watched us pass": Matthiessen, *The Snow Leopard*, 148.

280 "snow is hard": *SOS*, 229–30.

281 "this offer without thanks": Matthiessen, *The Snow Leopard*, 153–54.

281 "no basis for judging": *SOS*, 231.

281 "total increased expense": Matthiessen, *The Snow Leopard*, 160.

281 "supposedly impossible route": *SOS*, 231.

281 "a savage joy": *SOS*, 231.

281 began the descent: Journalist Kathryn Schulz has criticized the "recklessness" of this trip, its "bad equipment and intermittent but unmistakable bad leadership." Kathryn Schulz, "What Do We Hope to Find When We Look for a Snow Leopard?," *The New Yorker*, July 5, 2021, newyorker.com/magazine/2021/07/12/what-do-we-hope-to-find-when-we-look-for-a-snow-leopard.

283 "Top of Pass": Matthiessen, *The Snow Leopard*, 156.

283 "twenty-five dollars' worth": Matthiessen, *The Snow Leopard*, 162.

283 "unable to contain": Peter Matthiessen Papers, box 45; Matthiessen, *The Snow Leopard*, 160.

283 "come to my senses": Matthiessen, *The Snow Leopard*, 167.

283 what George was up to: This account comes from Schaller's journal; in *Stones*, he slides so quickly past his day of rest the reader barely notices.

283 "floundering through soft snow": Matthiessen, *The Snow Leopard*, 166.

284 *both* were suffering: *SOS*, 234.

284 "could not write": Matthiessen, *The Snow Leopard*, 164.

285 "a remorseless sleeper": Matthiessen, *The Snow Leopard*, 175.

285 "Never move more": Matthiessen, *The Snow Leopard*, 173.

285 impacts of tireless human: Villagers barred from taking live firs, for instance, girdled the trees so they would die. In twenty-five years, Nepal lost half its forest. See "Forests," in *Nepal: A Country Study*, ed. Andrea Matles Savada (Library of Congress, 1991), countrystudies.us/nepal/43.htm.

287 "'Kay gets wild'": Matthiessen, *The Snow Leopard*, 33.

287 "fine, old-fashioned qualities": Matthiessen, *The Snow Leopard*, 43.

288 Phu-Tsering, hunkered: Matthiessen, *The Snow Leopard*, 157.

288 "accept his reprimands": Matthiessen, *The Snow Leopard*, 94.

288 "a broken cricket": Matthiessen, *The Snow Leopard*, 30.

288 "freedom of the snow": Matthiessen, *The Snow Leopard*, 157.

288 "upon the wall": Matthiessen, *The Snow Leopard*, 42. Asked forty-five years later about this incident, George agreed: "That's true; it was careless of me."

289 "how to write": Matthiessen, *The Snow Leopard*, 33.

289 "learning about people": George also, in this unguarded journal entry, dispensed with an insight of Peter's that does seem far off the mark. Peter would make much of what he called George's "insecurity": his "paranoid" fear that his colleagues were just waiting for him to fail. After two meager and disheartening years, George did worry that he might again come home without the data that "justified" his existence. But that had little to do with what his colleagues might think. Even as a graduate student, he was more inclined to deflate the bestowers of accolades than to seek their favor. And it was Peter, he noted, who had "collected cannabis along the way to smoke [and] commented that he used to drink too much." Whatever insecurity George might suffer, at least "it does not rely on drugs for alleviation."

289 with "no company": Matthiessen, *The Snow Leopard*, 2.

289 "a stern pragmatist": Matthiessen, *The Snow Leopard*, 14. Matthiessen is quoting *The Tree Where Man Was Born*, 126.

289 fearful Phu-Tsering chanted: Matthiessen, *The Snow Leopard*, 102, 63–64.

290 "GS refuses to believe": Matthiessen, *The Snow Leopard*, 58.

290 "a dove calls": Matthiessen, *The Snow Leopard*, 34. In a marvelous piece for *The Hopkins Review*, historian of science Lily Huang noted that the Romantics had honored the "scientist-poet," and seen fieldwork as an "explicitly religious" pursuit of revelation. But in the twentieth century "you were no longer allowed to mingle divinatory intuition with crystalline fact-gathering." She saw in Schaller a repair of that rift. Lily Huang, "The Mystic Science: George Schaller in the Field," *The Hopkins Review* 9, no. 3 (Summer 2016): 368–70.

290 "unweave a rainbow": John Keats, "Lamia," part 2, (1819; Project Gutenberg, 2008), gutenberg.org/files/2490/2490-h/2490-h.htm.

290 "under [a] wintry light": Ralph Waldo Emerson, "Nature," in *The Complete Works of Ralph Waldo Emerson: Nature Addresses and Lectures*, vol. 1 (originally published 1876), University of Michigan Library Digital Collections, accessed April 3, 2024, name.umdl.umich.edu/4957107.0001.001.

290 Max Weber saw: The opposition of science and enchantment surfaces still: Writing of his 2018 reprise with photographer Vincent Munier of the search for the snow leopard, French author Sylvain Tesson derided the "number crunchers" who "ringed hummingbirds and disemboweled seagulls to take samples of bile. . . . Did their work contribute to the sum of knowledge?" He quoted Munier: "A yak is a lord. . . . What do I care whether he's ruminated twelve times this morning!" Sylvain Tesson, *The Art of Patience: Seeking the Snow Leopard in Tibet* (Penguin, 2021), 24–25. And again, in the September 2022 Voyages issue of the *New York Times Magazine*: "Even Schaller the hardened scientist," wrote Sam Anderson in a piece about *The Snow Leopard*, "summons a bit of poetry." Contradicting himself, Munier did deem Schaller—the great number cruncher and disemboweler—"the master" and "king."

290 seems "less dogmatic": Matthiessen, *The Snow Leopard*, 223, 225.

290 he begins writing haiku: Matthiessen, *The Snow Leopard*, 62.

290 a kind of Zen koan: Matthiessen, *The Snow Leopard*, 223, 225.

292 "'That's not what you meant'": Matthiessen, *The Snow Leopard*, 230.

293 "I lie buried": *SOS*, 117, 252.

293 "moves steadily ahead": Matthiessen, *The Snow Leopard*, 97.

293 "easy, ethereal lightness": When George says to him, as he squeezes past, "This is the first really interesting stretch of trail we've had so far," Peter thinks "how easy it would be to push him over." Matthiessen, *The Snow Leopard*, 142–43.

293 "this scientist has": SueEllen Campbell, "Science and Mysticism in the Himalayas: The Philosophical Journey of Peter Matthiessen and George Schaller," *Environmental Review* 12, no. 2 (Summer 1988): 127–42. This is one of many essays that have used this journey to explore the counterpoint, or convergence, between mysticism and science.

293 "strangers in nature": Emerson, "Nature."

293 All of us, except George: Reviewing *The Snow Leopard* together with Schaller's scientific monograph *Mountain Monarchs* for *The New York Review of Books*, Robert Adams (founding editor of the Norton Anthology of English Literature) found Matthiessen's Zen exposition "not very new" and a "bit watery" in the wake of such books as Robert M. Pirsig's *Zen and the Art of Motorcycle Maintenance.* Adams preferred Schaller's "nineteen closely printed pages of references," adding that for all his graphs and charts, "this man is no less a poet in his feeling for the high hills and their animal populations than his Zen companion." Robert M. Adams, "Blue Sheep Zen," *The New York Review of Books*, September 28, 1978, nybooks.com/articles/1978/09/28/blue-sheep-zen.

294 "separating the sheep": *The Dick Cavett Show*, episode 67, aired May 8, 1980, on ABC, Library of Congress.

294 "proper names . . . poetry": W. H. Auden, *A Certain World: A Commonplace Book* (Viking, 1970), 223.

294 "animals in Tibet": Vladimir Nabokov, *The Gift* (Random House, 1991), 68.

294 "incompletely named world": Nabokov, *The Gift*, 117.

294 "accustomed to speaking Latin": Tesson, *The Art of Patience*, 30.

295 "understand a frog": *SOS*, 243

295 "a good trip": *SOS*, 254.

295 seeing wolf tracks: Matthiessen, *The Snow Leopard*, 242.

295 "Half and half": *SOS*, 255.

296 "luxury of warmth": *SOS*, 260.

298 "like a schoolboy": *SOS*, 270.

298 "dispersing a gathering": *SOS*, 271.

302 idol and inspiration: Benson, *Wired Wilderness*, 104, 106.

303 "open to misinterpretation": Peter Matthiessen Papers, box 48.

303 "This letter is long overdue": Aerogramme from Kay Schaller to Peter Matthiessen, Peter Matthiessen Papers, box 48.

305 "the late Miocene": The rare Schaller mistake: Twenty million years ago is the early Miocene.

305 "wanderer or evil monk": Matthiessen, *The Snow Leopard*, 86.

305 a "Boddhisatva smile": Matthiessen, *The Snow Leopard*, 221.

305 "Tukten's spiritual propensities": Matthiessen, *The Snow Leopard*, 230.

305 "hurls goggles through": Matthiessen, *The Snow Leopard*, 96.

305 "intense respect of a private soul": Matthiessen, *The Snow Leopard*, 156–57.

306 Peter's notes on George's manuscript: Peter Matthiessen Papers, box 58.

306 "determined blue pencil": Peter Matthiessen Papers, box 58

306 George's one-sentence account: Peter described the beast that had "selected [him] for attack. . . . Snarling in a low, ugly way [it] lunged back and forth at the tip of my stick in a horrid fury." When GS hurled "a heavy split of wood . . . with all his force. The brute dodged, then sprang after it, sinking its teeth deep into the wood." Peter then marveled at whatever guidance had led him to cut his first stave an hour earlier. Matthiessen, *The Snow Leopard*, 47.

306 told *The Paris Review*: "Peter Matthiessen, The Art of Fiction No. 157," interview by Howard Norman, *The Paris Review*, no. 150 (Spring 1999), theparisreview.org/interviews/985/the-art-of-fiction-no-157-peter-matthiessen.

307 "partly to appease Peter": *SOS*, 251.

308 "harsh high-pitched" voice: Matthiessen, *The Snow Leopard*, 212.

308 "a penis-lick": Matthiessen, *The Snow Leopard*, 246.

308 most literary work: *SOS*, 250, 70, 66, 203, 217, 220.

308 at its most mystical: *SOS*, 215, 263, 32, 252, 111–12.

309 "The heights glimmer": *SOS*, 243, 248.

309 "some centuries hence": *SOS*, 254.

309 returns to the concrete: Writing four decades later, Indigenous poet Jidi Majia, a member of the Yi-Nuosu people who live alongside snow leopards on the Tibetan Plateau, would make clear whose account he preferred. In a long poem dedicated to Schaller, written in the voice of the snow leopard, the cat chides those who would make him an instrument of their inner quest.

> You may imagine I am . . .
> your refuge, a benign remedy, as if
> a creature in a long-forgotten
> experience or a Dream Vision.
>
> But I appear to you now
> incarnate, glorious
> as sunlight shattering
> on a dish of silver coins . . .

Jidi Majia, "I, Snow Leopard," translated from the Chinese by Frank Stewart, with an introduction by Barry Lopez, *Orion* 25, no. 4 and 5 (July–Sept 2016).

309 "*Alticola stoliczkanus* probably": *SOS*, 262–63.

309 Noting its similarity: A later paper by McNeely cited Schaller as a reference. See J. A. McNeely, E. W. Cronin, and H. B. Emery, "The Yeti—Not a Snowman," *Oryx*

12, no. 1 (May 1973): 65–73. In his 1986 fantasy novella *Escape from Kathmandu*, Kim Stanley Robinson also cites Schaller giving credence to the Yeti story.

310 George reported soberly: Peter Matthiessen Papers, box 175. See also James Shreeve, "Oliver's Travels," *Atlantic Magazine* 292, no. 3 (October 2003): 94–102.

310 "not much to tell": Mel Sunquist, email correspondence with the author, May 10, 2022.

311 George's acknowledgment of Kay: *SOS*, vi.

313 "survival and death": *SOS*, 42.

313 "very definition of invisible": In 1981, Rodney Jackson and Dara Hillard became the first to radio-collar snow leopards. Schaller protégé Tom McCarthy also managed to trap and collar snow leopards, but still saw them just five times in six years.

313 species of wild cat: Eric Dinerstein, "I ♡ Wild Cats," review of *Wild Cats of the World*, by Fiona Sunquist and Mel Sunquist, *Conservation Biology* 11, no. 6 (August 6, 2009): 1853–55.

314 an intestinal parasite: Fiona Sunquist and Mel Sunquist, *Tiger Moon: Tracking the Great Cats in Nepal*, rev. ed. (University of Chicago Press, 2002), 1.

314 No snow leopard: Sher Panah made this report in a letter to Schaller. *SOS*, 49.

315 no foreigner might enter: *SOS*, 79.

315 a warning shout: *SOS*, 86

316 quit crossing nearby: *SOS*, 89.

317 an ideal Nature: Amitav Ghosh, "Wild Fictions," 2008, amitavghosh.com/docs/Wild%20Fictions.pdf. Ghosh argues that stories like the Bon Bibi tales told in the Sundarbans are far more effective than state sanctions at constraining human heedlessness. For this critique, Ghosh drew from a 1999 article in the *Journal of Political Ecology* by anthropologist Are Knudsen, "Conservation and Controversy in the Karakoram: Khunjerab National Park."

317 "exclusion of all human": Trinity College anthropologist Shafqat Hussain has levied similar critiques, in *The Snow Leopard and the Goat: Politics of Conservation in the Western Himalaya* (University of Washington Press, 2019), 48–50, and *Remoteness and Modernity: Transformation and Continuity in Northern Pakistan* (Yale University Press, 2015), 189–99—though he, too, assailed recommendations Schaller had not in fact made. That jumbled history has been recycled to undercut conservation; see Shabbir Mir, "Khunjerab National Park: A Failed Project," *The Express Tribune*, June 27, 2011, tribune.com.pk/story/197118/khunjerab-national-park-a-failed-project.

318 Such dispossession had in fact: For literature on this topic, see "Decolonizing Conservation: A Reading List," curated by Sara Cannon, notion.so/c175fb8fa923469d840dda6555fc355a.

318 how everyone—human and nonhuman: Murie's vision was itself presaged in 1831 by American painter George Catlin, who appealed for a "*nation's Park*, containing man and beast, in all the wild and freshness of their nature's beauty!" In 1899, Kenya became the first to protect both wildlife and traditional people in its Southern Reserve. In 1969 Western proposed a similar model, a Maasai park, for Amboseli. George Catlin, *Letters and Notes on the Manners, Customs, and Condition of the North American Indians*, vol. 1 (London: published by the author, 1841), 262; Jonah Western, correspondence with the author, February 14, 2025.

318 now similarly recommended: Reproduced in his grant report "Ecology and Behavior of High-Altitude Mammals in South Asia" to the *National Geographic Society, Research Reports* 11 (1979): 461–78.

318 definition for a national park: For a discussion of the ongoing evolution of defining criteria, as the field followed Schaller in recognizing humans' place in these ecosystems and the ethical and practical need for local leadership, see Adrian Phillips, "The History of the International System of Protected Area Management Categories," *Parks* 14, no. 3, (2004): 3–14; see also, "What Is the Whakatane Mechanism?," https://www.worldparkscongress.org/sites/wpc/files/sessrep/453_5_Whakatane%20A%20rights-based%20tool%20to%20improve%20governance%20and%20equity%20in%20protected%20areas.pdf.

318 That no-grazing zone: Knudsen and then Ghosh cited biologist Per Wegge to suggest that livestock posed no threat to wild sheep because they did not compete for forage. Per Wegge, "Assessment of Khunjerab National Park and Environs, Pakistan, Survey 16 October–17 November 1988," unpublished report, 1988, portals.iucn.org/library/node/11723. This is the only part of this story that really riled Schaller, who noted that Wegge based his claim on a single visit to Khunjerab in mid-October 1988, well past the time when villagers had moved their livestock to lower ground. And that the damage done by such competition has been proven by Indian scientists "in years of actually measuring livestock impact in similar landscapes."

318 "George saw that before": Mirza, personal correspondence and interview.

319 minister of Sindh: As Hussain fairly notes, conservation has often been, here as elsewhere, the province of elites.

319 Rowell later recalled: Galen Rowell, *In the Throne Room of the Mountain Gods* (Sierra Club Books, 1986), 117; Schaller would write the foreword, though he was critical of the military scale of such expeditions and of climbers' willingness to risk so many Sherpas' lives for their "hobby." Around the time of Schaller's first trip to Nepal, ten Sherpas were killed on a South Korean expedition on Manaslu.

320 "He was so gentle": Bill Leacock, interview with the author, August 8, 2023.

321 Mount Drum "persists": *SOS*, 104.

322 monograph Schaller published: George B. Schaller, *Mountain Monarchs: Wild Sheep and Goats of the Himalaya* (University of Chicago Press, 1977), xiii, xv, 335.

Chapter 7

325 to George's desk: An oak monstrosity, which had been the great William Beebe's before him.

325 "memento[s] of exploration": *TW*, 1.

326 the federal agency: Instituto Brasileiro de Desenvolvimento Florestal (Brazilian Institute of Forestry Development).

327 small, medium, and large: XL, too, though Schaller saw a giant armadillo just once.

328 "think of nothing else": Peter Crawshaw, unpublished manuscript, June 2020, shared with the author. Also Crawshaw correspondence and interviews with the author, April, May, and September 2020. Further details are drawn from two unpublished Schaller manuscripts: "Realm of the Jaguar" and "Roosevelt's Ghosts."

329 The parks director: Celso Soares de Castro.

331 "a strayed Lilliputian": Schaller, *A Naturalist and Other Beasts*, 56–59, originally published as "The Mouse That Barks," in *International Wildlife* 6, no. 5 (1976): 12–16.

331 "bats on velvet wings": Schaller, *A Naturalist and Other Beasts*, 47, originally published as a foreword to *Rainforests of the World: Water, Fire, Earth, Air*, by Sir Ghillean Prance (Crown Publishers, 1998), 10.

333 "Life . . . smothers one": *SOS*, 6.

334 this primal force: Schaller, *A Naturalist and Other Beasts*, 68, originally published as "Epitaph for a Jaguar," in *Animal Kingdom* (April/May 1980): 4–11.

334 six bears had died: Benson, *Wired Wilderness*, 54, 56.

334 the Sucostrin Kid: Benson, *Wired Wilderness*, 60.

334 Monique the Space Elk: Benson, *Wired Wilderness*, 80–81.

334 lodged by Adolph Murie: Benson, *Wired Wilderness*, 69.

334 seconded by Olaus: Olaus had from the outset instilled in George a belief in research dependent not on equipment but on "close observation [and] thinking."

335 with foot snares: When an animal steps on a hidden trigger, a spring flips up and tightens a cable around its foot. The cable is attached to a tree or other immovable object. Fish or meat might be hung above as bait, or a pathway of twigs built to guide the animal in.

335 computer "so complicated": Email to the author, December 13, 2021.

336 But "that hunt": This is from an email George sent Peter Crawshaw on June 9, 2020, when Peter was working on his own memoir—which Crawshaw in turn shared with the author.

337 "the professor is there": Schaller email to Peter Crawshaw, June 9, 2020.

337 pig called Fuxica: Though he initially misunderstood the name she'd been given as Kushika, Schaller later concluded with Crawshaw's help that it was in fact short for "fuxiqueira": someone who pokes her nose into everything.

341 "the whole world was the whale's": Herman Melville, *Moby Dick: Or, the White Whale* (L. C. Page, 1926), 429–30.

342 "dad said, 'Cut' ": Eric Schaller, interview with the author at his parents' home and Eric's office at Dartmouth College, September 2019.

342 "jaguar had killed it": Mark Schaller, interview with the author.

345 "Reading this account again": George B. Schaller, foreword to *White Waters and Black*, by Gordon MacCreagh (University of Chicago Press, 2001), xi, xii.

346 collection of Yanomami: The Yanomami became emblems of the uncorrupted hunter-gatherer thanks to studies by Napoleon Chagnon and his critics; they inspired Matthiessen's fictional Niaruna tribe in *At Play in the Fields of the Lord*, a novel out of which Schaller's missionaries and mercenary pilots sometimes seem to have stepped.

354 Almeida's book on: Tony de Almeida, *Jaguar Hunting in Mato Grosso and Bolivia* (Stanwill Press, 1976).

356 "you found the jaguars": The account of the hunt in which Gigante died, and of the execution of the jaguars, is drawn in part from "Epitaph for a Jaguar," collected in Schaller, *A Naturalist and Other Beasts*, 64–79.

357 the US imported: William Conway, "Species in Peril," *The Britannica Yearbook of Science and the Future* (University of Chicago, 1973), 17.

357 "The hide with": George Schaller, "Realm of the Jaguar" (undated unpublished manuscript), George and Kay Schaller private collection, 20.

357 friend David Cardoso: See "David Cardoso," The Movie Database (TMDB), accessed on April 22, 2024, themoviedb.org/person/931100-david-cardoso.

358 "We stayed on together": Schaller, "Realm of the Jaguar," 20.

358 "as sporting as executing": Schaller, "Realm of the Jaguar," 2.

358 find a new home: Parque de Exposições Clube Cidade Rosa.

358 George stopped at Acurizal: He also stopped briefly at the Dutch embassy in Brasília, to receive the Chivalrous Order of the Golden Ark, in a ceremony he found a bit overwrought ("the ambassador pinned the medal on me like a war decoration").

362 "You're lucky we broke down": Kent Redford, interview with the author, August 29, 2022.

365 at the University of Tennessee: Howard Quigley, interview with the author, July 16, 2020.

365 dozen black bears: Quigley fitted radio collars on twenty-two black bears between June 1978 and December 1979. See Howard Quigley, "Activity Patterns, Movement Ecology, and Habitat Utilization of Black Bears in the Great Smoky Mountains National Park, Tennessee" (master's thesis, University of Tennessee—Knoxville, 1982), trace.tennessee.edu/utk_gradthes/2497.

367 "understanding of nature": Alan Rabinowitz, *Jaguar: One Man's Struggle to Establish the World's First Jaguar Preserve* (Arbor House, 1986; Island Press, 2000), 6. Citations refer to the Island Press edition.

367 "I said, 'Tell him yes'": Alan Rabinowitz, interview by Tom Veltre, December 19, 2008.

367 "always a minimalist": Alan Rabinowitz, interview by Tom Veltre, December 19, 2008.

367 Schaller would later explain: In a sad, strange twist of fate, Rabinowitz, Crawshaw, and Quigley all died young; Rabinowitz of cancer in 2018, Crawshaw of COVID in 2021, and Quigley of cancer in 2022. Richard Sandomir, "Alan Rabinowitz, Conservationist of Wild Cats, Dies at 64," *The New York Times*, August 8, 2018, nytimes.com/2018/08/08/obituaries/alan-rabinowitz-conservationist-of-wild-cats-dies-at-64.html.

Chapter 8

369 twenty-five WCS projects: Emily Hahn, "Symbol," Talk of the Town, *The New Yorker*, May 28, 1984.

370 "the real George": Western, interview.

370 a kind of dream tour: Western did give George more unintended thrills when he flew him and Moss to see the total eclipse over Tsavo National Park. Twenty minutes out, the plane began sputtering and oil spurted across the windscreen. Jonah had to land on yet another abandoned airstrip with his head stuck out the side to see.

371 in US-China relations: On April 10, 1971, the US table tennis team became the first American delegation to travel to the Chinese mainland since 1949; that led to both the lifting of an embargo in place since the Korean War and Nixon's 1972 visit to China.

371 sultry picture of George: Dawn Maria Clayton, "Zoologist George Schaller Heads for China to Help Save the Imperiled Giant Panda," *People*, September 15, 1980, 113–14.

371 "envy of zoologists": Andree Brooks, "A Trip to Panda Land for Black and White Answers," *The New York Times*, November 18, 1979.

371 had shot a panda: In 1975 Sheldon published an account of that expedition; in 1990, the museum moved the stuffed panda into the endangered species case in its Hall of Biodiversity.

372 "not have my name": Schaller was angrier still to learn that Scott was pressuring WCS to give WWF half the proceeds from a fundraiser featuring Prince Philip, though Philip had committed to the WCS event long before agreeing to be WWF president. "The panda project does not bring out the best in people. I am so dejected that Peter would even consider such a move against another conservation organization."

372 Professor Pan Wenshi: Elena Songster, *Panda Nation: The Construction and Conservation of China's Modern Icon* (Oxford University Press, 2018), 73.

372 Professor Hu Jinchu: "We heard he had suffered tragedy." George B. Schaller, *The Last Panda* (University of Chicago Press, 1993), 61 (hereafter abbreviated as *TLP*).

372 "cordial yet so inarticulate": *TLP*, xiii.

373 first-ever surveys of panda: This history is from Songster's *Panda Nation*, chapters 4 and 6.

374 Madame Sun Yat-sen: Madame Sun Yat-sen held various high positions in the People's Republic of China, including vice president and honorary president. Her husband was called "the forerunner of the revolution" for his role in overthrowing the Qing dynasty.

375 sent an SOS: Mary Ellen Collins, "A Tribute to William Conway," The Association of Zoos and Aquariums, March 16, 2022, aza.org/connect-stories/stories/a-tribute-to-william-conway.

375 blocked the cell: *TLP*, 31.

375 alongside the Buddhist reverence: *TLP*, 32.

375 "What's so special": *TLP*, 61.

375 "it had apparently pleased him": *TLP*, 32.

376 "We were stunned": *TLP*, 12.

376 at their heels: To Schaller's ongoing dismay, the Smithsonian had also been given its choice of three sites while WWF had been offered just one small area in a place that Chinese experts knew was not particularly good.

376 thanks to a word: *TLP*, 14, 37, 87.

377 most were irrelevant: *TLP*, 39.

378 obtuse publicity seeking: *TLP*, 37.

378 "a Chinese delegation": David Attenborough, "Are They Really Bears?," review of *The Last Panda*, by George Schaller, *Literary Review* (July 1993), literaryreview.co.uk/back-issue/181.

378 they couldn't guarantee: *TLP*, 41.

379 biologist's greatest danger: *TLP*, 26.

379 "couldn't straighten up": Hu Jinchu, interview conducted and translated by Shibin Yuan on behalf of the author, December 2020.

380 a panda can assimilate: *TLP*, 103.

381 rest of the sins: In one of her marvelous essays, Amy Leach imagines pandas were once blood-and-berry-stained "bon vivants" like other bears, but after a conversion experience chose the "pure and noble" path: the "radical sect of bears, the Bambooists." Amy Leach, "Radical Bears in the Forest Delicious," *Things That Are* (Milkweed Editions, 2012) 41.

382 "who idolized George": Quigley, interview.

383 "Cat belled, Schaller": His monthly reports to WWF are in the Yale archives.

383 But "get back": *TLP*, 56.

384 "first-ever observations": Henry Nicholls, "The Truth About Pandas," BBC Earth 2015, repr. October 25, 2023, henrynicholls.com/the-truth-about-pandas/.

384 "the indescribably strange": Hahn, "Symbol."

385 tuned out everything: *TLP*, 68–69.

385 "way more sex": Nicholls, "The Truth About Pandas."

386 finally giving up: *TLP*, 116.

386 One male ejaculated: Schaller to de Haes, July 1981, Yale archives.

387 of artificial insemination: *TLP*, 63.

387 "Beautiful, just beautiful": Schaller, "Letter from Wuyipeng" to WWF, No. 4, April 18, 1981, Yale archives.

388 director, Devra Kleiman: *TLP*, 112.

389 "Fear of denunciation": *TLP*, 60.

389 "a foreigner arrives": Schaller to de Haes, July 1981, Yale archives.

390 "my interpretation of": *TLP*, 74–75.

390 "No one dared": *TLP*, 60, 59.

391 "every human virtue": Schaller to de Haes, July 1981, Yale archives.

391 "He is a Marxist!": *TLP*, 11.

392 "ate 41 percent": Hu Jinchu, interview.

392 *Giant Panda* magazine: "The Father of Giant Pandas," *Giant Panda*, no. 4 (April 2023), 20–40.

393 "Emerging from the Cultural Revolution": Yong Yange, interview with the author via WeChat, November 24, 2020, Dajun Wang translating.

394 And Janet Stover: One of the few women then in veterinary science, Stover had just successfully used a Holstein cow as a surrogate for an endangered gaur.

394 "share insignificant thoughts": *TLP*, 32.

394 ingredient list included: *TLP*, 29.

395 but "not lonely": *TLP*, 73.

395 "apart for several months": *TLP*, xx.

395 "scent of moss": *TLP*, 33.

395 "return to the tent": *TLP,* 58–59.

396 "snow slanted down": *TLP,* 92.

396 long "downward crescendo": *TLP,* 57.

396 "her apprehension and": *TLP,* 79.

398 their brinksmanship prevailed: *TLP,* 121–22.

398 make a panda film: *TLP,* 37.

399 *The Washington Post* reported: Tom Zito, "Panda Panic: Faked Pictures Cause Pandemonium at GEO," *The Washington Post,* September 2, 1981.

399 first *genuine* pictures: George B. Schaller, "For the First Time a Westerner Joins Chinese Scientists to Study Pandas in the Wild," *National Geographic* 160, no. 6 (December 1981): 735–49.

399 Their *Save the Panda*: *Save the Panda,* written and produced by Miriam Birch, National Geographic Society television special, first aired March 1983 on PBS.

399 the "startling self-assurance": *TLP,* 252.

399 if any biologists: Birds of prey have lifted some nonscientists into this experience of "unselfing," as Iris Murdoch calls it. In her case it was a kestrel; for J. A. Baker, a falcon. Iris Murdoch, *The Sovereignty of Good* (Routledge, 2001), 82; J. A. Baker, *The Peregrine* (Harper & Row, 1967).

399 "goes far beyond": Alan Rabinowitz, interview by Tom Veltre, December 19, 2008.

400 "panda is the answer": *TLP,* 79–80.

400 "like a full moon": *TLP,* 68.

400 "The panda is a panda": *TLP,* 262.

401 humanize another species: Though he doesn't reference it, Schaller would surely have read philosopher Thomas Nagel's 1974 essay "What Is It Like to Be a Bat" on the impossibility of imagining not just how it might feel to echolocate with sonar but how it would feel to a bat to use that instrument. For a marvelous tour of the range of animal capabilities—from robins navigating via magnetic fields to dogs reading multicharacter histories with their noses, see Ed Yong, *An Immense World: How Animal Senses Reveal the Hidden Realms Around Us* (Random House, 2022).

401 "window on reality": *TLP,* 68.

401 "lion could talk": *TLP,* 106. Schaller attributes the Wittgenstein quote to the talking panda, who is surprised by such "an intelligent remark for a human, and a foreign barbarian at that."

401 "fragments of existence": *TLP,* 106–7.

402 "its death so empty": *TLP,* 123–24.

403 newborn "savage pessimism": *TLP,* 124.

404 and "stout indifference": *TLP,* 126–29.

404 the "Catholic temple": *TLP,* 134.

406 bamboo had died: *TLP,* 137.

406 media frenzy ("panda-monium"): Vincent del Giudice, "Pennies for Pandas," United Press International, March 27, 1984, upi.com/Archives/1984/03/27/Pennies-For-Pandas/8585449211600; and Michael Parks, "China Trying Hard to Save Its Starving Giant Pandas," *Los Angeles Times,* June 17, 1984, quoted in *TLP,* 205.

406 ***Time* reported that:** "Science: Battling a Bamboo Crisis," *Time*, November 28, 1983, content.time.com/time/subscriber/article/0,33009,926403,00.html; *China Daily* cited in *TLP*, 211.

406 ten holding stations: Schaller learned of the construction plans at a meeting in Beijing with De Haes.

406 he was "elderly": *TLP*, 205.

407 fill those camps: *TLP*, 201.

407 "No panda could starve": *TLP*, 212.

407 "my work here is finished": This is how he tells Tang-Tang's story in his journal. In *TLP* he tells it with slightly more diplomatic language (187).

407 few wild pandas were: *The MacNeil/Lehrer NewsHour*, NewsHour Productions, aired January 19, 1984, on PBS, American Archive of Public Broadcasting, accessed May 9, 2024, americanarchive.org/catalog/cpb-aacip-507-r785h7cp0z.

407 "can restore perspective": Donald Allan (WWF information director) to Schaller, August 31, 1984, Yale archives.

408 "Pandas Saved, China Reports": Schaller included a clipping of the UPI article in his scrapbook.

408 Forty were locked: *TLP*, 224.

408 leader, Wang Pengyan: Schaller tells the damning Wang Pengyan story in detail in *TLP*, 155–62.

408 failed to win: *TLP*, 156.

410 Pengyan still spent: *TLP*, 162.

411 "principles like slogans": *TLP*, 170, 173, 178.

411 "trust that George": Kay Schaller, undated interview by Tiffany Trent, shared with the author by Kay.

411 boy would repeat it: *TLP*, 178.

411 he quit collecting: *TLP*, 22.

412 rare snub-nosed monkeys: One threat to the snub-nosed monkey's survival, ironically, was their sacred status: Qiang priests often carried a monkey skull, liver, or fingernails wrapped in white paper.

412 shook the treetops: *TLP*, 71.

412 "like a bee-stung": George B. Schaller, "A Face Like a Bee-Stung Moose," *International Wildlife* 16, no. 6 (1986): 36–39.

412 "starts its existence": *TLP*, 179–80.

413 "One abandons pity": *TLP*, 165.

414 a *Xinhua* reporter: *TLP*, 167–68.

414 *The Giant Pandas of Wolong*: Published in the US by the University of Chicago Press.

415 second cover story: George Schaller, "Secrets of the Giant Panda," *National Geographic* 169, no. 3 (March 1986): 284–309.

415 "a grave disservice": *TLP*, xv, 250.

415 "hard and efficient worker": *TLP*, 32.

416 latest menace, foreign: Pandas had long been used to grease global diplomacy—gifted by China to Japan, Mexico, North Korea, and most famously the US following Nixon's visit in 1972.

416 "display of greed": Geoffrey C. Ward, "Loved to Death," review of *The Last Panda*, by George Schaller, *The New York Times*, March 28, 1993, nytimes.com/1993/03/28/books/loved-to-death.html.

416 revealed "ignoble traits": *TLP*, xv, 248.

416 the "relentless" efforts: *TLP*, 237.

416 Is a panda in a zoo: Devra G. Kleiman, "Killing with Kindness," review of *The Last Panda*, by George B. Schaller, *The Washington Post*, April 18, 1993.

417 a "clownish face": *TLP*, 97.

417 reference the Cultural Revolution: *TLP*, 119, 149–50.

417 "Telling and retelling": *TLP*, xvi.

417 "A blueprint for": George Schaller, foreword to *Giant Pandas: Biology and Conservation*, ed. Don Lindburg and Karen Baragona (University of California Press, 2004), xi.

417 Spurious "rescues" had: *TLP*, 224.

418 WWF undercover buyers: *TLP*, 225–27.

418 failure to act: *TLP*, 251, 233.

418 "crimes against nature": Ward, "Loved to Death."

418 "blame on every institution": Sharon Begley, "Killed by Kindness," *Newsweek*, April 11, 1993, newsweek.com/killed-kindness-191668.

418 British *Literary Review*: Then edited by Auberon Waugh, the *Literary Review* cover illustration was exceedingly strange: A skinned-up Attenborough sits in an embrace with a nonplussed-looking panda, while behind them Chinese peasants in Mao suits, long beards, and conical hats hold up empty rice bowls, forks, and knives. A cleaver-bearing chef seems inspired by Mickey Rooney in *Breakfast at Tiffany's*, another reminder of the glib racism often still in the air. Attenborough, "Are They Really Bears?"

419 "drained and unsettled": John Terborgh, "Solitary Enigma," review of *The Last Panda*, by George Schaller, *The New York Review of Books*, September 23, 1993, 12.

419 On *Day One* with: Kay described the *Day One* segment in a letter to George, June 28, 1993.

419 "after the pandas": *Charlie Rose*, "George Schaller," aired April 22, 1993, on PBS, charlierose.com/videos/14242.

419 *A River Runs Through It*: Kay to George, July 9, 1993. The news, Kay wrote, had come from his editor Susan Abrams, who had called to celebrate and report that the stellar sales had covered his advance and then some. "Such a good provider," Kay teased.

419 San Diego's waiver: *TLP*, 282; San Diego had concluded secret negotiations with China, guaranteeing $1 million a year for an adult pair plus a $100,000 monthly bonus for a cub. The zoo did get a pair on a research loan in 1996 and worked fruitfully with Wolong to develop pregnancy tests, infant formula, and a twin-swapping strategy that dramatically increased success rates of captive reproduction.

419 "But I was wrong": Conway, interview. Yange, interview.

420 "banned further capture": Alas, aggressive and sometimes injurious captive breeding to supply zoos had still not ended thirty years later. See Mara Hvistendahl and Joy Dong, "The Panda Factories," *The New York Times*, October 15, 2024, nytimes.com/interactive/2024/10/15/world/asia/pandas-zoo-breeding-death-captivity.html.

420 $77 million conservation plan: *TLP*, 281. Of this, $13 million would be provided internally. Wolong alone got forty antipoaching patrols.

420 "It is a challenge": Hu Jinchu, interview.

420 named the baby: Schaller recounts this in the afterword in the 1994 edition of *TLP*; additional details come from his journal.

Chapter 9

423 its "totemic loneliness": *TW*, 2.

423 "homesick for a world unknown": George B. Schaller, *Tibet's Hidden Wilderness: Wildlife and Nomads of the Chang Tang Reserve* (Harry N. Abrams, 1997), 16 (hereafter abbreviated as *THW*).

424 dirty and bearded: *TW*, 130–31.

424 White Russians fleeing: Bruce Weber, "Frank Bessac, Scholar and Adventurer, Dies at 88," *The New York Times*, December 25, 2010, nytimes.com/2010/12/26/world/americas/26bessac.html.

424 the surviving American: *TW*, 34.

424 "utterly God-forsaken region": Hedin quoted in *THW*, 33.

425 brief trip into Tibet: This account is from Schaller's journals, but see also Peter Jackson, "On the Tibetan Plateau," *Nature* 287, no. 5782 (October 9, 1980): 486–87.

426 "still antique people": *THW*, 38.

428 "diurnal and acute observers": *TW*, 109.

428 "fit of excess energy": George Schaller, "Wildlife in the Middle Kingdom," *Defenders* 60, no. 3 (May/June 1985): 10–15.

428 a dozen Aesopian fables: A few of the pika fables appear in *TW*, 119–23; as of 2025, an effort was underway to publish the whole collection in English.

429 Rolex was by now: Rolex, advertisement, *Country Life*, February 7, 1991, 56.

432 "into the white void": *TW*, 21.

433 "floats like an iceberg": *TW*, 15, 159.

435 receive another award: The Bruno H. Schubert award, with Bernhard Grzimek in attendance.

437 *News from Tartary*: An account of Fleming's 3,500-mile journey in 1935 from Peking (Beijing) to Kashmir.

439 "kindness to all living things": *THW*, 117.

440 "To all life is dear": *TW*, 73.

440 "glitter of broken glass": *THW*, 59.

440 a pound of *dongchongziazhow*: Sold today online as Dong Chong Xia Cao.

441 beautiful but barbed: George B. Schaller, "Tibet's Remote Chang Tang," *National Geographic* 184, no. 2 (August 1993): 68.

441 "you will not come out": *THW*, 28.

441 Mark got a sustained taste: These recollections are from Mark's journals, Mark Schaller private collection.

445 "the way to travel": *THW*, 28.

448 "an animal's artifacts": *THW*, 34–35.

449 "impervious to herbivory": George B. Schaller, *Wildlife of the Tibetan Steppe* (University of Chicago Press, 1998), 222 (hereafter abbreviated as *WTS*).

450 who "with imperial": *THW*, 22.

451 Still jet-lagged a week: Kay to George, in a long letter written over five days, September 22–26, 1988, Yale archives.

452 At the center of the story: Kim Masters, "Wildlife Scientist Investigated for Hunts," *The Washington Post*, December 13, 1990, front page; Kim Masters, "Smithsonian Spanked for Controversial Legal Fees," *The Washington Post*, March 5, 1992; Kim Masters, "Caught in the Snare of a Hunt," *The Washington Post*, March 31, 1992. Also *TW*, 262–69, and *WTS*, 80–83.

452 founded an NGO: The American "Ecological" Union.

453 included James Conklin: "Dr. James Conklin," The Conklin Foundation, theconklinfoundation.com/dr-james-conklin.

455 "One bull collapsed": *TW*, 270–71.

455 decimated the animals: *WTS*, 92.

456 "Where precisely has": George B. Schaller, *Into Wild Mongolia* (Yale University Press, 2020), 42.

456 supporter Edith McBean: Edith McBean, interview with the author, January 24, 2019.

457 produce valuable penises: *TLP*, 257.

458 important party leader: *THW*, 147.

459 George buried it: *THW*, 40.

461 "never, ever, seen": Kay Schaller, quoted in *THW*, 46.

462 fifth of Tibet: "Wildlife Reserve on the Tibetan Plateau," *National Geographic* magazine 178, no. 2 (August 1990).

462 the herds' migrations: Because of badly drawn boundaries, Serengeti wildebeest and Yellowstone bison both range far outside their parks; in April 2023, 1,150 of the latter were "culled" on the grounds they might spread disease. Jim Robbins, "Mass Yellowstone Hunt Kills 1,150 Bison," *The New York Times*, April 4, 2023, https://www.nytimes.com/2023/04/04/science/bison-hunt-yellowstone-native-americans.html.

463 "the Imperialist American": *THW*, 62.

465 Alec Le Sueur's bestselling: Alec Le Sueur, *Running a Hotel on the Roof of the World: Five Years in Tibet* (Summersdale Publishers, 1998), 165.

466 the chiru's undercoat: George returns to this story often: Chapter 4, "A Deadly Fashion," in *TW*, 67–94; *THW*, 91–95; and George B. Schaller, "Drop Dead Gorgeous," *National Geographic* 203, no. 4 (April 2003): 124–25.

467 as antifur activism: Bob Colacello, "O.K., Lady, Drop the Shawl," *Vanity Fair*, June 28, 2011, vanityfair.com/news/1999/11/shahtoosh-furs-199911.

468 set out that summer: *TW*, 131.

470 second-largest protected area: With three contiguous reserves (Kekexili, Arjin Shan, Mid-Kunlun) in Qinghai and Xinjiang, the Chang Tang reserve now protects an area larger than California.

470 the following summer: George Schaller, "In a High and Sacred Realm," *National Geographic* 184, no. 2 (August 1993): 62–87.

470 "suits my inner landscape": *TW*, 138.

470 what Lewis Hyde calls: W. Lewis Hyde, *The Gift: How the Creative Spirit Transforms the World* (Vintage, 2019).

471 into "small-time ranchers": George Schaller, "In Tibet, Change Comes to the Once-Pristine Roof of the World," *Yale Environment 360*, February 26, 2013, e360.yale.edu/features/in_tibet_change_comes_to_the_once-pristine_roof_of_the_world. In *WTS*, 288–303, Schaller draws on the work of Melvyn Goldstein and Cynthia Beall to summarize the nomads' complex history: from their subjection to a feudal system with mandated rotations, to collectivization, to sedentarization, often in places they'd not historically frequented.

471 His modern guns: *THW*, 63, 66, 70.

472 "empty sacks that can't stand": *TW*, 154.

472 came the famine: *TLP*, 151, 253.

472 Tibetan Medical College: *WTS*, 297.

472 faced ever more competition: Dan Miller and George Schaller, "Rangelands of the Chang Tang Wildlife Reserve in Tibet," *Rangelands* 18, no. 3 (June 1996); *WTS*, 309.

473 "capitalism foisted on": Western, interview.

473 Schaller found 681: Schaller reports to WCI, 1986–1992, WCS; Rick Ridgeway, *The Big Open: On Foot Across Tibet's Chang Tang* (National Geographic, 2004), 46.

473 the final pages: *TLP*, appendix A, 253–60.

474 Eric called excited: Scientist Sarah Harding is reassuring thirteen-year-old brainiac Kelly Malcolm that her mom might be wrong when she says boys don't like smart girls. "'It's like everything else in the world. You know about George Schaller?'

"'Sure. He studied pandas.'

"'Right. . . . He's the most important animal researcher in the twentieth century—and you know how he works? . . .

"'Before he goes into the field, George reads everything that's ever been written about the animal he's going to study. Popular books, newspaper accounts, scientific papers, everything. Then he goes out and observes the animal for himself. And you know what he usually finds? . . .

"'That nearly everything that's been written or said is wrong. Like the gorilla. George studied mountain gorillas ten years before Dian Fossey ever thought of it. And he found that what was believed about gorillas was exaggerated, or misunderstood, or just plain fantasy—like the idea that you couldn't take women on gorilla expeditions, because the gorillas would rape them. Wrong. Everything . . . just . . . wrong.'" Crichton, *The Lost World*, 232.

478 team at WCS: In 1993, New York Zoological Society became Wildlife Conservation Society, underlining its transformed mission.

478 "breaking several regulations": Schaller to Bleisch, October 29, 1994, Yale archives.

479 "together 40 years": C. Jackson Selsor had studied chimps in Uganda in the years just before the Schallers arrived in Congo; Ken Stott had edited the San Diego *Zoonooz* and worked on mammals in Central Africa since the late 1940s.

482 story on "the man": Michael Ryan, "We Have to Protect What We Have," *Parade* magazine, February 2, 1997, 17.

482 stunning *Tibet's Hidden Wilderness*: *THW*; the Dalai Lama quote is two brief paragraphs on page 159.

486 Kokoxili calving ground: George tells some of this story in *TW*, 35–42.

490 his 1998 monograph: *WTS*, 280; the epigraph is from Chinese poet Li Po, 701–762 AD.

492 they visited several: *TW*, 217, 233.

492 and were reviled: *TW*, 211, 245–46. In his study here, Qiu Mingjiang had learned that in one year, villagers had lost 140 cattle and twenty-seven horses and killed four tigers in return. Zhang Ming, Qiu Mingjiang, Li Shouchang, "The Present Status, Number and Distribution of Tigers in Jinzhu Tsangpo Valley, Southeastern Tibet," *Acta Theriologica Sinica*, 2006, mammal.cn/EN/Y1998/V18/I2/81.

492 2009, the last: George B. Schaller, "The Natural History of the Field Biologist," in *Saving Endangered Species: Lessons in Wildlife Conservation from Indianapolis Prize Winners*, ed. Robert W. Shumaker (Johns Hopkins University Press, 2020), 48–79.

493 "Whenever one works": *THW*, 23.

494 mountaineer David Breashears: Famous for transmitting the first live TV on Everest, Breshears stood George up twice, and was forever unforgiven.

494 about Baker's "discovery": In *TW* (211, 219), Schaller devotes several paragraphs to the prior discovery of the falls, and the entanglements with massive hydroelectric developments that by 2025 were under construction. Projected to cost $167 billion, it is reportedly the largest dam project in the world. "China Breaks Ground on Colossal Dam Project in Asia's Grand Canyon," YaleEnvironment 360, July 22, 2025, e360.yale.edu/digest/chinachina-tibet-yarlung-tsangpo-dam-construction.

495 the deity Vajrayogini: *TW*, 215; George B. Schaller, "Reflections in a Hidden Land," in *Extreme Landscape: The Lure of Mountain Spaces*, ed. Bernadette McDonald (National Geographic Society, 2002), 50–59.

496 "Spirits soar in": Schaller, "Reflections in a Hidden Land," 52.

496 "hidden and intangible": Schaller, "Reflections in a Hidden Land," 53.

496 banked "in winged": This last detail is from *TW*, 225.

497 "between Buddhist precepts": Schaller, "Reflections in a Hidden Land," 58, 55.

497 "when staff cheats": *TW*, 42.

497 You are selling: Julie K. L. Dam, Karen Grigsby Bates, J. Todd Foster, Bob Meadows, "Fashion Victims," *People*, June 14, 1999, 133; *TW*, 81.

497 "New Age Luxury": Suzy Menkes, "New Age Luxury Isn't So 'Correct,'" *The New York Times*, April 4, 1995, nytimes.com/1995/04/04/style/IHT-new-age-luxury-isnt-so-correct.html.

498 Jammu and Kashmir: *TW*, 82.

498 WPSI published an exposé: *Fashioned for Extinction: An Exposé of the Shahtoosh Trade* (Wildlife Protection Society of India, 1997), wpsi-india.org/images/fashioned_for_extinction.pdf.

498 "It was hysterical": Colacello, "O.K., Lady, Drop the Shawl."

499 *People, Time, The Wall Street Journal*: *TW*, 85; Dam et. al, "Fashion Victims"; Nadya Labi, "Soft, Warm and Illegal: Shahtooshes Become an Elegant Fashion Accessory at a Grim Price," *Time*, October 18, 1999, content.time.com/time/sub scriber/article/0,33009,992255,00.html; Matt Forney, "The Veil of Kashmir: How Many Chirus Died for That Shawl?," *The Wall Street Journal*, October 25, 1999, wsj .com/articles/SB940801160788497951.

499 the "Fashion Herd": Valli Herman-Cohen, "Fashion Herd Urged to Save Tibetan Beauties," *Los Angeles Times*, October 22, 1999, latimes.com/archives/la-xpm-1999 -oct-22-cl-24884-story.html.

501 broken his own taboo: *TW*, 236, 45.

503 his Biblical master: "What have I done to you," the ass asks, "that you have struck me three times?"

503 "newly opened to . . .": Milan Kundera, *The Unbearable Lightness of Being* (Harper and Row, 1984), 282; Joseph Conrad, *The Secret Agent* (Everyman Paperback, 1997), 127. See also Chris Townsend, "Nietzsche's Horse," *Los Angeles Review of Books*, April 25, 2017, blog.lareviewofbooks.org/essays/nietzsches-horse; and Michael John DiSanto, "'Dramas of Fallen Horses': Conrad, Dostoevsky, and Nietzsche," *Conradiana* 42, no. 3 (Fall 2010): 45–68, jstor.org/stable/24635332.

503 "a good donkey": *TW*, 51.

504 legend Rick Ridgeway: In 1978 Ridgeway was on the first American team to summit K2; he would go on to lead sustainability work at the Patagonia company and across the industry.

504 "to be open projects": Ridgeway, *The Big Open*, 181.

504 Ridgeway refused to: Ridgeway, *The Big Open*, 155, 68.

505 "the wildest places": Ridgeway, *The Big Open*, 91.

505 "Where are the chiru?": *TW*, 54.

505 photos and video: Rick Ridgeway, "275 Miles on Foot Through the Remote Chang Tang: A Team of Elite Mountaineers Sets Out to Witness Birth at the Calving Grounds of the Elusive Tibetan Chiru," *National Geographic* 203, no. 4 (April 2003) 104–8; Ridgeway, *The Big Open*, 205; Lisa Donchak, "Three Hundred Miles Later," *City on a Hill Press* (Santa Cruz, CA), April 10, 2008, cityonahillpress.com/2008/04 /10/three-hundred-miles-later.

505 persuade China to add: *TW*, 57, 58.

505 he had warned: In a report to the International Union for the Conservation of Nature. Ridgeway, *The Big Open*, 302. See also *Shahtoosh: The Illegal Trade* (Wildlife Protection Society of India, 2006), wpsi-india.org/images/shahtoosh_brochure-29 -08.pdf.

505 "I'd rather have": *Galen Rowell: A Retrospective* (Sierra Club Books, 2006), 117.

506 with fifty thousand inhabitants: *TW*, 316.

506 "together to create fertility": *WTS*, 331.

507 Tibetans "are not": *THW*, 130.

507 a 2005 paper: George Schaller, Lü Zhi, Wang Hao, and Su Tie, "Wildlife and Nomads in the Eastern Chang Tang Reserve, Tibet," *Memorie della Societa Civico di Storia Naturale di Milano* 33, no. 1 (January 2005): 59–67. Lü Zhi would soon found both the Center for Nature and Society at Peking University and an NGO, Shan Shui Conservation Center, dedicated to advancing "the equality and coexistence of humans and animals, through reliance on local guardians."

507 ***Wildlife Conservation in China*:** George Schaller, foreword to *Wildlife Conservation in China: Preserving the Habitat of China's Wild West*, by Richard Harris (M. E. Sharpe: 2008), xiii.

507 commercial chiru harvest: *THW*, 158.

508 And those that worked: Xinjiang statistic is from his journal; Pakistan from *TW*, 336.

509 "I'm not a person": Aili Kang, interview with the author, Center for Global Conservation, Bronx Zoo, December 2018.

509 " 'Leave it in peace' ": *TW*, 103.

510 A skilled mountaineer: Yanlin also had experienced tragedy; in the summer of 2002, the year before he met George, he'd led the Peking University climbing team up Shishapangma, the fourteenth-highest peak in the world. Hit by an avalanche, five of his teammates had died. In a dark coincidence, two and a half years before joining Ridgeway on their search for the chiru calving grounds, Conrad Anker had lost two climbing companions on the same mountain, a story Jimmy Chin would recall in his 2015 film *Meru*.

510 "at seventy-two he insisted": Yanlin Liu, interview with the author via WeChat, November and December 2019, May 2021, January–February 2022.

510 "small black circle": From an unpublished Schaller tribute Miller wrote in 2009, shared with the author in March 2021.

511 Wellby and Malcolm in 1896**:** *TW*, 125–66; Alex Chadwick, host, *Day to Day*, podcast, "A Resurgence of Wildlife in Northern Tibet," interview with George Schaller, February 21, 2007, npr.org/2007/02/21/7316356/a-resurgence-of-wildlife-in-northern-tibet; George Schaller, "The Great Chang Tang Traverse," report to National Geographic Society, December 2006.

512 went to mainland China: Karma Damchen, interview with the author, March 20, 2022.

512 already critical for: Schaller, "In Tibet, Change Comes to the Once-Pristine Roof of the World."

512 Zhaxi Duojie, called Zhaduo: *TW*, 127; Ching-Ching Ni, "Heroes of China's Wasteland," *Los Angeles Times*, August 29, 2000, latimes.com/archives/la-xpm-2000-aug-29-mn-11912-story.html

513 Global Environment Fund: Marc Foggin offers an excellent summary of the ecological significance of the plateau, the changes in nomads' lives, and the continuing shift to integrating conservation and socioeconomic development, including through local comanagement, in J. Marc Foggin, "Environmental Conservation in the Tibetan Plateau Region: Lessons for China's Belt and Road Initiative in the Mountains of Central Asia," *Land* 7, no. 2 (April 2018): 52, doi.org/10.3390/land7020052.

515 Several monasteries had: This tour of monasteries, dark and light, is chronicled in *TW*, 341–54.

515 where Buddhist beliefs: Lopez, introduction to "I, Snow Leopard" by Jidi Majia, dedicated to Schaller.

515 "an ideal vision": *TW*, caption to plate 32.

515 since an earthquake: April 2010.

515 At Thartung monastery: *TW*, 342–43.

516 the conservation NGO: See en.shanshui.org; Xiaoli Shen and Jiaxin Tan, "Ecological Conservation, Cultural Preservation, and a Bridge Between: The Journey of Shanshui Conservation Center in the Sanjiangyuan Region, Qinghai-Tibetan Plateau, China," *Ecology and Society* 17, no. 4 (September 2016), dx.doi.org/10.5751/ES-05345-170438.

516 "roadhouse," George nicknamed: *TW*, 320.

516 twenty-five-year-old Juan Li: Juan Li's study built on research by Lü Zhi on how the governance of parks resembled that of sacred lands: Both reserved core zones for non-human beings, ringed by buffers within which human activities were closely regulated.

517 Tibetan Buddhist regions: Juan Li, Dajun Wang, Hang Yin, Duojie Zhaxi, Zhala Jiagong, George B. Schaller, et al., "Role of Tibetan Buddhist Monasteries in Snow Leopard Conservation," *Conservation Biology* 28, no. 1 (2014): 87–94, jstor.org/stable/24479503.

517 culmination of this work: He described this in his 2013 annual report to WCS.

517 George gathered his thoughts: George B. Schaller, *Deki: The Adventures of a Dog and a Boy in Tibet* (Hachette India, 2014).

517 two nonfiction books: Pam Turner's charming *A Life in the Wild*, in which wild yaks run like "galloping sofas," and *The Chiru of High Tibet: A True Story* (2010) by Caldecott winner Jacqueline Briggs Martin.

517 Left to guard the tent: Deki's ambivalence recalls a Buddhist precept Schaller first encountered in an 1894 translation by his great-great-uncle Paul Carus: "The fowl in the coop has her food, but she will soon be boiled in the pot. / No provisions are given to the wild crane, but the heavens and the earth are his." Paul Carus, *The Gospel of Buddha According to Old Records* (1894; repr. Open Court Publishing Company, 1917), 122.

518 When Ridgeway had: Ridgeway, *The Big Open*, 76.

518 "Read the directions": Much of this is from a compilation assembled by Schaller of proposals and correspondence from 2010 through 2013, held at Yale. Also *TW*, 311–24, and his field journals.

520 "was really scary": Dajun Wang, interview with the author, November 14, 2019.

521 had trapped cats: Then a program coordinator for Panthera, the conservation group founded by Rabinowitz, Dickerson went on to run a tourist safari park and produce faux furs for use in ceremonial regalia in place of leopard skins. See *To Skin a Cat*, directed by Greg Lomas and Colwyn Thomas, starring Tristan Dickerson (Scholars and Gentlemen, 2016).

521 and spouting misinformation: Schaller wanted the whole scat to see all the animal had eaten; Dickerson said he should leave part as a scent post; Schaller said that, scattered about, it had already lost that function.

522 "The Chinese, they were told": Peter Hopkirk, *Foreign Devils on the Silk Road* (University of Massachusetts Press, 1984), 227.

523 Yanlin never forgot that moment: Liu, interview.

525 "Dr. Schaller took": "Biologist George B. Schaller in China," Sichuan Fun, June 21, 2015, sichuanfun.com/george-schaller-giant-panda.

525 a national hero: Wang, interview.

525 the Year Xi Zhinong: "Photographer Xi Zhinong Uses Camera to Protect Wildlife," CGTN America, May 14, 2015, america.cgtn.com/2015/05/14/photographer-xi-zhinong-uses-camera-to-protect-wildlife. See also "Zhinong Xi: Wildlife Photographer," National Geographic Explorers Directory, explorers.nationalgeographic.org/directory/zhinong-xi.

526 top environmental award: China Baosteel Environmental Prize, baosteel.com/group_en/contents/2863/39857.html.

526 foreword for *Snow Leopards*: Philip J. Nyhus, Tom McCarthy, and David Mallon, eds., *Snow Leopards: Biodiversity of the World: Conservation from Genes to Landscapes* (Elsevier, 2016).

526 beyond single species: Environmental historians Songster and Lewis argue that the snow leopard catalyzed this move from top-down to bottom-up conservation. Unlike the panda, a national symbol spurring mandates from Beijing, the transnational cat drew less state investment and therefore more space for work at both the local and transboundary scale. Michael Lewis and Elena Songster, "Studying the Snow Leopard: Reconceptualizing Conservation Across the China-India Border," *British Journal for the History of Science* 1 (May 2016): 169–98, doi.org/10.1017/bjt.2016.8.

527 new studies of signposts: Juan Li, George B. Schaller, Thomas M. McCarthy, Dajun Wang, Zhala Jiagong, Ping Cai, Lamao Basang, Zhi Lü, "A Communal Sign Post of Snow Leopards (*Panthera uncia*) and Other Species on the Tibetan Plateau, China," *International Journal of Biodiversity* 2013 (December 2013), doi.org/10.1155/2013/370905.

527 anthropologist Shafqat Hussain: Hussain is the one who mischaracterized George's work in Pakistan to critique him as neocolonial; it's unclear whether George knew that.

527 In a second foreword: Foreword to *Tales of the Snow Leopard* on the Ladakhi model of conservation. Philomene Verlaan, *Tales of the Snow Leopard* (CreateSpace Independent Publishing Platform, 2016).

527 Ethnography is becoming "multispecies": See, for example, S. Eben Kirksey and Stefan Helmreich, "The Emergence of Multispecies Ethnography," *Cultural Anthropology* 25, no. 4 (November 2010): 545–76, anthropology.mit.edu/sites/default/files/documents/helmreich_multispecies_ethnography.pdf; Danielle Celermajer and Sophie Chao, eds., "Multispecies Justice," special issue, *Cultural Politics* 19, no. 1 (March 2023), dukeupress.edu/multispecies-justice.

527 "I have had my hour": *TW*, 7; Scott Horton, "Dryden's 'Happy the Man,'" *Harper's Magazine*, November 28, 2007, harpers.org/2007/11/drydens-happy-the-man.

527 he lived T. S. Eliot: T. S. Eliot, "East Coker," *The Four Quartets* (Harcourt Brace, 1971), 32.

Epilogue

529 "The maestro," as colleague Joel Berger: Joel Berger, interview with the author, December 2020.

530 George told UNDP: Schaller stepped in again on behalf of Mongolian biologists when one was challenged on a staffing issue by George's young apprentice Kirk Olson. George gave Olson a list of "ten things to remember" when working in a foreign country. "You don't know the reason for others' behavior. (Maybe he was ordered.) Outsiders are tolerated, not wanted. If something goes wrong, the fault is yours, or will be seen that way." Kirk Olson, interview with the author, August 7, 2020.

530 "I'd never seen": Zamba Batjargal, interview with the author, October 18, 2020.

530 the first scientific paper: George Schaller, Ravdangiin Tulgat, and B. Navantsatsvalt, "Observations on the Gobi Brown Bear in Mongolia," in *Bears of Russia and Adjacent Countries—Status of Populations* (Moscow: Ministry of Environmental Protection, 1993), 110–25.

532 "splendour in the grass": William Wordsworth, "Ode on Intimations of Immortality from Recollections of Early Childhood," Academy of American Poets, poets.org/poem/ode-intimations-immortality-recollections-early-childhood.

532 In Vietnam, he: George Schaller, Nguyen Xuan Dang, Le Dinh Thuy, and Vo Thanhson, "Javan Rhinoceros in Vietnam," *Oryx* 24, no. 2 (April 1990): 77–80.

533 "I had to pretend": William Robichaud, interview with the author, April 6, 2020.

534 The Indochinese warty pig: Colin P. Groves, George B. Schaller, George Amato, Khamkhoun Khounboline, "Rediscovery of the Wild Pig *Sus bucculentus*," *Nature* 386 (March 1997): 335.

538 But he also knew: Landmark works include William Cronon, *Changes in the Land: Indians, Colonists, and the Ecology of New England* (Hill and Wang, 1983); William M. Denevan, "The Pristine Myth: The Landscape of the Americas in 1492," *Annals of the Association of American Geographers* 82, no. 3 (1992): 369–85; Charles C. Mann, *1491: New Revelations of the Americas Before Columbus* (Knopf, 2005). For a summary of studies see N. L. Boivin, M. A. Zeder, D. Q. Fuller, A. Crowther, G. Larson, J. M. Erlandson, T. Denham, and M. D. Petraglia, "Ecological Consequences of Human Niche Construction: Examining Long-Term Anthropogenic Shaping of Global Species Distributions," *PNAS* 113, no. 23 (2016): 6388–96.

538 "various island birds": George Schaller, "Expelled from the Garden of Eden," March 10, 1995, Commonwealth Club of California records, Hoover Institution Library & Archives, digitalcollections.hoover.org/objects/60412/expelled-from-the-garden-of-eden.

538 In an essay: Kent H. Redford, "The Empty Forest: Many Large Animals Are Already Ecologically Extinct in Vast Areas of Neotropical Forest Where the Vegetation Still Appears Intact," *BioScience* 42, no. 6 (June 1992): 412–22. The earliest rounds of battle spanned several issues of *Conservation Biology*, with Redford, Steve Sanderson, and John Terborgh arrayed against Stephan Schwartzman, Daniel Nepstad, and Adriana Moreira. See Schwartzman, Nepstad, and Moreira, "Arguing Tropical Forest Conservation: People Versus Parks," *Conservation Biology* 14, no. 5 (2000): 1370–74; and following issues. While both sides agreed that local sovereignty is the best defense against the ultimate destroyers—illegal loggers, miners, ranchers, and poaching syndicates—they had very different definitions of ecological integrity. "Even if [Indigenous hunters] do drive animals to local extinction," wrote Schwartzman et al., "there is no evidence that affects higher-level criteria of tropical forest integrity: vulnerability to fire, soil fertility, carbon sequestration and hydrological function." Schwartzman himself later documented the lack of animals (tapir, pigs, monkeys, armadillos) in Brazil's Xingu territories, and a shift by the Panará to hunting species they'd not historically eaten. Stephan Schwartzman, "Nature and Culture in Central Brazil: Panará Natural Resource Concepts and Tropical Forest Conservation," *Journal of Sustainable Forestry* 29, no. 2 (2010): 302–3. Updates on the empty forest include: Carlos Peres, "Effects of Subsistence Hunting on Vertebrate Community Structure in Amazonian Forests," *Conservation Biology* 14, no. 1 (February 2000): 240–53. Ana Benítez-López, Rob Alkemade, Aafke Schipper, Daniel Ingram, Pita Verweij, Jasper Eikelboom, et al., "The Impact of Hunting on Tropical Mammal and Bird Populations," *Science* 356, no. 6334 (April 2017): 180–83, doi.org/10.1126/science.aaj1891; Ana Benítez-López, Luca Santini, Aafke Schipper, Michela Busana, and Mark Huijbregts, "Intact but Empty Forests? Patterns of Hunting-Induced Mammal Defaunation in the Tropics," *PLoS Biology* 17, no. 5 (May 2019): e3000247, doi.org/10.1371/journal.pbio.3000247; Gustavo Canale, Carlos Peres, Carlos Guidorizzi, Cassiano Ferreira Gatto, and Maria Cecília Kierulff, "Pervasive Defaunation of Forest Remnants in a Tropical Biodiversity Hotspot," *PLoS One* 7, no. 8 (August 2012): e41671, doi.org/10.1371/journal.pone.0041671.

538 ***Rainforests of the World*****:** *Rainforests of the World: Water, Fire, Earth and Air*, photography by Art Wolfe, text by Ghillean T. Prance, foreword by George B. Schaller (Crown Publishers, 1998).

539 "hitched to everything else": John Muir, *My First Summer in the Sierra* (Houghton Mifflin, 1911), 211.

539 Lose the fruit eaters: Even losing diverse personalities within a species, Darwin argued, had costs: Timid seed-dispersers stick with the tried and true; bolder, more curious ones interact with novel seeds, enhancing the adaptation of fauna to changed circumstances. But it's those bolder ones who are more likely to be killed.

539 "The average American": George B. Schaller, foreword to *Hunting for Sustainability in Tropical Forests*, ed. John Robinson and Elizabeth Bennett (Columbia University Press, 2000), xv–xviii.

539 the nomenclature itself: Manvir Singh, "It's Time to Rethink the Idea of the 'Indigenous,'" *The New Yorker*, February 20, 2023, newyorker.com/magazine/2023/02/27/its-time-to-rethink-the-idea-of-the-indigenous.

540 "Yes, if there are": Schaller, "Expelled from the Garden of Eden."

540 "ecologically noble savage": Kent H. Redford, "The Ecologically Noble Savage," *Cultural Survival Quarterly* 15, no. 1 (January 1991): 46–48. "Precontact Indians were not 'ecosystem men'; they were not just another species of animal, largely incapable of altering the environment," Redford wrote in the essay that gave the debate its name. Nor were their interventions always a boon to other species. Precontact forests burned into prairies, as in North America, allowed grassland species like buffalo to thrive, but at obvious cost to forest species.

540 They were not sylvan: As Charles C. Mann put it, "the Americas before Columbus were filled with a staggering variety of cultures. . . . You can't say much about Indians in general." Charles C. Mann, "The Pristine Myth," interview by Katie Bacon, *Atlantic Unbound*, March 7, 2002, theatlantic.com/past/docs/unbound/interviews/int2002-03-07.htm.

540 Some Alaskan Natives: Gregory Scruggs, "Polar Opposites: The Remote Alaskan Village Divided over Oil Drilling," Reuters, May 1, 2019, reuters.com/article/us-usa-environment-oil-feature/polar-opposites-the-remote-alaskan-village-divided-over-oil-drilling-idUSKCN1S01B5; "The Refuge," *Threshold*, podcast, season 3, Auricle Productions, 2019, thresholdpodcast.org/season03. The Gwich'in, who George grew close to in his Alaska years, oppose drilling because it will harm the Porcupine caribou herd.

540 *Hunting for Sustainability*: Schaller, foreword to *Hunting for Sustainability*, xv.

541 "By the end I knew": Tshiring Lhamu Lama, interview with the author, May 31, 2020.

541 a most disturbing film: The film captures the brutal treatment of snow leopards by men who've come to harvest the valuable caterpillar fungus. When they find a dead baby yak, the men lace the carcass with rat poison (they sometimes embed razor blades) to punish the cats that killed it. They then drag two half-conscious snow leopard cubs by their tails, cans tied about their necks, finally kicking them into a ditch to die. Sonam Choekyi Lama, dir., *The Snow Leopard Calling* (Snow Leopard Journeys, 2020), vimeo.com/873214619. Tshiring's story and her success at helping protect her neighbors' goats and sheep and inspiring and training local snow leopard conservationists is captured in depth in Sonam Choekyi Lama, Ben Ayers, and Andrew Lynch, dirs., *Snow Leopard Sisters* (Noah Media Group, 2025).

541 "rightful claims to land": Schaller, foreword to *Hunting for Sustainability*, xvii.

541 Contrary to a persistent critique: Prerna Singh Bindra, who wrote her PhD thesis on relocations of villagers out of India's reserves, detailed the slipshod nature of much of the contemporary "reporting" on ostensibly forced dispossessions. See: Prerna Singh Bindra and Krithika Sampath, "Response to the Article 'Madia Gond Tribes Forced to Leave Ancestral Land, as Human-Animal Conflict Increases' Published on March 14, 2023," *Mongabay*, May 16, 2023, india.mongabay.com/2023/05/response-to-the-article-madia-gond-tribes-forced-to-leave-ancestral-land-as-human-animal-conflict-increases-published-on-march-14-2023; David Quammen also took on what he deemed a straw man. "Most people and organizations involved in [conservation biology] are not arguing for the preservation of some 'evolutionary Eden.' They're saying: Let's protect biological diversity . . . [which is] not a subjective matter. It can be measured. It can even be counted." David Quammen, "The Brilliant Plodder," *The New York Review of Books*, April 23, 2020, nybooks.com/articles/2020/04/23/charles-darwin-brilliant-plodder.

541 George's own influence: Human rights–based conservation, focused on helping IPLCs (Indigenous People and Local Communities) secure formal and inviolate recognition of their land and resources, is now the foundation of WCS's global work (https://www.wcs.org/our-work/communities). That includes ensuring that local people can continue to hunt and eat wild meat. Sustainable Wildlife Management Programme, Food and Agriculture Organization of the United Nations, 2025, swm-programme.info/homepage.

541 But it also holds: Alan Rabinowitz, *Beyond the Last Village: A Journey of Discovery in Asia's Forbidden Wilderness* (Island Press, 2001), 3.

542 the notorious SLORC: SLORC had annulled the 1990 elections, arrested activist and opposition leader Aung San Suu Kyi, and punished resisters with exile, torture, and execution. See Rabinowitz, *Beyond the Last Village*, 8.

542 Despite the heavy: Rabinowitz, *Beyond the Last Village*, 14.

543 "my mentor and friend": Alan Rabinowitz, *Life in the Valley of Death: The Fight to Save Tigers in a Land of Guns, Gold, and Greed* (Island Press, 2008), 85.

543 both criticized and praised: Zao Noam, "The Greening of a Dictatorship," *The Irrawaddy* 12, no. 9 (October 2004), www2.irrawaddy.com/article.php?art_id=4122; Zao

Noam, "Taming the Generals to Save the Tigers," *The Irrawaddy* 16, no. 3 (March 2008), www2.irrawaddy.com/print_article.php?art_id=10618; "Myanmar Officially Designates World's Largest Tiger Reserve in the Hukaung Valley," WCS Newsroom, August 3, 2010, newsroom.wcs.org/News-Releases/articleType/ArticleView/articleId/5354/Myanmar-Officially-Designates-Worlds-Largest-Tiger-Reserve-in-the-Hukaung-Valley.aspx; Charles Schmidt, "As Myanmar Opens to World, Fate of Its Forests Is on the Line," *Yale Environment 360*, November 15, 2012, e360.yale.edu/features/as_myanmar_opens_to_world_fate_of_its_forests_is_on_the_line.

544 "If you waited": Scott Wallace, "The Megafauna Man," *National Geographic Adventure* 8, no. 10 (December 2006/January 2007): 71.

545 "Why are we watching": Zalmai Moheb, interview with the author, July 24, 2020.

545 That again and again: More than 80 percent between 1950 and 2000. Thor Hanson, Thomas Brooks, Gustavo da Fonseca, Michael Hoffman, John Lamoreux, Gary Machlis, et al., "Warfare in Biodiversity Hotspots," *Conservation Biology* 23, no. 3 (June 2009): 578–87, doi.org/10.1111/j.1523-1739.2009.01166.x.

546 And rather than await: The creation of Wakhan National Park seemed to bear out the possibility for conservation as a strategy for building peace and democracy. Alex Dehgan, who set up and led the first WCS office in Kabul, met local Wakhi and Kyrgyz leaders who regularly walked hours to join in crafting rules for the new protected areas, planting the seeds of democratic governance. For the millions of Afghans who had been refugees, protecting their wildlife became a way to reclaim their national identity. Alex Dehgan, *The Snow Leopard Project and Other Adventures in Warzone Conservation* (Public Affairs, 2019), 5.

546 "flourishing . . . international diplomacy": Peter Zahler and George Schaller, "Saving More Than Just Snow Leopards," *The New York Times*, February 1, 2014, nytimes.com/2014/02/02/opinion/saving-more-than-just-snow-leopards.html.

547 In 2006 all four: George B. Schaller, "A Proposal for a Pamir International Peace Park," in *Science and Stewardship to Protect and Sustain Wilderness Values: Eighth World Wilderness Congress Symposium*, comp. Alan Watson, Janet Sproull, and Liese Dean, September 30–October 6, 2005, Anchorage, AK, Proceedings RMRS-P-49, (US Department of Agriculture, Forest Service, Rocky Mountain Research Station, 2007), 227–31.

548 coveted by royals: In addition to Schaller's journals, much of this is from his unpublished manuscript on Iran, shared with the author.

548 "that he was credible": Luke Hunter, interview with the author, April 8, 2020.

548 between 2001 and 2012: Mohammad Farhadinia, Luke Hunter, Houman Jowkar, George Schaller, and Stephanie Ostrowski, "Asiatic Cheetahs in Iran: Decline, Current Status, and Threats," in *Cheetahs: Biology and Conservation*, ed., Laurie Marker, Lorraine K. Boast, and Anne Schmidt-Küntzel (Academic Press, 2018), 55–69.

548 With camera traps: Mohammad Farhadinia, Luke Hunter, Alireza Jourabchian, Fetemeh Hosseini-Zavarei, Hasan Akbari, Hoosgand Ziaie, et al., "The Critically Endangered Asiatic Cheetah *Acinonyx jubatus venaticaus* in Iran: A Review of Recent Distribution and Conservation Status," *Biodiversity and Conservation* 26 (February 2017): 1027–46, doi.org/10.1007/s10531-017-1298-8. Some have argued that given the permanence of extinction and ecosystem destruction versus the impermanence of regimes, conservation science and tools should be exempt from sanctions. In her November 20, 2019, piece for *National Geographic*, Kayleigh Long quoted an email from Sarah Durant, cheetah researcher and professor of conservation science at the Zoological Society of London. "The protection of threatened biodiversity . . . should, and must, transcend politics," Durant wrote, adding that access to international support and expertise for the purpose of conservation should be exempt from sanctions. Long also linked to this piece: researchgate.net/publication/329644264_Sanctioning_to_extinction_in_Iran.

549 "superstar . . . a hero": Filmmaker Mani Mirsadeghi, interview by Beth Wald, September 2017, shared with the author. For more on Mirsadeghi, see "Mani Mirsadeghi: Documentary Filmmaker," National Geographic Explorers Directory, explorer-directory.nationalgeographic.org/mani-mirsadeghi and Mani Film, manifilm.com.

549 "the best cat": Hunter, interview.

549 They published together: Luke Hunter, Houman Jowkar, Hooshang Ziaie, George Schaller, Guy Balme, Chris Walzer, et al., "Conserving the Asiatic Cheetah in Iran: Launching the First Radio-Telemetry Study," *Cat News* 46 (Spring 2007): 8–11.

549 including a chapter: Farhadinia, Hunter, Jowkar, Schaller, and Ostrowski, "Asiatic Cheetahs in Iran."

549 On June 29, 2009: "Iran: Violent Crackdown on Protesters Widens," Human Rights Watch, June 23, 2009, hrw.org/news/2009/06/23/iran-violent-crackdown-protesters-widens.

550 billionaire Thomas Kaplan: Murtaza Hussain, "Did an American Billionaire Philanthropist Play a Role in the Imprisonment of Iranian Environmentalists?," *The Intercept*, November 27 2019, theintercept.com/2019/11/27/iran-environmentalists-panthera-thomas-kaplan.

550 "although we avoid": A screenshot of this correspondence was posted to Twitter by conservationist Tanya Rosen (@NarynTRosen) on November 15, 2019 (since deleted).

550 In early 2018: Kayleigh E. Long, "Cheetah Researchers Accused of Spying Sentenced in Iran," *National Geographic*, November 20, 2019, nationalgeographic.com/animals/article/jailed-cheetah-researchers-sentenced-iran.

550 "skin attracts atomic waves": Agence France-Presse and Michael Bachner, "Iran Accuses West of Using Lizards for Nuclear Spying," *The Times of Israel*, February 13, 2018, timesofisrael.com/iran-accuses-west-of-using-lizards-for-nuclear-spying/.

550 Seyed-Emami died: Patrick Wintour, "'Loss for Iran's Wildlife': Woman Jailed in Tehran Calls for Environmentalists' Release," *The Guardian*, May 31, 2023, theguardian.com/world/2023/jun/01/loss-for-irans-wildlife-woman-jailed-in-tehran-calls-for-environmentalists-release.

550 Despite appeals from: "IUCN Renews Call for Release of Iran Conservationists," IUCN, February 14, 2023, iucn.org/iucn-statement/202302/iucn-renews-call-release-iran-conservationists.

550 "contacts with the US enemy": Jon Gambrell, "Activists: Iran Conservationists Get Prison Time Amid Unrest," Associated Press, November 21, 2019, apnews.com/general-news-4c7125f9ffda485abc7f7063d0d22bd2.

550 dedicated conservationists, the kind: Rosen (@NarynTRosen) Twitter post, November 15, 2019. Schaller's letter is also mentioned in Long, "Cheetah Researchers Accused of Spying."

551 As the prisoners': Peter Schwartzstein, "How Iran Is Destroying Its Once Thriving Environmental Movement," *National Geographic*, November 12, 2020, nationalgeo graphic.com/environment/article/how-iran-destroying-once-thriving-environmen tal-movement.

551 A lone reported: Wintour, "'Loss for Iran's wildlife.'"

551 Of the migratory: Frances Davis, Andrew Szopa-Comley, Sarah Rouse, Aude Caromel, Andy Arnell, Saloni Basrur, et al., *State of the World's Migratory Species* (UNEP-WCMC, 2024), cms.int/sites/default/files/publication/State%20of%20the%20 Worlds%20Migratory%20Species%20report_E.pdf.

552 "George could say": Olson, interview.

552 "simply no longer there": Conway, interview.

552 "Will it skirt that meadow?": Alan Rabinowitz, interview by Tom Veltre, December 19, 2008.

552 "He showed us where": Kang, interview.

552 "works where you": From a ninetieth-birthday book WCS made for George in May 2023. Schaller personal archive, shared with the author.

553 "His appreciation of": Felix Ndagijimana, interview with the author, February 16, 2012.

553 "Schaller established the foundation": Yange, interview.

553 Schaller protégés support: His development of American protégés also continued. Kirk Olson, who at twenty-three blundered into Mongolia clueless enough to go into the steppe with a one-to-one-millionth scale map, nonetheless got the crucial Schaller nudge. "I knew nothing. But 'Would I mind,' he asked, 'if he joined my project?' Which was his way of offering help without taking over." Completing a PhD with Schaller's support, Kirk married a Mongolian woman, raised a family, and with twenty nationals launched WCS Mongolia. Bill Leacock, after working with him in Laos, asked George for help pursuing wildlife science and was soon in Kamchatka studying bears; he wound up as the mammal biologist for the Arctic National Wildlife Refuge.

553 "His terrible prerogative": Percy Bysshe Shelley, *Queen Mab* (Mrs. Carlile and Sons, 1832), 117–18.

553 "The most telling expression": Don Goddard, *Saving Wildlife, A Century of Conservation* (Harry Abrams, 1995), 158, 159.

554 "Where I lie": Alexander Sergeyevich Pushkin, *The Poems, Prose and Plays of Pushkin* (Modern Library, 1943), 71.

ILLUSTRATION CREDITS

FRONT COVER

George Schaller, age twenty-three, 1956 Sheenjek Expedition, Brooks Range, Alaska. Courtesy of George B. Schaller Archive.

CHAPTER OPENER PAGES

p. 14: Schaller, age eight, Copenhagen, fall 1941, just weeks before the US declared war against the country his mother came from and his father served. Courtesy of George B. Schaller Archive.

p. 40: Summer of '54, age twenty-one, after a failed attempt at a first ascent of Alaska's Mount Drum. A week later, with Austrian alpinist Heinrich Harrer, he would succeed. Courtesy of George B. Schaller Archive.

p. 82: By the second year of Schaller's gorilla study, his visas no longer came from Brussels but from the newly independent République du Congo. Courtesy of George B. Schaller Archive.

p. 144: The complex interdependencies among predators and prey became central to Schaller's work, following on his first carnivore study, in India in the early sixties. Courtesy of George B. Schaller Archive.

p. 180: To watch lions for days on end posed an unexpected challenge. "They spend masses of time doing absolutely nothing." Courtesy of George B. Schaller Archive.

p. 232: In December 1970, in northwest Pakistan, Schaller took the world's first pictures of a snow leopard. She and her cub stayed for a week, letting him sleep just 150 feet away. Courtesy of George B. Schaller Archive.

p. 324: Though it broke his heart, Schaller's work in the 1970s in Brazil's Pantanal began the recovery of the New World's only great cat. Courtesy of George B. Schaller Archive.

p. 368: *National Geographic* put Schaller's first-in-the-world pictures of pandas in the wild on the cover in December 1981. Courtesy of George B. Schaller Archive.

p. 422: Schaller's "covenant" with the threatened chiru called him back to the Tibetan Plateau many dozens of times. Courtesy of George B. Schaller Archive.

p. 528: At seventy-one, Schaller crossed thirteen valleys and passes above 14,000 feet in Afghanistan's Wakhan Corridor, in pursuit of Marco Polo sheep and an international peace park. Copyright © 2004 by Beth Wald.

INTERIOR ILLUSTRATIONS

p. 25: Courtesy of George B. Schaller Archive.

p. 57: Courtesy of the Peabody Museum of Natural History, Yale University, New Haven, Connecticut, USA.

pp. 104, 136, 185: Courtesy of George B. Schaller Archive.

p. 199: Courtesy of the Peabody Museum of Natural History, Yale University, New Haven, Connecticut, USA.

pp. 339, 361, 432, 485, 492: Courtesy of George B. Schaller Archive.

ILLUSTRATION INSERT

p. 1, top, middle, and bottom: Courtesy of George B. Schaller Archive.

p. 2, top and bottom: Courtesy of George B. Schaller Archive.

p. 3, top: Robert Krear, "Brina Kessel, George Schaller, Don MacLeod, Mardy and Olaus Murie, Sheenjek Expedition, 1956." The Murie Family photograph collection, The Murie Ranch, Teton Science School.

p. 3, bottom: Courtesy of George B. Schaller Archive.

pp. 4–9: Courtesy of George B. Schaller Archive.

p. 10, top: Courtesy of Peter Matthiessen Archive.

p. 10, middle and bottom: Courtesy of George B. Schaller Archive.

pp. 11–16: Courtesy of George B. Schaller Archive.

INDEX